Progress in Mathematics
Volume 106

Series Editors
J. Oesterlé
A. Weinstein

Peter Buser

Geometry and Spectra of Compact Riemann Surfaces

with 135 illustrations

Birkhäuser
Boston · Basel · Berlin

Peter Buser
Département de Mathématiques
Ecole Polytechnique Fédérale de Lausanne
CH-1015 Lausanne-Ecublens
Switzerland

Library of Congress Cataloging-in-Publication Data

Buser, Peter, 1946-
   Geometry and spectra of compact Riemann surfaces / Peter Buser.
     p.   cm. -- (Progress in mathematics ; v. 106)
   Includes bibliographical and references and index.
   ISBN 0-8176-3406-1 (acid free)
   1. Riemann surfaces. I. Title. II. Series: Progress in
mathematics (Boston, Mass.) ; vol. 106.
   QA333.B87   1992   92-23803
   515' .223--dc20   CIP

Printed on acid-free paper

© Birkhäuser Boston 1992

Copyright is not claimed for works of U.S. Government employees.
All rights reserved. No part of this publication may be reproduced, stored in a retrieval system, or transmitted, in any form or by any means, electronic, mechanical, photocopying, recording, or otherwise, without prior permission of the copyright owner.

Permission to photocopy for internal or personal use of specific clients is granted by Birkhäuser Boston for libraries and other users registered with the Copyright Clearance Center (CCC), provided that the base fee of $5.00 per copy, plus $0.20 per page is paid directly to CCC, 21 Congress Street, Salem, MA 01970, U.S.A. Special requests should be addressed directly to Birkhäuser Boston, 675 Massachusetts Avenue, Cambridge, MA 02139, U.S.A.

ISBN 0-8176-3406-1
ISBN 3-7643-3406-1

Camera-ready copy prepared by the Authors in TeX.
Printed and bound by Quinn-Woodbine, Woodbine, NJ.
Printed in the U.S.A.

9 8 7 6 5 4 3 2 1

*To
Barbara
Caroline
Patrick
Regula*

Preface

This book deals with two subjects. The first subject is the geometric theory of compact Riemann surfaces of genus greater than one, the second subject is the Laplace operator and its relationship with the geometry of compact Riemann surfaces.

The book grew out of the idea, a long time ago, to publish a Habilitationsschrift, a thesis, in which I studied Bers' pants decomposition theorem and its applications to the spectrum of a compact Riemann surface. A basic tool in the thesis was cutting and pasting in connection with the trigonometry of hyperbolic geodesic polygons. As this approach to the geometry of a compact Riemann surface did not exist in book form, I took this book as an occasion to carry out the geometry in detail, and so it grew by several chapters. Also, while I was writing things up there was much progress in the field, and some of the new results were too challenging to be left out of the book. For instance, Sunada's construction of isospectral manifolds was fascinating, and I got hooked on constructing examples for quite a while. So time went on and the book kept growing. Fortunately, the interest in existence proofs also kept growing. The editor, for instance, was interested, and so was my family. And so the book finally assumed its present form. Many of the proofs given here are new, and there are also results which appear for the first time in print.

Introductory remarks and some history about the individual subjects are given at the beginning of each chapter. I shall therefore use this place to add a few global remarks.

The book has two parts. The first part consists of Chapters 1 through 6 and is an introduction to the geometry of compact Riemann surfaces based on hyperbolic geometry and on cutting and pasting. This part is in textbook form at about graduate level. The prerequisites are kept to a minimum, but I assume that the reader has a background either in differential geometry or in complex Riemann surface theory. Consequently, the standard introductory material which belongs to the intersection of these fields is not treated here. In particular, the fundamental group, the universal covering and the topological classification of compact surfaces are assumed to be known. The theorems about isotopies of curves on surfaces, on the other hand, are less standard. Since they are basic for Teichmüller theory, they are treated in the Appendix.

Chapter 1 deals with the general properties of surfaces which are obtainable by pasting together geodesic polygons from the hyperbolic plane. Chapter 2 is an account of hyperbolic trigonometry, the basic computational tool in this book. This chapter also contains an account of two less familiar models of the hyperbolic plane, the hyperboloid model and the quaternion model. The reader may skip this chapter, though, as only the formulae will be needed later on.

Chapters 3 and 6 describe the construction of compact Riemann surfaces based on the pasting of geodesic hexagons and lead to the Fenchel-Nielsen model of Teichmüller space. The chapters may be read in this order. Chapter 6 is organized in such a way that it may also be used as a starting point for further reading in Teichmüller theory. Chapters 4 and 5 contain the basic qualitative geometric results about Riemann surfaces: the collar theorem and Bers' theorem on length controlled pants decompositions. In these chapters we also briefly consider surfaces of variable curvature.

The second part of the book starts with a fairly self-contained introduction to the spectrum of the Laplace operator based on the heat kernel. This approach is particularly suitable in the case of a Riemann surface because the heat kernel of the hyperbolic plane is known explicitly. After a brief look at isoperimetric techniques and the famous small eigenvalues in Chapter 8 we devote the rest of the book to the question of how far and to what extent the geometry of a compact Riemann surface is reflected in the spectrum of the Laplacian.

Many years ago Huber [2] proved that two compact Riemann surfaces have the same sequence of eigenvalues of the Laplace operator if and only if they have the same sequence of lengths of the closed geodesics. This theorem does not only show that the eigenvalues contain a great deal of geometric information, it also indicates that spectral problems may be approached by geometric methods such as those developed in the first part of the book. This is important because the computation of the individual eigenfunctions and eigenvalues is a very difficult matter and practically unsolved, whereas the computation of the closed geodesics can be carried out explicitly and we may focus our attention right away on the global problems. These so-called inverse spectral problems have been studied quite successfully in recent years (not only in the case of Riemann surfaces, of course), and we are now in a position to present a number of global results within a common framework. This will be carried out in Chapters 10 through 14. Huber's theorem and related results will be proved in Chapter 9 where we shall use trace formula techniques.

During the course of the years I was writing this book up I profited from innumerable discussions with friends and colleagues to whom I should like to express my warmest thanks. I am particularly indebted to Philippe Anker, Colette Anné, Pierre Bérard, Gérard Besson, Leesa Brieger, Isaac Chavel, Bruno Colbois, Gilles Courtois, Jozef Dodziuk, Patrick Eberlein, Edgar Feldman, Burton Randol, Paul Schmutz and Klaus-Dieter Semmler for their advice and encouragement, for teaching me special subjects, for reading drafts and lecturing on various chapters, and, last but not least, for their efforts to make this text look more English. My thanks also go to Françoise Achermann for typesetting the first version of the book.

And, finally, I am most indebted to my family for their love and patience.

Echandens, Peter Buser
May 1992

Contents

Chapter 1
Hyperbolic Structures 1

1.1	The Hyperbolic Plane	1
1.2	Hyperbolic Structures	5
1.3	Pasting	8
1.4	The Universal Covering	15
1.5	Perpendiculars	17
1.6	Closed Geodesics	21
1.7	The Fenchel-Nielsen Parameters	27

Chapter 2
Trigonometry 31

2.1	The Hyperboloid Model	31
2.2	Triangles	33
2.3	Trirectangles and Pentagons	37
2.4	Hexagons	40
2.5	Variable Curvature	43
2.6	Appendix: The Hyperboloid Model Revisited	49
	The Quaternion Model	49
	A Trace Relation	55
	The General Sine and Cosine Formula	57

Chapter 3
Y-Pieces and Twist Parameters 63

3.1	Y-Pieces	63
3.2	Marked Y-Pieces	67
3.3	Twist Parameters	69
3.4	Signature (1, 1)	76
3.5	Cubic Graphs	78
3.6	The Compact Riemann Surfaces	81

3.7	Appendix: The Length Spectrum Is of Unbounded Multiplicity	84
	Geometric Approach	85
	Algebraic Approach	89

Chapter 4
The Collar Theorem — 94

4.1	Collars	94
4.2	Non-Simple Closed Geodesics	98
4.3	Variable Curvature	104
4.4	Cusps	108
4.5	Triangulations of Controlled Size	116

Chapter 5
Bers' Constant and the Hairy Torus — 122

5.1	Bers' Theorem	123
5.2	Partitions	124
5.3	The Hairy Torus	130
5.4	Bers' Constant Without Curvature Bounds	133

Chapter 6
The Teichmüller Space — 138

6.1	Marked Riemann Surfaces	138
6.2	Models of Teichmüller Space	142
6.3	The Real Analytic Structure of $\mathcal{T}_g$	147
6.4	Distances in $\mathcal{T}_g$	152
6.5	The Teichmüller Modular Group	154
6.6	A Rough Fundamental Domain	160
6.7	The Coordinates of Zieschang-Vogt-Coldewey	164
6.8	Fuchsian Groups and Bers' Coordinates	170

Chapter 7
The Spectrum of the Laplacian — 182

7.1	Introduction	182
7.2	The Spectrum and the Heat Equation	184
7.3	The Abel Transform	194
7.4	The Heat Kernel of the Hyperbolic Plane	197
7.5	The Heat Kernel of $\Gamma \backslash H$	205

Chapter 8
Small Eigenvalues — 210

8.1	The Interval $[0, \frac{1}{4}]$	210
8.2	The Minimax Principles	213
8.3	Cheeger's Inequality	215
8.4	Eigenvalue Estimates	218

Contents   xiii

Chapter 9
Closed Geodesics and Huber's Theorem   224

9.1	The Origin of the Length Spectrum	225
9.2	Summation over the Lengths	227
9.3	Summation over the Eigenvalues	235
9.4	The Prime Number Theorem	241
9.5	Selberg's Trace Formula	252
9.6	The Prime Number Theorem with Error Terms	256
9.7	Lattice Points	261

Chapter 10
Wolpert's Theorem   268

10.1	Introduction	268
10.2	Curve Systems	270
10.3	Finitely Many Lengths Determine the Length Spectrum	273
10.4	Generic Surfaces Are Determined by Their Spectrum	275
10.5	Decoding the Moduli	278

Chapter 11
Sunada's Theorem   283

11.1	Some History	283
11.2	Examples of Almost Conjugate Groups	285
11.3	Proof of Sunada's Theorem	291
11.4	Cayley Graphs	296
11.5	Transplantation of Eigenfunctions	304
11.6	Transplantation of Closed Geodesics	307

Chapter 12
Examples of Isospectral Riemann Surfaces   311

12.1	Cayley Graphs and Hyperbolic Polygons	311
12.2	Smoothness	313
12.3	Examples over $\mathbf{Z}_8^* \ltimes \mathbf{Z}_8$	318
12.4	Examples over $SL(3, 2)$	321
12.5	Genus 6	325
12.6	Large Families	332
12.7	Criteria For Non-Isometry	333

Chapter 13
The Size of Isospectral Families   340

13.1	Finiteness	340
13.2	Parameter Geodesics of Length $> \exp(-4g)$	344
13.3	Measuring the Twist Parameters	347
13.4	Parameter Geodesics of Length $\leq \exp(-4g)$	355

Chapter 14
Perturbations of the Laplacian in Hilbert Space — 362

14.1 The Hilbert Spaces H_0 and H_1 — 362
14.2 The Friedrichs Extension of the Laplacian — 366
14.3 A Representation Theorem — 370
14.4 Resolvents and Projectors — 373
14.5 Holomorphic Families — 380
14.6 A Model of Teichmüller Space — 382
14.7 Reduction to Finite Dimension — 388
14.8 Holomorphic Families of Laplacians — 397
14.9 Analytic Properties of the Eigenvalues — 399
14.10 Finite Parts of the Spectrum — 406

Appendix
Curves and Isotopies — 409

The Theorems of Baer-Epstein-Zieschang — 409
An Application to the 3-Holed Sphere — 424
Length-Decreasing Homotopies — 428

Bibliography — 433

Index — 448

Glossary — 454

Chapter 1

Hyperbolic Structures

In this chapter we collect a number of geometric properties of surfaces which are modelled over the hyperbolic plane. The way of construction of these surfaces is cutting and pasting. For this we shall include surfaces with piecewise geodesic boundary and prove existence and uniqueness theorems for geodesics in various types of homotopy classes. The chapter ends with an outline of the Fenchel-Nielsen parameters.

1.1 The Hyperbolic Plane

In this section we fix the notation, recall a few properties of hyperbolic geometry and introduce various types of coordinates. We use Beardon [1] as a reference for hyperbolic geometry, Klingenberg [1, 2] as a reference for the differential geometry of surfaces and Macbeath [3] for the topology of coverings. For a short history of hyperbolic geometry we suggest Milnor [2].

N, Z, Q, R and **C** denote respectively the natural, integer, rational, real and complex numbers.

For any Riemannian manifold M we use the following notation.

$\quad$ dist(,) $\quad$ is the distance function,
$\quad$ $\ell(c)$ $\quad\;\;$ is the length of a curve c,
$\quad$ Is(M) $\quad$ is the isometry group of M,
$\quad$ Is$^+$(M) $\quad$ is the subgroup of all orientation-preserving isometries.

The *Poincaré model* of the hyperbolic plane is the following subset of the complex plane **C**,

$$\mathbf{H} = \{ z = x + iy \in \mathbf{C} \mid y > 0 \}$$

with the hyperbolic metric

(1.1.1) $$ds^2 = \frac{1}{y^2}(dx^2 + dy^2).$$

The distance in this model is given by the following formula (Beardon [1], p. 130)

(1.1.2) $$\cosh \operatorname{dist}(z, w) = 1 + \frac{|z - w|^2}{2 \, \operatorname{Im} z \, \operatorname{Im} w},$$

where cosh is the hyperbolic cosine function, and *Im* denotes the imaginary part of a complex number. The group

$$\mathrm{PSL}(2, \mathbf{R}) = \left\{ \begin{pmatrix} a & b \\ c & d \end{pmatrix} \mid a, b, c, d \in \mathbf{R}; \, ad - bc = 1 \right\} / \{\pm 1\}$$

acts biholomorphically on **H** via the mappings

$$z \mapsto \frac{az + b}{cz + d}$$

and leaves the metric (1.1.1) invariant. Moreover, PSL(2, **R**) is the full group of orientation-preserving isometries of **H**. The mapping

$$z \mapsto \frac{z - i}{z + i}, \quad z \in \mathbf{H},$$

maps **H** biholomorphically onto the unit disk

$$\mathbf{D} = \{ w = u + iv \in \mathbf{C} \mid u^2 + v^2 < 1 \}.$$

The induced metric is

(1.1.3) $$ds^2 = \frac{4(du^2 + dv^2)}{(1 - (u^2 + v^2))^2}.$$

This is the *disk model* of the hyperbolic plane. In both models the geodesics are the generalized circles which meet the boundary orthogonally. Yet another model, the hyperboloid, will be considered in Chapter 2 in connection with trigonometry. When no confusion arises, we will also use **H** to denote the abstract hyperbolic plane.

The following theorems for the hyperbolic plane can immediately be seen in the Poincaré model.

1.1.4 Theorem. *There is a unique geodesic through any two distinct points.* ◇

1.1.5 Theorem. *If a is a geodesic and p is a point, then there exists through p a unique geodesic perpendicular to a.* ◊

1.1.6 Theorem. *If a and b are geodesics with positive distance, then there exists a unique geodesic perpendicular to a and b.* ◊

A compact domain $P \subset \mathbf{H}$ is called a *geodesic polygon* if its boundary is a piecewise geodesic closed Jordan curve. Sides and interior angles of a polygon are defined in the usual way. An *n-gon* is a geodesic polygon with n sides.

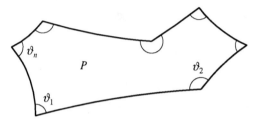

Figure 1.1.1

1.1.7 Theorem. *The area of an n-gon P with interior angles $\vartheta_1, \ldots, \vartheta_n$ is given by the following formula,*

$$\text{area } P = (n-2)\pi - (\vartheta_1 + \ldots + \vartheta_n).$$

Proof. Beardon [1], p.153. ◊

The hyperbolic trigonometry will follow in Chapter 2. The remainder of this section deals with geodesic coordinates.

Polar Coordinates. Let $p_0 \in \mathbf{H}$ be an arbitrary base point and let A be a unit tangent vector at p_0. For every point $p \in \mathbf{H} - \{p_0\}$ there exists a unique unit speed geodesic $\gamma : [0, \infty[\to \mathbf{H}$ with $\gamma(0) = p_0$ and passing through p. We let $\sigma = \sigma(p) \in [-\pi, \pi[$ be the directed angle from A to the initial tangent vector of γ and let $\rho = \rho(p)$ be the distance from p_0 to p such that $\gamma(\rho(p)) = p$. Then $(\rho, \sigma) = (\rho(p), \sigma(p))$ is the pair of *polar coordinates* of p with respect to p_0 and A. In polar coordinates the hyperbolic metric has the following expression.

(1.1.8) $$ds^2 = d\rho^2 + \sinh^2\rho \, d\sigma^2.$$

This is easily obtained in the unit disk model with p_0 at the origin (or Klingenberg [1, 2], Proposition 4.3.8). Instead of taking $\sigma(p)$ in the interval $[-\pi, \pi[$, we may consider $\sigma(p)$ an element of the unit circle

$$S^1 = \mathbf{R}/[s \mapsto s + 2\pi].$$

Fermi Coordinates. If we replace the above base point by a base line, we get Fermi coordinates. They are defined as follows. Let η be a geodesic in the hyperbolic plane, parametrized with unit speed in the form

$$t \mapsto \eta(t) \in \mathbf{H}, \qquad t \in \mathbf{R}.$$

Then η separates $\mathbf{H}$ into two half-planes: a left-hand side and a right-hand side of η. For each $p \in \mathbf{H}$ we have the *directed distance* ρ from p to η: positive on one side and negative on the other. There exists a unique t such that the perpendicular from p to η meets η at $\eta(t)$. Now (ρ, t) is the pair of *Fermi coordinates* of p with respect to η. The metric tensor becomes

(1.1.9) $$ds^2 = d\rho^2 + \cosh^2\rho\, dt^2.$$

This may be computed in the upper half-plane when η is the positive imaginary axis (or Klingenberg [1, 2], Proposition 4.4.6).

For convenience, and to avoid difficulties with minus signs we adopt the following *sign conventions* in all figures. If η is oriented we take the oriented distance ρ from a point p to η to be negative on the left-hand side of η and positive on the right-hand side. This fits with the convention that for a positively (= counterclockwise) oriented circle in the plane the points on the left-hand side lie in the interior of the circle and have smaller values of ρ than the points on the right-hand side.

Horocyclic Coordinates. For completeness we mention a third type of geodesic coordinate in the hyperbolic plane, useful for the study of cusps. We work in the upper half-plane. Let p_0 be a point in the extended complex plane $\mathbf{C} \cup \{\infty\}$ which lies either on the real axis or is the point at infinity. We call such a point a *boundary point* or *point at infinity* of $\mathbf{H}$. The generalized (Euclidean) circles which are contained in $\mathbf{H} \cup \{p_0\}$ passing through p_0

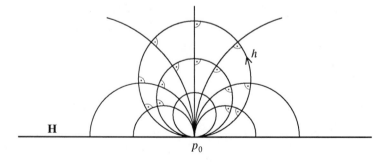

Figure 1.1.2

(cf. Fig. 1.1.2) are called the *horocycles with center* p_0.

The transformations of PSL(2, **R**) map horocycles onto horocycles. In the particular case $p_0 = \infty$, the horocycles are the horizontal straight lines, and the vertical straight lines are geodesics with p_0 an endpoint at infinity. The horizontal and vertical straight lines intersect each other orthogonally. It follows that for *any* position of p_0, the geodesics through p_0 and the horocycles at p_0 form an orthogonal family.

Now let h be a horocycle parametrized with unit speed in the form $t \mapsto h(t) \in \mathbf{H}$, $t \in \mathbf{R}$. The hyperbolic plane is again separated into a left-hand side and a right-hand side. We let the parametrization of h be such that p_0 lies on the left-hand side of h, and we take the oriented distance ρ from a point p to h to be negative on the left and positive on the right. This fits with the sign convention for the Fermi coordinates.

Again, there exists a unique t such that the perpendicular geodesic from p to h meets h at $h(t)$, and (ρ, t) is by definition the pair of *horocyclic coordinates* with respect to h. Using the particular example $t \mapsto h(t) = x(t) + iy(t) = t + i$, $t \in \mathbf{R}$, (and $p_0 = \infty$), we find the following formula for the metric tensor.

(1.1.10) $$ds^2 = d\rho^2 + e^{2\rho} dt^2.$$

1.2 Hyperbolic Structures

A Riemannian manifold of constant curvature -1 is locally isometric to hyperbolic space. A more self-contained approach to such manifolds is obtained if we define them through a hyperbolic atlas in which the coordinate neighborhoods are mapped into hyperbolic space and all overlap maps are local isometries. This approach is particularly useful with respect to cutting and pasting and has been most successful in dimensions two and three (see Benedetti-Dedo [1], Epstein [2], Sullivan [1], Thurston [1] for some literature in dimension 3). We use this approach for the compact Riemann surfaces.

In the following, a *surface* will always be a smooth *orientable* two dimensional manifold, possibly with piecewise smooth boundary. An atlas with coordinate systems (U, φ), where U is the coordinate neighborhood and φ the coordinate map, will always be a subatlas of the differentiable structure. The surfaces need not be connected, although later we shall restrict ourselves to connected ones. Since the definition for bordered surfaces is somewhat lengthy, we first consider the unbordered case.

1.2.1 Definition. Let S be a surface without boundary. An atlas $\mathcal{A}$ of S is called *hyperbolic* if it has the following properties.
  (i) $\varphi(U) \subset \mathbf{H}$, for all $(U, \varphi) \in \mathcal{A}$.
  (ii) If $(U, \varphi) \in \mathcal{A}$ and $(U', \varphi') \in \mathcal{A}$, then for each connected component V of $U \cap U'$ there exists an isometry $m \in \text{Is}^+(\mathbf{H})$ such that $\varphi' \circ \varphi^{-1}$ coincides with m on $\varphi(V)$.

As for atlases of other types, every hyperbolic atlas extends to a unique maximal one.

Every hyperbolic atlas induces a Riemannian metric of constant curvature -1 on S. Conversely, every Riemannian metric of constant curvature -1 on S is obtained in this way because any such metric is locally isometric to the hyperbolic plane.

In order to define hyperbolic atlases for surfaces with boundary we use the following terminology. Let $\delta > 0$ and let $0 < \vartheta < 2\pi$. A subset V of $\mathbf{H}$ is called a *circle sector* of radius δ and angle ϑ at $p_0 \in \mathbf{H}$ if it has the following form, where (ρ, σ) are polar coordinates based at p_0.

$$V = \{(\rho, \sigma) \in \mathbf{H} \mid 0 < \rho < \delta,\ 0 \leq \sigma \leq \vartheta\} \cup \{p_0\}.$$

A *half-disk* is a circle sector of angle π.

Now let S be a surface with non-empty piecewise smooth boundary ∂S. The set

$$\text{int } S := S - \partial S$$

is called the *interior*, and its points are the *interior points* of S. The boundary ∂S of S is composed of smooth arcs, the *sides* of S. A side may have 0, 1 or 2 endpoints. A smooth closed boundary curve is a side with no endpoints. Every endpoint of a side is the endpoint of an adjacent side and is called a *vertex* of S. A point of ∂S which is not a vertex is called an *ordinary boundary point*. In the following we shall allow that the boundary be smooth at some of the vertices, but the number of vertices must be locally finite.

1.2.2 Definition. Let S have non-empty boundary. An atlas $\mathcal{A}$ of S is called *hyperbolic* if it has the following properties.
  (i) For each $p \in S$ there exists a coordinate system $(U, \varphi) \in \mathcal{A}$ with $p \in U$ such that $\varphi(U) \subset \mathbf{H}$ is
    – a circle sector of angle $\vartheta \leq \pi$ at $\varphi(p)$, if p is a vertex,
    – a half-disk at $\varphi(p)$, if p is an ordinary boundary point,
    – an open disk with center $\varphi(p)$, if p is an interior point.
  (ii) If $(U, \varphi) \in \mathcal{A}$ and $(U', \varphi') \in \mathcal{A}$, then for each connected component V of $U \cap U'$ there exists an isometry $m \in \text{Is}^+(\mathbf{H})$ such that $\varphi' \circ \varphi^{-1}$ coincides

with m on $\varphi(V)$.

The restriction to angles $\vartheta \leq \pi$ is for convenience. We might as well extend Definition 1.2.2 to arbitrary $\vartheta \geq 0$. However, this causes the distinction of numerous additional cases in Theorems 1.5.2, 1.5.3 and 1.6.6 below.

1.2.3 Definition. Let S be as above. A maximal hyperbolic atlas on S is called a *hyperbolic structure*. The hyperbolic structure is *complete* if its induced metric on S is complete (in the sense of metric spaces). A *connected* surface together with a complete hyperbolic structure is called a *hyperbolic surface*.

In Theorem 1.4.1 we shall prove that every bordered hyperbolic surface is a subdomain of an unbordered hyperbolic surface and has a piecewise geodesic boundary.

Typical examples of hyperbolic surfaces are the hyperbolic plane, the half-planes, the hyperbolic polygons and the compact Riemann surfaces of signature (g, n) which will be defined in Section 1.7.

Note that we have restricted ourselves to orientable surfaces. If a hyperbolic atlas is given, the surface will be considered *oriented* with the orientation induced by the atlas.

On a surface with hyperbolic structure the definitions of *geodesics*, *surface area*, *angles*, etc. carry over from the hyperbolic plane in an obvious way. Note that the angle ϑ occurring in Definition 1.2.2 is the interior angle at vertex p of S.

1.2.4 Theorem. *Let S be a hyperbolic surface and let $\Gamma \subset \text{Is}^+(S)$ be a subgroup which acts properly discontinuously and without fixed points on S. Then the quotient $\Gamma \backslash S$ carries a uniquely determined complete hyperbolic structure such that the natural projection $\pi : S \to \Gamma \backslash S$ is a local isometry.*

Proof. We define the atlas $\mathcal{A}'$ for $\Gamma \backslash S$ using the atlas $\mathcal{A}$ of S. For all $q \in \Gamma \backslash S$ we select $q' \in \pi^{-1}(q)$ and let $(U, \varphi) \in \mathcal{A}$ be a coordinate system as in Definition 1.2.2, with radius δ so small that the restriction $\pi \,|\, U$ of π to U is a homeomorphism. We define $((\pi(U), \varphi \circ (\pi \,|\, U)^{-1})$ to be an element of $\mathcal{A}'$. This atlas defines the differentiable structure of $\Gamma \backslash S$ and it satisfies the conditions of Definition 1.2.2. The natural projection $\pi : S \to \Gamma \backslash S$ is a Riemannian covering map and hence $\Gamma \backslash S$ is complete. The uniqueness is clear. ◊

1.3 Pasting

If two sheets of paper are pasted together along a straight boundary, we get a smooth sheet of paper. Since the hyperbolic plane has a twice transitive isometry group, a similar fact holds for hyperbolic polygons: let S and S' be disjoint convex geodesic polygons in $\mathbf{H}$ such that all interior angles are not greater than π. Use an isometry $m \in \text{Is}^+(\mathbf{H})$ to move them together as in Fig.1.3.1. Then $S^* := m(S) \cup S'$ is again a hyperbolic geodesic polygon.

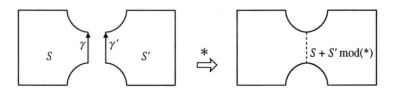

Figure 1.3.1

Now observe that in order to construct S^* we need not *move* S. We can *paste* S and S' together along sides γ and γ' as follows. Assume S and S' are disjoint. Let $\gamma : [0, 1] \to S$ and $\gamma' : [0, 1] \to S'$ be a parametrization of the sides with the same constant speed and with opposite boundary orientation (the sides are assumed to have the same length). Then there exists an isometry $m \in \text{Is}^+(\mathbf{H})$ such that

$$m(\gamma(t)) = \gamma'(t), \quad t \in [0, 1].$$

Define an equivalence relation on the disjoint union $S \cup S'$ as follows. For every point $p = \gamma(t)$, $t \in [0, 1]$, the equivalence class consists of the two points $\gamma(t)$ and $\gamma'(t)$. For every point $p \notin \gamma \cup \gamma'$ the equivalence class consists of the single point p. We shall say that the equivalence relation thus defined is determined by the *pasting condition*

(*) $\qquad \gamma(t) = \gamma'(t), \quad t \in [0, 1],$

and denote by

$$F = S + S' \bmod(*)$$

the quotient space of $S \cup S'$ with respect to this equivalence relation. If we project the Riemannian metric from $S \cup S'$ to F, then F is isometric to the domain $S^* := m(S) \cup S'$.

In a similar manner we paste together more general surfaces. Assume that $S_1, \ldots, S_m$ are pairwise disjoint hyperbolic surfaces and let $\gamma_1, \gamma_1', \gamma_2, \gamma_2', \ldots, \gamma_n, \gamma_n'$ be pairwise distinct sides of $S := S_1 \cup \ldots \cup S_m$. Assume that

Ch.1, §3] Pasting

for each k, the sides γ_k and γ'_k are parametrized in the form $\gamma_k : I_k \to S$, $\gamma'_k : I_k \to S$ with the same constant speed. Here I_k is an interval. When γ_k and γ'_k are smooth closed geodesics, then we also allow I_k to be the circle $\mathbf{R}/[t \mapsto t + a_k]$ for some a_k, or we may take $I_k =]-\infty, +\infty[$ and parametrize the geodesics periodically with period a_k.

In addition, we assume that the orientations of the sides are such that the quotient surface below will again be orientable.

We now define an equivalence relation on S by the *pasting condition* or *pasting scheme*:

$$(\wp) \qquad \gamma_k(t) = \gamma'_k(t), \quad t \in I_k, \quad k = 1, \ldots, n.$$

1.3.1 Definition. Under the above hypotheses,

$$F = S_1 + \ldots + S_m \bmod(\wp)$$

is the quotient space of the disjoint union $S = S_1 \cup \ldots \cup S_m$ with respect to the equivalence relation defined by $(\wp)$.

F is a separable Hausdorff space, and if certain general conditions are satisfied, then F is a hyperbolic surface in a natural way. Before we prove this, we give a few examples.

1.3.2 Example. (Hyperbolic cylinders).

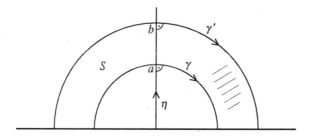

Figure 1.3.2

Consider the geodesic $\tau \mapsto \eta(\tau) = ie^\tau \in \mathbf{H}$, $\tau \in \mathbf{R}$. Let γ and γ' be geodesics intersecting η perpendicularly in a and b, as in Fig 1.3.2, where $|a| < |b|$. The closed strip S included between γ and γ' is a hyperbolic surface with atlas $\mathcal{A} = \{(S, id)\}$.

Parametrize γ and γ' with unit speed and opposite boundary orientation such that $\gamma(0) = a$ and $\gamma'(0) = b$. The isometry m given by

$$m(z) = \frac{bz}{a}, \quad z \in \mathbf{H},$$

maps η onto itself and satisfies the equation $m(\gamma(t)) = \gamma'(t)$, $t \in \mathbf{R}$. The pasting condition

(1) $$\gamma(t) = \gamma'(t), \quad t \in \mathbf{R},$$

yields the hyperbolic surface

(2) $$C = S \bmod(1).$$

The geodesic arc $\eta \,|\, [\log|a|, \log|b|]$ projects onto a closed geodesic η_1 on C of length

(3) $$\ell = \ell(\eta_1) = \log\left|\frac{b}{a}\right|.$$

If $\Gamma = \{m^k \mid k \in \mathbf{Z}\} \subset \mathrm{PSL}(2, \mathbf{R})$ denotes the cyclic subgroup generated by m, then the quotient manifold $\Gamma\backslash\mathbf{H}$ carries a natural hyperbolic structure such that the covering $\mathbf{H} \to \Gamma\backslash\mathbf{H}$ is a local isometry (Theorem 1.2.4). We therefore identify $\Gamma\backslash\mathbf{H}$ with C.

Introducing Fermi coordinates based on η (cf. above (1.1.9)), we obtain a description of C as the surface

(4) $$C = \mathbf{R} \times \mathbf{R}/[t \mapsto t + \ell],$$

where ℓ is as in (3), with the Riemannian metric

(5) $$ds^2 = d\rho^2 + \cosh^2\rho\, dt^2.$$

In any of these forms, C is called a *hyperbolic cylinder*. The coordinates used in (5) are called the *Fermi coordinates* for the cylinder. Observe that η_1 is parametrized with unit speed. If η_1 is reparametrized with speed ℓ instead, then (5) is replaced by

(6) $$ds^2 = d\rho^2 + \ell^2 \cosh^2\rho\, dt^2.$$

1.3.3 Example. (The pasting of closed geodesics).

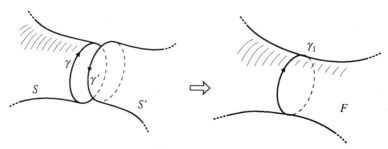

Figure 1.3.3

Let S, S' be hyperbolic surfaces (not necessarily distinct) and assume that γ on S and γ' on S' are closed boundary geodesics with the same length, for example ℓ. We parametrize γ and γ' periodically on $\mathbf{R}$ with period 1 and speed ℓ. If S coincides with S', then we assume that γ and γ' are distinct and have the *same boundary orientation*, that is, S and S' are either both on the left hand side or both on the right hand side of γ and γ'. Let

(1) $$F = S + S' \bmod (\gamma(t) = \gamma'(-t),\ t \in \mathbf{R}).$$

The two geodesics γ and γ' project to a simple closed geodesic γ_1 of length ℓ on F. For sufficiently small $\varepsilon > 0$ the following tubular neighborhoods are isometric

$$\{p \in F \mid \mathrm{dist}(p, \gamma_1) < \varepsilon\} \quad \text{and} \quad \{p \in C \mid \mathrm{dist}(p, \eta_1) < \varepsilon\},$$

where C is from Example 1.3.2 with $\ell(\eta_1) = \ell(\gamma_1)$. The pasting condition in (1) may be replaced by

(2) $$\gamma(t) = \gamma'(\alpha - t), \quad t \in \mathbf{R},$$

with an arbitrary so-called *twist parameter* $\alpha \in \mathbf{R}$. The above ε-neighborhood of γ_1 remains geometrically the same, but globally, the surfaces arising from different α are in general not isometric (cf. the examples in (3.6.6)).

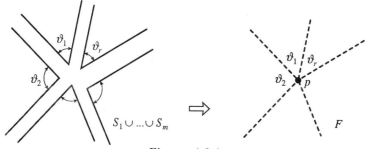

Figure 1.3.4

Not every pasting of hyperbolic surfaces yields a hyperbolic surface, even if the orientability condition is satisfied. For instance, if several vertices with interior angles $\vartheta_1, \ldots, \vartheta_r$ together define an interior point of the quotient F as shown in Fig. 1.3.4, then an obvious condition is that $\vartheta_1 + \ldots + \vartheta_r = 2\pi$. A more subtle problem is contained in the following example which we adopt from Benedetti-Dedo [1] (cf. also de Rham[1]).

1.3.4 Example. (A counterexample). Let α and β be the following geodesics in the upper half-plane: $\alpha(t) = 1 + ie^t$, $\beta(t) = 2 + ie^t$, $t \in \mathbf{R}$. The closed strip S included between α and β is a hyperbolic surface. The two sides of S

do not meet, but they have zero distance since $\text{dist}(\alpha(t), \beta(t)) \to 0$ as $t \to \infty$. Now paste together the two sides via the isometry m given by $m(z) = 2z$, $z \in \mathbf{H}$. The corresponding pasting condition is

(1) $$\alpha(t) = \beta(t + \log 2), \quad t \in \mathbf{R}.$$

The surface $F = S \bmod(1)$ has a hyperbolic structure such that the natural projection $S \to F$ is a local isometry, but the structure is not complete: consider the sequence of points $a_n = 1 + i2^n$, $b_n = 2 + i2^{n+1} = m(a_n)$, $n = 0, 1, \ldots$. Then a_n and b_n project to the same point p_n on F. The horocyclic arc (horizontal line) from a_n to b_{n-1} has length 2^{-n} and consequently we have $\text{dist}(p_{n-1}, p_n) < 2^{-n}$ on F, i.e. the p_n form a Cauchy sequence. This sequence does not converge.

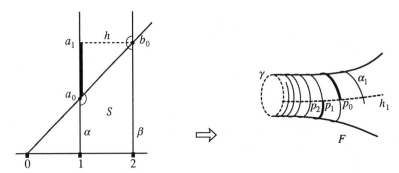

Figure 1.3.5

Fig. 1.3.5 shows S and the surface F obtained by this pasting. F is a half cylinder with boundary geodesic γ of length $\log 2$. However, γ does not belong to F. The boundary of S projects to an infinite geodesic α_1 on F for which γ is an asymptote. The points p_n lie on a horocyclic curve h_1 which intersects α_1 and γ orthogonally. The missing limit point of the sequence $\{p_n\}$ is the intersection point of h_1 and γ.

In the next theorem we consider again m distinct surfaces $S_1, \ldots, S_m$ with hyperbolic structures together with a choice of distinct sides $\gamma_1, \gamma_1', \gamma_2, \gamma_2', \ldots, \gamma_n, \gamma_n'$ as in Definition 1.3.1. The pasting scheme

($\wp$) $$\gamma_k(t) = \gamma_k'(t), \quad t \in I_k, \quad k = 1, \ldots, n,$$

gives rise to the orientable surface

$$F = S_1 + \ldots + S_m \bmod(\wp).$$

In the following, a *vertex cycle* is the set of all vertices of $S_1, \ldots, S_m$ which together define a single point of F (this point may be an interior point or a boundary point of F).

1.3.5 Theorem. *Assume that the following conditions hold.*

(i) *For each vertex cycle which yields an interior point of F the sum of the interior angles at the vertices is 2π,*

(ii) *for each vertex cycle which yields a boundary point of F the sum of the interior angles is $\leq \pi$.*

Then F carries a unique hyperbolic structure such that the natural projection $\sigma : S_1 \cup \ldots \cup S_m \to F$ is a local isometry.

Now assume that the following additional condition is satisfied.

(iii) *F is connected, the hyperbolic structures of $S_1, \ldots, S_m$ are complete, and any pair of non-adjacent sides in the list $\gamma_1, \ldots, \gamma_n'$ lying on the same surface S_μ has positive distance.*

Then the hyperbolic structure of F is complete.

Proof. The construction of the hyperbolic atlas $\mathcal{A}$ is straightforward and we restrict ourselves to a particular case.

Let $\mathcal{A}_1, \ldots, \mathcal{A}_m$ be the hyperbolic structures of $S_1, \ldots, S_m$ and let $p \in F$ be a boundary point whose inverse image under σ is the vertex cycle $p_1, \ldots, p_r$ with corresponding interior angles $\vartheta_1, \ldots, \vartheta_r$ whose sum is $\vartheta = \vartheta_1 + \ldots + \vartheta_r \leq \pi$. For sufficiently small $\varepsilon > 0$ the distance neighborhoods

$$U_i = \{ x \in S_1 \cup \ldots \cup S_m \mid \operatorname{dist}(x, p_i) < \varepsilon \}$$

are pairwise disjoint for $i = 1, \ldots, r$, and we obtain coordinate systems $(U_i, \varphi_i) \in \mathcal{A}_1 \cup \ldots \cup \mathcal{A}_m$ such that $\varphi_i(U_i)$ is a circle sector V_i in $\mathbf{H}$ of radius ε and angle ϑ_i. We suppose that the cycle is numbered in such a way that the left-hand side of U_i is identified with the right-hand side of U_{i+1} for $i = 1, \ldots, r-1$.

We then find orientation-preserving isometries $m_1, \ldots, m_r$ of $\mathbf{H}$ such that $V := m_1(V_1) \cup \ldots \cup m_r(V_r)$ is a circle sector of angle ϑ and such that for $i = 1, \ldots, r-1$ the left-hand side of $m_i(V_i)$ coincides with the right-hand side of $m_{i+1}(V_{i+1})$.

It follows that $x \in U_i$ and $y \in U_j$ are equivalent $\operatorname{mod}(\wp)$ if and only if $m_i(\varphi_i(x)) = m_j(\varphi_j(y))$. Hence, we have a well defined mapping

$$\varphi := m_1 \circ \varphi_1 + \ldots + m_r \circ \varphi_r \operatorname{mod}(\wp)$$

from $U := U_1 + \ldots + U_r \operatorname{mod}(\wp)$ onto V, and we let (U, φ) be an element of $\mathcal{A}$. In a similar manner we obtain coordinate neighborhoods in all other cases, and we check that the atlas $\mathcal{A}$ thus defined is hyperbolic and that the mapping $\sigma : S \to F$ is a local isometry. The uniqueness is clear.

The completeness of F in case (iii) is clear if the surfaces $S_1, \ldots, S_m$ are compact (which is the main case considered in this book). However, Example 1.3.4 shows that the non-compact case needs an argument. Rename the sides which are involved in the pasting as follows: $\gamma_k = \beta_{2k-1}$, $\gamma'_k = \beta_{2k}$, $k = 1, \ldots, n$. Denote by B the union $B = \beta_1 \cup \ldots \cup \beta_{2n}$ and let P be the set of all endpoints of $\beta_1, \ldots, \beta_{2n}$. There exists $\varepsilon > 0$ such that the distance sets $U_p^\varepsilon = \{x \in S \mid \text{dist}(x, p) < \varepsilon\}$, $p \in P$ are pairwise disjoint coordinate neighborhoods. Denote by U^ε the union of all U_p^ε, $p \in P$. Condition (iii) implies that for some positive $\delta < \varepsilon/2$ we have

(1) $$\text{dist}(\beta_i - U^{\varepsilon/2}, \beta_k) \geq \delta, \quad i, k = 1, \ldots, 2n, i \neq k.$$

Now let $x_1, x_2, \ldots$ be an infinite sequence in $S = S_1 \cup \ldots \cup S_m$ whose image $\sigma(x_1), \sigma(x_2), \ldots$ in F is a Cauchy sequence. To prove the completeness of F we must show that some infinite subsequence of $x_1, x_2, \ldots$ converges in S. We distinguish two cases.

Case 1. $\text{dist}(x_n, B) > \delta_2$ for some positive $\delta_2 < \delta$ and almost all n. Here we take N so large that for all $n \geq N$, $\text{dist}(\sigma(x_n), \sigma(x_N)) < \delta_2/2$ and such that $\text{dist}(x_N, B) > \delta_2$. The distance set

$$W = \{z \in S \mid \text{dist}(z, x_N) \leq \delta_2\}$$

is compact, connected and isometric under σ to the distance set

$$W' = \{z' \in F \mid \text{dist}(z', \sigma(x_N)) \leq \delta_2\}.$$

It follows that the x_n, $n \geq N$, form a converging Cauchy sequence in W.

Case 2. $\text{dist}(x_{n_k}, B)$ converges to 0 for some subsequence $\{x_{n_k}\}$, $k = 1, 2, \ldots$. We may assume that this is the sequence $\{x_n\}$ itself. We take N so large that for $n \geq N$, $\text{dist}(\sigma(x_n), \sigma(x_N)) < \delta/2$. We let $z_N \in B$ be the point which lies closest to x_N.

If $\text{dist}(z_N, P) < \varepsilon/2$, then $x_n \in U^\varepsilon$ for all $n \geq N$, and we are done, since U^ε has compact closure.

If $\text{dist}(z_N, P) > \varepsilon/2$, then the equivalence class of z_N consists of exactly two points, z_N and, say, w_N, where $\text{dist}(w_N, P) > \varepsilon/2$. Let

$$W_1 = \{z \in S \mid \text{dist}(z, z_N) < \delta\}, \quad W_2 = \{z \in S \mid \text{dist}(z, w_N) < \delta\}.$$

By (1), W_1 and W_2 are compact and connected. Moreover, $W_1 \cup W_2$ is the inverse image under σ of

$$W' := \{z' \in F \mid \text{dist}(z', \sigma(z_N)) < \delta\}.$$

As in case 1, it now follows that the x_n, $n \geq N$, form a converging Cauchy sequence in $W_1 \cup W_2$. ◇

To conclude this section we mention the inverse process of pasting.

1.3.6 Definition. Let $F = S_1 + \ldots + S_m \bmod(\wp)$ be as above with the natural projection $\sigma : S_1 \cup \ldots \cup S_m \mapsto F$, and let

$$C = \sigma(\gamma_1) \cup \ldots \cup \sigma(\gamma_n) = \sigma(\gamma_1') \cup \ldots \cup \sigma(\gamma_n').$$

Then we say that $S_1, \ldots, S_m$ are obtained by *cutting F open along C*.

It is not difficult to see, for example, that if C is a set of disjoint simple closed geodesics on a hyperbolic surface S then S can be cut open along C.

In a similar way we shall frequently cut open surfaces along sets of piecewise geodesic curves.

1.4 The Universal Covering

Every complete unbordered surface of constant curvature -1 is universally covered by the hyperbolic plane (Cheeger-Ebin [1] or Klingenberg [1, 2]). In this section we adapt this to the complete hyperbolic surfaces with boundary. This is made easy due to the following theorem.

1.4.1 Theorem. *Every hyperbolic surface S with boundary is isometrically embedded in a hyperbolic surface S^* without boundary such that S is a deformation retract of S^*.*

Proof. We paste additional pieces to the boundary as follows. Along each closed boundary geodesic γ of length ℓ we paste a half-cylinder

$$[0, \infty[\times \mathbf{R}/[t \mapsto t + 1]$$

with the metric $ds^2 = d\rho^2 + \ell^2 \cosh^2\rho\, dt^2$ as in Example 1.3.3. Along each non-closed side of length a we paste a strip

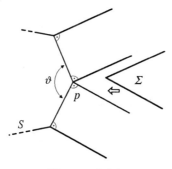

Figure 1.4.1

$$[0, \infty[\times [0, a]$$

with metric $ds^2 = d\rho^2 + \cosh^2\rho\, dt^2$. For sides of infinite length we proceed similarly. Note that the strips are right-angled. Finally, for each vertex p of S with interior angle ϑ we paste an infinite circle sector

$$\Sigma = \{p\} \cup \,]0, \infty[\times [0, \pi - \vartheta]$$

with metric $ds^2 = d\rho^2 + \sinh^2\rho\, d\sigma^2$ (cf. (1.1.8)) between the previously attached adjacent right-angled strips, as shown in Fig. 1.4.1. By Theorem 1.3.5, the surface S^* obtained in this way is hyperbolic, and we check that it has the required properties. ◇

In the following, the universal covering of a hyperbolic surface will be understood to carry the lifted structure so that it is again a hyperbolic surface. We recall that a hyperbolic surface is connected (cf. Definition 1.2.3).

1.4.2 Theorem. *Let S be a hyperbolic surface. The universal covering $\tilde{S}$ of S is isometric to a convex domain in $\mathbf{H}$ with piecewise geodesic boundary. If S is unbordered, then $\tilde{S}$ is isometric to $\mathbf{H}$.*

Proof. Let S^* be the unbordered hyperbolic surface as in Theorem 1.4.1 containing S as a subdomain. Then there exists a covering map $\pi : \mathbf{H} \to S^*$. Let $\tilde{S}$ be a connected component of $\pi^{-1}(S)$. Since all interior angles are less than or equal to π, $\tilde{S}$ is convex and, in particular, simply connected (cf. Beardon [1], p. 140). Hence, $\tilde{S}$ is a universal covering surface. Since all universal covering surfaces are isometric, this proves the theorem. ◇

1.4.3 Corollary. *Every simply connected hyperbolic surface is isometric to a convex domain in the hyperbolic plane with piecewise geodesic boundary.* ◇

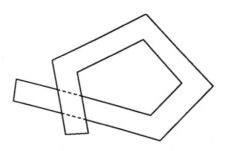

Figure 1.4.2

If we drop the condition that the interior angles of a hyperbolic surface be less than or equal to π, then the corollary is no longer true, as one can see in the example of Fig. 1.4.2. It shows a simply connected domain with piecewise geodesic boundary which is isometrically immersed in **H** but is not isometrically embeddable in **H**.

For better reference we note the following theorem in which $\pi : \tilde{S} \to S$ is the universal covering of a hyperbolic surface S, p is a point of S and $\tilde{p}$ is a covering point of p.

1.4.4 Theorem. (Unique lifting property). *Let I be an interval or a product of intervals and let $x \in I$. For every continuous mapping $\phi : I \to S$ satisfying $\phi(x) = p$, there exists a unique mapping $\tilde{\phi} : I \to \tilde{S}$ such that $\tilde{\phi}(x) = \tilde{p}$ and $\pi \circ \tilde{\phi} = \phi$.*

Proof. Macbeath [3], Section 2.3. ◇

1.5 Perpendiculars

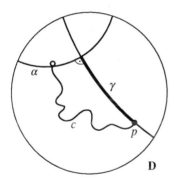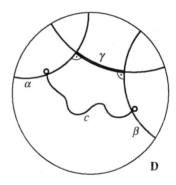

Figure 1.5.1

In the hyperbolic plane there is always a unique perpendicular from a point to a geodesic and a unique common perpendicular between two geodesics if the latter have positive distance. In this section we prove similar theorems for hyperbolic surfaces and various types of homotopy classes of curves.

In the following definition, S is a smooth surface with piecewise smooth boundary.

1.5.1 Definition. Let $A, B \subset S$ be closed connected subsets. Assume that $c, \gamma : [a, b] \to S$ are continuous curves with initial points $c(a), \gamma(a) \in A$ and endpoints $c(b)$, $\gamma(b) \in B$. We say that c is *homotopic to γ with endpoints gliding on A and B* if there exists a continuous map $\phi : [0, 1] \times [a, b] \to S$ such that

$$\phi(0, t) = c(t) \qquad \phi(s, a) \in A$$
$$\phi(1, t) = \gamma(t), \qquad \phi(s, b) \in B,$$
$$a \le t \le b, \qquad 0 \le s \le 1.$$

A typical example in the hyperbolic plane is given in Fig. 1.5.1 with $A = \alpha$ and $B = \{p\}$, respectively with $A = \alpha$ and $B = \beta$, where α and β are geodesics. If c is a rectifiable curve from p to α, respectively from β to α, and if we let c go (that is, we let it vary) in the defined homotopy classes, then c seeks its state of minimal energy and homotopes itself into the given perpendicular. (The verb *homotope* means to perform or to undergo a homotopy.)

In the following theorem and proof a similar situation is given on a surface.

1.5.2 Theorem. *Let S be a hyperbolic surface and let $c : [a, b] \to S$ be a curve with $c(a) \in A$ and $c(b) \in B$, where A and B are disjoint closed boundary geodesics of S. In the homotopy class of c with endpoints gliding on A and B there exists a unique geodesic γ. At its endpoints, γ meets A and B perpendicularly. All other points of γ lie in the interior of S.*

Proof. Let C be the given homotopy class of c. Then C contains smooth curves, and we let $\{\gamma_n\}_{n=1}^{\infty} \subset C$ be a sequence with lengths $\ell(\gamma_n)$ converging to the infimum L as $n \to \infty$. If we parametrize each curve in the interval $[a, b]$ with constant speed, then the family $\{\gamma_n\}_{n=1}^{\infty}$ is equicontinuous and the Arzelà-Ascoli theorem (Theorem A.19) yields a subsequence which converges to a rectifiable curve $\gamma : [a, b] \to S$ of length $\ell(\gamma) = L$.

Since the boundary of S is piecewise smooth, it is straightforward to see that γ also belongs to C. Since the interior angles of S are less than or equal to π, and since γ has minimal length, γ is a geodesic arc with only the endpoints on ∂S. Since A and B are closed geodesics and since γ has minimal length, γ meets A and B perpendicularly. (A variant of this method which is more elementary is sketched in the last section of the Appendix.)

To prove uniqueness, we consider the universal covering $\pi : \tilde{S} \to S$, where $\tilde{S} \subset \mathbf{H}$ (Theorem 1.4.2). We let $\tilde{A}$, $\tilde{B}$ and $\tilde{\gamma}$ be lifts of A, B and γ in $\mathbf{H}$ such that $\tilde{\gamma}$ is the common perpendicular of $\tilde{A}$ and $\tilde{B}$. Every homotopy of γ with endpoints gliding on A and B lifts to a homotopy of $\tilde{\gamma}$ with endpoints gliding on $\tilde{A}$ and $\tilde{B}$ (Theorem 1.4.4). The uniqueness of γ in S follows now from the uniqueness of the common perpendicular of $\tilde{A}$ and $\tilde{B}$ in $\mathbf{H}$. ◇

The first example in Fig. 1.5.2 shows two instances of Theorem 1.5.2. The next theorem is a completed version in which we collect a number of similar properties in various situations. In order to simplify the language we shall say that a closed connected subset A of a hyperbolic surface S is *admissible*, if either A consists of exactly one point or else A is a compact connected subset of ∂S. Observe the subtlety of some of the hypotheses.

1.5.3 Theorem. *Let S be a hyperbolic surface, let $A, B \subset S$ be admissible subsets and let $c : [a, b] \to S$ with $c(a) \in A$, $c(b) \in B$ be a curve from A to B (A and B need not be different or disjoint). Then the following hold.*

(i) *In the homotopy class of c with endpoints gliding on A and B there exists a shortest curve γ. This curve is a geodesic arc.*

(ii) *If γ is not contained in ∂S, then γ meets ∂S at most at its endpoints.*

(iii) *If $A = \{p\}$, $p \in S$, and if B is from one of the following categories:*
– *a smooth closed boundary geodesic,*
– *a side of S which meets its adjacent sides under an angle $\leq \pi/2$,*

then γ is either a point or a geodesic arc from p to B which meets B perpendicularly. In the latter case γ is the unique perpendicular geodesic from p to B in the homotopy class of c.

(iv) *If A and B are from the two categories in (iii) (but not necessarily from the same), and if γ is not a point, then γ is the unique common perpendicular from A to B in the homotopy class of c.*

(v) *If c in case (iv) is simple and γ is not a point, then γ is simple.*

(vi) *If A and B are points, then γ is unique.*

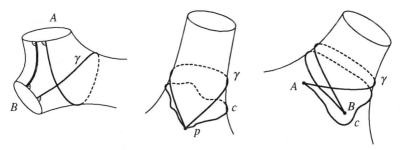

Figure 1.5.2

In (v) a curve $c : [a, b] \to S$ is called *simple* if c is an injective mapping. In Theorem 1.5.3 we have made an attempt to cover the most frequent cases. Fig. 1.5.2 illustrates some aspects. It shows that different homotopy classes may yield different common perpendiculars between boundary geodesics. It also gives an example where c is simple but γ is not.

Proof of Theorem 1.5.3. Points (i) - (iv) and (vi) have the same proof as Theorem 1.5.2. For the proof of (v), let us suppose that γ has a self-intersection. Then, since γ is not a closed geodesic, the self-intersection is transversal, and we find two lifts $\tilde{\gamma}$ and γ^* in the universal covering $\tilde{S}$ of S which intersect each other transversally. We let $\tilde{A}$ and $\tilde{B}$ be the lifts of A and B which contain the endpoints of $\tilde{\gamma}$. Similarly we let A^* and B^* be the lifts of A and B containing the endpoints of γ^* (Fig. 1.5.3). By Theorem 1.4.2, $\tilde{S}$ is a convex domain in, say, the unit disk **D**, and $\tilde{A}$, $\tilde{B}$, A^* and B^* are on the boundary of $\tilde{S}$. Lifting the homotopy between c and γ from S to $\tilde{S}$, we obtain covering curves $\tilde{c}$ homotopic to $\tilde{\gamma}$ and c^* homotopic to γ^* with endpoints on $\tilde{A}$ and $\tilde{B}$ respectively on A^* and B^*. By gliding the endpoints of c, if necessary, via a small homotopy which keeps c simple, we may assume that the endpoints of $\tilde{c}$ and c^* are not endpoints of $\tilde{A}$, A^*, $\tilde{B}$ and B^*.

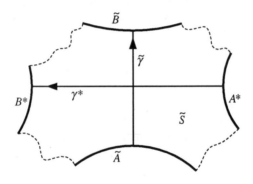

Figure 1.5.3

The interior of $\tilde{S} \subset \mathbf{D} \subset \mathbf{R}^2$ is a topological disk whose boundary (with respect to the topology of $\mathbf{R}^2$) is homeomorphic to a circle. Since $\tilde{\gamma}$ and γ^* intersect each other, the boundary arcs $\tilde{A}$ and $\tilde{B}$ separate the boundary arcs A^* and B^*. By the Jordan curve theorem, every curve in $\tilde{S}$ from A^* to B^* intersects the curves from $\tilde{A}$ to $\tilde{B}$. In particular, we find $x, y \in [a, b]$ such that $\tilde{c}(x) = c^*(y)$. Since c is simple, the relation $c(x) = \pi\tilde{c}(x) = \pi c^*(y) = c(y)$, where $\pi : \tilde{S} \to S$ is the covering map, implies that $x = y$, and hence $\tilde{c} = c^*$ by the unique lifting property (Theorem 1.4.4). Since the endpoints of $\tilde{c}$ are not endpoints of $\tilde{A}$, A^*, $\tilde{B}$ and B^*, it follows that $\tilde{A} = A^*$, and $\tilde{B} = B^*$, a contradiction. Theorem 1.5.3 is now proved. ◊

1.5.4 Example. A *loop*, that is, a curve $c : [a, b] \mapsto S$ such that $c(a) = c(b)$ is called *simple* if $c \,|\, [a, b[$ is injective. The following example shows that point (v) of the preceding theorem cannot be extended to simple loops. The domain shown in Fig. 1.5.4 is a geodesic rectangle with a hole (for further applications of this surface see Section 5.3). Curve c is a simple closed loop, but not simple in the sense of condition (v). If we let c go with endpoints gliding on A and B, we obtain a non-simple perpendicular γ from A to B. If we apply the preceding proof to this case, we will find that the covering curves $\tilde{c}$ and c^* coincide as point sets.

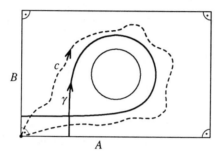

Figure 1.5.4

1.6 Closed Geodesics

We continue with the preceding theorems but now for the closed geodesics. The parametrization of closed curves in this section will be either on the real line, periodically, with period 1, or on the quotient $\boldsymbol{S}^1$:

(1.6.1) $$\boldsymbol{S}^1 = \mathbf{R}/[t \mapsto t + 1].$$

1.6.2 Definition. Let M be a topological space. Two closed curves $c, \gamma : \boldsymbol{S}^1 \to M$ are called *freely homotopic*, if there exists a continuous map $\phi : [0, 1] \times \boldsymbol{S}^1 \to M$ such that

$$\phi(0, t) = c(t), \quad \phi(1, t) = \gamma(t), \quad t \in \boldsymbol{S}^1.$$

Curves which are freely homotopic to a point are called *homotopically trivial*.

On compact hyperbolic surfaces or, more generally, on compact manifolds with negative curvature, the free homotopy classes and the closed geodesics are in one-to-one correspondence. To make this precise we adopt the following convention.

1.6.3 Definition. Two parametrized closed geodesics $\gamma, \gamma' : \boldsymbol{S}^1 \to M$ (of constant speed) are *equivalent* if there exists a homeomorphism $h : \boldsymbol{S}^1 \to$

S^1 of the form $h(t) = t + \text{const}$ such that $\gamma' = \gamma \circ h$. A *closed geodesic* is an equivalence class of closed parametrized geodesics.

Note that equivalent closed geodesics are always freely homotopic. Length, orientation and the property of being homotopic to other closed curves carry over from the parametrized closed geodesics in a natural way since these quantities are invariant under the above homeomorphisms. If no confusion arises, we identify closed geodesics with their parametrized representatives.

In Chapter 10, parametrized closed geodesics will also be considered equivalent if they differ by a parameter change of type $t \mapsto h(t) = \varepsilon t + c$, where $\varepsilon \in \{1, -1\}$ and c is a constant. In that case our present closed geodesics will be *oriented closed geodesics*.

1.6.4 Definition. Let γ and δ be closed geodesics and let $m \in \mathbf{Z} - \{0\}$. We say that γ is the *m-fold iterate* of δ, in symbols $\gamma = \delta^m$, if (for suitable parametrizations) $\gamma(t) = \delta(mt)$, $t \in S^1$. A closed geodesic (oriented or non-oriented) is *primitive* if it is not the m-fold iterate of another closed geodesic for some $m \geq 2$. An oriented primitive closed geodesic is called *prime geodesic*.

In this section we shall work in the disk model **D** of the hyperbolic plane. The points at the boundary of **D** are the *points at infinity*. If $c : \,]a, b[\, \to \mathbf{D}$ is a curve for which the limits

$$\lim_{t \downarrow a} c(t) \text{ and/or } \lim_{t \uparrow b} c(t)$$

exist and are points at infinity, then these points will be called the *endpoints at infinity* of c. Every geodesic in **D** has two endpoints at infinity.

1.6.5 Lemma. *Let S be a hyperbolic surface. Let c be a closed curve in the free homotopy class of a closed geodesic γ on S and let $\tilde{c}$, $\tilde{\gamma}$ be homotopic lifts in the universal covering $\tilde{S} \subset \mathbf{D}$. Then $\tilde{c}$ and $\tilde{\gamma}$ have the same endpoints at infinity.*

Proof. By Theorem 1.4.1 we may assume without loss of generality that S has empty boundary so that $\tilde{S} = \mathbf{D}$, and we apply the standard argument: the cyclic subgroup of the covering transformation group which leaves $\tilde{\gamma}$ invariant also leaves $\tilde{c}$ invariant. Hence, there exists a constant $d > 0$ such that $\text{dist}(\tilde{c}(t), \tilde{\gamma}(t)) \leq d$ for all $t \in \mathbf{R}$, and the lemma follows from (1.1.3).   ◇

In the next theorem we assume that every *closed* boundary component of S is parametrized as a *primitive* closed curve, that is, it cannot be written as the m-fold iterate of another closed curve with $m \geq 2$.

1.6.6 Theorem. *Let S be a compact hyperbolic surface and let c be a homotopically non-trivial closed curve on S. Then the following hold.*
  (i) *c is freely homotopic to a unique closed geodesic γ.*
  (ii) *γ is either contained in ∂S or $\gamma \cap \partial S = \emptyset$.*
  (iii) *If c is simple then γ is simple.*
  (iv) *If c is a non-smooth boundary component, then γ and c bound an embedded annulus.*

In the next theorem, intersection points are counted with multiplicities. That is, if c_1 and c_2 are curves intersecting each other at point p, and if p is a self-intersection point of c_i with multiplicity m_i, $i = 1, 2$, then p counts as $m_1 m_2$ intersection points of c_1 and c_2.

1.6.7 Theorem. *Let $n \geq 0$ be an integer and consider two non-trivial closed curves c and c' on a compact hyperbolic surface S which intersect each other in n points. If γ and γ' are the closed geodesics in their free homotopy classes, then either γ and γ' coincide as point sets or they intersect each other in at most n points.*

Proof. We prove the first theorem and leave the proof of the second theorem as an exercise. The existence of γ is proved as in Theorem 1.5.2 with the Arzelà-Ascoli argument. The compactness of S is needed to assure that the sequence $\{\gamma_n\}_{n=1}^{\infty}$ stays within bounded distance.

As in Theorem 1.5.2, property (ii) follows from the fact that γ has minimal length and that all vertices of S have interior angles $\leq \pi$. For the uniqueness in (i) we consider a universal covering $\pi : \tilde{S} \to S \subset \mathbf{D}$ (Theorem 1.4.2). Let δ be another closed geodesic in the free homotopy class of c and γ. Lift the homotopy to $\tilde{S}$ to obtain homotopic lifts $\tilde{\gamma}$, $\tilde{\delta}$ in $S \subset \mathbf{D}$ (Theorem 1.4.4). They have the same endpoints at infinity (Lemma 1.6.5). This proves that $\tilde{\gamma}$ and $\tilde{\delta}$ coincide as point sets. As in the proof of Lemma 1.6.5, there exists a constant $d > 0$ such that $\mathrm{dist}(\tilde{\gamma}(\tau), \tilde{\delta}(\tau)) \leq d$ for all $\tau \in \mathbf{R}$. It follows that $\tilde{\delta}(\tau) = \tilde{\gamma}(\tau + \omega)$ for some constant ω. (Recall that both curves are parametrized with constant speed.) Hence, $\delta(\tau) = \gamma(\tau + \omega)$, $\tau \in \mathbf{S}^1$ so that δ and γ are equal in the sense of Definition 1.6.3. This proves the uniqueness in (i).

The proof of (iii) is of considerable length; Example 1.6.9 below may explain the difficulty. Let $\tilde{c}$ and $\tilde{\gamma}$ be homotopic lifts of c and γ with the same endpoints at infinity. In view of Theorem 1.4.1 we may again assume that $\partial S = \emptyset$ so that $\tilde{S} = \mathbf{D}$. If γ is not simple, then it either has a transversal self-intersection or it is the k-fold iterate of a primitive closed geodesic γ_1 for some $k \geq 2$.

Let us begin with the second case. Let $T : \mathbf{D} \to \mathbf{D}$ be the covering transformation satisfying $T(\tilde{\gamma}(\tau)) = \tilde{\gamma}(\tau + 1)$, $\tau \in \mathbf{R}$. The relation $\gamma = \gamma_1^k$ implies that there is a covering transformation T_1 satisfying $T_1(\tilde{\gamma}(\tau)) = \tilde{\gamma}(\tau + 1/k)$ and $T = T_1^k$. We introduce Fermi coordinates (ρ, t) based on the axis $\tilde{\gamma}$ (cf. Section 1.1) with ρ the directed distance to $\tilde{\gamma}$. Since S is an orientable surface, T_1 preserves orientation and thus has the following form

$$T_1(\rho, t) = (\rho, t + \ell(\gamma)/k).$$

For the curves $\tau \mapsto \tilde{c}(\tau) = (\rho(\tau), t(\tau))$, and $\tau \mapsto T_1(\tilde{c}(\tau)) = (\rho_1(\tau), t_1(\tau))$, $\tau \in \mathbf{R}$, we find therefore parameter values τ' and τ'' satisfying

$$\rho(\tau') \leq \rho_1(\tau) \quad \text{for all } \tau \in \mathbf{R},$$
$$\rho(\tau'') \geq \rho_1(\tau) \quad \text{for all } \tau \in \mathbf{R}.$$

By the Jordan curve theorem, $T_1(\tilde{c})$ intersects $\tilde{c}$, i.e. there exists τ_0 such that $T_1(\tilde{c}(\tau_0)) \in \tilde{c}$. Now $\tilde{c}(\tau_0)$ and $T_1(\tilde{c}(\tau_0))$ are covering points of the same point $p \in c$. On the other hand, since c is simple, the only covering points of p which lie on $\tilde{c}$ are the points $T^m(\tilde{c}(\tau_0))$, $m \in \mathbf{Z}$. Hence, $T_1(\tilde{c}(\tau_0)) = T^m(\tilde{c}(\tau_0))$ for some m. Now $T = T_1^k$ so that $T_1(\tilde{c}(\tau_0)) = T_1^{mk}(\tilde{c}(\tau_0))$. Since T_1 has no fixed point this implies $k = 1$, a contradiction.

Next, assume that γ has a transversal self-intersection. Then the arguments are essentially as in the proof of (v) of Theorem 1.5.3. Namely: let p be an intersection point, let $\tilde{\gamma} \subset \tilde{S} = \mathbf{D}$ be a lift of γ and let $\tilde{p} \in \tilde{\gamma}$ be a lift of p.

There exists a covering transformation R such that $R(\tilde{\gamma})$ intersects $\tilde{\gamma}$ under some positive angle at $\tilde{p}$ as shown in Fig. 1.6.1. Lift the homotopy between γ and c to $\mathbf{D}$. The resulting covering curves $\tilde{c}$ and $R(\tilde{c})$ have the same endpoints at infinity as $\tilde{\gamma}$ and $R(\tilde{\gamma})$, (Lemma 1.6.5). The Jordan curve theorem

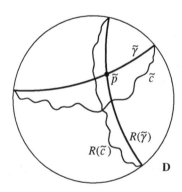

Figure 1.6.1

implies that $\tilde{c}$ intersects $R(\tilde{c})$. Since c is simple, the unique lifting property implies that for some constant ω, $R(\tilde{c}(\tau)) = \tilde{c}(\tau + \omega)$, a contradiction. A similar argument yields the proof of Theorem 1.6.7.

(iv). Since c is not smooth, $\gamma \cap c = \emptyset$, by (ii). By (iii), γ is simple. The proof that γ and c bound an embedded annulus is of purely topological nature and is postponed to the Appendix (Proposition A.11). ◇

The following examples show that in the non-compact case and also on non-orientable surfaces, some of the statements in Theorem 1.6.6 no longer hold.

1.6.8 Example. (Cusps). Let S be a non-compact hyperbolic surface and let $\mathscr{E} \subset S$ be a domain which is isometric to the surface

(1) $$]-\infty, 0] \times \boldsymbol{S}^1 =]-\infty, 0] \times \mathbf{R}/[t \mapsto t+1]$$

with the Riemannian metric

(2) $$ds^2 = d\rho^2 + e^{2\rho} dt^2$$

(cf. (1.1.10)). Such a domain is called a *cusp*. Likewise we may describe $\mathscr{E}$ (up to isometry) as the quotient $\mathscr{E} = \Gamma \backslash \tilde{\mathscr{E}}$, where

$$\tilde{\mathscr{E}} = \{ x + iy \in \mathbf{H} \mid y \geq 1 \},$$

and Γ is the cyclic subgroup $\Gamma = \{ T^m \mid m \in \mathbf{Z} \} \subset \mathrm{Is}^+(\mathbf{H})$ with $T(z) = z + 1$ for all $z \in \mathbf{H}$.

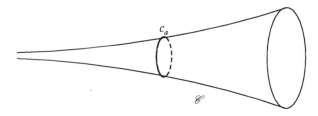

Figure 1.6.2

The horocycles $h_a = \{ x + iy \in \mathbf{H} \mid y = a \}$, $a \geq 1$, project to curves c_a of length $1/a$ on $\mathscr{E}$. They all belong to the same non-trivial free homotopy class. This class contains no closed geodesic.

1.6.9 Example. (Non-orientable surfaces). Let S be a compact hyperbolic surface whose boundary is a closed geodesic γ. Parametrize γ in the form $\gamma : \boldsymbol{S}^1 \to S$. Identify opposite points on γ via the pasting scheme

(1) $$\gamma(\tau) = \gamma(\tau + \tfrac{1}{2}), \qquad \tau \in \boldsymbol{S}^1.$$

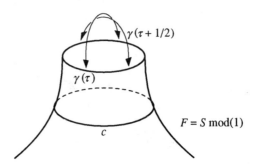

Figure 1.6.3

The quotient $F = S \bmod(1)$ is a compact non-orientable surface of constant curvature -1. The hyperbolic atlas is constructed as in Section 1.3 except that we now admit orientation reversing overlap maps. (The construction is similar to that of the projective plane.)

If we interpret (1) as the definition of a curve in F, then γ is the *2-fold iterate* of another closed geodesic. Curve c in Fig. 1.6.3 is the example of a simple closed curve which is homotopic to a non-simple closed geodesic.

We may use the two-fold orientable covering surface to prove that a non-trivial simple closed curve on a non-orientable compact hyperbolic surface is homotopic to either a simple closed geodesic or a 2-fold iterate of a simple closed so-called one-sided geodesic.

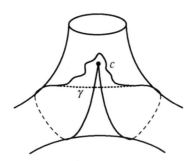

Figure 1.6.4

1.6.10 Example. (Geodesic loops). A geodesic arc $\gamma : [a, b] \to S$ is called a *geodesic loop* if $\gamma(a) = \gamma(b)$. By Theorem 1.5.3(vi), if $c : [a, b] \to S$ with $c(a) = c(b)$ is a closed curve, then the homotopy class of c with fixed endpoints contains a unique geodesic loop. Fig. 1.6.4 shows an example where c is simple and γ has self-intersections.

As a final result in this section we show that (in contrast for instance to the flat torus) the closed geodesics on a compact hyperbolic surface form a discrete set. Recall that closed geodesics which differ by a parameter change of type $t \mapsto t + \text{const}$ are considered equal (Definition 1.6.3).

1.6.11 Theorem. *Let S be a compact hyperbolic surface and let $L > 0$. Only finitely many closed geodesics on S have length $\leq L$.*

Proof. By compactness we may cover S with finitely many coordinate neighborhoods. Hence, there exists a constant $r > 0$ such that for any $x \in S$ the points at distance smaller than $4r$ form a convex neighborhood.

Assume now that there exists an infinite sequence of pairwise different closed geodesics $\gamma_1, \gamma_2, \ldots$ on S of length $\ell \leq L$, say parametrized on the interval $[0, 1]$. Then we may extract a subsequence such that the initial points, the initial tangent vectors and the lengths converge. We find therefore $n \neq k$ such that $\text{dist}(\gamma_n(t), \gamma_k(t)) \leq r$ for all $t \in [0, 1]$. It follows that γ_n and γ_k are homotopic. By Theorem 1.6.6, $\gamma_n = \gamma_k$, a contradiction. ◊

An estimate of the number of closed geodesics of length $\ell \leq L$ will follow in Lemma 6.6.4, and an asymptotic formula (for S without boundary) will be given in Theorem 9.4.14.

1.7 The Fenchel-Nielsen Parameters

In this section we give an overview of the Fenchel-Nielsen parameters and the corresponding construction of the compact Riemann surfaces. Fenchel and Nielsen used this construction in their celebrated manuscript [1], but we already find it in the work of Koebe [1], Löbell [1], and, implicitly, in Fricke-Klein [1] (see Keen [1, 2, 3]). More recently, the Fenchel-Nielsen parameters have been brought to new life through Thurston's work on hyperbolic manifolds (Thurston [1], Fathi-Laudenbach-Poenaru [1]).

The construction is based on the pasting of geodesic hexagons. Let a, b, c be arbitrary positive real numbers.

1.7.1 Lemma. *There exists a right-angled geodesic hexagon in the hyperbolic plane with pairwise non-adjacent sides of length a, b, c.*

Proof. Consider the perpendicular geodesics β, α and γ in **H** as in Fig. 1.7.1. Let B be the line of all points in **H** at distance c from β which lie on

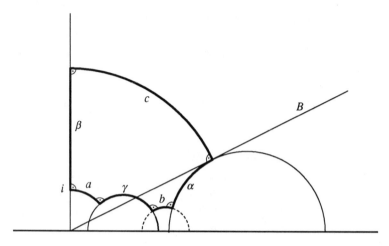

Figure 1.7.1

the same side of β as γ. In the upper half-plane model, B is a (Euclidean) straight line (because the mappings $z \mapsto \lambda z$, $\lambda > 0$ are isometries). Now consider a moving geodesic α tangent to B. Move α along B until dist$(\gamma, \alpha) = b$, then draw the common perpendicular of γ and α. This yields the hexagon. ◇

In the first step of the construction we paste two copies of such a hexagon together along the remaining three sides to obtain a complete hyperbolic surface Y with three closed boundary geodesics. (For the parametrization of the sides and the boundary geodesics, see Section 3.1.) By Lemma 1.7.1, the lengths of the boundary geodesics may be prescribed arbitrarily. We shall prove in Section 3.1 that they determine Y up to isometry (Theorem 3.1.7).

In the next step we let G be a cubic graph with $2g - 2$ vertices and $3g - 3$ edges, i.e. a combinatorial pattern such as that in Fig. 1.7.3 (cf. Section 3.5). To every vertex y of G we associate one of the above surfaces Y and paste these surfaces together according to G as shown in Fig. 1.7.3.

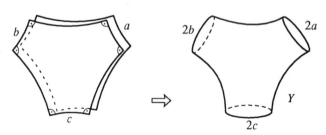

Figure 1.7.2

Boundary geodesics which come together must have the same length. On the resulting surface S we obtain $3g - 3$ closed geodesics which correspond to the $3g - 3$ edges of G. By the earlier remarks, the lengths $\ell_1, \ldots, \ell_{3g-3}$ of these geodesics may be prescribed arbitrarily. In addition to these lengths, we may prescribe twist parameters $\alpha_1, \ldots, \alpha_{3g-3}$ for the pasting of the closed geodesics as in Example 1.3.3. This yields a $(6g - 6)$-dimensional space $\mathcal{T}_G$ of examples of compact hyperbolic surfaces of genus g, modelled over the graph G. We shall see in Theorem 3.6.4 that every compact unbordered hyperbolic surface may be obtained in this way, and in Theorem 6.2.7 we shall see that $\mathcal{T}_G$ is a natural model of the so-called Teichmüller space. The parameters $\ell_1, \ldots, \ell_{3g-3}$ and $\alpha_1, \ldots, \alpha_{3g-3}$ are the *Fenchel-Nielsen parameters* of such constructed surfaces.

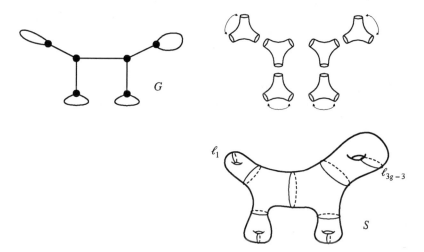

Figure 1.7.3

We add a remark concerning the conformal structures. By definition, a *conformal structure* on an orientable two-dimensional manifold is a maximal atlas in which all overlap maps are holomorphic. Since the overlap maps in a hyperbolic atlas are from $PSL(2, \mathbf{R})$ (if we use the upper half plane $\mathbf{H}$), every hyperbolic structure induces a conformal structure. Hence, the hyperbolic surfaces without boundary are Riemann surfaces in the sense of complex analysis.

Conversely, let F be a compact Riemann surface of genus $g \geq 2$. By the uniformization theorem (e.g Beardon [2], Farkas-Kra [1], Strebel [1], etc.), there exists a conformal universal covering map $\pi : \mathbf{H} \to F$. The covering transformations are conformal self mappings of $\mathbf{H}$ and belong therefore to $PSL(2, \mathbf{R})$. Hence, the local inverses of π define a subatlas of the conformal structure of F in which all overlap mappings are restrictions of elements of PSL(2, **R**). It follows that every compact Riemann surface in the sense of complex analysis is a compact unbordered hyperbolic surface. Therefore, the Fenchel-Nielsen construction also yields an overview of the conformal structures. This leads us to the following definitions.

1.7.2 Definition. A compact hyperbolic surface without boundary is called a *compact Riemann surface*.

For the bordered case we adopt the following terminology from the topology of surfaces. A compact connected orientable topological surface with boundary is said to have *signature* (g, n) if it is obtained from a closed surface of genus g by cutting away the interior of n disjoint closed topological disks. For example, the above building blocks Y have signature $(0, 3)$. A surface of signature $(g, 0)$ is, by definition, a compact surface of genus g without boundary.

1.7.3 Definition. A compact hyperbolic surface of signature (g, n) is a *Riemann surface of signature* (g, n) if every boundary component is a smooth closed geodesic.

Chapter 2

Trigonometry

Hyperbolic trigonometry is a basic tool in various studies, and various approaches are known (cf. e.g. Beardon [1], Fenchel [1], Meschkowski [1], Perron [1], Rees [1], Thurston [1]). This chapter gives an account based on the hyperboloid model which has grown out of discussions with Patrick Eberlein and Klaus-Dieter Semmler. In the first part we use the isometry group acting on the hyperboloid model to obtain the trigonometric formulae of the triangle by comparing matrix elements. In the second part, beginning with Section 4, we generalize this to hexagons and similar configurations. Then we briefly sidestep to variable curvature. In the final part, Section 6, we describe a variant of the approach which is from Semmler [1] and uses a vector product and quaternions.

The results of Sections 2.1 - 2.4 will be widely used in this book. Sections 2.5 and 2.6 are appendices and may be skipped in a first reading.

2.1 The Hyperboloid Model

The starting point is the bilinear form

(2.1.1) $\quad h(X, Y) = x_1 y_1 + x_2 y_2 - x_3 y_3, \quad X, Y \in \mathbf{R}^3,$

where x_i, y_i are the coordinates of X, Y. The linear mappings L_σ, $M_\rho \in \mathrm{GL}(3, \mathbf{R})$ given by

(2.1.2) $\quad L_\sigma = \begin{pmatrix} \cos\sigma & -\sin\sigma & 0 \\ \sin\sigma & \cos\sigma & 0 \\ 0 & 0 & 1 \end{pmatrix}, \quad M_\rho = \begin{pmatrix} \cosh\rho & 0 & \sinh\rho \\ 0 & 1 & 0 \\ \sinh\rho & 0 & \cosh\rho \end{pmatrix}$

(acting on column vectors) leave h and also the corresponding differential form

$$h^1 = dx_1^2 + dx_2^2 - dx_3^2$$

invariant. We let $\Omega \subset GL(3, \mathbf{R})$ be the subgroup generated by all L_σ and M_ρ; $\sigma, \rho \in \mathbf{R}$. Now consider the hyperboloid

(2.1.3) $\qquad H = \{X \in \mathbf{R}^3 \mid h(X, X) = -1, x_3 > 0\}$

and let g be the quadratic differential form on H which is obtained by restricting h^1 to the tangent vectors of H. Then H and g are invariant under Ω. Moreover, we have

2.1.4 Lemma. *Ω acts twice transitively on H.* ◇

Since g is positive definite on the tangent plane of the point

$$p_0 = \begin{pmatrix} 0 \\ 0 \\ 1 \end{pmatrix} \in H,$$

the two-fold transitivity implies that $H = (H, g)$ is a complete two-point homogeneous Riemannian manifold, and $\Omega = \mathrm{Is}^+(H)$. We could now use the constant curvature of H to prove that (possibly up to a scaling factor) H is a model of the hyperbolic plane, but it is easy to give a direct proof. The proof runs as follows. For fixed σ, the curve

$$\rho \mapsto L_\sigma M_\rho(p_0), \qquad \rho \geq 0,$$

has unit speed on H, and for every fixed $\rho > 0$ the curve

$$\sigma \mapsto L_\sigma M_\rho(p_0), \qquad \sigma \in \mathbf{R},$$

has constant speed $\sinh \rho$ and intersects orthogonally the curves of the first type. If we introduce therefore (ρ, σ) as the pair of coordinates of the point $p = L_\sigma M_\rho(p_0)$, for $0 < \rho < \infty$ and $-\pi \leq \sigma < \pi$, then the coordinate system obtained in this way covers $H - \{p_0\}$, and the tensor g has the expression $g = ds^2 = d\rho^2 + \sinh^2\rho \, d\sigma^2$. This proves that H is isometric to $\mathbf{H}$ and that (ρ, σ) is the pair of geodesic polar coordinates centered at p_0. For later application we shall denote by μ the geodesic

(2.1.5) $\qquad r \mapsto \mu(r) := M_r(p_0), \qquad -\infty < r < \infty.$

Observe that μ is invariant under M_ρ for all $\rho \in \mathbf{R}$.

At some places we shall be obliged to calculate in Ω. Since every $A \in \Omega$ leaves the quadratic form h invariant, we have that

(2.1.6) $$A^{-1} = SA^tS, \quad A \in \Omega,$$

where A^t is the transpose of A, and S denotes the diagonal matrix with diagonal elements $1, 1, -1$.

2.2 Triangles

For geodesic triangles in the hyperbolic plane, we denote by a, b, c the sides and by α, β, γ the corresponding opposite angles. These letters also denote the side-length and the angular measure. If the triangle is right-angled, we usually let γ be the right angle.

2.2.1 Theorem. (Ordinary triangles). *The following formulae hold for geodesic triangles in the hyperbolic plane.*

(i) $\quad \cosh c = -\sinh a \sinh b \cos \gamma + \cosh a \cosh b,$
(ii) $\quad \cos \gamma = \sin \alpha \sin \beta \cosh c - \cos \alpha \cos \beta,$
(iii) $\quad \sinh a : \sin \alpha = \sinh b : \sin \beta = \sinh c : \sin \gamma.$

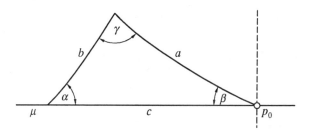

Figure 2.2.1

Proof. Let the triangle T, say, be placed in H such that β is at p_0, the "origin" (cf. Section 2.1), α is at $M_{-c}(p_0)$ and γ is at $L_{\pi-\beta}(M_a(p_0))$, as shown in Fig. 2.2.1. The isometry $L_{\pi-\alpha}M_c$ first parallel translates T along μ, bringing α to p_0, and then rotates T about p_0 by an angle $\pi - \alpha$. Hence, $L_{\pi-\alpha}M_c$ brings T back into a position like that of Fig. 2.2.1, but now with side b, rather than side c, on μ, and with α, rather than β, at p_0. Clearly, the product $L_{\pi-\beta}M_aL_{\pi-\gamma}M_bL_{\pi-\alpha}M_c$ brings T back into its original position. Hence, this product is the identity, and the following relationship holds:

(1) $$M_aL_{\pi-\gamma}M_b = L_{\pi-\beta}M_{-c}L_{\alpha-\pi}.$$

Computing the components in (1) we obtain nine identities, four of which are those of the theorem. ◇

2.2.2 Theorem. (Right-angled triangles). *For any hyperbolic triangle with right angle γ, the following hold.*

(i) $\cosh c = \cosh a \cosh b$,
(ii) $\cosh c = \cot \alpha \cot \beta$,
(iii) $\sinh a = \sin \alpha \sinh c$,
(iv) $\sinh a = \cot \beta \tanh b$,
(v) $\cos \alpha = \cosh a \sin \beta$,
(vi) $\cos \alpha = \tanh b \coth c$.

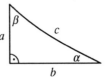

Proof. The first three identities are the restatement of Theorem 2.2.1 in the case $\gamma = \pi/2$. The remaining three identities are obtained via cyclic permutation and elementary computations   ◊

In hyperbolic geometry, angles and perpendiculars are, in some sense, the same thing, and there are various configurations whose trigonometry is similar to that of the triangle. The right-angled hexagon of Lemma 1.7.1 is an example. In the remainder of this section we give a general definition of this configuration - the generalized triangle - and prove a relation which will later imply all trigonometric formulae. For another configuration which unites triangles and hexagons we refer to the book of Fenchel [1]. The uniformizing configuration in Fenchel's book is the general right-angled geodesic hexagon in hyperbolic three space (any pair of consecutive sides is orthogonal, but the sides of the hexagon are in general not coplanar).

In what follows, a *polygon P* is a piecewise geodesic *oriented* closed *curve*, possibly with self-intersections. The geodesic arcs of P are the *sides*. We do not admit sides of length zero.

Let (u, w) be an ordered pair of consecutive sides of P with respect to the given orientation, and let p be the common vertex of u and w. The *angle* of P at p is defined to be the rotation $v \in \text{Is}^+(\mathbf{H})$ which fixes p and rotates w to u. We shall say that v is the *subsequent angle* of side u, and w is the *subsequent side* of angle v.

With v we also denote the corresponding *angle of rotation*. To allow addition and subtraction, angles of rotation are considered elements of the

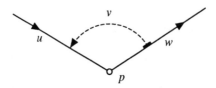

Figure 2.2.2

group $S^1 = \mathbf{R}/[s \mapsto s + 2\pi]$. Our next aim is to define a concept which unites the various configurations of Fig. 2.2.3.

2.2.3 Definition. Let x and y belong to the set of sides and angles of polygon P. The ordered pair (x, y) is said to be of *angle type* if one of the following conditions holds:
  (i) y is the subsequent angle of side x,
  (ii) (x, y) is a pair of consecutive sides, and y is orthogonal to x,
  (iii) y is the subsequent side of angle x.

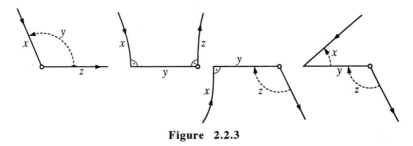

Figure 2.2.3

In Fig. 2.2.3 all pairs (x, y) and (y, z) are of angle type. To each pair (x, y) of angle type we associate an isometry $N_y \in \text{Is}^+(\mathbf{H})$ as follows (L and M are as in (2.1.2)).

2.2.4 Definition. (i) If y is the subsequent angle of side x, then we define

$$N_y := L_{\pi-y} = \begin{pmatrix} -\cos y & -\sin y & 0 \\ \sin y & -\cos y & 0 \\ 0 & 0 & 1 \end{pmatrix}.$$

(ii) If (x, y) is a pair of consecutive sides with angle $\varepsilon\pi/2$, where $\varepsilon = \pm 1$, then we define

$$N_y := L_{\varepsilon\pi/2} M_y = \begin{pmatrix} 0 & -\varepsilon & 0 \\ \varepsilon \cosh y & 0 & \varepsilon \sinh y \\ \sinh y & 0 & \cosh y \end{pmatrix}.$$

(iii) If y is the subsequent side of angle x, then we define

$$N_y := M_y = \begin{pmatrix} \cosh y & 0 & \sinh y \\ 0 & 1 & 0 \\ \sinh y & 0 & \cosh y \end{pmatrix}.$$

2.2.5 Definition. A *generalized triangle* is a closed oriented hyperbolic geodesic polygon P together with a cycle $a, \gamma, b, \alpha, c, \beta$ of consecutive sides and angles of P in which all pairs (a, γ), (γ, b), ..., (c, β), (β, a) are of angle type.

The polygons in Figures 2.2.1, 2.2.4, 2.3.1, 2.4.1, 2.4.3 and 2.6.2 are generalized triangles.

We adopt the convention that all sides of P have *positive* length, independently of the orientation, whereas the angles have values in $\mathbf{S}^1$. If angular measures with values in $[0, \pi]$ are preferred, as e.g. in Section 2.6, then the signs in the formulae which follow must be adjusted accordingly.

If P is the boundary of a convex domain in $\mathbf{H}$, we shall always orient P positively, so that the angles of rotation will be the interior angles of P.

2.2.6 Theorem. *For every generalized triangle* $a, \gamma, b, \alpha, c, \beta$ *we have*
$$N_a N_\gamma N_b = (N_\alpha N_c N_\beta)^{-1}.$$

Proof. A pair (x, y) of angle type in the generalized triangle T, say, is said to have *standard position* if it is placed as in Fig. 2.2.3. That is, if y is an angle, then the subsequent side z lies on the geodesic μ (cf. (2.1.5)) with initial point at p_0 and endpoint at $M_z(p_0)$; if y is a side, then y lies on μ with initial point at $M_{-y}(p_0)$ and endpoint at p_0. One checks with Definition 2.2.4 that if (x, y) has standard position and if N_y is applied to T, then the preceding pair, say (w, x), moves into standard position. Therefore, if T is placed such that (α, c) has standard position (as in Fig. 2.2.4 and Fig. 2.2.1) and if we successively apply $N_c, N_\alpha, \ldots, N_\beta$, then T returns to its original position. Hence, $N_\beta N_a N_\gamma N_b N_\alpha N_c = id$. ◇

2.2.7 Example. In the generalized triangles of Fig. 2.2.4, we have in both cases $N_a N_\gamma N_b = L_{\pi/2} M_a L_{\pi-\gamma} M_b$. Notice the difference between the two examples for the product $N_\alpha N_c N_\beta$.

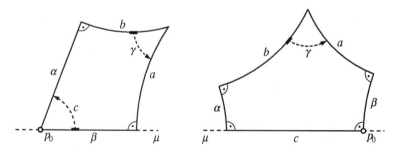

Figure 2.2.4

In the first example,
$$N_\alpha N_c N_\beta = L_{\pi/2} M_\alpha L_{\pi-c} M_\beta,$$
while in the second example,
$$N_\alpha N_c N_\beta = L_{\pi/2} M_\alpha L_{\pi/2} M_c L_{\pi/2} M_\beta.$$
Theorem 2.2.6 yields for the first example

(i) $\quad \cos c = -\cosh a \cosh b \cos\gamma + \sinh a \sinh b,$
(ii) $\quad \cosh a : \cosh\alpha = \cosh b : \cosh\beta = \sin c : \sin\gamma.$

The corresponding identities for the second example are

(iii) $\quad \cosh c = -\cosh a \cosh b \cos\gamma + \sinh a \sinh b,$
(iv) $\quad \cosh a : \sinh\alpha = \cosh b : \sinh\beta = \sinh c : \sin\gamma,$
(v) $\quad \cos\gamma = \sinh\alpha \sinh\beta \cosh c - \cosh\alpha \cosh\beta.$

2.3 Trirectangles and Pentagons

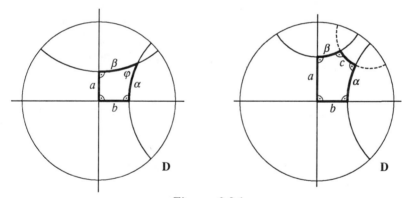

Figure 2.3.1

Consider two perpendicular geodesic arcs, a and b, in **D** (the disk model of the hyperbolic plane) with a common vertex at the origin as in Fig. 2.3.1. Let α and β be the perpendicular geodesics at the endpoints of b and a. If a and b are small, then α and β intersect each other at an acute angle φ and we obtain a geodesic quadrilateral with three right angles. Following Coxeter [1] we call this configuration a *trirectangle*. By abuse of notation we also denote by α and β the sides of the trirectangle.

Trirectangles have played a decisive role in the discovery of non-Euclidean geometry (see for instance the very early approach of Saccheri [1]). They occur in many geometric constructions and are useful for computations. The following formulae, similar to those of the right-angled triangle in Theorem 2.2.2, hold.

2.3.1 Theorem. (Trirectangles). *For every trirectangle with sides labelled as in Fig. 2.3.1 the following relations are true*:

(i) $\cos\varphi = \sinh a \sinh b$,
(ii) $\cos\varphi = \tanh\alpha \tanh\beta$,
(iii) $\cosh a = \cosh\alpha \sin\varphi$,
(iv) $\cosh a = \tanh\beta \coth b$,
(v) $\sinh\alpha = \sinh a \cosh\beta$,
(vi) $\sinh\alpha = \coth b \cot\varphi$.

Proof. The first three identities are the restatement of formulae (i) and (ii) of Example 2.2.7 for the case $\gamma = \pi/2$. The remaining ones are obtained via cyclic permutation and elementary computations. ◇

As an example we prove the distance formula for Fermi coordinates. We let (ρ, t) denote the Fermi coordinates with respect to a fixed base line η in **H**, (cf. Section 1.1), and let $p_1 = (\rho_1, t_1)$ and $p_2 = (\rho_2, t_2)$ be points in **H**. Then the following formula holds. Note that in this formula the quantities ρ_1 and ρ_2 are *oriented* lengths and have opposite signs when the corresponding sides lie on opposite sides of η.

(2.3.2) $\cosh \text{dist}(p_1, p_2) = \cosh\rho_1 \cosh\rho_2 \cosh(t_2 - t_1) - \sinh\rho_1 \sinh\rho_2$.

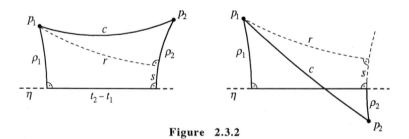

Figure 2.3.2

Proof. We give a proof which is based on a decomposition into trirectangles and right-angled triangles. Another proof would consist of understanding the figure as a generalized triangle and to apply Theorem 2.2.6 as in Example 2.2.7.

We may assume that $\rho_1 \geq 0$. Dropping the perpendicular r from p_1 to that geodesic through p_2 which is orthogonal to η, we obtain a trirectangle with sides ρ_1, $|t_2 - t_1|$, s, say, and r and also a right-angled triangle with sides r, $|\rho_2 - s|$ and c, whose hypothenuse c has length $\text{dist}(p_1, p_2)$. The trirectangle is again decomposed into two right-angled triangles with common hypothenuse. Formula (i) of Theorem 2.2.2 and formula (v) of Theorem 2.3.1 yield the relations $\cosh r \cosh s = \cosh \rho_1 \cosh(t_2 - t_1)$, and $\cosh r \sinh s = \sinh \rho_1$. Hence,

$$\begin{aligned} \cosh \text{dist}(p_1, p_2) &= \cosh r \cosh(\rho_2 - s) \\ &= \cosh r \cosh \rho_2 \cosh s - \cosh r \sinh \rho_2 \sinh s \\ &= \cosh \rho_1 \cosh \rho_2 \cosh(t_2 - t_1) - \sinh \rho_1 \sinh \rho_2. \end{aligned}$$ ◇

We return to the trirectangle of Fig. 2.3.1. If side a grows continuously, the vertex at angle φ moves towards the endpoint at infinity of geodesic α. Let the limiting position be obtained for $a = a_b$. As $a \to a_b$, $\varphi \to 0$. This follows from Theorem 2.3.1(vi) or by a glance at Fig. 2.3.1. From Theorem 2.3.1(i) we obtain, by continuity,

(2.3.3) $$\sinh a_b \sinh b = 1.$$

If a grows beyond a_b, the angle φ disappears and is replaced by c, the common perpendicular of α and β. We obtain a *right-angled pentagon*.

2.3.4 Theorem. (Right-angled pentagons). *For any right-angled pentagon with consecutive sides a, b, α, c, β we have*:

(i) $\cosh c = \sinh a \sinh b$,
(ii) $\cosh c = \coth \alpha \coth \beta$.

Proof. This is Example 2.2.7(iii) - (v) in the particular case $\gamma = \pi/2$. ◇

2.3.5 Lemma. *Let a and b be any positive real numbers satisfying $\sinh a \sinh b > 1$. Then there exists a unique right-angled geodesic pentagon with two consecutive sides of lengths a and b.*

Proof. a and b are the consecutive orthogonal sides of either a trirectangle, a limiting trirectangle with a vertex at infinity, or a right-angled pentagon. By Theorem 2.3.1(i) and by (2.3.3), the first two cases are excluded. The uniqueness follows from Theorem 2.3.4 (or Theorem 1.1.6). ◇

2.4 Hexagons

If we paste two right-angled pentagons together along a common side r, we obtain a right-angled hexagon. Every compact Riemann surface of genus greater than one can be obtained by pasting together such hexagons (Chapter 3). The boundary of a right-angled hexagon is a generalized triangle with consecutive sides $a, \gamma, b, \alpha, c, \beta$. We use the word *hexagon* for both the domain and its boundary. The hexagon is called *convex* if it is convex as a domain. Fig. 2.4.3 shows a right-angled hexagon which is not the boundary of a convex domain.

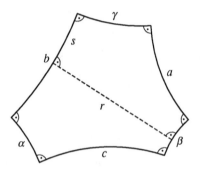

Figure 2.4.1

The pairs $(a, \gamma), \ldots, (\beta, a)$ are all of type (ii) from Definition 2.2.3, with the associated matrices $N_a = L_{\pi/2}M_a$, $N_\gamma = L_{\pi/2}M_\gamma$, ... (cf. (ii) in Definition 2.2.4). The next theorem follows from Theorem 2.2.6.

2.4.1 Theorem. (Right-angled hexagons). *For any convex right-angled geodesic hexagon with consecutive sides $a, \gamma, b, \alpha, c, \beta$, the following are true:*

(i) $\quad \cosh c = \sinh a \sinh b \cosh \gamma - \cosh a \cosh b,$
(ii) $\quad \sinh a : \sinh \alpha = \sinh b : \sinh \beta = \sinh c : \sinh \gamma,$
(iii) $\quad \coth \alpha \sinh \gamma = \cosh \gamma \cosh b - \coth a \sinh b.$ ◊

An alternative proof would be to use the decomposition into pentagons along r and to proceed as in the proof of (2.3.2), but now with the pentagon formulae.

2.4.2 Theorem. *Let x, y and z be any positive real numbers. Then there exists a unique convex right-angled geodesic hexagon $a, \gamma, b, \alpha, c, \beta$ such that $x = a$, $y = b$ and $z = c$.*

Proof. Uniqueness follows from Theorem 2.4.1. A construction has been described in Section 1.7. We also may paste together two pentagons as in Fig. 2.4.1. For this we first determine $s \in {]}0, b{[}$ by the equation

$$\frac{\sinh(b - s)}{\sinh s} = \frac{\cosh c}{\cosh a},$$

where $a, b, c = x, y, z$. After that we let $r > 0$ be defined by the equation $\sinh r \sinh s = \cosh a$. Then $\sinh r \sinh(b - s) = \cosh c$. By Lemma 2.3.5 and Theorem 2.3.4, the pentagons of Fig. 2.4.1 exist for our values of r, s, a, b and c. ◇

The trigonometric formulae show the strong analogy between triangles and right-angled hexagons (Theorems 2.2.1 and 2.4.1). Another analogy is the following.

2.4.3 Theorem. *In any convex right-angled geodesic hexagon the three altitudes are concurrent.*

Proof. By definition, an altitude is the common perpendicular of two opposite sides. By Theorem 1.5.3(iii), the three altitudes exist and are contained in the hexagon. We prove their concurrence as an application of trigonometry.

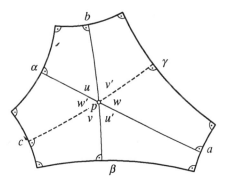

Figure 2.4.2

With the notation as in Fig. 2.4.2, we let p be the intersection of altitudes $a\alpha$ and $b\beta$. We drop the perpendiculars $p\gamma$ from p to γ and pc from p to c. By Theorem 2.3.1(ii), the acute angles satisfy the relation

$$\cos u \cos v \cos w = \cos u' \cos v' \cos w',$$

where $u = u'$. Hence, if $v < v'$, then $w > w'$ and vice versa. But $v + w' = v' + w$. This implies that $v = v'$ and $w = w'$. Hence $p\gamma$ and pc together form the remaining altitude $c\gamma$. This proves the theorem. ◇

While convex hexagons are important for the construction of Riemann surfaces, self-intersecting hexagons like that of Fig. 2.4.3 are important for the twist parameters (cf. Section 3.3).

2.4.4 Theorem. *In a right-angled geodesic hexagon a, γ, b, α, c, β with intersecting sides c and γ, we have*

$$\cosh c = \sinh a \sinh b \cosh \gamma + \cosh a \cosh b.$$

Proof. This follows from Theorem 2.2.6. Another proof, based on pentagons and trirectangles is as follows.

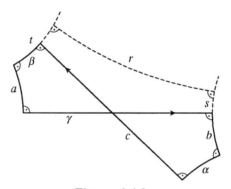

Figure 2.4.3

Assume first that the geodesic extensions of β and b have a common perpendicular r. Since c and r are orthogonal to β, they do not intersect. Hence, we have two right-angled pentagons, one with sides a, γ, s, r and t and the other with sides c, α, $(b + s)$, r and $(t - \beta)$. Theorem 2.3.4(i) (right-angled pentagons) yields:

$$\begin{aligned}\cosh c &= \sinh r \sinh(b + s) \\ &= \sinh r \sinh b \cosh s + \sinh r \cosh b \sinh s \\ &= \sinh a \sinh b \cosh \gamma + \cosh a \cosh b.\end{aligned}$$

Assume next that the extensions of β and b intersect. The pentagons are now replaced by two trirectangles, one with sides a, γ, s and t, the other with sides c, α, $(b + s)$ and $(t - \beta)$. The proof is the same, now with Theorem 2.3.1(iii) and (v) (trirectangles). The remaining case, where b and β meet at infinity, is filled-in by a continuity argument in the proof. ◇

2.5 Variable Curvature

This section belongs to a small subculture of the book consisting of Sections 2.5, 4.3 and 5.4. In these sections we extend some of the results to variable curvature. The results, in this form, will not be used in other parts of the book. We use two comparison theorems. The first one is Sturm's theorem, where we use Klingenberg [1, 2], Lemma 6.5.5 as a reference, the second one is Toponogov's theorem, for which we refer to Cheeger-Ebin [1].

Pentagons and hexagons are also useful on surfaces of variable curvature. In this section we show that they satisfy trigonometric inequalities. We derive them from the constant curvature case with Sturm's comparison theorem.

Throughout the section (with the exception of the final lemma) we make the following *assumption*. M is a complete simply connected two dimensional Riemannian manifold of negative curvature K with the following bounds

$$(2.5.1) \qquad -\kappa^2 \leq K \leq -\omega^2 < 0,$$

where κ and ω are positive constants.

Using Hadamard's theorem (Klingenberg [1, 2], Theorem 6.6.4), we introduce polar coordinates (ρ, σ) centered at some given point $p_0 \in M$. These coordinates are valid on $M - \{p_0\}$. We consider σ to be an element of $\mathbf{S}^1$. The metric tensor has the following form

$$(2.5.2) \qquad ds^2 = d\rho^2 + f^2(\rho, \sigma)\, d\sigma^2$$

with a smooth positive function $f :]0, \infty[\times \mathbf{S}^1 \mapsto \mathbf{R}$. Sturm's comparison theorem now states that

$$(2.5.3) \qquad \frac{1}{\kappa} \sinh \kappa\rho \geq f(\rho, \sigma) \geq \frac{1}{\omega} \sinh \omega\rho.$$

This suggests introducing the comparison metrics

$$(2.5.4) \qquad ds_\tau^2 = d\rho^2 + \frac{1}{\tau^2} \sinh^2 \tau\rho \, d\sigma^2$$

of constant curvature $-\tau^2$ for $\tau = \kappa$ and $\tau = \omega$. The three lengths of a smooth curve c on M with respect to ds_κ^2, ds^2 and ds_ω^2 will be denoted respectively by $\ell_\kappa(c)$, $\ell(c)$ and $\ell_\omega(c)$. By (2.5.3) we have the following inequalities, for any such curve,

$$(2.5.5) \qquad \ell_\kappa(c) \geq \ell(c) \geq \ell_\omega(c).$$

Note that for each σ the straight line

$$(2.5.6) \qquad \rho \mapsto (\rho, \sigma), \quad \rho \in \,]0, \infty[,$$

is a unit speed geodesic with respect to all three metrics. This will be used tacitly below.

In order to translate a trigonometric identity from curvature -1 to curvature $-\tau^2$, we replace every length x by τx (curvature $\times$ length^2 is a scaling invariant). Thus, Theorem 2.2.2(i) (right-angled triangles) on (M, ds_τ^2) reads as before, but with every argument x replaced by τx:

(2.5.7) $$\cosh \tau c = \cosh \tau a \cosh \tau b,$$

etc. If the curvature is non-constant, the identities can be replaced by inequalities. We give four examples.

2.5.8 Theorem. *For any right-angled geodesic triangle a, b, c in M with right angle γ we have*

$$\cosh \kappa a \cosh \kappa b \geq \cosh \kappa c,$$
$$\cosh \omega c \geq \cosh \omega a \cosh \omega b.$$

Proof. We introduce polar coordinates centered at the vertex of γ. By the above remark, a and b are the sides of a right-angled triangle with respect to all three metrics. The corresponding hypothenuses c_κ, c, and c_ω are distinct curves, in general. Using the fact that geodesics are shortest connecting curves, we obtain from (2.5.5)

$$\ell_\kappa(c_\kappa) \geq \ell(c_\kappa) \geq \ell(c) \geq \ell_\omega(c) \geq \ell_\omega(c_\omega),$$

where in our notation $\ell(c) = c$. The theorem follows now from (2.5.7). ◇

2.5.9 Theorem. *For any right-angled geodesic pentagon in M with consecutive sides a, b, α, c, β, we have*

$$\sinh \kappa a \sinh \kappa b \geq \cosh \kappa c,$$
$$\cosh \omega c \geq \sinh \omega a \sinh \omega b.$$

Proof. We use polar coordinates with the common vertex p_0 of sides a and b as center, and introduce the comparison metrics (2.5.4) for $\tau = \kappa$ and $\tau = \omega$. To prove the first inequality we draw in (M, ds_κ^2) the perpendiculars β' and α' at the endpoints a and b. In (M, ds^2) the curves β' and α' are generally not geodesics but they are orthogonal to a and b with respect to both metrics. This follows from (2.5.2).

We check that Fig. 2.5.1 is drawn correctly in the sense that β' and α' do not intersect the open strip between β and α. For this we let $p' \in \beta'$ (respectively, $p' \in \alpha'$) and consider the geodesic arc (with respect to either metric) w from p_0 to p'.

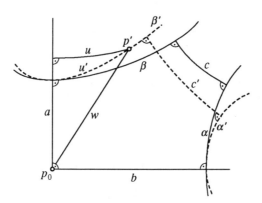

Figure 2.5.1

Dropping in (M, ds^2) the perpendicular u from p' to the geodesic extension of a, we obtain a right-angled triangle u, v, w, where v is on the extension of a and w is the hypotenuse. In (M, ds_κ^2), w is the hypotenuse of a right-angled triangle u', a, w, where u' lies on β'. Theorem 2.5.8 yields

$$\cosh \kappa v \cosh \kappa \ell(u) \geq \cosh \kappa w = \cosh \kappa a \cosh \kappa \ell_\kappa(u').$$

Since in negative curvature the perpendiculars are shortest connecting curves, we have by (2.5.5)

$$\ell_\kappa(u') \geq \ell(u') \geq \ell(u).$$

With this and the preceding inequality, $v \geq a$ follows. Since (M, ds^2) has negative curvature, the perpendiculars u and β either coincide or else are disjoint. This proves that p' is not contained in the open strip between β and α.

We conclude that β' and α' have positive distance in (M, ds_κ^2). Namely, any connecting curve c' from β' to α' intersects β and α so that, by (2.5.5),

$$\ell_\kappa(c') \geq \ell(c') \geq \ell(c) = c.$$

In (M, ds_κ^2), β' and α' therefore have a common perpendicular. We let c' be this perpendicular. It follows that

$$\sinh \kappa a \sinh \kappa b = \cosh \kappa \ell_\kappa(c') \geq \cosh \kappa c.$$

The second inequality of the theorem is proved in a similar way, with β' and α' in (M, ds_ω^2), where now β' and α' do not intersect the exterior of the strip between β and α. It possible that β' and α' intersect or have a common endpoint at infinity. In this case we have $\sinh \omega a \sinh \omega b \leq 1$, and the second inequality of the theorem is trivially true. ◇

2.5.10 Corollary. *Under the hypothesis of Theorem 2.5.9, we have*

$$\cosh \kappa c \geq \coth \kappa \alpha \coth \kappa \beta,$$
$$\cosh \omega c \leq \coth \omega \alpha \coth \omega \beta.$$

Proof. Twice applying the first inequality of Theorem 2.5.9 (with cyclic permutation) yields

$$\begin{aligned}\cosh^2 \kappa \beta &\leq \sinh^2 \kappa \alpha \, (\cosh^2 \kappa b - 1) \\ &\leq \sinh^2 \kappa \alpha \, (\sinh^2 \kappa \beta \sinh^2 \kappa c - 1) \\ &= \sinh^2 \kappa \alpha \, (\sinh^2 \kappa \beta \cosh^2 \kappa c - \cosh^2 \kappa \beta).\end{aligned}$$

This gives the first inequality. The second inequality is proved in the same way. ◇

Our final example will be applied in Section 4.2 to obtain a sharp lower bound for the length of a closed geodesic with transversal self-intersections.

2.5.11 Theorem. *For any right-angled convex geodesic hexagon in* $M = (M, ds^2)$ *with consecutive sides* $a, \gamma, b, \alpha, c, \beta$, *the following hold.*

$$\sinh \kappa a \sinh \kappa b \cosh \kappa \gamma - \cosh \kappa a \cosh \kappa b \geq \cosh \kappa c,$$
$$\cosh \omega c \geq \sinh \omega a \sinh \omega b \cosh \omega \gamma - \cosh \omega a \cosh \omega b.$$

Proof. We begin with the second inequality. By the scaling invariance of curvature × length^2, we may assume that $\omega = 1$. The idea is to use a comparison hexagon in the hyperbolic plane. For this we draw the consecutive orthogonal geodesic arcs a, γ, b in **H** and extend this configuration into a hexagon. If this is impossible, then we are done because then

$$1 \geq \sinh a \sinh b \cosh \gamma - \cosh a \cosh b.$$

(Increase γ until the hexagon comes into existence, then use Theorem 2.4.1 and observe that the function $t \mapsto \sinh a \sinh b \cosh t - \cosh a \cosh b$ is

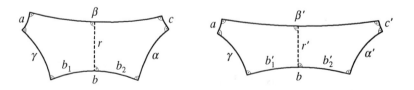

Figure 2.5.2

monotone increasing.)

Assume therefore that the completion, say a, γ, b, α', c', β', exists. Divide both hexagons into right-angled pentagons as shown in Fig. 2.5.2 (the existence of r is proved as in the case of constant curvature with the Arzelà-Ascoli theorem). By the second inequality in Theorem 2.5.9, applied to the pentagon on the left-hand side of the first hexagon, we have $r \geq r'$. Since $\sinh r' \sinh b_1' = \cosh a \geq \sinh r \sinh b_1$, we find that $b_1' \geq b_1$ and $b_2 \geq b_2'$. Hence,

$$\cosh c \geq \sinh r \sinh b_2 \geq \sinh r' \sinh b_2' = \cosh c',$$

where $\cosh c' = \sinh a \sinh b \cosh \gamma - \cosh a \cosh b$.

The first inequality of the theorem will be proved in a similar way, except that the inequalities obtained at intermediate steps will at the same time be used to prove that the comparison hexagon exists. For this we assume that $\kappa = 1$, that is, we assume the curvature bounds $-1 \leq K < 0$.

Draw orthogonal sides a, γ, in **H**. This configuration can be completed into a right-angled pentagon a, γ, b_1'', r'', β_1''. (This has been shown in the proof of Theorem 2.5.9, see Fig. 2.5.1, where a and b play the roles of a and γ of Fig. 2.5.2). By Theorem 2.5.9 we have $r \leq r''$, $b_1'' \leq b_1$, and $b_2 \leq b_2''$, where $b_1'' + b_2'' = b$. For the same reason, there exists a pentagon with sides r, b_2 in **H**. Hence, we complete the bigger sides r'' and b_2'' to find the right-angled pentagon r'', b_2'', α'', c'', β_2'', where, by Theorem 2.5.9 and by the monotonicity of the involved trigonometric functions, $c'' \geq c$. Pasting the two pentagons together along r'', we get the desired comparison hexagon. ◇

For better reference we restate the last part as a corollary.

2.5.12 Corollary. *Let a, γ, b, α, c, β be a right-angled convex geodesic hexagon in M, where M is a complete surface of curvature K satisfying $-1 \leq K < 0$. Then there exists a right-angled convex geodesic hexagon a, γ, b, α'', c'', β'' in **H** and this hexagon satisfies $c'' \geq c$.* ◇

We do not know whether the inequalities above, which involve the lower curvature bound, remain valid if the curvature is allowed to assume positive values. In the proofs we needed the non-positive curvature assumption in order to work with polar coordinates. In some cases however, stronger methods can be applied which are not restricted to non-positive curvature. As an illustration, we apply Toponogov's theorem to prove the following lemma. (The lemma will be needed in Section 4.3 to extend the collar theorem to variable curvature.)

2.5.13 Lemma. *Let G be a (simply connected) right-angled geodesic pentagon in a surface of arbitrary curvature $K \geq -1$. Then any pair of adjacent sides a_1, a_2 satisfies the relation* $\sinh a_1 \sinh a_2 > 1$.

Proof. We use the notation of Fig. 2.5.3. A new feature is that we no longer have the uniqueness theorems for connecting geodesics arcs and perpendiculars. However, we still have the existence of geodesic arcs *in G* of minimal length in the various homotopy classes. This follows, as always, from the Arzelà-Ascoli theorem, now applied to the compact metric space G. In the arguments which follow, "minimal" is meant with respect to G. Also, all arcs considered will be arcs in G.

Replace each side a_i of G by a *shortest* geodesic arc a_i' in G having the same endpoints. Since the arcs have minimal length, they do not intersect each other, except for the common vertices, and we obtain a geodesic pentagon G' in G. We triangulate G' as shown in Fig. 2.5.3(a) with diagonals which again have minimal length in their homotopy classes. By Toponogov's comparison theorem (Cheeger-Ebin [1]), there exists a geodesic pentagon G'' in **H** with sides a_i'' of length $\ell(a_i'') = \ell(a_i')$, $i = 1, \ldots, 5$, such that all interior angles of G'' are less than or equal to the corresponding angles of G'. In particular, all interior angles of G'' are less than or equal to $\pi/2$.

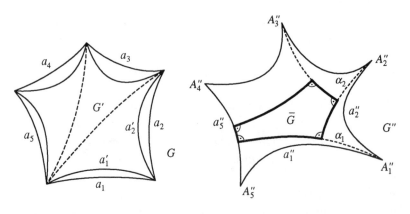

Figure 2.5.3(a)   **Figure 2.5.3(b)**

In **H** we drop the perpendiculars α_1 from A_1'' to a_5'', α_2 from A_2'' to α_1, α_3 from A_3'' to α_2 and α_4 from a_5'' to α_3 (Fig. 2.5.3(b)). We obtain a right-angled geodesic pentagon $\bar{G}$ in **H** with sides $\bar{a}_i$ of length $\ell(\bar{a}_i) \leq \ell(a_i'') = \ell(a_i') \leq \ell(a_i)$. By virtue of the pentagon formula (Theorem 2.3.4(i)) this proves the lemma. ◇

2.6 Appendix: The Hyperboloid Model Revisited

In this section we describe a more algebraic aspect of the hyperboloid model of the hyperbolic plane. The new feature is the vector product associated with the bilinear form h. This will allow the unification of the hyperbolic plane and its isometry group into a quaternion algebra which one may call the quaternion model of the hyperbolic plane. As an application we prove the triple trace theorem (Theorem 2.6.16 and Corollary 2.6.17) and an improved version of the general sine and cosine laws of Thurston [1] (Theorem 2.6.20).

The Quaternion Model

As in Section 2.1 we consider the bilinear form $h : \mathbf{R}^3 \times \mathbf{R}^3 \to \mathbf{R}$ given by

$$h(X, Y) = x_1 y_1 + x_2 y_2 - x_3 y_3.$$

h is a symmetric bilinear form of signature $(2, 1)$. To simplify the notation we shall write

$$h(X, Y) = XY.$$

$\mathbf{R}^3$ together with this product will be denoted by $\mathcal{H}_0$.

In Section 2.1 the hyperboloid model was introduced as the surface

$$H = \{ X \in \mathcal{H}_0 \mid h(X, X) = -1, x_3 > 0 \}.$$

The geodesics are the sets $H \cap \Gamma$, where Γ is a plane through the origin of $\mathcal{H}_0$. Using the action of Ω (cf. Lemma 2.1.4), we easily check that if $C \in \mathcal{H}_0 - \{0\}$ is a vector h-orthogonal to Γ, then $h(C, C) > 0$. Conversely, for every non-zero vector C with $h(C, C) > 0$, the h-orthogonal plane Γ of C intersects H in a geodesic. We may thus say that *C represents a geodesic* whenever $h(C, C) > 0$, and *C represents a point* whenever $h(C, C) < 0$. This leads to the following

2.6.1 Definition. A vector $X \in \mathcal{H}_0 - \{0\}$ is *point-like* if $h(X, X) < 0$, *geodesic-like* if $h(X, X) > 0$, and *infinity-like* if $h(X, X) = 0$.

Our point of view is to see $\mathcal{H}_0$ as an example of hyperbolic geometry with the basic geometric quantities: points and lines represented by point-like and geodesic-like vectors. For this we call vectors *equivalent* if and only if they differ by a non-zero factor. The equivalence classes of the point-like, geodesic-like and infinity-like vectors are respectively the *points*, the *geodesics* (or *lines*) and the *points at infinity*. As usual we shall identify equivalence classes with their representatives. The set of all points, lines and points at infinity is denoted by $\mathcal{H}'$.

2.6.2 Definition. Two vectors $X, Y \in \mathcal{H}_0 - \{0\}$, along with the equivalence classes represented by these vectors, are called *incident* iff $XY = 0$.

To compare this with the hyperboloid model H, we use the following notation. If X is point-like we denote by X^* the intersection point of H with the straight line spanned by X. If X is geodesic-like we denote by X^* the geodesic in H which is cut out by the h-orthogonal plane of X. In this way we have a natural one-to-one correspondence between the points and lines of $\mathcal{H}'$ and the points and geodesics of H. Finally, if X is infinity-like we denote by X^* the straight line spanned by X and call X^* a *point at infinity* of H.

We now interpret the above incidence relation. If $X \in \mathcal{H}_0$ is point-like, we may, by virtue of Lemma 2.1.4 (the twice transitivity of Ω), assume that $X = (0, 0, 1)$ and then easily check that X is incident with Y if and only if Y is geodesic-like with the corresponding geodesic Y^* containing the point X^*.

If X is infinity-like, then X can be incident with Y in exactly two cases: (i) $Y^* = X^*$, and (ii) Y is geodesic-like and X^* is an endpoint at infinity of the geodesic Y^*. (Again use Lemma 2.1.4.)

If both X and Y are geodesic-like, then X and Y are incident if and only if the geodesics X^* and Y^* are orthogonal.

With our definition of incidence we can restate and complete Theorems 1.1.4 - 1.1.6 as follows.

2.6.3 Theorem. *If α and β are distinct elements of $\mathcal{H}'$, then there exists a unique $\gamma \in \mathcal{H}'$ which is incident with α and β.*

Proof. Check the various cases, or wait until the end of the proof of Theorem 2.6.7. ◊

The theorem becomes more tangible if we introduce as a new tool the *vector product* associated with h.

Let $I = (1, 0, 0)$, $\mathcal{J} = (0, 1, 0)$ and $\mathcal{K} = (0, 0, 1)$. Then $\{I, \mathcal{J}, \mathcal{K}\}$ is an orthonormal basis of $\mathcal{H}_0$ in the sense that $I\mathcal{J} = I\mathcal{K} = \mathcal{J}\mathcal{K} = 0$ and

$$II = \mathcal{J}\mathcal{J} = -\mathcal{K}\mathcal{K} = 1.$$

2.6.4 Definition. The *vector product* associated with h is the unique antisymmetric bilinear mapping $\wedge : \mathcal{H}_0 \times \mathcal{H}_0 \to \mathcal{H}_0$ satisfying

$$I \wedge J = K, \qquad I \wedge K = J, \qquad J \wedge K = -I.$$

In the following lemma, det(A, B, C) is the determinant of the matrix formed by the coordinate vectors of A, B, C with respect to the basis $\{I, J, K\}$.

2.6.5 Lemma. *The vector product has the following properties*:

(i) $(A \wedge B)C = A(B \wedge C) = -\det(A, B, C)$,

(ii) $A \wedge (B \wedge C) = (AB)C - (AC)B$,

      $(A \wedge B) \wedge C = (BC)A - (AC)B$,

(iii) $(A \wedge B)(C \wedge D) = (AD)(BC) - (AC)(BD)$,

(iv) $(A \wedge B)$ *is h-orthogonal to A and B*.

Proof. By the linearity in all arguments it suffices to check (i) and the first identity in (ii) on the above orthonormal basis. The second identity in (ii) is a consequence of the first. (iii) is a consequence of (i) and the first identity in (ii). The final statement, (iv), is a consequence of (i). ◇

From the lemma follows that up to a factor ±1 the vector product $\wedge$ is independent of our particular choice of the orthonormal basis $\{I, J, K\}$. In fact, let A, B, C be h-orthonormal vectors. Then from the incidence relations (or by using Ω) we see that two of the vectors are geodesic-like and one is point-like. We may therefore assume that $AA = BB = -CC = 1$. By (iii) we have $(A \wedge B)(A \wedge B) = -1$. Therefore, by (iv), $A \wedge B = \varepsilon C$ with $\varepsilon = \pm 1$. Now (i) and (iv) imply $A \wedge C = \varepsilon B$ and $B \wedge C = -\varepsilon A$. Hence, the vector product based on $\{A, B, C\}$ differs from $\wedge$ only by the factor ε.

As in the Euclidean case we have the following lemma.

2.6.6 Lemma. $A \wedge B = 0$ *iff A and B are linearly dependent*.

Proof. For each $B = b_1 I + b_2 J + b_3 K$ the linear mapping $A \mapsto A \wedge B$ is represented by the following matrix (operating on column vectors) with respect to the basis $\{I, J, K\}$

$$\begin{pmatrix} 0 & -b_3 & b_2 \\ b_3 & 0 & -b_1 \\ b_2 & -b_1 & 0 \end{pmatrix}.$$

If $B \neq 0$, this matrix has rank 2, so the kernel of this mapping consists of the multiples of B. ◇

Here is a first application of the vector product.

2.6.7 Theorem. *If $A, B \in \mathcal{H}_0$ represent distinct elements of $\mathcal{H}'$, then $A \wedge B$ represents the unique element which is incident with A and B.*

Proof. By Lemma 2.6.6 we have $A \wedge B \neq 0$. By Lemma 2.6.5(iv) $A \wedge B$ is incident with A and B. If C is another vector incident with A and B, then by Lemma 2.6.5(ii), $(A \wedge B) \wedge C = 0$. By Lemma 2.6.6, C is a multiple of $A \wedge B$, i.e. C and $A \wedge B$ represent the same element of $\mathcal{H}'$. This proves the theorem. At the same time this provides a proof of Theorem 2.6.3. ◊

Let us next look at the isometries of $\mathcal{H}_0$, or, more precisely at the endomorphisms of $\mathcal{H}_0$ which preserve h. It is not difficult to check that the subgroup $\Omega \subset GL(3, \mathbf{R})$ of Section 2.1 acts by isometries and preserves the vector product. However, we want to redevelop the isometries in a more algebraic way which uses quaternions. We begin by adding an additional dimension to $\mathcal{H}_0$ by taking the direct sum of vector spaces:

$$\mathcal{H} = \{ \mathcal{A} = a + A \mid a \in \mathbf{R}, A \in \mathcal{H}_0 \}.$$

$\mathcal{H}$ is a vector space with basis $\{1, \mathcal{I}, \mathcal{J}, \mathcal{K}\}$. The vector product is extended to $\mathcal{H}$ as follows.

2.6.8 Definition. For $A, B \in \mathcal{H}_0$ we define

$$A * B = AB + A \wedge B,$$

and for $\mathcal{A} = a + A$, $\mathcal{B} = b + B \in \mathcal{H}$ we define

$$\mathcal{A} * \mathcal{B} = ab + aB + bA + A * B.$$

This is the distributive extension of all previous products and we check that $\mathcal{H}$ together with $+$ and $*$ is an algebra. More precisely, $\mathcal{H}$ is a quaternion algebra of type $(1, 1)$ (Vignéras [3]). $(\mathcal{H}, *)$ is the *quaternion model* of hyperbolic geometry.

The *hyperbolic plane* in this model is the set of all equivalence classes of point-like vectors of $\mathcal{H}_0 \subset \mathcal{H}$. The *geodesics* are the equivalence classes of the geodesic-like vectors. Since we identify equivalence classes by their representatives, we may rephrase this by saying that $\mathcal{H}$ *contains* the hyperbolic plane and that $\mathcal{H}$ also *contains* the geodesics. We next describe how the isometries sit in $\mathcal{H}$ and how the various relationships may be expressed in terms of the algebra.

The following are basic concepts of quaternion algebras (for a general introduction to quaternion algebras and applications to Riemann surfaces we refer to Vignéras [3]).

2.6.9 Definition. For every $\mathcal{A} = a + A \in \mathcal{H}$ we define the *quaternion conjugate* $\bar{\mathcal{A}} := a - A$, the *norm* $n(\mathcal{A}) := \mathcal{A} * \bar{\mathcal{A}}$, and the *trace* $\operatorname{tr}(\mathcal{A}) := a$.

Observe that the trace allows us to rewrite the product h on $\mathcal{H}_0$ in terms of the quaternion algebra:

$$h(A, B) = AB = \operatorname{tr}(A * B), \quad A, B \in \mathcal{H}_0.$$

2.6.10 Lemma. $\operatorname{tr} : \mathcal{H} \to \mathbf{R}$ *is a vector space homomorphism. Quaternion conjugation is a vector space homomorphism which satisfies the rule*

$$\overline{\mathcal{A} * \mathcal{B}} = \bar{\mathcal{B}} * \bar{\mathcal{A}},$$

and $n : \mathcal{H} \to \mathbf{R}$ *is a multiplicative homomorphism with respect to* $*$.

Proof. Clearly, trace and conjugation are vector space homomorphisms, and the rule is easily checked on the basis vectors 1, I, J and K. With this rule we compute $n(\mathcal{A} * \mathcal{B}) = \mathcal{A} * \mathcal{B} * \overline{\mathcal{A} * \mathcal{B}} = \mathcal{A} * \mathcal{B} * \bar{\mathcal{B}} * \bar{\mathcal{A}} = \mathcal{A} * n(\mathcal{B})\bar{\mathcal{A}} = n(\mathcal{A})n(\mathcal{B})$. ◊

We introduce the subsets $\mathcal{H}^1$ and $\mathcal{H}^{-1}$.

2.6.11 Definition.

$$\mathcal{H}^1 = \{a + A \in \mathcal{H} \mid a^2 - AA = 1\}, \quad \mathcal{H}^{-1} = \{a + A \in \mathcal{H} \mid a^2 - AA = -1\}.$$

Since n is a $*$-homomorphism, $\mathcal{H}^1$ and $\mathcal{H}^1 \cup \mathcal{H}^{-1}$ are groups with respect to $*$. For $\mathcal{A} = a + A \in \mathcal{H}^1 \cup \mathcal{H}^{-1}$ the inverse $\mathcal{A}^{-1}$ is given by

$$\mathcal{A}^{-1} = a - A \text{ if } \mathcal{A} \in \mathcal{H}^1, \quad \mathcal{A}^{-1} = -a + A \text{ if } \mathcal{A} \in \mathcal{H}^{-1}.$$

For $\mathcal{A} \in \mathcal{H}^1 \cup \mathcal{H}^{-1}$ and $\mathcal{B} \in \mathcal{H}$ we compute that

$$\operatorname{tr}(\mathcal{A}^{-1} * \mathcal{B} * \mathcal{A}) = \operatorname{tr}(\mathcal{B}).$$

Since $\mathcal{H}_0$ is the set of quaternions with trace zero, it follows that $\mathcal{H}^1$ and $\mathcal{H}^1 \cup \mathcal{H}^{-1}$ act on $\mathcal{H}_0$ by conjugation (conjugation in the sense of groups). Moreover, the following is also true.

2.6.12 Lemma. *For each* $\mathcal{A} \in \mathcal{H}^1 \cup \mathcal{H}^{-1}$ *the vector space endomorphism* $X \mapsto \mathcal{A}^{-1} * X * \mathcal{A}, X \in \mathcal{H}_0$ *is an isometry with respect to* h. *This isometry commutes with the vector product.*

Proof. We use the fact that for $X, Y \in \mathcal{H}_0$ the bilinear form h can be written as $h(X, Y) = \operatorname{tr}(X * Y)$ (see above). Hence $h(\mathcal{A}^{-1} * X * \mathcal{A}, \mathcal{A}^{-1} * Y * \mathcal{A}) = \operatorname{tr}(\mathcal{A}^{-1} * X * \mathcal{A} * \mathcal{A}^{-1} * Y * \mathcal{A}) = \operatorname{tr}(\mathcal{A}^{-1} * X * Y * \mathcal{A}) = h(X, Y)$. This proves the first claim. The equation $X \wedge Y = X * Y - h(X, Y)$ proves the second claim ◇

Note in particular that for $\mathcal{A} \in \mathcal{H}^{-1}$ conjugation with $\mathcal{A}$ is commutative and not, as on might expect, anticommutative with the vector product. For the action of $\mathcal{A} \in \mathcal{H}^1 \cup \mathcal{H}^{-1}$ on $\mathcal{H}_0$ we use the notation

$$\mathcal{A}^{-1} * X * \mathcal{A} = \mathcal{A}[X].$$

In axiomatic geometry the isometries are studied by looking at the half-turns (point symmetries) and the reflections (axial symmetries). This is easy to undertake in $\mathcal{H}$. In the following, δ is either 1 or -1.

2.6.13 Proposition. *Let $A \in \mathcal{H}_0$ with $AA = -\delta$. Set $\mathcal{A} = A$ and interpret $\mathcal{A}$ as an element of $\mathcal{H}^\delta$ acting on $\mathcal{H}_0$ by conjugation. Then*
  (i) $\mathcal{A}[A] = A$,
  (ii) $\mathcal{A}[X] = -X$ *for all $X \in \mathcal{H}_0$ which are incident with A.*

Proof. (i) is clear. For (ii), with Lemma 2.6.5 we compute that

(iii) $\qquad\qquad\qquad \mathcal{A}[X] = -2\delta(AX)A - X$

and recall that, by definition, X is incident with A if and only if $AX = 0$. ◇

If we interpret $\mathcal{A}$ as acting on $\mathcal{H}'$ and denote, as earlier, by A^*, X^*, etc. the elements in $\mathcal{H}'$ represented by A, X, etc., then (iii) shows that $\mathcal{A} \neq id$. Moreover, $\mathcal{A}$ is a half-turn about A^* if A is point-like (it fixes each geodesic X^* through A^*), and $\mathcal{A}$ is a symmetry with axis A^* if A is geodesic-like (it fixes the points X^* on A^* and the geodesics Y^* orthogonal to A^*).

The two-fold transitive action of $\mathcal{H}^1$ on $\mathcal{H}'$ can now be proved in the same way as in axiomatic geometry:

If X^* and Y^* are points, we can chose the representatives X and Y in $\mathcal{H}_0$ such that $XX = YY$ and such that $XY < 0$. This implies that $(X + Y)$ is again point-like and we can further normalize the representatives such that, in addition $(X + Y)(X + Y) = -1$. The proposition below will show that under these normalizations $(X + Y)^*$ is the mid-point of X^* and Y^* and, if $(X + Y)$ is interpreted as an element of $\mathcal{H}^1$, then $(X + Y)$ is the half-turn about $(X + Y)^*$ which interchanges X^* and Y^*.

Similarly, if X^* and Y^* are geodesics, we can choose the representatives X and Y such that $XX = YY$ and $(X + Y)(X + Y) = 1$. The following proposition will show that $(X + Y)^*$ is either an angle bisector or the median of X^* and Y^*, and that $(X + Y)$, if seen as element of $\mathcal{H}^{-1}$, is the symmetry with axis $(X + Y)^*$ and interchanges X^* and Y^*.

2.6.14 Proposition. *Let $X, Y \in \mathcal{H}_0$ be linearly independent such that $XX = YY$ and $(X + Y)(X + Y) = -\delta$, where $\delta = \pm 1$. Set $\mathcal{A} = (X + Y)$. Then $\mathcal{A}$, when seen as an isometry, satisfies the following relations.*

$$\mathcal{A}[X] = Y, \; \mathcal{A}[Y] = X \text{ and } \mathcal{A}[X \wedge Y] = -X \wedge Y.$$

Proof. This is obtained by a straightforward computation. The statement about $X \wedge Y$ can also be concluded from Proposition 2.6.13 because $X \wedge Y$ is incident with X and Y (cf. Theorem 2.6.7). ◇

For every $\mathcal{A} \in \mathcal{H}^1 \cup \mathcal{H}^{-1}$ the action of $\mathcal{A}$ on $\mathcal{H}'$ coincides with the action of $-\mathcal{A}$. By looking at the fixed points (respectively, the fixed geodesics) we see that the kernel of the action is $\{\pm 1\}$ and that $\mathcal{H}^1 / \{\pm 1\}$ can be identified with the group Ω of Section 2.1.

In order to see the relation between $\mathcal{H}^1$ and the isometries in the upper half-plane model we associate with each basis vector of $\mathcal{H}$ a (2×2) matrix as follows.

(2.6.15) $\quad 1 \mapsto \begin{pmatrix} 1 & 0 \\ 0 & 1 \end{pmatrix}, \; I \mapsto \begin{pmatrix} 1 & 0 \\ 0 & -1 \end{pmatrix}, \; \mathcal{J} \mapsto \begin{pmatrix} 0 & 1 \\ 1 & 0 \end{pmatrix}, \; \mathcal{K} \mapsto \begin{pmatrix} 0 & 1 \\ -1 & 0 \end{pmatrix}.$

This mapping extends to an isomorphism between the quaternion algebra $\mathcal{H}$ and $GL(2, \mathbf{R})$, and the restriction to $\mathcal{H}^1$ is an isomorphism between $\mathcal{H}^1$ and $SL(2, \mathbf{R})$.

Resumé. The quaternion algebra $\mathcal{H}$ contains the points, the lines and the points at infinity in the subset $\mathcal{H}_0$; the incidence of elements $X, Y \in \mathcal{H}_0$ is given by the relation $\text{tr}(X * Y) = 0$; and the isometries are contained in $\mathcal{H}^1 \cup \mathcal{H}^{-1}$ which acts twice transitively by conjugation.

A Trace Relation

We postpone the introduction of distances and angles to the end of the section and now prove, as a first application, a theorem about traces in finitely generated subgroups of $SL(2, \mathbf{R})$. We start with a more general version in $\mathcal{H}^1$.

Consider n distinct elements $X_1, \ldots, X_n \in \mathcal{H}^1$, (e.g. generators of a discrete subgroup). We can write *words* in $X_1, \ldots, X_n$ as follows. Let $m \in \mathbf{N}$, and consider a finite sequence

$$\omega = v_1, n_1, v_2, n_2, \ldots, v_m, n_m,$$

where every v_i is an index, $v_i \in \{1, \ldots, n\}$, and every n_i is an exponent,

$n_i \in \mathbf{Z}$, $i = 1, \ldots, m$. Then write

$$W = W_\omega(X_1, \ldots, X_n) = X_{v_1}^{n_1} * X_{v_2}^{n_2} * \ldots * X_{v_m}^{n_m}.$$

We want to express this word in a different form using the traces from the following list

$$\begin{array}{ll} \operatorname{tr} X_i, & i = 1, \ldots, n, \\ \operatorname{tr}(X_i * X_j), & 1 \leq i < j \leq n, \\ \operatorname{tr}(X_i * X_j * X_k), & 1 \leq i < j < k \leq n. \end{array}$$

For simplicity we rewrite this list as a sequence $\tau_1, \ldots, \tau_N$, where $\tau_\iota = \tau_\iota(X_1, \ldots, X_n)$, $\iota = 1, \ldots, N$. (Later on, the group elements $X_1, \ldots, X_n$ will be varied.) The following theorem has its roots in Fricke-Klein [1], p. 366.

2.6.16 Theorem. (Triple trace theorem I). *For every sequence* $\omega = v_1, n_1, v_2, n_2, \ldots, v_m, n_m$ *as defined above, there exists a sequence of real polynomial functions*

$$\begin{array}{ll} f_i = f_i(x_1, \ldots, x_N), & i = 0, \ldots, n, \\ f_{ij} = f_{ij}(x_1, \ldots, x_N), & 1 \leq i < j \leq n \end{array}$$

$((x_1, \ldots, x_N) \in \mathbf{R}^N)$ *which depend only on* ω *such that any word* $W = W_\omega(X_1, \ldots, X_n)$ *with* $X_1, \ldots, X_n \in \mathcal{H}^1$ *can be written in the form*

$$W = f_0(\tau_1, \ldots, \tau_N) + \sum_i f_i(\tau_1, \ldots, \tau_N) X_i + \sum_{i<j} f_{ij}(\tau_1, \ldots, \tau_N) X_i * X_j.$$

Proof. We first consider some elementary relations. If $X_i = x_i + \mathbf{X}_i$ with $x_i = \operatorname{tr}(X_i)$ and $\mathbf{X}_i \in \mathcal{H}_0$, then $X_i^{-1} = x_i - \mathbf{X}_i = 2x_i - X_i$. This yields the relation

(1) $$X_i^{-1} = 2\operatorname{tr} X_i - X_i,$$

and since $X_i \in \mathcal{H}^1$, we have

(2) $$\mathbf{X}_i \mathbf{X}_i = \operatorname{tr}^2(X_i) - 1.$$

If $X_j = x_j + \mathbf{X}_j$, then, by the definition of the $*$-product,

(3) $$\mathbf{X}_i \mathbf{X}_j = \operatorname{tr}(X_i * X_j) - \operatorname{tr}(X_i)\operatorname{tr}(X_j).$$

We also compute that

(4) $\mathbf{X}_i \wedge \mathbf{X}_j = X_i * X_j - \operatorname{tr}(X_i)X_j - \operatorname{tr}(X_j)X_i - \operatorname{tr}(X_i * X_j) + 2\operatorname{tr}(X_i)\operatorname{tr}(X_j).$

Finally, if $X_k = x_k + \mathbf{X}_k$, then

(5) $\operatorname{tr}(X_i * X_j * X_k) = x_i x_j x_k + x_k \mathbf{X}_i \mathbf{X}_j + x_i \mathbf{X}_j \mathbf{X}_k + x_j \mathbf{X}_i \mathbf{X}_k + (\mathbf{X}_i \wedge \mathbf{X}_j)\mathbf{X}_k.$

The theorem is now proved by a straightforward induction: if $W = X_i^{\pm 1}$, the claim follows from (1). Assume therefore that the claim holds for W and let $W' = W * X_k^{\pm 1}$ for some k. By (1) we may assume that $W' = W * X_k$. Then

$$W' = f_0 X_k + \sum_i f_i X_i * X_k + \sum_{i<j} f_{ij} X_i * X_j * X_k \, .$$

By (2), (3) and (4), we can rewrite each term $X_i * X_k$ in the form given in the theorem. The same is verified for the terms $X_i * X_j * X_k$ if we apply (5) to $(X_i \wedge X_j)X_k$ and use that $(X_i \wedge X_j) \wedge X_k = (X_j X_k)X_i - (X_i X_k)X_j$ (Lemma 2.6.5(ii)). Hence, W' can be rewritten in the desired form. ◇

With the isomorphism $\mathcal{H}^1 \to \mathrm{SL}(2, \mathbf{R})$ given by (2.6.15) we immediately obtain the following.

2.6.17 Corollary. (Triple trace theorem II). *Let $v_1, n_1, v_2, n_2, \ldots, v_m, n_m$, be a sequence as given above. Then there exists a polynomial function $f = f(x_1, \ldots, x_N)$ such that for any sequence $A_1, \ldots, A_n \in \mathrm{SL}(2, \mathbf{R})$ the word*

$$W = A_{v_1}^{n_1} A_{v_2}^{n_2} \ldots A_{v_m}^{n_m}$$

has trace $\mathrm{tr}\, W = f(\tau_1, \ldots, \tau_N)$, *where* $\tau_1, \ldots, \tau_N$ *is the sequence of all traces* $\mathrm{tr}\, A_i$, $\mathrm{tr}\, A_i A_j$, $\mathrm{tr}\, A_i A_j A_k$, $1 \leq i < j < k \leq n$.

Proof. Use the correspondence given by (2.6.15) and observe that the trace of each $\mathcal{A} \in \mathcal{H}^1$ is half the trace of its corresponding matrix in $\mathrm{SL}(2, \mathbf{R})$. ◇

The General Sine and Cosine Formula

The final step in this essay is to introduce length and angular measure and then to give a new proof of the general sine and cosine formulae of Thurston [1], p. 218.

Our point of view is that most frequently, a trigonometric formula has to be applied to a configuration given by a drawn figure. We have therefore tried to find a formula *together with an adjustment rule* so that a reader who is used to reading figures may conveniently adjust the general formula to the given configuration.

It turns out that there are quite tedious sign conventions. In order to make them more applicable, we have decided to introduce *oriented points* and *oriented arcs*, but work with non-oriented measures. The orientations will then be used in the adjustment rule in order to rewrite the formula for the given case. The so adjusted formula then applies to the non-oriented lengths and angles.

The general sine and cosine law will be given in Theorem 2.6.20. Prior to it there will be a rule telling how to set arrows in the figure, and a list (Fig. 2.6.1) telling how to interpret the terms occurring in the general rule. The definitions which now follow are only needed for the proof.

For $U \in \mathcal{H}_0$ we denote again by U^* the element in the hyperboloid H represented by U (cf. the paragraph after Definition 2.6.2). Ω is the twice transitive group generated by the isometries L_σ and M_ρ as defined in (2.1.2).

To orient the geodesics we first consider the case $U = I$. Here I^* is a geodesic in H which passes through the point $p_0 = (0, 0, 1)$ in an h-orthogonal plane of I. We parametrize I^* with unit speed such that the tangent vector at p_0 becomes $-\mathcal{J}$. This introduces an orientation on I^* and we let $I°$ denote the geodesic I^* together with this orientation. If $U \in \mathcal{H}_0$ is an arbitrary geodesic-like vector, normalized so that $UU = 1$, then there exists an isometry $\phi \in \Omega$ which sends I to U and I^* to U^*, and we denote by $U°$ the geodesic U^* together with the orientation induced from $I°$ by ϕ. We abbreviate this by writing

$$U° = \phi(I°).$$

Note that with our convention the tangent vector of $\mathcal{J}°$ at p_0 is $I°$. Since every isometry of Ω which fixes I also fixes the orientation of $I°$, this definition is independent of the particular choice of ϕ, and the following compatibility with Ω holds:

$$\Psi(U°) = (\Psi(U))° \text{ for all } \Psi \in \Omega.$$

Finally, if U is geodesic-like satisfying $UU = a^2$, $a > 0$, then we define $U° = (U/a)°$. Note that $(-U)°$ has opposite orientation.

To define the orientation of a point, we let $U \in \mathcal{H}_0$ be a point-like vector and denote by $U°$ the point $U^* \in H$ together with an orientation which is defined to be *positive* if $U\mathcal{K} < 0$ (i.e. if U has a positive third component) and *negative* otherwise. Again $(-U)°$ has the opposite orientation of $U°$.

Points at infinity will not be considered.

In the figures which follow, we adopt the following *conventions*. The orientation of a geodesic will be marked by an arrow (indicating the sense of the parametrization), and the orientation of a point will be marked by an oriented curve which goes around the point in the counterclockwise sense if the orientation is positive and in the clockwise sense otherwise. This convention commutes with the action of Ω: if $V = \Psi(U)$ with $\Psi \in \Omega$, then marking V in the figure is the same as first marking $U°$ and then letting Ψ operate on the figure. Finally, we adopt the convention that a counterclockwise rotation of $\pi/2$ is needed to rotate $I°$ into $\mathcal{J}°$. (This is our definition of *counterclockwise*.)

In the next step we introduce a non-oriented measure and a *parity* for each of the following configurations (cf. Fig. 2.6.1 further down): (1) an angle flanked by two oriented geodesic arcs, (2) a geodesic arc flanked by two oriented perpendicular geodesics, (3) a geodesic arc flanked by two oriented points, (4) a geodesic arc flanked by an oriented point and an oriented perpendicular geodesic.

If the geodesic arcs in (1) intersect in opposite directions the curve marking the orientation of the vertex, then we define the *parity* δ of the configuration to be 1. If the intersections are in the same direction we set $\delta = -1$. For the configurations (2) - (4), the parity is defined in a similar way (cf. Fig. 2.6.1). Note that in case (3) the parity is 1 if both endpoints have the same orientation. We also remark that the parity is preserved under the action of Ω.

We next define the *non-oriented measures*. To simplify the notation we use the following abbreviation for $X \in \mathcal{H}_0$:

$$|X| = (|XX|)^{1/2}.$$

2.6.18 Definition and Remarks. (1) First let $U°$ and $V°$ with U, $V \in \mathcal{H}_0$ be the oriented geodesics intersecting at $p \in H$ and carrying the geodesic arcs of the first configuration. There exists a unique $\phi \in \Omega$ and a unique $\sigma \in [-\pi, \pi[$ such that $\phi(p) = p_0$, $\phi(U°) = I°$, $\phi((-\delta V)°) = L_\sigma(I°)$, where δ denotes the parity and L_σ is the rotation as in (2.1.2) with fixed point p_0 and angle σ. We define $a = |\sigma|$ to be the measure of the angle flanked by $U°$ and $V°$. This is just the ordinary angular measure and is independent of the orientations of the arcs. Using ϕ we easily check that $UV = -\delta \cos a\, |U||V|$ and $|U \wedge V| = \sin a\, |U||V|$. For this configuration we define

$$\mathbf{C}(a) := \cos a, \qquad \mathbf{S}(a) := \sin a.$$

(2) Now let $U°$ and $V°$ be oriented geodesics in H having a common perpendicular, say γ. There exists a unique $\phi \in \Omega$ and a unique $a \geq 0$ such that ϕ sends the intersection point of $U°$ and γ to p_0 and the intersection point of $V°$ and γ to $M_a(p_0)$, where M_a is the isometry from (2.1.2) with axis $\mathcal{J}°$ and displacement length a. We define a to be the distance from $U°$ to $V°$. Again using ϕ, we verify the two equations $UV = -\delta \cosh a\, |U||V|$ and $|U \wedge V| = \sinh a\, |U||V|$. Here we define

$$\mathbf{C}(a) := \cosh a, \qquad \mathbf{S}(a) := \sinh a.$$

(3) If $U°$ and $V°$ are the oriented endpoints of a geodesic arc, we have a unique $\phi \in \Omega$ and a unique $a \geq 0$ such that ϕ sends $U*$ to p_0 and $V*$ to $M_a(p_0)$, (disregarding the orientations). We define a to be the distance of $U°$ and $V°$, and check that $UV = -\delta \cosh a\, |U||V|$, $|U \wedge V| = \sinh a\, |U||V|$. Here we define

$$C(a) := \cosh a, \quad S(a) := \sinh a.$$

(4) The remaining case is that $U°$ is a point and $V°$ a geodesic (or vice-versa). We let γ be the (non-oriented) geodesic through $U°$ orthogonal to $V°$. There exist uniquely determined $\phi \in \Omega$ and $a \geq 0$, satisfying $\phi(U^*) = p_0$ and $\phi(V° \cap \gamma) = M_a(p_0)$. (Again we disregard the orientation of curves.) We define a to be the distance from the point to the geodesic. In contrast to the above cases (1) - (3) we now have $UV = -\delta \sinh a \, |U||V|$ and $|U \wedge V| = \cosh a \, |U||V|$, and we define

$$C(a) := \sinh a, \quad S(a) := \cosh a. \qquad \diamond$$

Using the above isometries $\phi \in \Omega$, we check that in cases (2) - (4) the arrow which marks the orientation of $(U \wedge V)°$ points from U to V if the parity is -1, and from V to U if the parity is $+1$. This is also true in case (1) if the arrow of the marking curve is drawn on the part which is inside the angular region of angle a flanked by the two geodesic arcs. We abbreviate this fact by saying that $(-\delta U \wedge V)°$ *points from* $U°$ *to* $V°$. The above arguments yield:

2.6.19 Lemma. *In each of the configurations* (1) - (4) *the following hold.*

$$UV = -\delta \, C(a) \, |U||V|,$$
$$|U \wedge V| = S(a) \, |U||V|,$$

and $(-\delta U \wedge V)°$ *points from* $U°$ *to* $V°$. $\qquad \diamond$

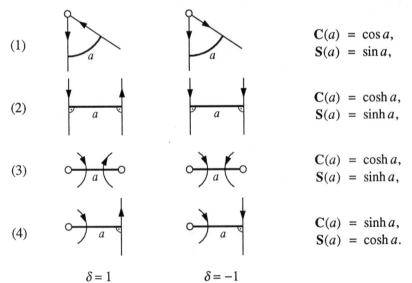

Figure 2.6.1

Ch.2, §6] The General Sine and Cosine Formula

Rule for setting the arrows. Let a generalized triangle $a, \gamma, b, \alpha, c, \beta$ in H be given. Mark each element of angle type by a small arc connecting the adjacent sides and contained in the sector of angle $\leq \pi$ as shown in Fig. 2.6.2. Then orient a, γ, etc. by setting arrows in such a way that a points from β to γ, γ points from a to b, and so on (cf. Fig. 2.6.2). The configurations $\beta a \gamma$, $a \gamma b$, $\gamma b \alpha$, etc. now have well defined parities which we denote by $\delta_a, \delta_\gamma, \delta_b$, etc. Finally, we define $\varepsilon_\gamma = 1$ if γ is a geodesic arc, and $\varepsilon_\gamma = -1$ if γ is an angle.

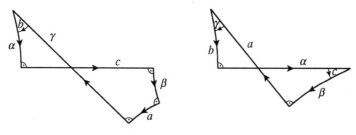

Figure 2.6.2

2.6.20 Theorem. *With the above definitions each generalized triangle a, $\gamma, b, \alpha, c, \beta$ satisfies the cosine law*

$$C(c) = \varepsilon_\gamma \delta_a \delta_b \delta_c [\delta_\gamma S(a) S(b) C(\gamma) - C(a) C(b)]$$

and the sine law

$$S(a) : S(\alpha) = S(b) : S(\beta) = S(c) : S(\gamma).$$

Proof. Choose $A, B, C \in \mathcal{H}_0$ with $|A| = |B| = |C| = 1$ such that with respect to the given orientations $A° = \alpha$, $B° = \beta$, $C° = \gamma$ (recall that $|X| = |XX|^{1/2}$). Lemma 2.6.5(iii) implies that

$$(AB)(CC) = (B \wedge C)(C \wedge A) + (BC)(CA).$$

From Lemma 2.6.19 we have

$$AB = -\delta_c C(c), \quad BC = -\delta_a C(a), \quad CA = -\delta_b C(b),$$
$$|B \wedge C| = S(a), \quad (-\delta_a B \wedge C)° = a°,$$
$$|C \wedge A| = S(b), \quad (-\delta_b C \wedge A)° = b°,$$

where $a°$ and $b°$ are the oriented geodesics carrying a and b and having the orientations of a and b. By the first relation of Lemma 2.6.19 we have therefore

$$\delta_a\delta_b(B \wedge C)(C \wedge A) = -\delta_\gamma C(\gamma)|B \wedge C||C \wedge A| = -\delta_\gamma C(\gamma)S(a)S(b),$$

and the cosine law follows. To prove the sine law we note from Lemma 2.6.19 and Lemma 2.6.5 that

$$S(\gamma)|B \wedge C||C \wedge A| = |(B \wedge C) \wedge (C \wedge A)| = |\det(A, B, C)|.$$

Hence $S(c) : S(\gamma) = |A \wedge B||B \wedge C||C \wedge A| : |\det(A, B, C)|$, where the right-hand side is invariant under cyclic permutation. ◇

As an example we compute the cosine formula for the generalized triangles in Fig. 2.6.2. In the first triangle the signs are $\varepsilon_\gamma = 1$, $\delta_a = \delta_b = 1$, $\delta_c = \delta_\gamma = -1$. The formula becomes

$$\cosh c = \sinh a \sin b \sinh \gamma + \cosh a \cos b.$$

In the second triangle the signs are $\varepsilon_\gamma = -1$, $\delta_b = \delta_c = \delta_\gamma = 1$, $\delta_a = -1$. The formula becomes

$$\cos c = \cosh a \cosh b \cos \gamma - \sinh a \sinh b.$$

One may now again compute the formulae for the configurations in the earlier sections. We recall that in the present section the angles and sides have non-oriented measures. When oriented measures are used, as for example in (2.3.2), then the signs in the formula must be adjusted accordingly.

Chapter 3

Y-Pieces and Twist Parameters

In this chapter we carry out the construction of the compact Riemann surfaces and introduce the Fenchel-Nielsen parameters as outlined in Section 1.7. The building blocks are the pairs of pants or Y-pieces, whose geometry will be studied in some detail in Section 3.1. We then give a first idea of Teichmüller space by considering the so-called marked Y-pieces. If two Y-pieces are pasted together, one obtains an X-piece, and an additional degree of freedom appears: the twist parameter. We show that the twist parameter can be computed in terms of the lengths of closed geodesics and vice-versa. In Section 3.5 we briefly consider cubic graphs which are the underlying combinatorial structure in the construction of the surfaces. We also give an estimate of the number of such possible graphs. The construction then follows in Section 3.6 at the end of which we give a large number of pairwise non-isometric examples. The final section is an essay on the multiplicity of the length spectrum and is considered as a first application. This section is an appendix and may be skipped in a first reading.

3.1 Y-Pieces

3.1.1 Definition. A compact Riemann surface of signature (0, 3) is called a *Y-piece* or a *pair of pants*.

Pants are building blocks for all compact Riemann surfaces of genus greater than one. They are obtained as follows.

Let G be a right-angled geodesic hexagon in the hyperbolic plane with consecutive sides $\alpha_1, c_3, \alpha_2, c_1, \alpha_3, c_2$, and let G' be a copy of G with corresponding sides α_1', c_3', α_2', c_1', α_3', c_2'.

We parametrize all sides on the interval [0, 1] with *constant speed*:

(3.1.2) $\quad t \mapsto \alpha_i(t), \quad t \mapsto \alpha_i'(t),$
$\qquad\qquad\qquad\qquad\qquad\qquad t \in [0, 1], \quad i = 1, 2, 3,$
$\quad t \mapsto c_i(t), \quad t \mapsto c_i'(t),$

and such that the sides of G and G' together form a closed boundary curve (Fig. 3.1.1). The hexagons are assumed to be disjoint. The *pasting condition*

(3.1.3) $\quad \alpha_i(t) = \alpha_i'(t) := a_i(t), \quad t \in [0, 1], \quad i = 1, 2, 3,$

(cf. Section 1.3) defines a 3-holed sphere

$$Y = G + G' \mod(3.1.3)$$

which inherits the hyperbolic structures of G and G'. Since all angles are right angles, the boundary curves

(3.1.4) $\quad t \mapsto \gamma_i(t) := \begin{cases} c_i(2t) & \text{if } 0 \le t \le 1/2 \\ c_i'(2-2t) & \text{if } 1/2 \le t \le 1, \end{cases} \quad i = 1, 2, 3.$

are closed geodesics. Thus, Y is a pair of pants.

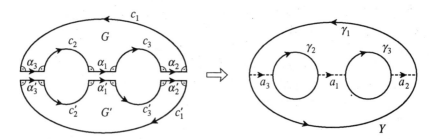

Figure 3.1.1

Equation (3.1.4) will also be interpreted as a parametrization of γ_i on $S^1 = \mathbf{R}/[t \mapsto t+1]$ instead of on [0, 1]. With either interpretation, (3.1.4) is called the *standard parametrization* of the boundary, and Y as described here is said to be given in *standard form*. All Y-pieces are obtained in this way:

3.1.5 Proposition. *Let S be an arbitrary Y-piece. For every pair of boundary geodesics of S there exists a unique simple common perpendicular. The three perpendiculars together decompose S into two isometric right-angled geodesic hexagons.*

Proof. Let a be a simple curve on S which connects, for example, boundary components γ_2 and γ_3. By surface topology (or Proposition A.18), any other simple curve from γ_2 to γ_3 is homotopic to a with endpoints gliding on γ_2 and γ_3 (see Definition 1.5.1 for homotopies with gliding endpoints). By Theorem 1.5.3, a is homotopic to a unique common perpendicular a_1 and this perpendicular is simple. Hence, we have the three unique simple perpendiculars a_1, a_2, a_3 between the boundary geodesics.

To see that the perpendiculars are pairwise disjoint, we cut S open along a_3, for example, to obtain a hyperbolic surface A of signature $(0, 2)$ with piecewise geodesic boundary. By Theorem 1.5.3 again, a simple perpendicular a_1' between γ_2, γ_3 exists on A. (Fig. 3.1.2 shows a curve α_1' in the homotopy class of a_1'.) By the uniqueness of a_1 on S, we have $a_1' = a_1$. Hence a_1 does not intersect a_3, and the same holds for the other pairs of perpendiculars. If we cut S open along a_1, a_2, a_3, we obtain two simply connected right-angled geodesic hexagons G and G'. By Corollary 1.4.3, G and G' are hexagons in **H**. By the uniqueness in Theorem 2.4.2 they are isometric. ◇

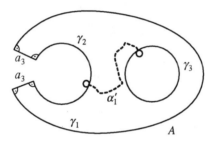

Figure 3.1.2

3.1.6 Corollary. *The endpoints of the perpendiculars a_1, a_2, a_3 divide each boundary geodesic of S into two arcs of the same length.* ◇

3.1.7 Theorem. *For any triple of positive real numbers ℓ_1, ℓ_2, ℓ_3 there exists a unique pair of pants Y with boundary geodesics γ_1, γ_2, γ_3 of lengths $\ell(\gamma_i) = \ell_i$, $i = 1, 2, 3$.*

Proof. The existence follows from the existence of hexagons (Lemma 1.7.1 or Theorem 2.4.2) and from the above construction. Now let Y, $\tilde{Y}$ with boundary components γ_i, $\tilde{\gamma}_i$, $i = 1, 2, 3$, be two such pairs of pants. Let a_i, $\tilde{a}_i$ be the perpendiculars as in Proposition 3.1.5 which divide Y and $\tilde{Y}$ into hexagons G, G' and $\tilde{G}, \tilde{G}'$. By Corollary 3.1.6 and by the uniqueness of the hexagons in Theorem 2.4.2, there exist isometries from G to $\tilde{G}$ and from G' to $\tilde{G}'$ which both send a_i to $\tilde{a}_i$, $i = 1, 2, 3$. The two isometries together define an isometry from Y to $\tilde{Y}$ which sends γ_i to $\tilde{\gamma}_i$, $i = 1, 2, 3$. ◇

3.1.8 Proposition. (Half-collars). *Let Y be a pair of pants with boundary geodesics γ_1, γ_2, γ_3. The sets*

$$\mathscr{C}^*[\gamma_i] = \{p \in Y \mid \sinh(\mathrm{dist}(p, \gamma_i)) \sinh \tfrac{1}{2}\ell(\gamma_i) \leq 1\}, \quad i = 1, 2, 3,$$

are pairwise disjoint and each is homeomorphic to $[0, 1] \times S^1$. For any $p \in \mathscr{C}^*[\gamma_i]$ there exists in $\mathscr{C}^*[\gamma_i]$ a unique perpendicular to γ_i, $i = 1, 2, 3$.

Proof. We decompose Y into right-angled geodesic hexagons G and G' with sides labelled and parametrized as in (3.1.2). We then have $c_i = \tfrac{1}{2}\ell(\gamma_i) = c_i'$, $i = 1, 2, 3$.

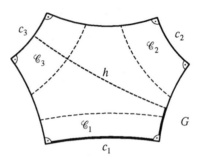

Figure 3.1.3

On G (and similarly on G') we define

$$\mathscr{C}_i = \{p \in G \mid \sinh(\mathrm{dist}(p, c_i)) \sinh c_i \leq 1\}, \quad i = 1, 2, 3.$$

To see that the $\mathscr{C}_i$ are pairwise disjoint, we drop the common perpendiculars between opposite sides, for example h from side c_3 to the side between c_1 and c_2 (Fig. 3.1.3). The pentagon formula, Theorem 2.3.4(i), yields

$$\sinh c_k \sinh \mathrm{dist}(c_k, h) > 1; \quad k = 1, 2.$$

Hence, h separates $\mathscr{C}_1$ and $\mathscr{C}_2$, and similarly $\mathscr{C}_1$, $\mathscr{C}_3$ and $\mathscr{C}_2$, $\mathscr{C}_3$ are separated. Moreover, we see from this separation that each $\mathscr{C}_i$ is swept out by pairwise disjoint geodesic arcs of length d_i emanating perpendicularly from c_i, where $\sinh c_i \sinh d_i = 1$, $i = 1, 2, 3$. The same holds for the corresponding domains on G', and the proposition follows. ◇

3.1.9 Example. The *width* of the half-collars in Proposition 3.1.8 is maximal in the following sense. Let a sequence of Y-pieces be given in which the lengths $\ell(\gamma_1)$ and $\ell(\gamma_2)$ are kept fixed but $\ell(\gamma_3) \to 0$. With the arguments of the preceding proof we immediately see that $\mathrm{dist}(\mathscr{C}^*[\gamma_1], \mathscr{C}^*[\gamma_2]) \to 0$. Hence, the first conclusion of Proposition 3.1.8 may fail if we increase the width of the half-collars only by little. In connection with the collar theorem we shall see that the given width is a natural quantity also from another point of view (cf. Example 4.1.3).

3.2 Marked Y-Pieces

The survey over the Y-pieces given by Theorem 3.1.7 allows us to give a first idea of Teichmüller space. The *moduli problem* for a class of Riemann surfaces is to find a set of isometry invariants which determine the surface up to isometry. In the case of a Y-piece, it is natural to choose the boundary lengths as moduli. The parameter space then becomes $\mathbf{R}_+^3$. The set $\mathcal{R}$ of all Y-pieces, however, is smaller. In order to obtain a one-to-one correspondence with $\mathbf{R}_+^3$, one introduces an additional structure, a *marking*, which in our case consists of labelling the boundary components. In this way, a pair of pants Y together with the labelled boundary geodesics γ_1, γ_2, γ_3, and a pair of pants $\tilde{Y}$ with $\tilde{\gamma}_1$, $\tilde{\gamma}_2$, $\tilde{\gamma}_3$ are said to be *marking equivalent* if and only if there exists an isometry $\phi: Y \to \tilde{Y}$ satisfying $\phi(\gamma_i) = \tilde{\gamma}_i$, $i = 1, 2, 3$. The set $\mathcal{T}$ of all such defined marking equivalence classes is now in the desired one-to-one correspondence with $\mathbf{R}_+^3$, and the set $\mathcal{R}$ of all unmarked Y-pieces is the quotient

(3.2.1) $$\mathcal{R} = \mathcal{T}/\mathfrak{S}_3$$

of $\mathcal{T}$ by the action of the permutation group $\mathfrak{S}_3$ which acts by permutation of the labels of the boundary geodesics. The moduli problem is now to find a fundamental domain for the action of $\mathfrak{S}_3$ on $\mathbf{R}_+^3$. In the present case the problem has an easy solution. A fundamental domain is e.g. the set

(3.2.2) $$\mathcal{F} = \{(x, y, z) \in \mathbf{R}_+^3 \mid 0 < x \leq y \leq z\}.$$

$\mathcal{T}$ is called the *Teichmüller space* of signature (0, 3), $\mathfrak{S}_3$ is the *Teichmüller modular group* and $\mathcal{R} = \mathcal{T}/\mathfrak{S}_3$ is the *Riemann space*.

For arbitrary signature the moduli problem is much more complicated. Linda Keen [5] has given a solution in the case of signature (1, 1), and Semmler [1] has recently given an explicit fundamental domain for the compact Riemann surfaces of genus 2. For higher genus the problem is unsolved. The Teichmüller space, however, is well defined. The definition will follow in Section 6.1. The marking will be given by a homeomorphism from a base surface to the given surface, mapping building blocks onto building blocks. For this it is convenient to use certain standard mappings which we now define.

3.2.3 Definition. A homeomorphism $\phi : A \to B$ between two metric spaces A and B is a *quasi isometry*, more precisely a *q-quasi isometry* ($q \geq 1$) if

$$\frac{1}{q} \operatorname{dist}(x, y) \leq \operatorname{dist}(\phi x, \phi y) \leq q \operatorname{dist}(x, y)$$

for all $x, y \in A$. For a quasi isometry ϕ we denote by $q[\phi]$ the infimum over all q' for which ϕ is a q'-quasi isometry. $q[\phi]$ is sometimes called the *maximal length distortion of* ϕ.

In order to fix certain basic quasi isometries between pants let G and $\tilde{G}$ be right-angled geodesic hexagons in **H**, with sides labelled and parametrized as in (3.1.2).

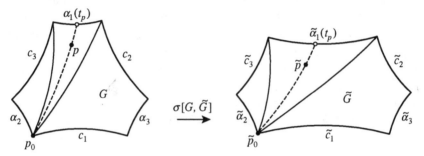

Figure 3.2.1

We let p_0 be the common vertex of sides α_2 and c_1 in G and $\tilde{p}_0$ the common vertex of sides $\tilde{\alpha}_2$ and $\tilde{c}_1$ in $\tilde{G}$. The diagonals at p_0 and $\tilde{p}_0$ triangulate the hexagons as shown in Fig. 3.2.1.

We define a mapping

$$\sigma[G, \tilde{G}] : G \to \tilde{G}$$

(a "stretch") as follows. For each point $p \in G$, $p \neq p_0$, there exists a unique side α_i (respectively, c_i) and a unique parameter $t_p \in [0, 1]$ such that p lies on the geodesic ray from p_0 to $p_* = \alpha_i(t_p)$ (from p_0 to $p_* = c_i(t_p)$). We let $\sigma[G, \tilde{G}](p)$ be the point $\tilde{p} \in \tilde{G}$ which lies on the geodesic ray from $\tilde{p}_0$ to $\tilde{p}_* = \tilde{\alpha}_i(t_p)$ (from $\tilde{p}_0$ to $\tilde{p}_* = \tilde{c}_i(t_p)$) satisfying

$$\frac{\operatorname{dist}(\tilde{p}_0, \tilde{p})}{\operatorname{dist}(\tilde{p}_0, \tilde{p}_*)} = \frac{\operatorname{dist}(p_0, p)}{\operatorname{dist}(p_0, p_*)}.$$

We complete the definition by setting $\sigma[G, \tilde{G}](p_0) = \tilde{p}_0$. Clearly, $\sigma[G, \tilde{G}]$ is a quasi isometry which preserves the boundary parametrization.

3.2.4 Definition. Let Y and $\tilde{Y}$ be arbitrary Y-pieces in standard form (Proposition 3.1.5) and let G and G' ($\tilde{G}$ and $\tilde{G}'$) be the corresponding hexagons. We define a mapping $\sigma[Y, \tilde{Y}] : Y \to \tilde{Y}$ by setting

$$\sigma[Y, \tilde{Y}] = \begin{cases} \sigma[G, \tilde{G}] & \text{on } G \\ \sigma[G', \tilde{G}'] & \text{on } G'. \end{cases}$$

Observe that $\sigma[G, \tilde{G}]$ and $\sigma[G', \tilde{G}']$ coincide on $G \cap G'$ since $\sigma[G, \tilde{G}]$ and $\sigma[G', \tilde{G}']$ preserve the boundary parametrization. For the same reason we have the following.

3.2.5 Lemma. *$\sigma[Y, \tilde{Y}]$ preserves the parametrization (3.1.4) of the boundary geodesics.* ◊

The following lemma will be needed for the topology of Teichmüller space.

3.2.6 Lemma. *Each $\sigma[Y, \tilde{Y}]$ is a quasi isometry. If $\{Y_n\}_{n=1}^{\infty}$ is a sequence of marked pairs of pants converging to a given pair of pants Y in the sense that $\ell(\gamma_{i,n}) \to \ell(\gamma_i)$ as $n \to \infty$, $i = 1, 2, 3$, then $q[\sigma[Y, Y_n]] \to 1$.*

Proof. This is a standard application of trigonometry. Since rather rough distance estimates are sufficient, we omit the details. ◊

3.3 Twist Parameters

3.3.1 Definition. A compact Riemann surface of signature $(0, 4)$ is called an *X-piece*.

X-pieces are obtained by pasting together two Y-pieces along two boundary geodesics of the same length. An additional parameter then appears: the *twist parameter*. The aim of this section is to compute twist parameters in terms of lengths of closed geodesics and vice versa.

Let Y and Y' be two Y-pieces given in standard form (Section 3.1) with boundary geodesics γ_i, γ_i', parametrized on $\boldsymbol{S}^1$, $i = 1, 2, 3$, and suppose $\ell(\gamma_1) = \ell(\gamma_1')$. Then for any real number α we obtain an X-piece via the identification

(3.3.2) $\qquad \gamma_1(t) = \gamma_1'(\alpha - t) =: \gamma^\alpha(t), \qquad t \in \boldsymbol{S}^1.$

The number α is called the *twist parameter*. With the notation of Section 1.3

(pasting) we set

$$X^\alpha := Y + Y' \mod(3.3.2).$$

In order to distinguish the elements in the family $\{X^\alpha\}$, $\alpha \in \mathbf{R}$ from each other, we use X^α to designate the ordered pair $(\alpha, Y + Y' \mod(3.3.2))$: the *X-piece marked* with α.

3.3.3 Proposition. *Every X-piece can be obtained by the above construction.*

Proof. Let X be the given X-piece. As is known from surface topology, X contains homotopically non-trivial simple closed curves which are not homotopic to a boundary curve. Taking one such curve and replacing it by the simple closed geodesic in its free homotopy class (Theorem 1.6.6), we obtain a decomposition of X into two Y-pieces. ◊

3.3.4 Definition. Let X^α with geodesic γ^α be as above. The set

$$\mathscr{C}[\gamma^\alpha] = \{p \in X^\alpha \mid \text{dist}(p, \gamma^\alpha) \leq w\},$$

where

$$w = \text{arcsinh}\{1/\sinh \tfrac{1}{2}\ell(\gamma^\alpha)\}$$

is called the *collar around* γ, and w is its *width*.

By Proposition 3.1.8, for $p \in \mathscr{C}[\gamma^\alpha]$, there exists the unique perpendicular in $\mathscr{C}[\gamma^\alpha]$ from p to γ^α which meets γ^α at the point $\gamma^\alpha(t_p)$ for a uniquely determined $t_p \in \mathbf{S}^1$. We let $\rho_p = \varepsilon \, \text{dist}(p, \gamma^\alpha)$ be the directed distance with $\varepsilon = -1$ if $p \in Y$ and $\varepsilon = +1$ if $p \in Y'$.

3.3.5 Definition. For $p \in \mathscr{C}[\gamma^\alpha]$, the above

$$(\rho, t) = (\rho_p, t_p) \in [-w, w] \times \mathbf{S}^1$$

is the pair of *Fermi coordinates* of p with respect to γ^α.

We point out that the sign convention in Definition 3.3.5 is compatible with the sign convention for Fermi coordinates in the hyperbolic plane (Section 1.1): the orientation of the boundary geodesics (3.1.4) of a Y-piece in standard form is such that the Y-piece lies on the left-hand side of each boundary geodesic. In the definition (3.3.2) of γ^α the orientation is that of γ_1 so that Y lies on the left-hand side of γ^α. Accordingly, the points in $\mathscr{C}[\gamma^\alpha]$ belonging to Y have $\rho \leq 0$ and those belonging to Y' have $\rho \geq 0$.

By Proposition 3.1.8, these coordinates define a homeomorphism from $\mathscr{C}[\gamma^\alpha]$ to the annulus $[-w, w] \times \mathsf{S}^1$. Moreover, since γ^α has speed $\ell(\gamma^\alpha)$, we can give the expression for the metric tensor from Example 1.3.2(6), namely

(3.3.6) $$ds^2 = d\rho^2 + \ell^2(\gamma^\alpha) \cosh^2\rho \, dt^2.$$

In the family $\{X^\alpha\}$, $\alpha \in \mathbf{R}$ we regard X^0 as a base surface and write $X^0 = X$, $\gamma^0 = \gamma$. The next step is to define a homeomorphism $\tau^\alpha : X \to X^\alpha$ as shown in Fig. 3.3.1. This mapping will later be used to define the marking homeomorphisms for Teichmüller space.

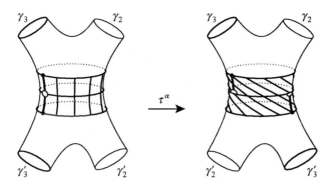

Figure 3.3.1

In what follows, $\pi^\alpha : Y \cup Y' \to X^\alpha = Y + Y' \bmod (3.3.2)$ denotes the natural projection, and for $\alpha = 0$, we write $\pi^0 = \pi$.

Observe what the pasting condition (3.3.2) means for the Fermi coordinates: if $p \in \mathscr{C}[\gamma]$ has coordinates (ρ, t) and $\rho \leq 0$, then, by the above sign convention, $(\pi)^{-1}(p) \in Y$. It follows that $\pi^\alpha \circ (\pi)^{-1}(p) \in \mathscr{C}[\gamma^\alpha]$ has exactly the same coordinates (ρ, t) with respect to γ^α (because $\gamma^\alpha(t) = \gamma_1(t)$). If, however, $\rho \geq 0$, then $(\pi)^{-1}(p) \in Y'$, and it follows that $\pi^\alpha \circ (\pi)^{-1}(p)$ has coordinates $(\rho, \alpha + t)$. It follows that the mapping $T^\alpha : \mathscr{C}[\gamma] \to \mathscr{C}[\gamma^\alpha]$ defined by

$$T^\alpha(\rho, t) = (\rho, t + \alpha \frac{w + \rho}{2w})$$

(w is the common width of the collars $\mathscr{C}[\gamma]$ and $\mathscr{C}[\gamma^\alpha]$) coincides with $\pi^\alpha \circ (\pi)^{-1}$ on the boundary of $\mathscr{C}[\gamma]$. With

(3.3.7) $$\tau^\alpha := \begin{cases} T^\alpha & \text{on } \mathscr{C}[\gamma] \\ \pi^\alpha \circ (\pi)^{-1} & \text{on } X - \mathscr{C}[\gamma] \end{cases}$$

we obtain therefore a homeomorphism

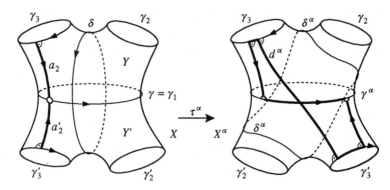

Figure 3.3.2

$$\tau^\alpha : X \to X^\alpha.$$

The mappings T^α and τ^α are sometimes called *twist homeomorphisms*.

The following analogue of Lemma 3.2.6 is clear and will be used in Section 6.4 to define a distance function in Teichmüller space. (q is as in Definition 3.2.3.)

3.3.8 Lemma. *Each τ^α is a quasi isometry. If $\{\alpha_n\}_{n=1}^\infty$ is a sequence converging to α, then $q[\tau^{\alpha_n} \circ (\tau^\alpha)^{-1}] \to 1$ as $n \to \infty$.* ◇

The next step is to introduce mappings analogous to the above, but now from X^α onto itself: the Dehn twists. We start in $\mathscr{C}[\gamma^\alpha]$ with the mapping $\mathscr{T} : \mathscr{C}[\gamma^\alpha] \to \mathscr{C}[\gamma^\alpha]$ defined by

$$\mathscr{T}(\rho, t) = (\rho, t + \frac{w + \rho}{2w}).$$

3.3.9 Definition. The mapping $\mathscr{D} : X^\alpha \to X^\alpha$ defined by

$$\mathscr{D} := \begin{cases} \mathscr{T} & \text{on } \mathscr{C}[\gamma^\alpha] \\ \text{id} & \text{on } X^\alpha - \mathscr{C}[\gamma^\alpha] \end{cases}$$

is called an *elementary Dehn twist*. A *Dehn twist* of X^α of order m, $m \in \mathbf{Z}$, is a homeomorphism of X^α isotopic to $\mathscr{D}^m$ which fixes the boundary ∂X pointwise.

(We refer to Definition A.1 for the isotopy of homeomorphisms, and to Definition A.15 for Dehn twists in general.) Note that if $\mathscr{D}^m$ is a Dehn twist of $X = X^0$, then $\tau^\alpha \circ \mathscr{D}^m \circ (\tau^\alpha)^{-1}$ is a Dehn twist of X^α.

It is an important fact that the twist parameter α in the family $\{X^\alpha\}$, $\alpha \in \mathbf{R}$, can be computed in terms of lengths of closed geodesics and vice versa. For this we consider the following geodesics.

On X we let $d = a_2' a_2^{-1}$ (the parametrized curve a_2' followed by the inversely parametrized a_2). Then d is a perpendicular from γ_3' to γ_3. In the free homotopy class of the curve $d\gamma_3 d^{-1} \gamma_3'$, there are simple closed curves so that, by Theorem 1.6.6, $d\gamma_3 d^{-1} \gamma_3'$ is homotopic to a simple closed geodesic δ which separates γ_2 and γ_2' from γ_3 and γ_3' (Fig. 3.3.2). As a second curve of this sort we let η be the geodesic in the homotopy class of $\mathscr{D}(\delta)$, where $\mathscr{D} : X \to X$ is the Dehn twist of Definition 3.3.9. We memorize the following.

$$\delta \text{ is homotopic to } d\gamma_3 d^{-1} \gamma_3', \quad \text{where } d = a_2' a_2^{-1},$$
$$\eta \text{ is homotopic to } \bar{d}\gamma_3 \bar{d}^{-1} \gamma_3', \quad \text{where } \bar{d} = a_2' \gamma_1 a_2^{-1}.$$

In the following definition, $\tau^\alpha : X \to X^\alpha$ is again the (marking-) homeomorphism from (3.3.7)

3.3.10 Definition. On X^α we let δ^α and η^α be the simple closed geodesics in the free homotopy classes of $\tau^\alpha(\delta)$ and $\tau^\alpha(\eta)$.

We now show that the length of δ^α determines $|\alpha|$, and that the lengths of δ^α and η^α together determine α.

3.3.11 Proposition. *For the above family* $\{X^\alpha\}$, $\alpha \in \mathbf{R}$ *we let F be the function*

$$F(\alpha) = \sinh \tfrac{1}{2}\gamma_3 \sinh \tfrac{1}{2}\gamma_3' \{ \sinh a_2 \sinh a_2' \cosh(\alpha\gamma) + \cosh a_2 \cosh a_2' \}$$
$$- \cosh \tfrac{1}{2}\gamma_3 \cosh \tfrac{1}{2}\gamma_3',$$

where we use the notational convention $\gamma_3 = \ell(\gamma_3)$, *etc. Then* $\cosh \tfrac{1}{2}\delta^\alpha = F(\alpha)$ *and* $\cosh \tfrac{1}{2}\eta^\alpha = F(\alpha + 1)$.

Proof. By the definition of τ^α, the curve $\tau^\alpha(d)$ is homotopic (with endpoints gliding on γ_3' and γ_3) to the curve $a_2' b a_2^{-1}$ in X^α, where b is a parametrized arc on γ^α of length $|\alpha|\ell(\gamma^\alpha) = |\alpha\ell(\gamma)| = |\alpha\gamma|$. Using Theorem 1.5.3 we *homotope* (= to perform a homotopy) this curve into the perpendicular d^α in its homotopy class. (Fig. 3.3.2.) Lifting the curves into the universal covering of X^α (without changing the notation), we obtain a crossed right-angled hexagon as shown in Fig. 3.3.3, and Theorem 2.4.4 yields

(1) $\qquad \cosh d^\alpha = \sinh a_2 \sinh a_2' \cosh(\alpha\gamma) + \cosh a_2 \cosh a_2'$.

This relates $|\alpha|$ to the length of d^α. In order to relate it to the length of the closed geodesic δ^α, we observe that δ^α is freely homotopic to the closed curve $d^\alpha \gamma_3 \, (d^\alpha)^{-1} \, \gamma_3'$. This curve does not intersect d^α (by Theorem 1.6.6: cut X^α open along d^α, homotope $d^\alpha \gamma_3 \, (d^\alpha)^{-1} \, \gamma_3'$ into a closed geodesic in the *interior* of the new surface; by the uniqueness in X^α this geodesic is δ^α). Hence, d^α is the simple perpendicular from γ_3' to γ_3 in the pair of pants $\mathcal{Y}$ in X^α, whose boundary geodesics are γ_3', γ_3 and δ^α. We may therefore decompose $\mathcal{Y}$ into two right-angled hexagons and apply Theorem 2.4.1(i) to obtain

(2) $\qquad \cosh \tfrac{1}{2}\delta^\alpha = \sinh \tfrac{1}{2}\gamma_3 \sinh \tfrac{1}{2}\gamma_3' \cosh d^\alpha - \cosh \tfrac{1}{2}\gamma_3 \cosh \tfrac{1}{2}\gamma_3'$.

This proves that $\cosh \tfrac{1}{2}\delta^\alpha = F(\alpha)$. To prove the formula for η, it suffices to observe that η^α on X^α has the same length as $\delta^{\alpha+1}$ on $X^{\alpha+1}$. $\diamond$

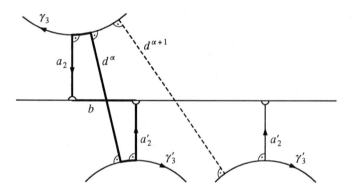

Figure 3.3.3

Proposition 3.3.11 will be used in the proof of Wolpert's theorem and also in the finiteness theorem of Chapter 13. To prove the real analytic equivalence of the Fenchel-Nielsen coordinates with respect to the various underlying graphs, we restate Proposition 3.3.11 in yet another form.

We can express a_2 and a_2' in terms of the lengths of γ and the boundary geodesics, which, together with the above, gives the following result.

3.3.12 Proposition. *In the family* $\{X^\alpha\}$, $\alpha \in \mathbf{R}$ *we have*

$$\cosh \tfrac{1}{2}\delta^\alpha = u + v \cosh(\alpha\gamma)$$
$$\cosh \tfrac{1}{2}\eta^\alpha = u + v \cosh((\alpha+1)\gamma),$$

where the coefficients u and v are real analytic functions of the lengths of γ, γ_2, γ_3, γ_2' and γ_3' but do not depend on α. Moreover, $v > 0$. $\diamond$

(We use again the notational convention $\ell(\delta^\alpha) = \delta^\alpha$, etc.)

So far we have kept fixed the lengths of γ and the boundary geodesics. If we let them vary also, then each X^α determines a vector

(3.3.13) $\quad L = L[\alpha, \gamma, \gamma_2, \gamma_3, \gamma_2', \gamma_3'] := (\gamma^\alpha, \delta^\alpha, \eta^\alpha, \gamma_2, \gamma_3, \gamma_2', \gamma_3') \in \mathbf{R}^7,$

where we also use the notation $\gamma = \ell(\gamma)$, etc. We recall that

$$\ell(\gamma^\alpha) = \ell(\gamma) = \ell(\gamma_1)$$

for all α. The parameters $\alpha, \ldots, \gamma_3'$ fill out $\mathbf{R} \times \mathbf{R}_+^5$, and the corresponding values of L fill out a closed real analytic subvariety $\mathcal{L}$ of dimension 6 in $\mathbf{R}^7$. The function $\alpha \mapsto \delta^\alpha$ itself is not invertible, but the following important lemma holds.

3.3.14 Lemma. *There exists an open neighborhood U_L of $\mathcal{L}$ in $\mathbf{R}^7$ and a real analytic function $\mathcal{A} : U_L \to \mathbf{R}$ such that $\alpha = \mathcal{A}(L[\alpha, x_1, \ldots, x_5])$ for all $(\alpha, x_1, \ldots, x_5) \in \mathbf{R} \times \mathbf{R}_+^5$.*

Proof. Recall that cosh is an even function, so that $z \mapsto \cosh\sqrt{z}$, $z \in \mathbf{C}$ is a holomorphic function on the entire complex plane. Its restriction to the real line has positive first derivatives on the interval $]-\pi^2, \infty[$. Therefore, its inverse

(1) $\quad g(t) = \begin{cases} -(\arccos t)^2 & \text{if } -1 < t \le 1 \\ (\operatorname{arccosh} t)^2 & \text{if } t \ge 1 \end{cases}$

is real analytic on the interval $]-1, \infty[$. With Proposition 3.3.12 we have the explicit solution

(2) $\quad \alpha = \dfrac{1}{2\ell^2(\gamma)} [g(\tfrac{1}{v}(\cosh(\tfrac{1}{2}\eta^\alpha) - u)) - g(\tfrac{1}{v}(\cosh(\tfrac{1}{2}\delta^\alpha) - u))] - \tfrac{1}{2},$

where u and v are real analytic and v is positive. If $L \in \mathcal{L}$, then both arguments of g in the formula are positive. Since g is real analytic in an open interval which contains $[0, \infty[$, the above expression is well defined and real analytic in an open neighborhood of $\mathcal{L}$. ◇

From Lemma 3.3.14 we note in particular that

(3.3.15) $\quad\quad \alpha = \mathcal{A}(\gamma^\alpha, \delta^\alpha, \eta^\alpha, \gamma_2, \gamma_3, \gamma_2', \gamma_3'),$

where we again recall that $\ell(\gamma^\alpha) = \ell(\gamma) = \ell(\gamma_1)$.

3.4 Signature (1, 1)

Let Y be a pair of pants, given in normal form, with boundary geodesics γ_1, γ_2, γ_3, and suppose that γ_1 and γ_2 have the same length. Then the pasting condition

(3.4.1) $$\gamma_1(t) = \gamma_2(\alpha - t) =: \gamma_\alpha(t), \quad t \in \mathsf{S}^1,$$

gives rise to a compact Riemann surface of signature $(1, 1)$:

$$Q_\alpha := Y \bmod (3.4.1).$$

As in the preceding section, we distinguish the members in the family $\{Q_\alpha\}$, $\alpha \in \mathbf{R}$ from each other by formally understanding them as ordered pairs $(\alpha, Y \bmod (3.4.1))$. We also adopt the notational convention that we omit the subscript α if $\alpha = 0$.

Although the topology is now different, the statements of the preceding section still hold. The aim of this section is to translate these statements, using two-fold coverings $X^\alpha \to Q_\alpha$, thereby avoiding extra considerations. These coverings are a useful tool also in Chapter 6, and one may understand the present section as an introductory example of the application of such coverings.

To define the covering surface X^α, we let Y' be a copy of Y. In order to avoid a change of notation we relabel the boundary geodesics γ_i' of Y' in such a way that we have an isometry $m : Y' \to Y$ satisfying

$$m(\gamma_1'(t)) = \gamma_2(t), \quad m(\gamma_2'(t)) = \gamma_3(t), \quad m(\gamma_3'(t)) = \gamma_1(t)$$

for $t \in \mathsf{S}^1$. The pasting condition is now identical with (3.3.2),

(3.4.2) $$\gamma_1(t) = \gamma_1'(\alpha - t) =: \tilde{\gamma}^\alpha(t), \quad t \in \mathsf{S}^1.$$

The covering surface is defined as

$$X^\alpha := Y + Y' \bmod (3.4.2).$$

We denote by $\pi_\alpha : Y \to Q_\alpha$ the natural projection and define $J_\alpha : X^\alpha \to Q_\alpha$ as follows.

(3.4.3) $$J_\alpha = \begin{cases} \pi_\alpha & \text{on } Y \\ \pi_\alpha \circ m & \text{on } Y' \end{cases}$$

(Y and Y' are interpreted as subsets of X^α). This is an isometric immersion. We call X^α a *two-fold covering* of Q_α, although J_α is not exactly a covering map. Fig. 3.4.1 shows the situation in the case $\alpha = 1/2$.

By Proposition 3.1.8, the half-collars $\mathscr{C}^*[\gamma_1]$, $\mathscr{C}^*[\gamma_2]$ on Y formed by the

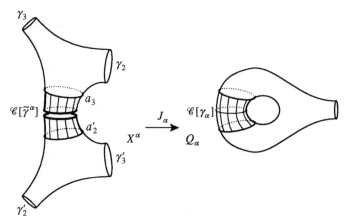

Figure 3.4.1

points at distance $\leq w = \operatorname{arcsinh}\{1/\sinh \frac{1}{2}\gamma_1\}$ from γ_1 (respectively, γ_2) are disjoint. It follows that the collars

$$\mathscr{C}[\tilde{\gamma}^\alpha] = \{p \in X^\alpha \mid \operatorname{dist}(p, \tilde{\gamma}^\alpha) \leq w\}, \mathscr{C}[\gamma_\alpha] = \{p \in Q_\alpha \mid \operatorname{dist}(p, \gamma_\alpha) \leq w\}$$

are isometric, where $J_\alpha : \mathscr{C}[\tilde{\gamma}^\alpha] \to \mathscr{C}[\gamma_\alpha]$ is the isometry. In all collars we denote by (ρ, t) the Fermi coordinates.

Similarly to the preceding section we let $T_\alpha : \mathscr{C}[\gamma] \to \mathscr{C}[\gamma_\alpha]$ be given by

$$T_\alpha(\rho, t) = (\rho, t + \alpha \frac{w+\rho}{2w})$$

and define a twist homeomorphism $\tau_\alpha : Q \to Q_\alpha$ as follows

(3.4.4) $$\tau_\alpha = \begin{cases} T_\alpha & \text{on } \mathscr{C}[\gamma] \\ \pi_\alpha \circ (\pi)^{-1} & \text{on } Q - \mathscr{C}[\gamma]. \end{cases}$$

Let $\tau^\alpha : X \to X^\alpha$ be the corresponding homeomorphism from (3.3.7). The following lemma is then easily checked.

3.4.5 Lemma. *Let μ be a closed curve on X. Then the curves $\tau_\alpha \circ J(\mu)$ and $J_\alpha \circ \tau^\alpha(\mu)$ on Q_α are homotopic.* ◊

A possible way to compute the twist parameter α in terms of closed geodesics is now as follows. We denote by $\tilde{\delta}^\alpha$ and $\tilde{\eta}^\alpha$ on X^α the geodesics from Definition 3.3.10 which determine the twist parameter of X^α, and define the following curves on Q_α

(3.4.6) $$\delta_\alpha = J_\alpha(\tilde{\delta}^\alpha), \quad \eta_\alpha = J_\alpha(\tilde{\eta}^\alpha).$$

By Lemma 3.4.5, δ_α and η_α are the closed geodesics in the free homotopy classes of the curves $\tau_\alpha(\delta)$ and $\tau_\alpha(\eta)$. Since Q_α and X^α have the same twist parameter, and since J_α is length-preserving, we have the analog of (3.3.15) (again with the notation $\gamma = \ell(\gamma)$, etc.)

(3.4.7) $\qquad \alpha = \mathscr{A}(\gamma_\alpha, \delta_\alpha, \eta_\alpha, \gamma_2, \gamma_3, \gamma_3, \gamma_2),$

where $\mathscr{A}$ is the function as in Lemma 3.3.14. We recall that $\ell(\gamma_\alpha) = \ell(\gamma_2) = \ell(\gamma_1)$.

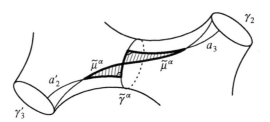

Figure 3.4.2

In certain cases another geodesic is useful for the computation of α. Consulting Fig. 3.4.1 or 3.4.2 (it shows X^α rather than Q_α), we observe that for $\alpha = 1/2$ the endpoints of the perpendicular a_3 meet in $Q^{1/2}$ and form a simple closed geodesic $\mu_{1/2}$. We define μ_α to be the closed geodesic in the free homotopy class of the curve $\tau_\alpha \circ \tau_{1/2}^{-1}(\mu_{1/2})$. It is easy to see that μ_α and γ_α form two isometrically immersed right-angled geodesic triangles with sides of length $\tfrac{1}{2}\mu_\alpha$, $\tfrac{1}{2}a_3$ and $\tfrac{1}{2}\gamma_\alpha|\alpha - \tfrac{1}{2}|$. From formula (i) of Theorem 2.2.2 for right-angled triangles we obtain

(3.4.8) $\qquad \cosh \tfrac{1}{2}\mu_\alpha = \cosh \tfrac{1}{2}a_3 \cosh(\tfrac{1}{2}\gamma_\alpha(\alpha - \tfrac{1}{2})).$

Note that the twist parameter is again determined by $\ell(\mu_\alpha)$ and $\ell(\mu_{\alpha+1})$.

3.5 Cubic Graphs

A *cubic graph* is a finite 3-regular connected graph. For an introduction to graph theory we refer to Bollobás [1]. We use cubic graphs as the combinatorial skeleton for the pasting of pairs of pants, and for this purpose the very rudiments of graph theory, as given below, are sufficient.

A graph G consists of a set of *vertices* and a set of *edges*. We denote by $\#G$ the cardinality of the vertex set. In the figures the vertices will be drawn as dots and the edges as lines. Each edge connects two vertices; an edge may also connect a vertex with itself.

For our purposes it is useful to view each edge as the union of two *half-edges* with each half-edge emanating from one of the two connected vertices. G is called 3-*regular*, if every vertex has three emanating half-edges. In the construction of compact Riemann surfaces each Y-piece with its three boundary geodesics will be interpreted as a vertex with its three half-edges. G is *connected*, if for each pair of distinct vertices x and y we find a sequence $x_1, \ldots, x_n$ of vertices with $x_1 = x$ and $x_n = y$ such that each pair x_i, x_{i+1} is connected by an edge, $i = 1, \ldots, n-1$.

Let now G be a fixed cubic graph. Then $\#G$ is an even number, and we write it in the form

$$\#G = 2g - 2,$$

where $g \geq 2$. Since G is 3-regular, it must have $3g - 3$ edges. We use the following notation. The vertices and edges of G are denoted by

$$y_1, \ldots, y_{2g-2} \quad \text{and} \quad c_1, \ldots, c_{3g-3}.$$

For each y_i, the three half-edges are denoted by $c_{i\mu}$, $\mu = 1, 2, 3$. If $c_{i\mu}$ and c_{jv} are the two half-edges of edge c_k, we shall write $c_k = (c_{i\mu}, c_{jv})$. In this way the graph G is fully described by the list

(3.5.1) $\qquad c_k = (c_{i\mu}, c_{jv}), \qquad k = 1, \ldots, 3g - 3.$

For the construction of Riemann surfaces it is practical to view the list (3.5.1) itself as the graph. In fact, assume that symbols $c_{i\mu}$ are given, with $i = 1, \ldots, 2g - 2$ and $\mu = 1, 2, 3$. Then write a list (3.5.1) of ordered pairs in which each symbol occurs exactly once, and set $y_i = \{c_{i1}, c_{i2}, c_{i3}\}$ for $i = 1, \ldots, 2g - 2$ (the triplets are not ordered). This defines a 3-regular graph. Call the list *admissible* if the graph is connected.

3.5.2 Definition. An admissible list as in (3.5.1) is called a *marked cubic graph*.

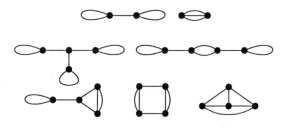

Figure 3.5.1

Two graphs, marked or unmarked, are called *isomorphic* if there exists a bijection of the vertex sets with the property that whenever two vertices are connected by k edges, then the image vertices are also connected by k edges.

Fig. 3.5.1 shows the cubic graphs with two and four vertices. For an arbitrary cardinality we have the following estimate.

3.5.3 Theorem. *Let $\#(g)$ denote the number of pairwise non-isomorphic cubic graphs with $2g - 2$ vertices. Then*

$$2^{g-3} \leq \#(g) \leq g^{3g}.$$

Proof. For the lower bound in the theorem we have to produce exponentially many examples. Consider first the example in Fig. 3.5.2.

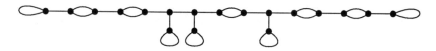

Figure 3.5.2

We may read this example as the word "$a\,a\,b\,b\,b\,a\,b\,a\,a$", where ⬤─⬤ is the letter a and ⬤ with loop is the letter b. The loops at either end are used to mark the beginning and the end. In a similar way we can give examples for any word with letters a, b. Since two loops are "lost" at the ends, and since every letter needs 2 vertices, these words have $g - 2$ letters, and there are 2^{g-2} words. Two such graphs are isomorphic only if the words are the same or if one word is the mirror image of the other. Hence we obtain something over 2^{g-3} graphs in this way.

A simple upper bound for $\#(g)$ is obtained by counting the number of different lists 3.5.1: There are at most $(6g - 6)^2$ ways of writing a line and there are $3g - 3$ lines. This yields the estimate

$$\#(g) \leq (6g - 6)^{6g-6}.$$

A better bound is obtained be considering the following way of constructing cubic graphs. Start with two vertices which are connected by an edge.

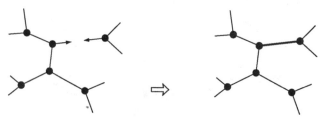

Figure 3.5.3

Then add successively more vertices such that at each step a new vertex is attached to the previous ones along exactly one of its half-edges.

Initially, the number of free half-edges is four. This number increases by 1 at each step. After $2g - 4$ steps, all vertices are used up and we shall have to paste together the remaining $2g$ half-edges. The number of different possibilities up to this point is at most

$$4 \bullet 5 \bullet \ldots \bullet (2g - 1) = \frac{1}{6}(2g - 1)! \ .$$

For each of these there are at most $(2g - 1)(2g - 3) \bullet \ldots \bullet 3 \bullet 1$ ways to combine the $2g$ half-edges. Thus

$$\#(g) \leq \frac{1}{6}(2g - 1)! \frac{(2g - 1)!}{2^{g-1}(g - 1)!} \ .$$

The upper bound in Theorem 3.5.3 is now obtained by simplification. ◇

For sharper estimates we refer to Bollobás [1, 2], Read [1], Wormwald [1, 2] and others. For large g these references yield the following inequalities:

$$\alpha_g \leq \#(g) \leq 3(g - 1) \, 2^{g-1} \alpha_g,$$

where

$$\alpha_g = \frac{(6g - 6)!}{e^2 \bullet (3g - 3)! \bullet (2g - 2)! \bullet 288^{g-1}} \ .$$

Hence, the true order of magnitude of $\#(g)$ is roughly $g!$.

3.6 The Compact Riemann Surfaces

We now construct Riemann surfaces using Y-pieces as building blocks and marked cubic graphs as combinatorial skeletons. In Theorem 3.6.4 we show that all compact Riemann surfaces are obtained in this way, and at the end of the section we give a large family of pairwise non-isometric examples.

First we fix a marked cubic graph G with vertices $y_1, \ldots, y_{2g-2}$ and edges $c_1, \ldots, c_{3g-3}$ defined by the relations

$$c_k = (c_{i\mu}, c_{j\nu}), \quad k = 1, \ldots, 3g - 3,$$

(cf.(3.5.1)). Then we choose

$$L = (\ell_1, \ldots, \ell_{3g-3}) \in \mathbf{R}_+^{3g-3}, \quad A = (\alpha_1, \ldots, \alpha_{3g-3}) \in \mathbf{R}^{3g-3}$$

and define a compact Riemann surface $F(G, L, A)$ as follows. To each vertex

y_i with emanating half-edges c_{i1}, c_{i2}, c_{i3}, we associate a pair of pants Y_i with boundary geodesics γ_{i1}, γ_{i2}, γ_{i3} parametrized on $\mathbf{S}^1$ in standard form (see the text prior to Proposition 3.1.5) such that for the pairs in the above list we have

$$\ell_k = \ell(\gamma_{i\mu}) = \ell(\gamma_{jv}), \qquad k = 1, \ldots, 3g-3.$$

This is possible by Theorem 3.1.7. For each pair in the list we then paste Y_i and Y_j together along these geodesics via the *pasting scheme*

(3.6.1) $\qquad \gamma_{i\mu}(t) = \gamma_{jv}(\alpha_k - t) := \gamma_k(t); \qquad t \in \mathbf{S}^1.$

With the notation of Section 1.3 (pasting) we let

(3.6.2) $\qquad F = F(G, L, A) = Y_1 + \ldots + Y_{2g-2} \mod(3.6.1).$

Since the boundary geodesics of $Y_1, \ldots, Y_{2g-2}$ are in standard form, they are oriented in such a way that F is orientable. Since G is a connected graph, F is connected. F satisfies now all conditions of Theorem 1.3.5 and is therefore a compact Riemann surface. Computing the Euler characteristic, we find that F has genus g. The construction of F and the relation with G is illustrated in Fig. 3.6.1 and Fig. 1.7.3.

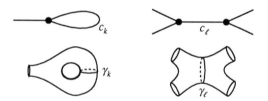

Figure 3.6.1

3.6.3 Definition. (L, A) is the sequence of *Fenchel-Nielsen coordinates* of the compact Riemann surface $F(G, L, A)$.

The geodesics $\gamma_1, \ldots, \gamma_{3g-3}$ on F corresponding to the edges $c_1, \ldots, c_{3g-3}$ of the graph will be called the *parameter geodesics* of F.

We now give an overview of the surfaces $F(G, L, A)$. The first remark is that they cover all compact Riemann surfaces of genus $g \geq 2$. More precisely, the following holds.

3.6.4 Theorem. *Let G be a fixed marked cubic graph with $2g-2$ vertices. Then $F(G, L, A)$ runs through all compact Riemann surfaces of genus g.*

Ch.3, §6]   The Compact Riemann Surfaces   83

Proof. Let $F_0 = F(G, L_0, A_0)$ be some fixed base surface. For every compact Riemann surface S of genus g there exists a homomorphism $\phi : F_0 \to S$. By the theorem below, ϕ can be chosen such that the images $\phi(\gamma_i)$ of the parameter geodesics $\gamma_1, \ldots, \gamma_{3g-3}$ of F_0 form a system of pairwise disjoint simple closed geodesics on S. It follows that ϕ maps the pants of F_0 onto pants on S. These define S as a surface $S = F(G, L, A)$ for suitable L and A. Since ϕ is a homeomorphism, the graph G remains the same. ◇

We recall that homeomorphisms $\phi_0, \phi_1 : A \to B$ of topological spaces A, B are called *isotopic* if there exists a continuous mapping $J : [0, 1] \times A \to B$ such that $J(0, \ ) = \phi_0$, $J(1, \ ) = \phi_1$ and such that $J(s, \ ) : A \to B$ is a homeomorphism for each $s \in [0, 1]$ (Definition A.1). A homeomorphism $h : A \to A$ which is isotopic to the identity is called a 1-*homeomorphism*. In the proof of the next theorem we shall use the theorem of Baer-Zieschang (Theorem A.3). It states that if c and γ are homotopically non-trivial simple closed curves on a surface F, and if c is homotopic to γ, then there exists a 1-homeomorphism $h : F \to F$ sending c to γ.

3.6.5 Theorem. *Let $\phi : S \to R$ be a homeomorphism of compact Riemann surfaces, and let $\gamma_1, \ldots, \gamma_N$ be pairwise disjoint, simple closed geodesics on S. Then there exists a homeomorphism ϕ' isotopic to ϕ such that the curves $\phi'(\gamma_1), \ldots, \phi'(\gamma_N)$ are closed geodesics on R.*

Proof. We successively add 1-homeomorphisms $R \to R$ using the theorem of Baer-Zieschang and the fact that non-trivial simple closed curves are homotopic to simple closed geodesics (Theorem 1.6.6).

Suppose that for a given $k \in \{1, \ldots, N\}$, ϕ has already been isotoped such that all $\phi(\gamma_i)$ are closed geodesics, for $i = 1, \ldots, k-1$. Cut R open along these geodesics and let R' be the connected component containing $\phi(\gamma_k)$.

By Theorem 1.6.6, there exists a simple closed geodesic γ_k' in the free homotopy class of $\phi(\gamma_k)$ in R'. By the uniqueness of the closed geodesic in a free homotopy class, the $\gamma_1, \ldots, \gamma_N$ on S are pairwise non-homotopic, and so are $\phi(\gamma_1), \ldots, \phi(\gamma_N)$ on R. Hence, γ_k' cannot be one of the boundary geodesics of R. By Theorem 1.6.6 again, γ_k' lies in the interior of R. Now the theorem of Baer-Zieschang (Theorem A.3) provides a 1-homeomorphism $h : R' \to R'$ which leaves the boundary of R' pointwise fixed and maps $\phi(\gamma_k)$ onto γ_k'. Since the boundary of R' is pointwise fixed, h extends to a 1-homeomorphism of R which fixes $\phi(\gamma_i)$ for $i = 1, \ldots, k-1$. We now replace ϕ by $h \circ \phi$ and continue. ◇

We finish this section by giving a large family of pairwise non-isometric examples. Let $\mathcal{G}$ be the set of all pairwise non-isomorphic marked cubic graphs with $2g - 2$ vertices. Thus, each $G \in \mathcal{G}$ is a list as in (3.5.1). For fixed $G \in \mathcal{G}$ we define

(3.6.6) $\quad \mathcal{F}(G) = \{F(G, L, A) \mid 0 < \ell_1 < \ldots < \ell_{3g-3} < \operatorname{arccosh} 3\,;$
$$0 < \alpha_1, \ldots, \alpha_{3g-3} < \tfrac{1}{4}\},$$

where $L = (\ell_1, \ldots, \ell_{3g-3})$, $A = (\alpha_1, \ldots, \alpha_{3g-3})$, and we define

$$\mathcal{F}_g = \bigcup_{G \in \mathcal{G}} \mathcal{F}(G).$$

We claim that *the surfaces in $\mathcal{F}_g$ are pairwise non-isometric.*

Proof. The strategy is to recover G, L and A out of the intrinsic geometry of $F = F(G, L, A)$, under the assumption that we already know that F belongs to $\mathcal{F}_g$.

By the collar theorem which we anticipate from the next chapter, the parameter geodesics $\gamma_1, \ldots, \gamma_{3g-3}$ are the only geodesics of length less than $\operatorname{arccosh} 3 = 2 \operatorname{arcsinh} 1$. Hence, we find the parameter geodesics by sorting out *all* simple closed geodesics of length $< \operatorname{arccosh} 3$ on F. The labelling can be found by arranging these geodesics in increasing order. After these steps, we have L and also the graph G.

Consider now a particular γ_k. In each of the adjacent Y-pieces of γ_k we drop the unique simple common perpendiculars between γ_k and the remaining two boundary geodesics (cf. Proposition 3.1.5). By Corollary 3.1.6, these perpendiculars are antipodal on γ_k. It follows that the smallest distance among the feet of the four perpendiculars determines α_k because by hypothesis $0 < \alpha_k < \tfrac{1}{4}$. Hence we have A. $\quad\diamond$

3.7 Appendix: The Length Spectrum Is of Unbounded Multiplicity

As an application of the geometry of 1-holed tori (= signature $(1, 1)$) we prove here a first spectral result which has to do with the distribution of the lengths of the closed geodesics on a compact Riemann surface. The result itself will not be used in this book. We recall that geodesics which differ only by a parameter change are considered equal and that the closed geodesics on a compact Riemann surface of genus ≥ 2 are discrete in the sense that for any $L > 0$ only finitely many closed geodesics have length $\ell \leq L$ (Theorem 1.6.11). We also mention that this holds for any compact Riemannian manifold of negative curvature.

Guillemin and Kazhdan [1] found an inverse spectral result for surfaces of variable negative curvature for which they had to assume that the length spectrum (cf. Section 9.2) is *simple*, that is, different closed geodesics have different lengths. For surfaces of variable negative curvature this is not a severe restriction since in this case the simplicity of the length spectrum is a generic property. However, in the case of constant curvature Randol [5] has observed that the complete contrary is true.

3.7.1 Theorem. *On every compact Riemann surface the length spectrum has arbitrarily high multiplicities.*

Randol's note [5] is based on a result of Horowitz [1] which concerns characters of representations of free groups in SL(2, **R**). We want to *visualize* the occurrence of high multiplicities geometrically, by immersing certain 1-holed tori into others, using cutting and pasting. However, the geometric method has its limits, and we give a second proof of Theorem 3.7.1 which is a modification of the first, but this time in algebraic language, similar to that in the paper of Horowitz.

Geometric Approach

Our first remark is that Theorem 3.7.1 is already valid on Riemann surfaces of signature (1, 1) and also on Y-pieces (cf. the end of this section). Since every compact Riemann surface of genus $g \geq 2$ may be decomposed into a Riemann surface of signature $(g - 1, 1)$ and one of signature (1, 1), we shall restrict ourselves to 1-holed tori for the rest of this section.

We let

(3.7.2) $\qquad Q = Y \bmod (\gamma_2(t) = \gamma_3(\alpha - t) =: \gamma(t),\ t \in S^1)$

be a 1-holed torus, where Y is pair of pants in normal form with boundary geodesics $\gamma_1 = \eta$, γ_2, γ_3 of length $\ell(\gamma_2) = \ell(\gamma_3)$, and α is a twist parameter which is kept fixed. To simplify the notation, the boundary geodesic of Q has been denoted by η instead of γ_1.

Since $\ell(\gamma_2) = \ell(\gamma_3)$, there exists an orientation-preserving isometry $\rho : Y \to Y$ which fixes η and interchanges γ_2 with γ_3. Since $\rho(\gamma_2(t)) = \gamma_3(t + \frac{1}{2})$ for $t \in S^1$, ρ is compatible with the pasting condition in (3.7.2) and may therefore be interpreted as an isometry of the quotient :

$$\rho : Q \to Q.$$

It is with this ρ that we obtain families of geodesics of the same length on Q.

In the following we also use that Y has an orientation *reversing* isometry $\sigma : Y \to Y$ which fixes η and interchanges γ_2, γ_3. This isometry does, in general, not go to the quotient Q. If $\ell(\eta) \neq \ell(\gamma_2)$, then the isometry group of Y is exactly $\{id, \rho, \sigma, \rho\sigma\}$ with $\rho^2 = \sigma^2 = (\rho\sigma)^2 = id$ and $\rho\sigma = \sigma\rho$. The fixed points of σ form a geodesic arc h (the "symmetry axis" of σ) which stands orthogonally on η. The fixed point set of $\rho\sigma$ is the union of the common perpendiculars a_1, a_2, a_3 between the boundary geodesics.

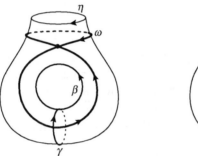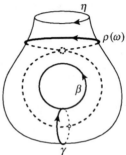

Figure 3.7.1

Now let ω be a figure eight geodesic on Q as shown in Fig. 3.7.1 with one self-intersection. The geodesic consists of two loops, where one of the loops is freely homotopic to the boundary η, and the other loop intersects γ in exactly one point. One finds such geodesics, for example, by cutting Q open into a different pair of pants Y' along a simple closed geodesic β which intersects γ once, and then by drawing ω on Y'.

We claim that $\rho(\omega)$ is different from ω (as a point set). In fact, if $\rho(\omega) = \omega$, then ρ fixes the self-intersection point of ω and is locally a rotation about this point, since ρ preserves orientation. But then ρ must interchange the two loops of ω. This is impossible because ρ fixes γ and only one of the loops intersects γ. Thus, we already have an example with multiplicity 2.

In the next step we construct an isometric immersion $J : Q^* \to Q$ of another 1-holed torus Q^* into Q. Once this is achieved, we show that the corresponding examples of multiplicity 2 in Q^* inject into examples in Q which will be different from their images under ρ, so that we obtain multiplicity 4, and so on.

In order to define the immersion J, we temporarily consider γ and ω as geodesic loops whose base point is the common intersection point. We then form the commutator $\hat{\mu} = \gamma \omega \gamma^{-1} \omega^{-1}$ (first along γ, then along ω, etc). A small homotopy moves $\hat{\mu}$ away from γ so that the closed geodesic μ in the homo-

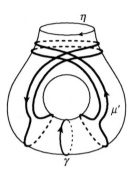

Figure 3.7.2

topy class of $\hat{\mu}$ lies on Y (cf. Theorem 1.6.6). Fig. 3.7.2 shows an example μ' of the homotopy between $\hat{\mu}$ and μ. The reader who has difficulties in seeing such homotopies may take these lines as a pictorial guide to the algebraic proof which will follow later.

Since $\sigma(\mu)$ is homotopic to μ (except for the orientation of the curves, which is irrelevant), the uniqueness of a geodesic in its homotopy class implies that, up to a change of parametrization,

(3.7.3) $$\sigma(\mu) = \mu.$$

We split μ into two arcs μ_2 and μ_3 with endpoints on the "symmetry axis" h of σ and satisfying $\sigma(\mu_2) = \mu_3$. Figure 3.7.3 shows each of these arcs separately and drawn on Y rather than Q.

The initial and end segments of μ_2 which connect h with a_2, cut out a simply connected geodesic quadrangle B_2. The remainder of μ_2 together with γ_2 and an arc on a_2 bounds an annular domain A_2 (which may overlap B_2). Similarly, μ_3 yields the domains A_3, B_3, and we have

$$\sigma(B_2) = B_3, \qquad \sigma(A_2) = A_3.$$

Let $\tilde{Y}$ be a copy of Y with boundary geodesics $\tilde{\gamma}_2$, $\tilde{\gamma}_3$, $\tilde{\eta}$, and let $\tilde{\mu}_3$, $\tilde{B}_3$, $\tilde{A}_3$ be the corresponding subsets of $\tilde{Y}$. We paste B_2 and $\tilde{B}_3$ together along the

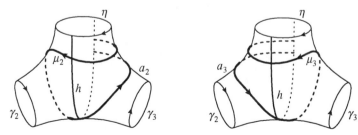

Figure 3.7.3

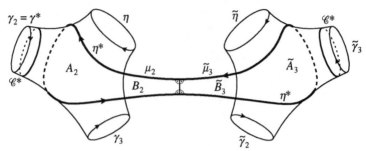

Figure 3.7.4

sides which lie on h (respectively, $\tilde{h}$) and we paste A_2 and $\tilde{A}_3$ together via the following pasting condition, where α is as in (3.7.2)

(3.7.4) $\qquad \gamma_2(t) = \tilde{\gamma}_3(\alpha - t) =: \gamma^*(t), \qquad t \in S^1.$

The surface Q^* thus obtained is a 1-holed torus (see Fig. 3.7.4), the curves μ_2 and $\tilde{\mu}_3$ together form a closed boundary geodesic η^*, and, since (3.7.4) is compatible with (3.7.2), we have a natural isometric immersion

(3.7.5) $\qquad\qquad\qquad J : Q^* \to Q$

with the property that the restriction $J \,|\, A_2$ of J to A_2 is the identity mapping. In particular, $J(\gamma^*) = \gamma$. Moreover, there exist tubular neighborhoods $\mathscr{C}^*$ of γ^* in Q^* and $\mathscr{C}$ of γ in Q such that

(3.7.6) $\qquad\qquad J^{-1}(\mathscr{C}) = \mathscr{C}^* \text{ and } J \,|\, \mathscr{C}^* \text{ is injective.}$

For the proof of Theorem 3.7.1 it remains to show that if g^* and g^{**} are different closed geodesics in Q^*, intersecting γ^* transversally, then $J(g^*)$, $J(g^{**})$, $\rho \circ J(g^*)$ and $\rho \circ J(g^{**})$ are different geodesics. (The condition that g^* and g^{**} intersect γ could be relaxed, cf. Remark 3.7.13 (iv).)

Thus, let $g = J(g^*)$. We have to show that $\rho(g)$ cannot be the image under J

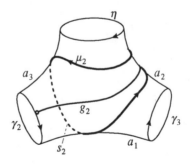

Figure 3.7.5

of a closed geodesic in Q^*. For the proof we use the fact that for $k = 2$ and 3, the geodesic g contains an arc g_k (Fig. 3.7.5) lying on Y, connecting γ_k with a_k and contained in A_k. The part of μ_k which lies on the boundary of A_k contains an arc s_k which connects a_{k-1} with a_{k+1} (dotted line in Fig. 3.7.5). We claim that for $k = 2$ and 3

(3.7.7) $$s_k \cap g_k = \varnothing, \qquad s_k \cap \rho\sigma(g_k) \neq \varnothing,$$

where $\rho\sigma$ is the isometry of Y which fixes a_1, a_2 and a_3 pointwise.

Let $k = 2$, for simplicity. If the first statement is valid, then the second statement is also valid, because s_2, $\rho\sigma(s_2)$ and γ_2 bound an annulus which does not meet a_2. For the first statement we note, by virtue of (3.7.6), that g_2 has a *unique* lift in Q^*. The injectivity of J on A_2 then implies that any intersection of g_2 with s_2 lifts to an intersection of g^* with $J^{-1}(s_2) \subset \eta^* = \partial Q^*$. This proves (3.7.7).

Now suppose that $\rho(g) = J(\bar{g})$ for some closed geodesic $\bar{g}$ in Q^*. Then (3.7.7) holds also if g_3 is replaced by $\rho(g_2)$ and g_2 by $\rho(g_3)$. Applying σ and observing that $\sigma(s_2) = s_3$ and $\sigma(s_3) = s_2$ (cf. (3.7.7)), we get for $k = 2$ and 3,

$$s_k \cap \rho\sigma(g_k) = \varnothing, \qquad s_k \cap g_k \neq \varnothing,$$

a contradiction.

It is thus proved that $J(g^*)$, $J(g^{**})$, $\rho \circ J(g^*)$ and $\rho \circ J(g^{**})$ are four distinct geodesics.

Taking g^* and g^{**} in Q^* such that they have the same length, we obtain an example of multiplicity 4 on Q. More generally, the preceding argument has shown that if Q^* has examples with multiplicity m, then Q has examples with multiplicity $2m$. Theorem 3.7.1 now follows by induction over m. ◇

Algebraic Approach

We now give an algebraic proof of Theorem 3.7.1 which is in terms of traces of matrices in $SL(2, \mathbf{R})$. This approach will lead to a more general construction of examples with high multiplicities. Before we start, however, we want to review the preceding proof more algebraically, in order to clarify the link between the two approaches. After that (that is, after Corollary 3.7.8), we shall begin anew in $SL(2, \mathbf{R})$.

The two geodesics γ and β in Fig. 3.7.1, if seen as loops at their common intersection point, are generators of the fundamental group of the 1-holed torus Q. The geodesic ω is homotopic to

$$v = v(\gamma, \beta) := \beta\gamma^{-1}\beta^{-1}\gamma\beta.$$

This is best checked by drawing some instances of the homotopy. The isometry $\rho : Q \to Q$ maps γ and β to their inverses, and consequently $\rho(\omega)$ is homotopic to

$$w = w(\gamma, \beta) := \beta^{-1}\gamma\beta\gamma^{-1}\beta^{-1}.$$

Now let $\tilde{Q} \to Q$ with $\tilde{Q} \subset \mathbf{H}$ be a universal covering of Q (cf. Theorem 1.4.2), and let Γ be the associated group of deck transformations, which we see as a discrete subgroup of $SL(2, \mathbf{R})$. Then Γ is canonically isomorphic to the fundamental group $\pi_1(Q)$ of Q, and we let $A, B \in \Gamma$ be generators of Γ corresponding to the generators γ, β of $\pi_1(Q)$ under this isomorphism (A goes to γ and B to β). It is well known that $\pi_1(Q)$ is a free group, and hence we have a free subgroup of two generators of $SL(2, \mathbf{R})$:

$$\Gamma = \langle A, B \rangle.$$

The closed geodesics of Q are in a natural one-to-one correspondence with the conjugacy classes in Γ, and their lengths are determined by the traces of the representatives of the corresponding conjugacy classes. Finding multiplicities in the length spectrum of Q is thus translated into the problem of finding non-conjugate words $W = W(A, B)$ in Γ having the same trace. The above proof of Theorem 3.7.1 has provided such words. They are iteratively defined as follows.

$$W_1(A, B) = v(A, B), \qquad W_2(A, B) = v(A^{-1}, B^{-1}),$$

where $v(A, B) = B A^{-1} B^{-1} A B$, and

$$W_{\sigma 1}(A, B) = W_\sigma(A, v(A, B)), \qquad W_{\sigma 2}(A, B) = W_\sigma(A^{-1}, v(A^{-1}, B^{-1})),$$

where the multi-indices σ are strings of 1-s and 2-s such as $\sigma = 12211$, $\sigma = 2212121$, etc. We denote by $\mathscr{S}_k$ the set of all 2^k strings of length k. The geometric proof then yields the following corollary in which k, A and B are fixed. (Another proof that the words $W_\sigma^{\pm 1}(A, B)$ in the corollary are pairwise non-conjugate will be obtained via Lemma 3.7.11 and Lemma 3.7.12.)

3.7.8 Corollary. *The words $W_\sigma^{\pm 1}(A, B)$, $\sigma \in \mathscr{S}_k$, all have the same trace and are pairwise non-conjugate.*

We now give an algebraic proof of this which is similar to that in Horowitz [1]. We start with the *trace relation* (cf. Horowitz [1] for a more general form). In the following, tr denotes the trace of a matrix, and a, b are variables. The next lemma is an immediate consequence of the triple trace theorem of Section 2.6 (Corollary 2.6.17). However, in order to keep the present section more independent, we shall give a direct proof.

3.7.9 Lemma. *For any word $W = W(a, b) = a^{m_1} b^{n_1} \ldots a^{m_s} b^{n_s}$ there exists a unique polynomial P_W such that*

$$\operatorname{tr}(W(A, B)) = P_W(\operatorname{tr} A, \operatorname{tr} B, \operatorname{tr} AB)$$

for all $A, B \in SL(2, \mathbf{R})$.

Proof. We first check the elementary identities

$$\operatorname{tr} A = \operatorname{tr} A^{-1}, \quad \operatorname{tr} A = \operatorname{tr}(BAB^{-1})$$
$$\operatorname{tr}(AB) = \operatorname{tr} A \operatorname{tr} B - \operatorname{tr}(AB^{-1}).$$

The lemma holds for the words $W = a^m b^n$ with $m, n \in \{-1, 0, 1\}$. The identity

$$\operatorname{tr}(A^m B^n) = \operatorname{tr}(A^m B^{n-1}) \operatorname{tr} B - \operatorname{tr}(A^m B^{n-2})$$

then allows this to be extended to words $W = a^m b^n$ with arbitrary $m, n \in \mathbf{Z}$. The identity

$$\operatorname{tr}(A^{m_1}B^{n_1}\ldots A^{m_s}B^{n_s}) = \operatorname{tr}(A^{m_1}B^{n_1} \ldots A^{m_{s-1}}B^{n_{s-1}})\operatorname{tr}(A^{m_s}B^{n_s}) - $$
$$\operatorname{tr}(A^{(m_1-m_s)}B^{n_1}A^{m_2}B^{n_2} \ldots A^{m_{s-1}}B^{(n_{s-1}-n_s)})$$

gives access to induction over s. In this way we obtain the polynomials. As for their uniqueness, we observe that the values of $\operatorname{tr} A$, $\operatorname{tr} B$ and $\operatorname{tr}(AB)$ fill out a region of $\mathbf{R}^3$ which contains an open neighborhood. The relation $\operatorname{tr}(W(A, B)) = P_W(\operatorname{tr} A, \operatorname{tr} B, \operatorname{tr} AB)$ implies therefore that two polynomials with this property must be identical. ◇

An immediate consequence of Lemma 3.7.9 is

3.7.10 Lemma. *Let $V(a, b)$, $W(a, b)$ and $u(a, b)$ be words in the variables a, b and set*

$$V'(a, b) = V(a, u(a, b)), \quad W'(a, b) = W(a, u(a, b))$$
$$V''(a, b) = V'(a^{-1}, b^{-1}), \quad W''(a, b) = W'(a^{-1}, b^{-1}).$$

If $P_V = P_W$, then $P_{V'} = P_{W'} = P_{V''} = P_{W''}$. ◇

Lemma 3.7.10 shows by induction that the 2^k words $W_\sigma(A, B)$ in Corollary 3.7.8, $\sigma \in \mathscr{S}_k$, have the same trace. Let us next show that the $W_\sigma^{\pm 1}(A, B)$, $\sigma \in \mathscr{S}_k$, are pairwise non-conjugate. For this we first make some general remarks about free groups. Let

$$G = \langle a, b \rangle$$

be the abstract free group in the two generators a and b. Every element in G

can be written as a *unique* reduced (that is, all xx^{-1} are cancelled) word $W = a^{m_1} b^{n_1} \ldots a^{m_s} b^{n_s}$ with all exponents non-zero except possibly for m_1 and n_s. If W_1 and W_2 are words, then $W_1 W_2$ and $W_2 W_1$ are conjugate. Conversely, if $W' = UWU^{-1}$, then induction over the word length of U shows that W' can be obtained by splitting up W into a product $W = W_1 W_2$ and then by putting $W' = W_2 W_1$. Conjugacy classes in G are therefore represented by *cyclic words*: we denote by

$$\Omega = [a^{m_1} b^{n_1} \ldots a^{m_s} b^{n_s}]$$

the equivalence class of words modulo cyclic permutations of the $2s$ "syllables" $a^{m_1}, b^{n_1}, \ldots, a^{m_s}, b^{n_s}$. The point of view here is that the last syllable is glued to the first. A more space consuming notation would be to write these words on cylindrical paper.

If we restrict ourselves to reduced cyclic words, then different words belong to different conjugacy classes. For better reference we state this as follows.

3.7.11 Lemma. *In the free group $G = \langle a, b \rangle$, two elements are conjugate if and only if their associated reduced cyclic words coincide.* ◊

In the next lemma, v is again the word

$$v(a, b) = b\, a^{-1} b^{-1} a\, b.$$

Lemma 3.7.12 together with Lemma 3.7.11 will yield a new proof that the words $W_\sigma^{\pm 1}$ occurring in Corollary 3.7.8 are non-conjugate. Observe below that $\Omega^{-1}(a, v(a, b)) = (\Omega(a, v(a, b)))^{-1}$, etc. (cf. also Remark 3.7.13 (v)).

3.7.12 Lemma. *Let $\Omega(a, b)$ and $\Omega_*(a, b)$ be reduced cyclic words in the free group $G = \langle a, b \rangle$, each with at least one occurrence of a syllable b^k, $k \neq \pm 1$. If $\Omega \neq \Omega_*$ and $\Omega \neq \Omega_*^{-1}$, then the eight words $\Omega^{\pm 1}(a, v(a, b))$, $\Omega^{\pm 1}(a^{-1}, v(a^{-1}, b^{-1}))$, $\Omega_*^{\pm 1}(a, v(a, b))$ and $\Omega_*^{\pm 1}(a^{-1}, v(a^{-1}, b^{-1}))$ are pairwise different.*

Proof. Let θ be one of these eight words. By reading θ we can tell whether $v(a, b)$ has been used or $v(a^{-1}, b^{-1})$: since both Ω and Ω_* have an occurrence of a syllable b^k with $k \neq \pm 1$, θ must have a syllable $b^{\pm 2}$. Take any such syllable in θ. The structure of the word v (all exponents of the letters b in v are ± 1) shows that this syllable has its occurrence in the center of a sequence

$$v^\delta(a^\varepsilon, b^\varepsilon)\, v^\delta(a^\varepsilon, b^\varepsilon) = b^{\delta\varepsilon} a^{-\varepsilon} b^{-\delta\varepsilon} a^\varepsilon b^{2\delta\varepsilon} a^{-\varepsilon} b^{-\delta\varepsilon} a^\varepsilon b^{\delta\varepsilon},$$

where $\delta, \varepsilon \in \{-1, 1\}$. By reading the four rightmost letters of our syllable, we determine the values of ε and δ, that is, we determine which of the four possible words $v^\delta(a^\varepsilon, b^\varepsilon)$ has been used. Once this is known, we easily read off which of the four words $\Omega^{\pm 1}$, $\Omega_*^{\pm 1}$ has been used to build θ. ◇

This finishes the algebraic proof of Theorem 3.7.1 and the proof of Corollary 3.7.8. We add several remarks.

3.7.13 Remarks. (i) The above multiplicities are *stable* in the sense that if we vary the values of the parameters $\ell(\eta)$, $\ell(\gamma)$ and α of the one-holed torus Q, then the 2^k geodesics determined by the words W_σ, $\sigma \in \mathcal{S}_k$, remain examples of multiplicity 2^k. The stability of the above multiplicities is due to the fact that they result from identities among traces.

(ii) The above construction yields 2^k geodesics of length of order $O((\ell(v(a,b))/\ell(b))^k)$, where ℓ denotes the length of the geodesic represented by the given word. If γ and β in Q are chosen such that $\ell(\gamma) \leq \ell(\beta)$, we obtain geodesics of length of order $O(5^k)$ with multiplicity 2^k as $k \to \infty$. This is still far from answering the question in Hejhal [2], vol I, p.321.

(iii) Instead of 1-holed tori we may use Y-pieces whose fundamental group is also the free group of two generators. This shows that Theorem 3.7.1. holds also on non-compact Riemann surfaces of finite area (cf. Randol [5]).

(iv) With additional effort one may improve Lemma 3.7.12 as follows. Assume that $\Omega(a, b)$ and $\Omega_*(a, b)$ are reduced cyclic words in the free group $G = \langle a, b \rangle$. If Ω and Ω_* are not of type $[a^n]$ or $[(abab^{-1})^n]$, $n \in \mathbb{Z}$, and if $\Omega \neq \Omega_*^{\pm 1}$, then the eight words $\Omega^{\pm 1}(a, v(a, b))$, $\Omega^{\pm 1}(a^{-1}, v(a^{-1}, b^{-1}))$, $\Omega_*^{\pm 1}(a, v(a, b))$ and $\Omega_*^{\pm 1}(a^{-1}, v(a^{-1}, b^{-1}))$ are pairwise different.

For the immersion $J: Q^* \to Q$ and the isometry $\rho: Q \to Q$ used earlier, it follows that if g^* is a closed geodesic in Q^*, then $\rho(J(g^*))$ is not the image under J of a closed geodesic, except possibly in the following two cases

$$J(g^*) = \gamma^n, \quad n \in \mathbb{Z}, \quad \text{and} \quad J(g^*) = (\gamma \beta \gamma \beta^{-1})^n, \quad n \in \mathbb{Z}.$$

One checks that these geodesics are invariant under ρ, so that these cases are, in fact, counterexamples.

(v) The proof of Lemma 3.7.12 shows that the construction is not restricted to the particular word $v = ba^{-1}b^{-1}ab$. We may for example replace it by any $v' = ba^r b^s a^t b$ with $r \neq s$ and $t \neq 2$. We may use any such words at different steps of the iteration. Since Lemma 3.7.10 remains valid, this gives numerous other examples, however, not sufficiently many to solve Hejhal's problem mentioned in (ii).

Chapter 4

The Collar Theorem

A fundamental property of Riemann surfaces, known as the collar theorem, is that the small closed geodesics have large tubular neighborhoods which are topological cylinders. This was first observed in Keen [4], and various versions have been proved in Basmajian [1], Buser [4], Chavel-Feldman [1], Halpern [1], Matelski [1], Randol [4], Seppälä-Sorvali [1, 3] and others. The collar theorem is a basic tool in different parts of this book. The proof is given in Section 4.1. In Section 4.2 we apply the collar theorem to obtain a lower bound for the lengths of closed geodesic with transversal self-intersections. Another application is the triangulation of controlled size in Section 4.5. In Section 4.3 we extend the collar theorem to surfaces with variable curvature, and in Section 4.4 we extend the collar theorem to non-compact Riemann surfaces of finite area.

Section 4.1 is the main part of this chapter. The remaining sections, except for Theorem 4.2.1, may be skipped in a first reading.

4.1 Collars

4.1.1 Theorem. (Collar theorem I). *Let S be a compact Riemann surface of genus $g \geq 2$, and let $\gamma_1, \ldots, \gamma_m$ be pairwise disjoint simple closed geodesics on S. Then the following hold.*
  (i) $m \leq 3g - 3$.
  (ii) *There exist simple closed geodesics $\gamma_{m+1}, \ldots, \gamma_{3g-3}$ which, together with $\gamma_1, \ldots, \gamma_m$, decompose S into pairs of pants.*
  (iii) *The collars*
$$\mathscr{C}(\gamma_i) = \{p \in S \mid \text{dist}(p, \gamma_i) \leq w(\gamma_i)\}$$

of widths

$$w(\gamma_i) = \operatorname{arcsinh}\{1/\sinh(\tfrac{1}{2}\ell(\gamma_i))\}$$

are pairwise disjoint for $i = 1, \ldots, 3g - 3$.

(iv) *Each $\mathscr{C}(\gamma_i)$ is isometric to the cylinder $[-w(\gamma_i), w(\gamma_i)] \times \mathbf{S}^1$ with the Riemannian metric $ds^2 = d\rho^2 + \ell^2(\gamma_i)\cosh^2\rho\, dt^2$.*

Proof. Cut S open along $\gamma_1, \ldots, \gamma_m$. Each connected component S' of the surface thus obtained is a hyperbolic surface of some signature (g', n'). As the γ_i are pairwise non-homotopic and each geodesic is homotopically non-trivial (Theorem 1.6.6), we either have $g' \geq 1$ or $n' \geq 3$ (if $g' = 0$).

Assume that S' does not have signature $(0, 3)$. Then S' contains a homotopically non-trivial simple closed curve, say γ_{m+1}, which is not homotopic to a boundary component of S'. By Theorem 1.6.6, we may assume that γ_{m+1} is a closed geodesic. Now cut S' open along γ_{m+1}, and continue. After finitely many steps, S is decomposed into pairs of pants. The Euler characteristic tells us that the number of pants is $2g - 2$. This gives (i) and (ii).

Since S is decomposed into pairs of pants, assertion (iii) and the first part of (iv) are immediate consequences of Proposition 3.1.8 (half-collars), and we can copy the expression for the metric tensor from Example 1.3.2.   ◇

4.1.2 Corollary. *Let γ, δ be closed geodesics on S which intersect each other transversally, and assume that γ is simple. Then*

$$\sinh \tfrac{1}{2}\ell(\gamma) \sinh \tfrac{1}{2}\ell(\delta) > 1.$$

Proof. Let $\tilde{\gamma}, \tilde{\delta}$ be intersecting lifts of γ, δ in the universal covering $\mathbf{H}$ of S. Then the point set $\tilde{\mathscr{C}}(\tilde{\gamma}) = \{z \in \mathbf{H} \mid \operatorname{dist}(z, \tilde{\gamma}) \leq w(\gamma)\}$ is a lift of $\mathscr{C}(\gamma)$, and $\tilde{\delta}$ connects the two boundary components of $\tilde{\mathscr{C}}(\tilde{\gamma})$. Hence, δ contains an arc in $\mathscr{C}(\gamma)$ which connects the boundary components of $\mathscr{C}(\gamma)$. This arc has length $\geq 2w(\gamma)$.   ◇

4.1.3 Example. The following example shows that the width $w(\gamma)$ of the collar of a closed geodesic γ is a natural quantity, and also that the bound in Corollary 4.1.2 is sharp. As in Example 3.1.9, we consider a pair of pants Y with $\ell(\gamma_3)$ arbitrarily small but now such that $\ell(\gamma_1) = \ell(\gamma_2)$. The distance on Y of the half-collars $\mathscr{C}^*(\gamma_1)$ and $\mathscr{C}^*(\gamma_2)$ is also arbitrarily small. With the pasting condition

(*) $\gamma_1(t) = \gamma_2(\tfrac{1}{2} - t) =: \gamma(t), t \in \mathbf{S}^1,$

we obtain a surface $Q = Y \operatorname{mod}(*)$ with a closed geodesic μ of length $\ell =$

$\ell(\mu)$ arbitrarily close to $2w(\gamma)$ and intersecting γ orthogonally. The distance set $\mathscr{C}' = \{ p \in Q \mid \mathrm{dist}(p, \gamma) \leq \ell/2 \}$ is no longer an annulus. ◇

The important implication of Corollary 4.1.2 is that the small simple closed geodesics on S, that is, those of length $< 2 \operatorname{arcsinh} 1$, are pairwise disjoint. By Theorem 4.1.1(i), the number of small simple closed geodesics on S cannot exceed $3g - 3$. We next show that outside the collars of the small geodesics, the injectivity radius of S cannot be too small. This gives rise to a "thick and thin" decomposition of S which is responsible for certain spectral phenomena of the Laplacian, such as the limitation of the number of small eigenvalues (Theorem 8.1.1). We remark also that, in modified form, a thick and thin decomposition with similar phenomena is possible on arbitrary manifolds with negative curvature and finite volume (see for instance Ballman-Gromov-Schroeder [1]).

In what follows, S is again a fixed compact Riemann surface of genus $g \geq 2$. For $p \in S$ we denote by U_p^r the distance set

$$U_p^r = \{ q \in S \mid \mathrm{dist}(p, q) < r \}.$$

For small $r > 0$, U_p^r is isometric to an open disk of radius r in $\mathbf{H}$.

4.1.4 Definition. The supremum of all r for which U_p^r is isometric to a disk is called the *injectivity radius of S at p*, and will be denoted by $r_p(S)$. The (over all) *injectivity radius* of S is defined as

$$r_{inj}(S) = \inf \{ r_p(S) \mid p \in S \}.$$

4.1.5 Lemma. $r_p(S) = \tfrac{1}{2}\ell(\mu_p)$, *where μ_p is the shortest geodesic loop at p, and $r_{inj}(S) = \tfrac{1}{2}\ell(\mu)$, where μ is the shortest closed geodesic on S. The curves μ_p and μ are simple.*

Proof. Since for $r < r_p(S)$, U_p^r is isometric to a disk in $\mathbf{H}$, and since there are no geodesic loops in $\mathbf{H}$, we have $\ell(\mu_p) \geq 2r_p(S)$. Let $r = r_p(S)$. Then the lifts of p in the universal covering $\mathbf{H}$ of S have pairwise distances of at least $2r$, and there are two such lifts, say p_1 and p_2 such that $\mathrm{dist}(p_1, p_2) = 2r$. Under the covering $\mathbf{H} \to S$, the geodesic arc from p_1 to p_2 is mapped to a geodesic loop of length $2r$ at p. Since U_p^r is still a disk, μ_p is a simple loop.

To prove the statement about μ, we observe that the function $p \mapsto \ell(\mu_p)$ is continuous (we have $r_q \geq r_p - \mathrm{dist}(p, q)$ for $p, q \in S$). By the compactness of S, $\ell(\mu_p)$ has a minimum, say at $p = m$. Since μ_m is homotopically non-trivial, it contains a unique shortest curve μ in its free homotopy class, and this μ is a closed geodesic (Theorem 1.6.6). Now, $r_{inj}(S) = \tfrac{1}{2}\ell(\mu_m) \geq \tfrac{1}{2}\ell(\mu) \geq r_{inj}(S)$.

Hence, $\mu = \mu_m$ and all statements are proved. ◊

4.1.6 Theorem. (Collar theorem II). *Let $\beta_1, \ldots, \beta_k$ be the set of all simple closed geodesics of length ≤ 2 arcsinh 1 on S. Then $k \leq 3g - 3$, and the following hold.*
  (i) *The geodesics $\beta_1, \ldots, \beta_k$ are pairwise disjoint.*
  (ii) *$r_p(S) > $ arcsinh 1 for all $p \in S - (\mathscr{C}(\beta_1) \cup \ldots \cup \mathscr{C}(\beta_k))$.*
  (iii) *If $p \in \mathscr{C}(\beta_i)$, and $d = \mathrm{dist}(p, \partial \mathscr{C}(\beta_i))$, then*

$$\sinh r_p(S) = \cosh \tfrac{1}{2}\ell(\beta_i) \cosh d - \sinh d.$$

Proof. Corollary 4.1.2 implies that $\beta_1, \ldots, \beta_k$ are pairwise disjoint, and the inequality $k \leq 3g - 3$ becomes a restatement of Theorem 4.1.1(i).

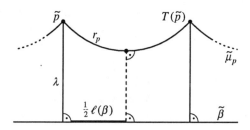

Figure 4.1.1

Now let $p \in S$ and assume that $\sinh r_p(S) \leq 1$. By Lemma 4.1.5, there exists a simple geodesic loop μ_p at p of length $2r_p(S)$. By Theorem 1.6.6, μ_p is freely homotopic to a simple closed geodesic β. Let $\tilde{\mu}_p$ and $\tilde{\beta}$ be homotopic lifts of μ_p and β in the universal covering of S, and let T be the covering transformation with axis $\tilde{\beta}$ which corresponds to μ_p. Since $\tilde{\mu}_p$ and $\tilde{\beta}$ are homotopic lifts, T maps $\tilde{\mu}_p$ onto itself. Let $\tilde{p} \in \tilde{\mu}_p$ be a lift of p and consider its image $T(\tilde{p})$ (Fig. 4.1.1). Since T preserves orientation, $\tilde{p}$ and $T(\tilde{p})$ are on the same side of $\tilde{\beta}$. The geodesic arc of length $2r_p(S)$ on $\tilde{\mu}_p$ from $\tilde{p}$ to $T(\tilde{p})$ does therefore not intersect $\tilde{\beta}$, and we have a geodesic quadrangle with sides of length λ, $\ell(\beta)$, λ, $2r_p(S)$, where $\lambda = \mathrm{dist}(\tilde{p}, \tilde{\beta}) = \mathrm{dist}(T(\tilde{p}), \tilde{\beta})$. Dropping the common perpendicular between this arc and $\tilde{\beta}$ (dotted line in Fig. 4.1.1), we obtain two isometric trirectangles. Theorem 2.3.1(v) (trirectangles) yields

(4.1.7) $\qquad \sinh r_p(S) = \sinh \tfrac{1}{2}\ell(\beta) \cosh \lambda > \sinh \tfrac{1}{2}\ell(\beta) \sinh \lambda.$

Since, by hypothesis, $\sinh r_p(S) \leq 1$, it follows that $\sinh \tfrac{1}{2}\ell(\beta) < 1$. We conclude that β is one of the geodesics $\beta_1, \ldots, \beta_k$. It also follows that $\mathrm{dist}(p, \beta) \leq \lambda < \mathrm{arcsinh}\{1/\sinh \tfrac{1}{2}\ell(\beta)\} = w(\beta)$, so that $p \in \mathscr{C}(\beta)$.

Now $\lambda = w - d$, where $w = w(\beta)$. From (4.1.7) we have

$$\sinh r_p(S) = \sinh \tfrac{1}{2}\ell(\beta) (\cosh w \cosh d - \sinh w \sinh d)$$
$$= \sinh \tfrac{1}{2}\ell(\beta) \cosh w \cosh d - \sinh d.$$

Since
$$\sinh^2 \tfrac{1}{2}\ell(\beta) \cosh^2 w = \sinh^2 \tfrac{1}{2}\ell(\beta) \sinh^2 w + \sinh^2 \tfrac{1}{2}\ell(\beta) = \cosh^2 \tfrac{1}{2}\ell(\beta),$$
we obtain
$$\sinh r_p(S) = \cosh \tfrac{1}{2}\ell(\beta) \cosh d - \sinh d. \qquad \diamond$$

4.1.8 Remarks. (i) We remark without proof that the collar theorem is not restricted to surfaces. It holds (with an appropriate definition of the width of the collar) on any Riemannian manifold with negative curvature and finite volume. This may be derived from Margulis' lemma. A reference for this is Ballmann-Gromov-Schroeder [1]. For hyperbolic 3-manifolds one has a sharper version of Margulis' lemma, known as Jørgensen's inequality. For the collar theorem on hyperbolic 3-manifolds we refer the reader to Beardon [1], Brooks-Matelski [1], Gilman [1] and Jørgensen [1].

(ii) A version of the collar theorem which holds for geodesics with self-intersections may be found in Basmajian [1].

(iii) A variety of other geometric inequalities and constraints for compact Riemann surfaces may be found in Beardon [1], chapter 11.

(iv) The antecedent and earliest version of all collar theorems seems to be Siegel [1].

4.2 Non-Simple Closed Geodesics

In Section 10.5 and again in Section 13.1 we shall need a lower bound for the length of a closed geodesic which has a transversal self-intersection. In the first part of the present section we obtain a bound which is sufficient for the needs of Sections 10.5 and 13.1. After that, we improve our efforts and obtain a sharp bound. Sharp lower bounds for the lengths of non-simple closed geodesics have been found by Hempel [1], Nakanishi [1], Yamada [1] and others for compact Riemann surfaces and also for quotients of Fuchsian groups of more general signatures. The proof given in this section is designed in such a way that it also works for surfaces with variable curvature. For a shorter proof we refer to Beardon [1], chapter 11. The main tool in the proof given here is that the shortest geodesic with transversal self-intersections has exactly one self-intersection (Theorem 4.2.4).

4.2.1 Theorem. *Let S be a compact Riemann surface of genus $g \geq 2$. Then every primitive non-simple closed geodesic on S has length greater than 1.*

Proof. Let δ be a primitive closed geodesic of length $\ell(\delta) \leq 1$ on S, and let $p \in \delta$. For the injectivity radius we then have $r_p(S) \leq \frac{1}{2}\ell(\delta) \leq \frac{1}{2}$. By Theorem 4.1.6, we have $p \in \mathscr{C}(\beta)$ for some simple closed geodesic β of length $\ell \leq 1$, and the identity (iii) of the same theorem implies that

$$\sinh \tfrac{1}{2} \geq \sinh r_p(S) > \cosh d - \sinh d = e^{-d},$$

where d is the distance from p to the boundary of $\mathscr{C}(\beta)$. From the inequality $\sinh \tfrac{1}{2} < e^{-1/2}$ follows that $d = \mathrm{dist}(p, \partial\mathscr{C}(\beta)) > \tfrac{1}{2}$. Hence, δ is contained in $\mathscr{C}(\beta)$, and therefore $\delta = \beta$ (up to a parameter change). ◇

In the remainder of the section we prove a sharp bound (Hempel [1], Nakanishi [1], Yamada [1]). The result itself will not be used later in the book. S is again a compact Riemann surface.

4.2.2 Theorem. *Every primitive non-simple closed geodesic on S has length greater than* $4 \operatorname{arcsinh} 1 = 3.52\dots$ *. This bound is sharp.*

Proof.

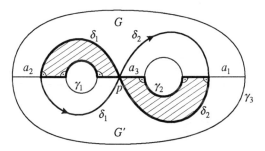

Figure 4.2.1

We first give an example to show that the bound is sharp. Since some of the arguments will be used later in the proof, we give the example in more general form than necessary. Let Y be an arbitrary pair of pants with boundary geodesics γ_1, γ_2, γ_3 and the perpendiculars a_1, a_2, a_3, which decompose Y into the right-angled hexagons G, G' as shown in Fig. 4.2.1 (cf. Proposition 3.1.5)

The figure eight geodesic $\delta = \delta_1 \delta_2$ (first along δ_1, then along δ_2) in Fig. 4.2.1, whose loops δ_1 and δ_2 are freely homotopic to γ_1 and γ_2, has its intersection point p on a_3. This follows from the fact that there exists an isometry from Y to itself with fixed point set $a_1 \cup a_2 \cup a_3$. For the same reason, δ intersects a_1 and a_2 orthogonally. Now consider the crossed right-angled hexagon formed by arcs on γ_1, γ_2, δ and arcs on a_1, a_2, a_3. Again by the symmetry of Y, the sides of this hexagon (with the short notation $\ell(\gamma_1) = \gamma_1$, etc.) are the following: $a_3, \frac{1}{2}\gamma_2, \bullet, \frac{1}{2}\delta, \bullet, \frac{1}{2}\gamma_1$. Theorem 2.4.4 yields

$$\cosh \tfrac{1}{2}\delta = \sinh \tfrac{1}{2}\gamma_1 \sinh \tfrac{1}{2}\gamma_2 \cosh a_3 + \cosh \tfrac{1}{2}\gamma_1 \cosh \tfrac{1}{2}\gamma_2.$$

Theorem 2.4.1(i) (right-angled hexagons) applied to G yields

$$\cosh \tfrac{1}{2}\gamma_3 = \sinh \tfrac{1}{2}\gamma_1 \sinh \tfrac{1}{2}\gamma_2 \cosh a_3 - \cosh \tfrac{1}{2}\gamma_1 \cosh \tfrac{1}{2}\gamma_2.$$

The two equations together yield

(4.2.3) $\qquad \cosh \tfrac{1}{2}\delta = \cosh \tfrac{1}{2}\gamma_3 + 2 \cosh \tfrac{1}{2}\gamma_1 \cosh \tfrac{1}{2}\gamma_2 > 3.$

Now consider a sequence of Y-pieces such that $\ell(\gamma_i) \to 0$, $i = 1, 2, 3$. Then

$$\ell(\delta) \to 2 \operatorname{arccosh} 3 = 4 \operatorname{arcsinh} 1.$$

Hence, the lower bound in Theorem 4.2.2 cannot be improved. Moreover, with the above considerations we have actually *proved* Theorem 4.2.2 for the particular case in which the geodesic is a figure eight on a pair of pants. In the next theorem we show that, on S again, the shortest geodesic with a transversal self-intersection is the image of a figure eight under a suitable isometric immersion $Y \to S$. This will prove Theorem 4.2.2 in the general case. ◇

The following theorem will also be used in the next section and is therefore stated for variable curvature. The proofs for constant curvature and for variable curvature are almost the same.

For simplicity, on any compact surface M of negative curvature a primitive closed geodesic with exactly one self-intersection will be called a *figure eight geodesic*. It is easy to see, and it will also come out in the proof of the next theorem, that a figure eight geodesic is always contained in a Y-piece (with geodesic boundary) whose interior is embedded in M.

4.2.4 Theorem. *Let M be a compact orientable surface of variable negative curvature and let δ be the shortest primitive non-simple closed geodesic on M. Then δ is a figure eight geodesic.*

Proof. In this proof "self-intersection" is the same as "transversal self-intersection". We assume that δ has more than one self-intersection. The strategy is to find a closed non-smooth path z on δ which is freely homotopic to a figure eight geodesic ζ. This ζ will then be shorter than δ.

We first remark that δ contains simple loops. Here a *loop in* δ is, by definition, a smooth arc on δ whose initial point coincides with the endpoint. For example, the curve cu in Fig. 4.2.2(b) is a loop in this sense, but the curve bd is not.

To find a simple loop we choose a parametrization $\delta : [0, 1] \to M$ and let t_2 be the supremum of all τ such that the restriction $\delta | [0, \tau]$ is a simple arc. Then there exists a unique $t_1 \in [0, t_2]$ with $\delta(t_1) = \delta(t_2)$, and it follows that $\delta | [t_1, t_2]$ is a simple loop in δ.

Now let η with base point A be the shortest loop in δ. The argument just given shows that η is simple. We choose our parametrization of δ anew so that δ now begins with η.

Since M is orientable, a small tubular neighborhood of η is an annulus and is decomposed by η into two open annuli, say L and R. We let L be the one containing the angle of η at A which is smaller than π. For small positive ε the segments $\delta | [t_2, t_2 + \varepsilon]$ and $\delta | [1 - \varepsilon, 1]$ are then contained in R. From this we see that δ can be written as

$$\delta = abcudw,$$

where the arcs a, b, c, u, d, w, fulfill the following requirements (the arcs need not all be simple and some may have zero length).

- $\eta = abc$,
- d has positive length,
- the intersection $d \cap \eta$ consists of the endpoints B and C of d (possibly with $B = C$).
- $d \cap L = \emptyset$.

To make this possible we must allow that $u = 0$ and/or $w = 0$. The following proof is valid also in these limiting cases so that they need no extra considerations. Note that if $u = 0$, then $A = B$ and $a = 0$. If $w = 0$, then $C = A$ and $c = 0$. If $u = w = 0$, then $A = B = C$ and $\delta = \eta d$, where d is a non-simple loop (because δ is assumed to have more than one self-intersection).

As a first case we assume that d is either a simple arc with $B \neq C$ or a simple loop with base point $B = C$. If $B \neq C$ the notation is such that b goes from B to C. The arc d then may go from B to C as in Fig. 4.2.2(a) or from C to B as in Fig. 4.2.2(b).

Let us first look at Fig. 4.2.2(a). Here wab with base point C is a loop in δ. This loop is non-trivial because if $w = 0$ then $c = 0$. By the minimal length of η we have $w + a + b \geq a + b + c$, that is, $w \geq c$. Similarly, bcu is a non-trivial loop in δ with base point B and so $u \geq a$. Altogether $u + w \geq a + c$, and we define

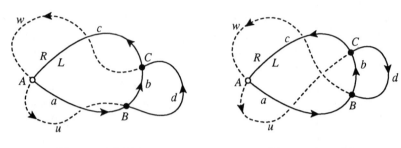

Figure 4.2.2(a)   **Figure 4.2.2(b)**

(a) $$z = abcadc,$$

where $\ell(z) \le \ell(\delta)$. In the present case we cannot have $u = w = 0$, for then $\delta = bd$ and δ would have only one self-intersection. The curve z is therefore not smooth and the closed geodesic ζ in the free homotopy class of z is strictly shorter than δ. That ζ is a figure eight geodesic will be proved together with the remaining cases.

Next let us look at Fig. 4.2.2(b). Since the limiting situations $u = 0$ and $a = 0$ are covered by the preceding case, we now have $u \ne 0$ and $a \ne 0$. There is the non-trivial loop cu in δ implying the inequality $c + u \ge a + b + c$ so that $u \ge a + b > b$. Here the closed curve

(b) $$z = abdbc$$

is strictly shorter than δ and we let again ζ be the closed geodesic in the free homotopy class of z.

The remaining case is that d has a transversal self-intersection different from the initial point of d. Here we may write d as a product $d = xvy$ where x is a simple arc and v is a simple loop. We may also write $d^{-1} = x'v'y'$ with a simple arc x' and a simple loop v', and observe that x' is contained in y and x

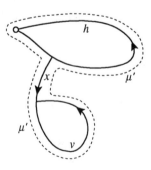

Figure 4.2.2(c)

is contained in y' so that either $x \leq y$ or $x' \leq y'$. We may therefore, without loss of generality, assume that $x \leq y$.

We let h be the closed curve η but now with a new parametrization such that the product $\mu = hxvx^{-1}$ is well defined. The orientation of h is to be chosen in such a way that an arbitrarily small homotopy deforms μ into a simple closed curve μ' which together with h and v bounds an open neighborhood of signature (0. 3). Since $x \leq y$ the closed curve

(c) $$z = hxv^{-1}x^{-1}$$

has length $\ell(z) \leq \ell(\delta)$. Again, z is not smooth, and the closed geodesic ζ in its free homotopy class satisfies $\ell(\zeta) < \ell(\delta)$.

We have now considered all possible cases, and it remains to prove that in each case z is not homotopic to a simple curve.

The curve η is not homotopic to a point because M has negative curvature. We also check that the following closed curves are homotopically non-trivial: bd^{-1} and adc in the case of Fig. 4.2.2(a); bd and $ad^{-1}c$ in the case of Fig. 4.2.2(b); v and μ in the case of Fig. 4.2.2(c). In fact, if any of these curves were homotopically trivial, then a lift of it in the universal covering of M would bound a simply connected geodesic polygon with, say, n sides in which the angle sum is greater than $(n-2)\pi$. This is impossible since M has negative curvature.

By the preceding statements, z has an open neighborhood W of signature (0, 3) where none of the boundary components is homotopically trivial. We now make use of Theorems 1.6.6 and 1.6.7. In order to make them applicable (rather than generalizing them to variable negative curvature, which would also be a possibility) we introduce an auxiliary hyperbolic metric on M and denote by S the surface with this new metric.

Assume first that, even after a possible change of orientation, the boundary components of W are pairwise non-homotopic. By Theorems 1.6.6 and 1.6.7, they are homotopic to pairwise disjoint simple closed geodesics in S. By Theorem A.3, there exists a 1-homeomorphism $\phi : S \to S$ mapping W onto a pair of pants Y bounded by the latter. On Y, $\phi(z)$ is homotopically non-trivial (with respect to Y) and not homotopic to a boundary component. Hence, $\phi(z)$ is homotopic to a closed geodesic ζ' in the interior of Y with a transversal self-intersection. By Theorem 1.6.6, the homotopy class of ζ' contains no simple curves. In particular, ζ is not simple.

Finally we assume that, possibly after a change of orientation two of the boundary components of W, say α and β are homotopic. By Proposition A.11, α and β bound an annulus $\mathcal{A}$ which is embedded in S. If the third component of W, say γ, were homotopic to α then, by Proposition A.11, W would be an annulus. Now let $\phi : S \to S$ be a 1-homeomorphism mapping α

and γ to the closed geodesics α' and γ' in their respective homotopy classes. Then $\phi(W \cup \mathcal{A})$ is a hyperbolic surface of signature (1, 1), and after cutting S open along α and γ we obtain a pair of pants Y whose interior is embedded in S and contains $\phi(z)$. As before, this implies that the homotopy class of $\phi(z)$ contains no simple curves. Theorem 4.2.4 is now proved. ◊

We remark that the preceding proof also shows that on a compact hyperbolic surface S any figure eight geodesic is contained in a pair of pants whose interior is embedded in S. The proof of this fact for variable negative curvature is obtained by generalizing Theorems 1.1.6 and 1.1.7 to variable negative curvature.

4.3 Variable Curvature

In this section we extend the collar theorem and the length estimate for the self-intersecting closed geodesics to variable curvature. For the sake of continuity we begin with the length estimate.

4.3.1 Theorem. *Let M be a compact orientable surface of genus $g \geq 2$ with curvature K such that $-\kappa^2 \leq K < 0$. Then every primitive non-simple closed geodesic on M has length $\geq \frac{4}{\kappa}$ arcsinh 1.*

Proof. Let δ be the geodesic. Since ($\kappa \times$ length) is a scaling invariant, we may assume that $\kappa = 1$. By Theorem 4.2.4, it suffices to consider the case where δ has exactly one self-intersection. Our first observation is that δ is contained in a pair of pants Y (with curvature satisfying $-1 \leq K < 0$). This follows from the remark at the end of Section 4.2 or from the following direct argument.

Let $\pi : H \to M$ be a universal covering with covering transformation group Γ. The geodesic δ is a product of two loops $\delta = \delta_1 \delta_2$. Let γ_1, γ_2 and γ_3 be the closed geodesics in the free homotopy classes of δ_1, δ_2 and $\delta_1 \delta_2^{-1}$ respectively. The homotopy between δ_1 and γ_1 can be lifted to a homotopy between lifts $\tilde{\delta}_i$ and $\tilde{\gamma}_1$. By the negative curvature of H these lifts do not intersect each other and bound therefore a strip S_1 in H. We let $T_1 \in \Gamma$ be the element with axis $\tilde{\gamma}_1$ which corresponds to δ_1. The cyclic group $[T_1] = \{T_1^n\}_{n \in \mathbb{Z}}$, acts on S_1, and the quotient $\mathcal{A}_1 = [T_1] \backslash S_1$ is an annulus. The loop δ_2 gives rise to a similar annulus $\mathcal{A}_2$. Since δ has only one self-intersection, any lift $\tilde{\mu}$ of $\delta_1 \delta_2^{-1}$ in H is a simple curve and has the property that one of the components of $H - \tilde{\mu}$ is convex. Therefore, if $\tilde{\mu}$ and $\tilde{\gamma}_3$ are homotopic lifts of $\delta_1 \delta_2^{-1}$ and

γ_3, then $\tilde{\mu}$ and $\tilde{\gamma}_3$ bound a strip S_3, that is, the closed curve $\delta_1\delta_2^{-1}$ gives rise to a third annulus $\mathcal{A}_3$. By pasting together $\mathcal{A}_1$, $\mathcal{A}_2$ and $\mathcal{A}_3$ along δ we obtain a pair of pants Y containing δ in its interior. It would not be difficult to see that the interior of Y is embedded in M, however, we shall not need this fact since all the remaining arguments in the proof of Theorem 4.3.1 take place on Y.

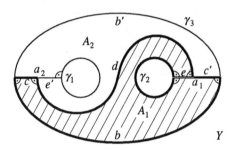

Figure 4.3.1

Fig. 4.3.1 shows Y with the boundary geodesics γ_1, γ_2, γ_3 labelled in such a way that the two loops δ_1 and δ_2 of $\delta = \delta_1\delta_2$ are freely homotopic to γ_1 and γ_2, respectively, exactly as in the preceding section. As in the case of constant curvature, one proves with the Arzelà-Ascoli argument that each pair of boundary geodesics of Y has a unique simple common perpendicular, and that the three perpendiculars a_1, a_2, a_3 decompose Y into right-angled hexagons. As remarked earlier, the hexagons are, of course, no longer isometric in general.

The geodesic δ consists of two arcs, each of which connects a_1 with a_2. Unlike in the case of constant curvature, these arcs need not meet a_1 or a_2 orthogonally. Now let δ' be the smaller of the two arcs. In the homotopy class of δ' with endpoints gliding on a_1 and a_2 (Definition 1.5.1) we let d be the shortest curve (Fig. 4.3.1). This curve is a geodesic arc in Y which meets a_1 and a_2 orthogonally at interior points of a_1 and a_2.

Now, d together with a_1 and a_2 decomposes Y into right-angled hexagons A_1 and A_2, where
$$A_1 = a_1bcde\gamma_2, \qquad A_2 = a_2b'c'de'\gamma_1,$$
and where (speaking of lengths) we have
$$e + c' = a_1, \qquad c + e' = a_2.$$
By Corollary 2.5.12, there exist two right-angled geodesic comparison hexagons in $\mathbf{H}$:
$$\bar{A}_1 = \bar{a}_1\bar{b}cde\bar{\gamma}_2, \qquad \bar{A}_2 = \bar{a}_2\bar{b}'c'de'\bar{\gamma}_1$$

such that $a_1 \leq \bar{a}_1$, $a_2 \leq \bar{a}_2$, that is,

$$e + c' \leq \bar{a}_1, \quad c + e' \leq \bar{a}_2.$$

In order to construct a hyperbolic comparison Y-piece, we fix the lengths of sides c and d of $\bar{A}_1$ and decrease e continuously. By Theorem 2.4.1(i), side $\bar{a}_1$ decreases also and converges to zero as e approaches some positive limiting value. Hence, there exists a value $\tilde{e}$ in between, such that the corresponding value $\tilde{a}_1$ satisfies the equation $\tilde{e} + c' = \tilde{a}_1$. Similarly, we fix c' and d of $\bar{A}_2$ and decrease e' until $\bar{a}_2$ reaches a value $\tilde{a}_2$ such that $c + \tilde{e}' = \tilde{a}_2$. Hence, there exist comparison hexagons in **H**:

$$\tilde{A}_1 = \tilde{a}_1 \tilde{b} c d \tilde{e} \tilde{\gamma}_2, \quad \tilde{A}_2 = \tilde{a}_2 \tilde{b}' c' d \tilde{e}' \tilde{\gamma}_1$$

satisfying

$$\tilde{e} + c' = \tilde{a}_1, \quad c + \tilde{e}' = \tilde{a}_2.$$

The last relation allows us to paste $\tilde{A}_1$ and $\tilde{A}_2$ together, with the same configuration as that in Fig. 4.3.1, to obtain a hyperbolic pair of pants $\tilde{Y}$ on which $\tilde{a}_1$ and $\tilde{a}_2$ are the unique simple common perpendiculars from $\tilde{\gamma}_2$ to $\tilde{\gamma}_3$ and from $\tilde{\gamma}_1$ to $\tilde{\gamma}_3$, respectively. By symmetry, d extends to a figure eight geodesic $\tilde{\delta}$ of length $\ell(\tilde{\delta}) = 2d \leq \ell(\delta)$. Now we can apply Theorem 4.2.2. ◊

We do not know whether Theorem 4.3.1 remains valid if the curvature is allowed to assume positive values. The collar theorem, however, holds for variable curvature (Buser [4], Chavel-Feldman [1]) without any assumption about the upper curvature bound.

4.3.2 Theorem. (Collar theorem for variable curvature). *Let M be a compact orientable 2-dimensional Riemannian manifold of genus ≥ 2 and curvature $K \geq -1$. For any homotopically non-trivial simple closed geodesic γ in M the collar*

$$\mathscr{C}(\gamma) = \{p \in M \mid \sinh(\text{dist}(p, \gamma)) \sinh \tfrac{1}{2}\ell(\gamma) \leq 1\}$$

is homeomorphic to $[0, 1] \times S^1$. If γ and β are disjoint simple closed geodesics from different non-trivial homotopy classes in M, then $\mathscr{C}(\gamma)$ and $\mathscr{C}(\beta)$ are disjoint.

Proof. We reduce the proof to a 3-holed sphere by cutting M open, first along γ and β, and then successively along further simple closed geodesics, which we find using length-decreasing homotopies. They are described in the last section of the Appendix.

Thus, let M be a geodesically bordered Riemannian manifold of signature (0, 3) with curvature $K \geq -1$ and boundary geodesics γ_1, γ_2, γ_3. We prove that the half-collars

$$\mathscr{C}^*(\gamma_i) = \{p \in M \mid \sinh(\text{dist}(p, \gamma_i)) \sinh \tfrac{1}{2}\ell(\gamma_i) \leq 1\}$$

are disjoint annuli. The notation is as in Fig. 4.3.2.

We cut open M as in Fig. 4.3.2(a) along a simple geodesic arc μ which is orthogonal to γ_3 and separates γ_1 from γ_2. To obtain such an arc we begin with a piecewise smooth arc with the given separation property, and then use length-decreasing homotopies. M after having been cut open along μ, is decomposed into two doubly connected domains Z_1, Z_2 with γ_i on the boundary of Z_i, $i = 1, 2$. We dissect Z_1 further into two right-angled geodesic pentagons G, G' along *shortest* perpendiculars from γ_1 to μ and from γ_1 to γ_3. We denote by c_1 the side of G which lies on γ_1 and let c_2 be the shorter of the two adjacent sides of c_1. We may assume that $c_1 \leq \tfrac{1}{2}\ell(\gamma_1)$ (otherwise we interchange G and G'). By Lemma 2.5.13, we have $\sinh c_1 \sinh c_2 > 1$. It follows that $\mathscr{C}^*(\gamma_1)$ is contained in Z_1, and similarly $\mathscr{C}^*(\gamma_2)$ is contained in Z_2. This proves that the half-collars in M are disjoint.

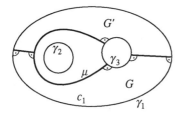 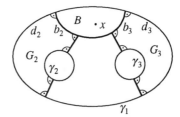

Figure 4.3.2(a) **Figure 4.3.2(b)**

Now suppose that $\mathscr{C}^*(\gamma_1)$ is not an annulus. More generally, let

$$r \leq \text{arcsinh}\{1/\sinh(\tfrac{1}{2}\ell(\gamma_1))\},$$

consider the set

$$\mathscr{C}^* = \{p \in Z_1 \mid \text{dist}(p, \gamma_1) \leq r\}$$

and suppose that $\mathscr{C}^*$ is not an annulus. We lead this to a contradiction. Since Z_1 is an annulus, the complement $Z_1 - \mathscr{C}^*$ has an open connected component D whose boundary does not intersect the boundary of Z_1. Pick $x \in D$. Then $\text{dist}(x, \gamma_1) > r$. There exists a simple arc β of length less than or equal to $2r$ in Z_1 with both endpoints on γ_1 which separates x from $(\partial Z_1 - \gamma_1)$. Now apply a length-decreasing homotopy to β. Since $\text{dist}(x, \gamma_1) > r$, this homotopy cannot sweep over x, and we obtain a *geodesic* arc b of length less than or equal to $2r$ orthogonal to γ_1 which cuts out a simply connected right-angled geodesic biangle $B \subset M$ as shown in Fig. 4.3.2(b).

On $M - \operatorname{int} B$ we drop *shortest* perpendiculars from γ_2 and γ_3 to b. Since they have minimal length, they do not intersect each other. In their complement we then find non-intersecting shortest perpendiculars from γ_2 and γ_3 to $(\gamma_1 - B)$ as shown in Fig. 4.3.2(b). We obtain two disjoint right-angled geodesic pentagons G_2, G_3, with pairs of adjacent sides b_2, d_2 and b_3, d_3, respectively, where b_2, b_3 are disjoint arcs on b, and d_2, d_3 are disjoint arcs on γ_1. Hence, $b_2 + b_3 < 2r$ and $d_2 + d_3 < \ell(\gamma_1)$. By the assumed upper bound of r we have therefore

$$\sinh(\tfrac{1}{2}(b_2 + b_3)) \sinh(\tfrac{1}{2}(d_2 + d_3)) < \sinh r \sinh \tfrac{1}{2}\ell(\gamma_1) \leq 1.$$

On the other hand, by virtue of Lemma 2.5.13,

$$\sinh b_2 \sinh d_2 > 1, \quad \sinh b_3 \sinh d_3 > 1.$$

By the inequality given below, this is a contradiction. Hence, $\mathscr{C}^*$ is an annulus, and Theorem 4.3.2 is proved. ◇

The inequality is the following, where we assume that $0 \leq |s| \leq x$, $0 \leq t \leq y$:

(4.3.3)
$$\sinh x \sinh y \geq \min\{\sinh(x - s) \sinh(y + t), \sinh(x + s) \sinh(y - t)\}.$$

Proof. The statement in non-trivial only for $s > 0$. We set $\vartheta = \sinh x \sinh y$, $s_1 = x - \operatorname{arcsinh}\{\vartheta/\sinh(y + t)\}$, $s_2 = -x + \operatorname{arcsinh}\{\vartheta/\sinh(y - t)\}$. For $s \in [0, s_2]$ we then have

$$\sinh(x + s) \sinh(y - t) \leq \sinh(x + s_2) \sinh(y - t) = \vartheta,$$

and for $s \in [s_1, x]$ we get

$$\sinh(x - s) \sinh(y + t) \leq \sinh(x - s_1) \sinh(y + t) = \vartheta.$$

It remains to show that $f(t) := s_2 - s_1 \geq 0$ for $t \in [0, y]$. But this is true because $f(0) = 0$ and $f'(t) > 0$ for $t \in \,]0, y[$. ◇

4.4 Cusps

In this section we consider non-compact hyperbolic surfaces with finite area. A new type of neighborhood occurs which is a limiting case of a half-collar: the *cusp*. Due to the importance of the thick and thin decomposition of hyperbolic manifolds, we include a short discussion of cusps for completeness. For the thick and thin decomposition in general we refer, for instance, to Ballmann-Gromov-Schroeder [1]. For some applications to the spectrum of the Laplacian we refer to Burger-Schroeder [1] and Dodziuk-Randol [1].

Let us first extend the definition of a Y-piece as follows. Instead of compact hexagons we now admit right-angled hyperbolic hexagons as shown in Fig. 4.4.1 in the unit disk **D**, where either one, two or three pairwise non-adjacent sides are degenerated into points at infinity. The points at infinity themselves are not considered elements of the hexagon.

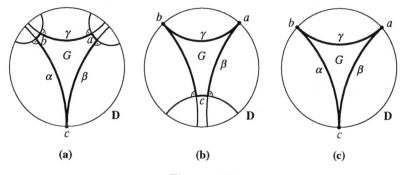

Figure 4.4.1

Two copies of the hexagon, say G and G', pasted together along the three remaining sides as in the non-degenerate case (in Fig. 4.4.1 along α, β, γ) yield a degenerate pair of pants Y which has either two, one or no boundary geodesics. From the geometry of hexagons we obtain

4.4.1 Lemma. *A degenerate pair of pants Y exists for any prescribed lengths of its boundary geodesics. These lengths determine Y up to isometry.* ◇

In particular, there exists exactly one pair of pants for which all three geodesics are degenerate. This is in accordance with the fact that the Möbius transformations which preserve **D** act three times transitively on $\partial \mathbf{D}$.

For the boundary geodesics of Y we define the half-collars in the same way as in the non-degenerate case. (Proposition 3.1.8). For the infinitely long ends, the analog of the half-collar is obtained by the following construction which we describe in the case of Fig. 4.4.2 (a). To simplify the language, geodesic arcs and their geodesic extensions will be given the same name. By "endpoints" of a geodesic are meant the endpoints at infinity.

Thus, let Y be obtained by pasting together the degenerate hexagons G and

G' with the notation of Fig. 4.4.2(a). A neighborhood $\mathscr{E} \subset G$ is defined as follows. We drop the perpendicular s from the endpoint p_β of β to point q on side α. Then there exists a unique horocycle h through q centered at c. We let $H \subset \mathbf{D} \subset \mathbf{R}^2$ denote the closed disk with boundary h and define

$$\mathscr{E} = G \cap H.$$

On Y we let $\mathscr{E}^*$ be the domain formed by $\mathscr{E}$ and its copy $\mathscr{E}' \subset G'$. Topologically, $\mathscr{E}^*$ is a cylinder. To compute the length of its boundary we use a limiting argument.

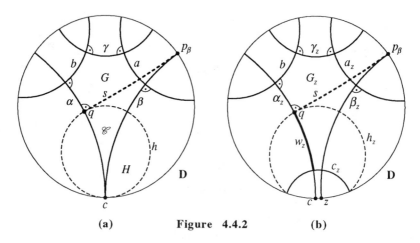

Figure 4.4.2

Let $z \in \partial \mathbf{D}$ be a point at infinity near c which will later converge to c. Replace β with the geodesic β_z from p_β to z, set $\alpha_z := \alpha$ and let c_z denote the common perpendicular of α_z and β_z. The hexagon G is thus approximated by a hexagon G_z with sides a_z, γ_z, b, α_z, c_z, β_z (Fig. 4.4.2(b)). The perpendicular s from p_β to $q \in \alpha_z$ has not moved. Set

$$w_z = \text{dist}(q, c_z)$$

and denote by h_z the parallel line formed by all those points at distance w_z from c_z and lying on the same side of c_z as q. In our model, h_z is a Euclidean circle through q and the endpoints of c_z.

Using Fermi coordinates based on c_z, we compute, using (1.1.9),

$$\ell(h_z \cap G_z) = c_z \cosh w_z.$$

Then, applying trigonometry to the degenerate trirectangle with vertex p_β, we have

$$\sinh c_z \sinh w_z = 1.$$

Now let z converge to c. Then h_z converges to h (in $\mathbf{R}^2$) and $\ell(h_z \cap G_z)$ converges to $\ell(h \cap G)$. On the other hand, we have $w_z \to \infty$ and $c_z \to 0$. Together with the above, this gives $\ell(h_z \cap G_z) \to 1$ and thus

$$\ell(\partial \mathscr{E}^*) = 2.$$

Using horocyclic coordinates at $c \in \partial \mathbf{D}$, we conclude, using (1.1.10), that $\mathscr{E}^*$ is isometric to the cylinder

(4.4.2) $\qquad]{-}\infty, \log 2] \times \mathbf{S}^1 \quad \text{with} \quad d\rho^2 + e^{2\rho} dt^2,$

where $-\infty < \rho \le \log 2$ and $t \in \mathbf{S}^1$.

4.4.3 Definition. *Any domain which is isometric to $\mathscr{E}^*$ or (4.4.2) will be called a cusp.*

The same construction also works in the cases of Fig. 4.4.1(b) and (c), in which the Y-pieces have two or three cusps. As in the non-degenerate case, the common perpendiculars in G and G' which go from a to α, b to β and c to γ, separate the half-collars and cusps. The following analog of Proposition 3.1.8 therefore follows.

4.4.4 Proposition. *In a degenerate Y-piece the cusps and half-collars are pairwise disjoint.* ◊

For the rest of this chapter, the Y-pieces which will be considered may be degenerate or non-degenerate.

A Y-piece with p boundary geodesics and q cusps is said to have *signature* $(0, p; q)$. In order to obtain hyperbolic surfaces of arbitrary signature $(g, p; q)$ we paste together Y-pieces along boundary geodesics just as in the compact case. The number of Y-pieces needed for signature $(g, p; q)$ is $2g - 2 + p + q$ and the number of pairs of geodesics involved in the pasting is $3g - 3 + p + q$. In view of Lemma 4.4.1, the parameter space for these surfaces has dimension $6g - 6 + 3p + 2q$. The only signatures excluded are those for which $2g + p + q < 3$.

Since only the cusps contain points with arbitrarily small injectivity radius, the cusps of a hyperbolic surface of signature $(g, p; q)$ obtained by the above construction are easily detected by means of the *intrinsic* geometry of the surface. If we cut off the cusps (along the boundaries of length 2), we are left with a compact topological surface of genus g with $p + q$ holes. This shows that the signature introduced above is independent of the particular description of the surface. Moreover, the next theorem shows that a signature is defined, intrinsically, for any hyperbolic surface of finite area whose boundary components are closed geodesics.

The next two theorems are the goal of this section. To keep the statements shorter, we have not included the bordered case in the collar theorem. We recall that the term "hyperbolic" includes that the metric is complete (Definition 1.2.3).

4.4.5 Theorem. *Let S be a hyperbolic surface of finite area, all of whose boundary components are closed geodesics. Then S is a surface of some signature $(g, p; q)$ obtained by the above construction.*

4.4.6 Theorem. (Collar theorem in the non-compact case). *Let S be a hyperbolic surface of signature $(g, 0; q)$. Then*

(i) *S has uniquely determined cusps $\mathscr{E}^1, \ldots, \mathscr{E}^q$. The cusps are pairwise disjoint.*

(ii) *If $\gamma_1, \ldots, \gamma_m$ are pairwise disjoint simple closed geodesics on S, then $m \leq 3g - 3 + q$, and there exist simple closed geodesics $\gamma_{m+1}, \ldots, \gamma_{3g-3+q}$ such that $\gamma_1, \ldots, \gamma_{3g-3+q}$ decompose S into Y-pieces.*

(iii) *The collars $\mathscr{C}(\gamma_i) = \{p \in S \mid \sinh(\text{dist}(p, \gamma_i)) \sinh(\frac{1}{2}\ell(\gamma_i)) \leq 1\}$, around the geodesics in (ii) are pairwise disjoint and do not intersect the cusps $\mathscr{E}^1, \ldots, \mathscr{E}^q$.*

(iv) *If $\beta_1, \ldots, \beta_k$ is the sequence of all simple closed geodesics of length $\leq 2 \operatorname{arcsinh} 1$ on S, then $\beta_1, \ldots, \beta_k$ are pairwise disjoint, and the injectivity radius $r_p(S)$ satisfies the inequality*

$$r_p(S) > \operatorname{arcsinh} 1$$

for any point $p \in S - (\mathscr{C}(\beta_1) \cup \ldots \cup \mathscr{C}(\beta_k) \cup \mathscr{E}^1 \cup \ldots \cup \mathscr{E}^q)$.

Proof of Theorem 4.4.6. The proof is similar to the compact case and we restrict ourselves to outlining the few modifications. It would also be possible to conclude Theorem 4.4.6 directly from Section 4.1 by a limiting argument.

By hypothesis, S is obtained by the above construction, and hence the cusps can be found by means of the intrinsic geometry, as remarked earlier. That the cusps are pairwise disjoint follows from Proposition 4.4.4, and (i) is clear.

For (ii) we note that S is homeomorphic to a compact surface of genus g with q punctures (q points removed). This yields the bound on m, and we find completing curves $c_{m+1}, \ldots, c_{3g-3+q}$ which together with $\gamma_1, \ldots, \gamma_m$ topologically decompose S into 3-holed spheres. Then we observe that none of the c_i can be homotoped into the interior of one of the cusps $\mathscr{E}^1, \ldots, \mathscr{E}^q$.

Hence, if for given i, $\{c_i^\nu\}_{\nu=1}^\infty$ is a sequence of homotopic curves with lengths converging to the infimum of the homotopy class of c_i, then the sequence stays in a compact subset of S, and the theorem of Arzelà-Ascoli can be applied in the same way as in Section 1.6. Similarly, the proof of Theorem 1.6.6(iii) goes through. We then get (ii) using Theorem A.3, as in the compact case.

(iii) follows from Proposition 4.4.4.

The geodesics $\beta_1, \ldots, \beta_k$ in (iv) are pairwise disjoint because the collars $\mathscr{C}(\beta_1), \ldots, \mathscr{C}(\beta_k)$ are, and the bound on $r_p(S)$ is obtained by the same arguments as in the compact case. ◇

Proof of Theorem 4.4.5. The strategy is to cut off Y-pieces from S. Since all Y-pieces have the same area, the procedure will end after finitely many steps. It suffices therefore to show that we can cut out *one* Y-piece.

We shall use covering arguments. For this we fix a universal covering $\pi : \tilde{S} \to S$, $\tilde{S} \subset \mathbf{D}$ (Theorem 1.4.2) with covering transformation group $\Gamma : \tilde{S} \to \tilde{S}$. Since Γ operates by isometries in $\tilde{S}$, we have $\Gamma \subset \mathrm{Is}^+(\mathbf{D})$. By Theorem 1.4.1, S is contained in an unbordered hyperbolic surface with the same fundamental group. This larger surface is $\Gamma \backslash \mathbf{D}$, and consequently Γ acts freely and properly discontinuously not only on $\tilde{S}$ but also on $\mathbf{D}$.

Consider first the case that S has a boundary geodesic, say γ. For small $r > 0$ the distance set $Z(r) = \{p \in S \mid \mathrm{dist}(p, \gamma) \leq r\}$ is an annulus isometric to $[0, r] \times \mathbf{S}^1$ with metric $d\rho^2 + \ell^2(\gamma) \cosh^2\rho \, dt^2$. As r grows, we find a first $r = r_0$, where this fails to be the case (here we use the fact that S has finite area). We then have one of the following two cases (Fig. 4.4.3).

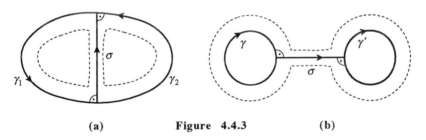

Figure 4.4.3

In the first case there is a geodesic arc σ of length $2r_0$ perpendicular to γ at both endpoints as in Fig. 4.4.3(a). In the second case there is a geodesic arc σ of length $2r_0$ which is the common perpendicular of γ and a second boundary geodesic γ' as in Fig. 4.4.3(b).

Case 1. The endpoints of σ decompose γ into two arcs γ_1, γ_2 and we can parametrize the curves in such a way that $\sigma\gamma_1$ and $\gamma_2\sigma^{-1}$ are simple closed curves. We let ω be a lift of $\sigma\gamma_1$ in $\tilde{S}$. Then ω is an infinite right-angled geodesic polygon whose sides alternately are lifts of the arcs σ and γ_1. We consider the perpendiculars at the midpoint of the sides of ω. By symmetry, these perpendiculars either meet each other in a point $c \in \mathbf{D} \cup \partial \mathbf{D}$, or they have a common perpendicular *geodesic c*. We let $\{T\} := \{T^n\}_{n \in \mathbb{Z}}$ be the maximal cyclic subgroup of Γ which leaves ω invariant. Then T leaves c invariant. Since Γ acts freely on $\mathbf{D}$, T is either parabolic and c its fixed point at infinity, or T is hyperbolic and c is the axis of T.

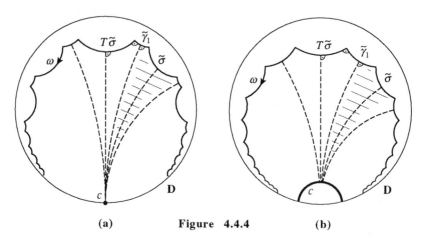

(a) Figure 4.4.4 (b)

Let us first look at the parabolic case (Fig: 4.4.4(a)). We want to show here that $\sigma\gamma_1$ bounds a cylindrical end. For this let $R \in \Gamma$, $R \notin \{T\}$. Since $\sigma\gamma_1$ is simple, its lift ω is simple, and, moreover, either $R\omega$ coincides with ω or $R\omega \cap \omega = \emptyset$. But $\{T\}$ is the maximal subgroup of Γ which fixes ω so that $R\omega \neq \omega$. We claim that c is not a fixed point of R.

Namely, if R fixes c, then, following from the fact that $R\omega \cap \omega = \emptyset$, c cannot be a parabolic fixed point of R. Hence R must be hyperbolic and c an endpoint at infinity of the axis of R. Replacing R by R^{-1} if necessary, we may assume that c is an attracting fixed point in the sense that $R^n p$ converges to c for $n \to \infty$, where p is a test point on the axis of R. For every $\varepsilon > 0$, there must exist n sufficiently large so that $0 < \text{dist}(p, R^{-n}TR^n p) < \varepsilon$. This contradicts the discreteness of the action of Γ. Hence, c cannot be a fixed point of R, as claimed.

Since $R\omega \cap \omega = \emptyset$, it follows next that $R\Omega \cap \Omega = \emptyset$, where Ω is the simply connected domain with boundary ω. Hence $\{T\}\backslash\Omega = \Gamma\backslash\Omega$. Let $\tilde{\sigma}$ be a side of ω which is a lift of σ. The perpendicular at the midpoint of $\tilde{\sigma}$ together with its image under T bounds a fundamental domain for $\{T\}$ on Ω. Its projection in S is an infinitely long cylinder with boundary curve $\sigma\gamma_1$.

Let us next look at the case of Fig. 4.4.4(b), where T is hyperbolic with axis c. We want to show that $\sigma\gamma_1$ bounds an annulus whose other boundary component is a closed geodesic. (This case has already been studied in the proof of Theorem 1.6.6(iv).) The proof is as before: if $R \in \Gamma$, $R \notin \{T\}$, then $R\omega \cap \omega = \emptyset$, hence $R \cap c = \emptyset$ and therefore $R\Omega \cap \Omega = \emptyset$, where now Ω is the domain enclosed by ω and c. A fundamental domain is obtained as before, and the desired annulus is $\{T\}\backslash\Omega$.

So far we have shown that $\sigma\gamma_1$ (Fig. 4.4.3(a)) cuts away an infinite cylinder or an annulus from the surface S. The same is true for the curve $\gamma_2\sigma^{-1}$. Since the interior angles at the boundary of the cut-off domains are right angles (acute would be sufficient), we see from Fig. 4.4.3 that the two domains are disjoint, and their union is cut away from S by the closed geodesic $\gamma = \gamma_1\gamma_2$. The two together therefore define the desired Y-piece.

Case 2. Next let us assume that the arc σ is the common perpendicular of γ and a second boundary geodesic γ' of S (Fig. 4.4.3(b)). Here we can parametrize the curves such that $\sigma\gamma'\sigma^{-1}\gamma$ contains simple curves in its homotopy class and we let ω be a lift of $\sigma\gamma'\sigma^{-1}\gamma$ in $\tilde{S} \subset \mathbf{D}$. Then ω is simple, and the same arguments as before show that $\sigma\gamma'\sigma^{-1}\gamma$ bounds a cylindrical neighborhood of finite or infinite length. This neighborhood together with its boundary forms the desired Y-piece in this second case.

So far we have assumed that S has at least one boundary geodesic. Now let the boundary of S be empty. (Points at infinity are not boundary points of S.) Choose $p \in S$ arbitrarily and let μ_p be the shortest geodesic loop at p. Such a loop exists because S is not the hyperbolic plane. The loop μ_p need not be homotopic to a closed geodesic. Hence we consider, once again, a lift ω of μ_p in $\tilde{S} \subset \mathbf{D}$. Then ω is a simple infinite geodesic polygon, and we have a maximal cyclic subgroup $\{T\} \subset \Gamma$ of covering transformations which leave ω invariant. Again the generator T may be hyperbolic or parabolic.

If T is hyperbolic, then, by the preceding arguments, μ_p is homotopic to a simple closed geodesic. We may then cut S open along this geodesic, bringing us back again to the preceding case.

If T is parabolic, μ_p bounds an infinite cylinder which contains a domain $\mathscr{E}_m$ isometric to $]-\infty, -m] \times \mathbf{S}^1$ with metric $d\rho^2 + e^{2\rho}dt^2$ for some $m > 0$. Here we let γ be the boundary of $\mathscr{E}_m$ and set

$$Z(r) := \{p \in S - \mathscr{E}_m \mid \text{dist}(p, \gamma) \leq r\}.$$

There exists a first r_0, where two of the geodesic segments emanating perpendicularly from γ into $S - \mathscr{E}_m$ meet and form a smooth arc σ. We are then in the situation of Fig. 4.4.3(a), except that $\gamma = \gamma_1\gamma_2$ is not a geodesic. However, this does not affect the arguments used earlier for $\sigma\gamma_1$ and $\gamma_2\sigma^{-1}$, and so these curves bound domains (cylinders and/or annuli) which together with $\mathscr{E}_m$ form a Y-piece. Theorem 4.4.5 is now proved. ◊

4.5 Triangulations of Controlled Size

It is sometimes desirable to have a triangulation of a compact Riemann surface with uniformly size-controlled triangles. However, at places where the injectivity radius is small, the triangles have to be small too. In this section we apply the collar theorem to bypass this difficulty using a more general definition of triangles. The result is from Buser [2]. It will be used in the proof of Theorem 8.1.4.

4.5.1 Definition. Let S be a compact Riemann surface of genus ≥ 2. A closed domain $D \subset S$ is called a *trigon* if it satisfies one of the following conditions.

(i) D is a simply connected embedded geodesic triangle.

(ii) D is an embedded doubly connected domain. One boundary component of D is a smooth closed geodesic and the other boundary component consists of two geodesic arcs. The closed geodesic and the two arcs are the *sides* of D.

Trigons of type (i) are ordinary geodesic triangles; examples of trigons of type (ii) are shown in Fig. 4.5.1.

In the following, a *triangulation* of S will be a triangulation with trigons. Hence, the union of all trigons is S, and the intersection of any two distinct trigons is either empty or a common vertex or a common side.

4.5.2 Theorem. *Any compact Riemann surface of genus ≥ 2 admits a triangulation such that all trigons have sides of length $\leq \log 4$ and area between 0.19 and 1.36.*

Proof. The idea is borrowed from an article by Fejes Tóth [1] on circle packings of the hyperbolic plane. To obtain a particularly dense packing of the hyperbolic plane, Fejes Tóth defines a maximal set of points at pairwise distances greater than or equal to some fixed constant. Then he considers those circles which pass through at least three of these points but contain none of these points in the interior. This can also be carried out on compact surfaces.

We begin with the collars. Let S be the compact Riemann surface and let $\gamma_1, \ldots, \gamma_m$ be all simple closed geodesics of length $\leq \log 4$ on S (if there are any). The constant $\log 4$ comes from the present proof and is not optimal. By Theorem 4.1.6, the collars $\mathscr{C}(\gamma_i)$, $i = 1, \ldots, m$ are pairwise disjoint.

For fixed $\gamma = \gamma_i$, we select four points A, B, A', B' in the interior of $\mathscr{C}(\gamma)$ as shown in Fig. 4.5.1, such that

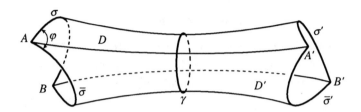

Figure 4.5.1

(i) $\text{dist}(A, \gamma) = \text{dist}(A', \gamma) = \text{dist}(B, \gamma) = \text{dist}(B', \gamma) = w(\gamma) - \frac{1}{2}\log 2$,

(ii) the geodesics in $\mathscr{C}(\gamma)$ from A to A' and from B to B' intersect γ perpendicularly and in opposite points.

Here $w(\gamma) = \operatorname{arcsinh}\{1/\sinh \frac{1}{2}\ell(\gamma)\}$ is the *width* of the collar. Observe that $w(\gamma) > 1$. The minimal geodesic arcs σ and $\bar{\sigma}$ from A to B and σ' and $\bar{\sigma}'$ from A' to B' together with γ form two isometric trigons D and D'. Each of these trigons consists of four isometric trirectangles. Theorem 2.3.1(v) and (vi) (trirectangles) yields the following equations, where φ is the interior angle of the trigons at the vertices A, B, A', B'.

$$\sinh \tfrac{1}{2}\sigma = \sinh \tfrac{1}{4}\ell(\gamma) \cosh(w(\gamma) - \tfrac{1}{2}\log 2)$$
$$= 2^{-1/2}\left(\tfrac{3}{2}\cosh \tfrac{1}{4}\ell(\gamma) - (\cosh \tfrac{1}{4}\ell(\gamma))^{-1}\right)$$
$$\cot \tfrac{1}{2}\varphi = \sinh(w(\gamma) - \tfrac{1}{2}\log 2) \tanh \tfrac{1}{4}\ell(\gamma)$$
$$= 2^{-1/2}\left((\cosh \tfrac{1}{4}\ell(\gamma))^{-2} - \tfrac{1}{2}\right).$$

From these equations we obtain the numerical bounds

(4.5.3) $1.07 \leq \text{area } D \leq 1.36, \quad \log 2 < \sigma \leq 0.89, \quad \varphi > 2\pi/3$.

After having introduced the trigons in $\mathscr{C}(\gamma_1), \ldots, \mathscr{C}(\gamma_m)$, we let S' be the closure of the complement of these trigons and define $\mathcal{P}_1$ to be the set of the $4m$ vertices in $\mathscr{C}(\gamma_1), \ldots, \mathscr{C}(\gamma_m)$ corresponding to A, B, A', B'. Since the collars are pairwise disjoint, it follows from (4.5.3) and from the inequality $w(\gamma_i) > 1$, $i = 1, \ldots, m$, that the points in $\mathcal{P}_1$ have pairwise distances $\geq \log 2$. We find therefore a finite set $\mathcal{P} \subset S$ with the following properties:

(i) $\mathcal{P}_1 \subset \mathcal{P}$,
(ii) $\text{dist}(p, q) \geq \log 2$ for all $p, q \in \mathcal{P}, p \neq q$,
(iii) for all $x \in S'$ there exists $p \in \mathcal{P}$ such that $\text{dist}(p, x) < \log 2$.

S' is the thick part of S in the following sense.

(4.5.4) *For all $x \in S'$ the injectivity radius of S at x satisfies*
$$r_x(S) \geq \min\{\log 2, \operatorname{dist}(x, \mathcal{P}_1)\}.$$

For the proof we recall from Theorem 4.1.6 that $r_x(S) > \log 2$ for all x in the complement of $\mathcal{C}(\gamma_1) \cup \ldots \cup \mathcal{C}(\gamma_m)$. Now fix the geodesic $\gamma = \gamma_i$ and let $x \in \mathcal{C}(\gamma) \cap S'$. We may assume that $\operatorname{dist}(x, \mathcal{P}_1) = \operatorname{dist}(x, A)$.

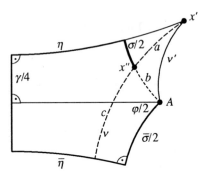

Figure 4.5.2

At vertex A we have two trirectangles with acute angle $\varphi/2$ as shown in Fig. 4.5.1 and Fig. 4.5.2. We denote by η and $\bar{\eta}$ the geodesic rays perpendicular to γ which form a side of the trirectangles. Now let $x' \in \eta$ be a point with $\operatorname{dist}(x', \gamma) = \operatorname{dist}(x, \gamma)$. Since $\operatorname{dist}(x, A) = \operatorname{dist}(x, \mathcal{P}_1) \leq \operatorname{dist}(x, B)$, we have $\operatorname{dist}(x, A) \leq \operatorname{dist}(x', A)$. Since $r_x(S) = r_{x'}(S')$, it suffices to prove (4.5.4) for x'.

Connect x' and A with a minimizing geodesic arc v' and drop the shortest perpendicular v from x' to $\bar{\eta}$ (Fig. 4.5.2). Then $v = r_{x'}(S)$, and we have to prove that $v \geq \min\{v', \log 2\}$. Assume first that v intersects side $\sigma/2$ (Fig. 4.5.2), say in x'', and abbreviate

$$a = \operatorname{dist}(x', x''), \quad b = \operatorname{dist}(x'', A), \quad c = \operatorname{dist}(x'', \bar{\eta}).$$

By (4.5.3), $\varphi > \pi/2$ so that $c \geq \bar{\sigma}/2 = \sigma/2 \geq b$ and therefore $v = a + c \geq a + b > v'$. Next assume that v does not intersect side $\sigma/2$. We shift x' along η towards γ until v contains A and thus becomes an extension of $\bar{\sigma}/2$. The length of v decreases in this procedure and has the lower bound $\bar{\sigma}/2 + \sigma/2 > \log 2$ (by (4.5.3)). This proves (4.5.4).

To define a triangulation of S with vertex set $\mathcal{P}$, we let $\mathcal{B}$ denote the set of all closed distance sets

$$B_x(\rho) = \{y \in S \mid \operatorname{dist}(x, y) \leq \rho\}$$

with the following properties (int denotes the interior).

(i) $x \in S'$,
(ii) int $B_x(\rho) \cap \mathcal{P} = \emptyset$,
(iii) $\partial B_x(\rho)$ contains at least 3 points of $\mathcal{P}$.

We shall see below that $\mathcal{B}$ is not empty.

The following is an immediate consequence of (4.5.4) and of the properties of $\mathcal{P}$ and $\mathcal{B}$.

(4.5.5) *If* $B_x(\rho) \in \mathcal{B}$, *then* $\frac{1}{2}\log 2 < \rho < \log 2$ *and* $r_x(S) > \rho$.

It follows that each $B \in \mathcal{B}$ is an *embedded* metric disk of the hyperbolic plane. For each $B \in \mathcal{B}$ we denote by G_B the geodesic polygon domain in B whose vertex set is $\partial B \cap \mathcal{P}$ (so that G_B is the convex hull in B of $\partial B \cap \mathcal{P}$).

(4.5.6) *The domains* G_B, $B \in \mathcal{B}$ *tessellate* S'.

This means, by definition, that the domains G_B fill out S', and for $B \neq B^* \in \mathcal{B}$ the intersection $G_B \cap G_{B^*}$ is either empty or is a common vertex or a common side.

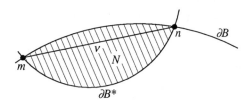

Figure 4.5.3

To prove (4.5.6), we first look at $G_B \cap G_{B^*}$. We may assume that $B \cap B^* \neq \emptyset$. Since int $B \cap \mathcal{P} = \emptyset$ and int $B^* \cap \mathcal{P} = \emptyset$, it is impossible that $B \subset B^*$ or that $B^* \subset B$. By (4.5.5), each connected component N of $B \cap B^*$ is the embedded intersection of two disks of the hyperbolic plane and contains exactly two points m and n of $\partial B \cap \partial B^*$ on its boundary. The geodesic arc v from m to n separates N into a left-hand side and a right-hand side. Since int$(B \cup B^*)$ does not intersect $\mathcal{P}$, neither side contains a vertex of G_B, except possibly for m and n. It follows that $G_B \cap N$ and $G_{B^*} \cap N$ are separated by v. Moreover, $(G_B \cap G_{B^*} \cap N)$ is either empty or is exactly one of the points m, n (which then is a common vertex), or the entire arc v (which then is a common side).

To prove (4.5.6) it remains to prove that the domains G_B, $B \in \mathcal{B}$, cover S'. We begin with the case where the above set $\gamma_1, \ldots, \gamma_m$ of closed geodesics of length $\leq \log 4$ is not empty, and let σ be a side of one of the trigons as in Fig. 4.5.1 or Fig. 4.5.2. We show that σ is the side of one of the polygons G_B. Let m be the midpoint of σ. Consider the unit speed geodesic $t \mapsto \eta(t)$, $t \geq 0$, $\eta(0) = m$, orthogonal to σ and pointing toward S'. For every $t \in [0, \log 2]$, we consider the closed metric disk $B(t)$ with center $\eta(t)$ and radius $\rho(t) = \operatorname{dist}(\eta(t), A) = \operatorname{dist}(\eta(t), B)$ (notation of Fig. 4.5.2). Since $\rho(0) = \sigma/2$, every $x \in B(0)$ satisfies $\cosh \operatorname{dist}(x, \{A, B\}) \leq \cosh^2 \sigma/2$. In conjunction with (4.5.3) this yields

(4.5.7) $\qquad \operatorname{dist}(x, \{A, B\}) < \log 2$ *for all* $x \in B(0)$.

Hence, $\operatorname{int} B(t) \cap \mathcal{P} = \emptyset$ for sufficiently small $t > 0$. We also have $\eta(t) \in S'$ for all $t \in [0, \log 2]$ and $\rho(\log 2) > \log 2$. By the definition of $\mathcal{P}$ we find therefore $t_0 \in \,]0, \log 2[$ such that $B := B(t_0) \in \mathcal{B}$. Now σ is a side of G_B. In particular, $\mathcal{B} \neq \emptyset$ if the geodesics $\gamma_1, \ldots, \gamma_m$ exist.

If there are no geodesics $\gamma_1, \ldots, \gamma_m$, that is, if the injectivity radius of S is everywhere greater than $\log 2$, then we let $B_x(\rho)$ be the distance set with the largest possible radius ρ that satisfies $\operatorname{int} B_x(\rho) \cap \mathcal{P} = \emptyset$. By the definition of $\mathcal{P}$, we have $\rho < \log 2$. Hence, $B_x(\rho)$ is again an embedded disk of the hyperbolic plane. An elementary argument now shows that $B_x(\rho)$ contains three points of $\mathcal{P}$ on its boundary.

To finish the proof of (4.5.6), it remains to prove that each side s of a domain G_B, $B \in \mathcal{B}$, which is not one of the above sides σ (Fig. 4.5.2), must be the side of a second domain G_{B*}. For this we let m be the midpoint of s and define the mappings $t \mapsto \eta(t)$, $t \mapsto \rho(t)$, $t \mapsto B(t)$ with $\eta(0) = m$, etc. as before. The center of the disk B itself is $\eta(t_0)$, say. We assume η so oriented that for small $\varepsilon > 0$, G_B is not contained in $B(t_0 + \varepsilon)$. We then find a smallest $t_1 > t_0$ such that $\partial B(t_1)$ contains a third point of $\mathcal{P}$. We claim that $\eta(t) \in S'$ for all $t \in [t_0, t_1]$.

Assume this is not the case. Then $\eta(t') \in \partial S'$ for some $t' \in [t_0, t_1[$, that is, $\eta(t') \in \sigma$ for one of the above sides σ. By the definition of t_1, the endpoints of σ cannot be contained in $B(t')$; hence $B(t') \subset B_\sigma$, where B_σ is the disk with center $\eta(t')$ and radius $\sigma/2$. In particular, $s \subset B_\sigma$. By (4.5.7) we have an endpoint $p \in s$ and an endpoint $q \in \sigma$, $p \neq q$, and such that $\operatorname{dist}(p, q) < \log 2$. This is impossible because $p, q \in \mathcal{P}$. Thus it is proved that $\eta(t) \in S'$ for all $t \in [t_0, t_1]$. In particular, $\eta(t_1) \in S'$, and therefore $B^* := B(t_1) \in \mathcal{B}$. The domain G_{B*} is the desired neighbor of G_B along side s, and (4.5.6) is proved.

To go from the tessellation to a triangulation, we dissect each G_B with more than three vertices into geodesic triangles using diagonals emanating from a fixed vertex. This yields a triangulation of S' and, together with the earlier trigons, a triangulation of S. It remains to estimate the size.

By (4.5.3) the doubly connected trigons satisfy the inequalities of Theorem 4.5.2. Also, all ordinary triangles have side length between $\log 2$

and $2\log 2$. An upper bound of the area of the ordinary triangles is given by the equilateral triangle in the hyperbolic plane which is inscribed in a circle of radius $\log 2$. To find the lower bound, we minimize the area of an inscribed triangle under the given constraints.

Let $\rho \in]\frac{1}{2}\log 2, \log 2[$ and consider triangles inscribed in the circle of radius ρ. We keep one side of the inscribed triangle fixed and shrink the shorter of the remaining sides (keeping the triangle inscribed). Then the area decreases. This shows that the smallest area in the given circle is obtained by the isosceles triangle with two sides of length $\log 2$. Now vary ρ and consider the area of this isosceles triangle as a function of the angle ϑ between the sides of length $\log 2$. This function has no local minimum for $\vartheta \in]0, \pi[$. We find that the smallest possible area is either achieved for $\rho = \log 2$ or for the equilateral triangle of side length $\log 2$. Comparing the two, we see that the latter has the smaller area. Altogether we obtain the following numerical bounds for the *ordinary* triangles D in the above triangulation of S'.

(4.5.8) $\qquad\qquad 0.196 \leq \text{area } D \leq 0.545.$

Theorem 4.5.2 is now proved. $\diamond$

4.5.9 Remark. Similar arguments yield estimates for the interior angles of the above triangles D. The lower bound is achieved for the isosceles triangle with basis of length $\log 2$ and inscribed in a circle of radius $\log 2$. The upper bound is achieved for the isosceles triangle in the same circle but with the two sides of length $\log 2$. The numerical bounds in degrees are 22.6° and 112.6°.

Chapter 5

Bers' Constant and the Hairy Torus

Every compact Riemann surface can be decomposed into Y-pieces. What can we say about the lengths of the geodesics involved in such a decomposition? Bers [3, 4] proved that there exists a decomposition with lengths less than some constant which depends only on the genus. Bers' theorem has numerous consequences for the geometry of compact Riemann surfaces (see for instance Abikoff [1], Bers [4], Seppälä [1]). In this book we shall give the following applications of Bers' theorem. In Chapter 6 it gives a rough fundamental domain for the Teichmüller modular group, in Chapter 10 it is used in the proof of Wolpert's theorem, and in Chapter 13 we apply Bers' theorem to estimate the number of pairwise non-isometric isospectral Riemann surfaces possible.

The chapter is organized as follows. In Section 5.1 we state various versions of Bers' theorem. Section 5.2 contains the proof for the case of a compact Riemann surface. In Section 5.3 a torus with "thin hairs" will be constructed in order to show that the upper bound in the theorem cannot be improved too much. For instance it is not possible to obtain a logarithmic bound. The final section deals with variable curvature. Although this section will not be applied elsewhere in the book, we included it for the following reasons.

While partitions of Riemann surfaces can be traced back to as far as Fricke-Klein [1], the possibility of a 3-holed sphere decomposition of an *arbitrary* compact orientable surface of genus ≥ 2 with minimal geodesics (i.e. shortest in their homotopy classes) follows from the far more recent work of Ballmann [1] and Freedmann-Hass-Scott [1]. (The length-decreasing homotopies of the last section of the Appendix do not produce *minimal* geodesics in general.) The problem of finding such decompositions with, in addition, length controlled geodesics is closely related to Loewner's problem (Berger [1], Blatter [1], Hebda [1], Gromov [1]), which is in the realm of curvature free geometric estimates. In Section 5.4 we therefore prove a version of Bers' theorem which holds without any curvature assumptions but uses area instead.

5.1 Bers' Theorem

5.1.1 Definition. Let S be a compact Riemann surface of genus $g \geq 2$. A *partition* on S is a set of $3g - 3$ pairwise disjoint simple closed geodesics.

A partition of such a surface is the same as a decomposition into Y-pieces. Bers [4] proved that every compact Riemann surface of genus $g \geq 2$ has a partition $\gamma_1, \ldots, \gamma_{3g-3}$ with geodesics of length

$$\ell(\gamma_1), \ldots, \ell(\gamma_{3g-3}) \leq L_g,$$

where L_g is a constant depending only on g. The best possible constant with this property, to be denoted L_g, will be called *Bers' constant*.

Bers' original proof is based on a compactness argument in Teichmüller space. A constructive proof based on the collar theorem is given in Abikoff [1]. The bound which results from Abikoff's proof is explicit but grows much more rapidly than exponentially in g. In Section 5.2 we shall use area estimates to prove the following quantitative version.

5.1.2 Theorem. *Every compact Riemann surface of genus $g \geq 2$ has a partition $\gamma_1, \ldots, \gamma_{3g-3}$ satisfying*

$$\ell(\gamma_k) \leq 4k \log \frac{8\pi(g-1)}{k}, \quad k = 1, \ldots, 3g - 3.$$

Theorem 5.1.2 yields the following upper bound on Bers' constant:

$$L_g \leq 26(g - 1).$$

How much can it bound be improved? Since many of the finiteness theorems and quantitative estimates which result from Bers' theorem involve bounds which themselves are rapidly growing functions of L_g, it would be desirable to have a logarithmic bound. Yet this is not possible. In fact, the hairy torus of Section 5.3 will give the following *lower* bound for L_g.

5.1.3 Theorem. $L_g \geq \sqrt{6g} - 2$ *for all $g \geq 2$.*

The problem of finding length controlled partitions without using curvature bounds has an antecedent in Loewner's problem, as mentioned in the introduction to this chapter. Loewner's problem is that of finding short homologically non-trivial cycles on compact surfaces, and various results have been proved by Berger [1], Blatter [1], Hebda [1], Loewner (unpublished) and others. The sharpest form together with a number of related results may be found in Gromov [1]. The result of Hebda [1] states that for any metric on a compact surface the globally shortest closed geodesic has length $\ell \leq \sqrt{2 \text{ area}}$. We reprove this result in Section 5.4 for genus $g \geq 2$ and extend it to partitions in the following form.

5.1.4 Theorem. *Let M be an arbitrary compact orientable two dimensional Riemannian manifold of genus $g \geq 2$. Then there exists a decomposition of M into 3-holed spheres along pairwise disjoint simple closed geodesics $\gamma_1, \ldots, \gamma_{3g-3}$ of length*

$$\ell(\gamma_k) \leq 3\sqrt{kA}, \quad k = 1, \ldots, 3g-3,$$

where A is the area of M.

5.2 Partitions

In this section S denotes a compact Riemann surface of genus $g \geq 2$. By the theorem of Gauss-Bonnet (or by Theorem 1.1.7), S has area

$$\text{area } S = 4\pi(g-1).$$

We shall use area estimates to find a length controlled partition on S, thereby proving Theorem 5.1.2.

5.2.1 Lemma. *On S there exists a simple closed geodesic γ of length $\ell(\gamma) \leq 2 \log(4g-2)$.*

Proof. Let γ be the shortest non-trivial closed geodesic on S and fix a point $p \in \gamma$. The distance set

$$U = \{q \in S \mid \text{dist}(p, q) < r\}$$

is a hyperbolic disk of radius r as long as r is smaller than $\ell(\gamma)/2$ (Lemma 4.1.5). Using polar coordinates we compute the area for these values of r as follows.

$$\text{area } U = 2\pi \int_0^r \sinh \rho \, d\rho = 2\pi(\cosh r - 1)$$

(cf. (1.1.8)). Since area $U \leq$ area $S = 4\pi(g-1)$, this proves the lemma. ◊

5.2.2 Remark. In Buser-Sarnak [1] there are examples of compact Riemann surfaces of genus g, for infinitely many values of g, where the shortest closed geodesic has length $\geq \frac{4}{3}\log g$. These examples are based on quaternion groups. In Buser [3] examples are obtained by pasting together pairs of pants with respect to cubic graphs with large girth. In that construction any $g \geq 2$ is obtained, but the lower bound for the length of the shortest closed geodesic is only $2(\log g)^{1/2}$.

We shall prove Theorem 5.1.2 in the following form which is tailored for later application in Section 6.6 and Chapter 13.

5.2.3 Theorem. *Let $\gamma_1, \ldots, \gamma_m$ be the set of all distinct simple closed geodesics on S of length $\ell \leq 2$ arcsinh 1. This system is extendable to a partition $\gamma_1, \ldots, \gamma_{3g-3}$ satisfying*

$$\ell(\gamma_k) \leq 4k \log \frac{8\pi(g-1)}{k}, \quad k = 1, \ldots, 3g-3.$$

Proof. We shall obtain these geodesics inductively. At the beginning we cut S open along the simple closed geodesics $\gamma_1, \ldots, \gamma_m$ of length ≤ 2 arcsinh 1, if there are any. By the collar theorem, these geodesics are pairwise disjoint and $m \leq 3g - 3$. If there are no such geodesics on S, then we set $m = 1$ and let γ_1 be the geodesic of Lemma 5.2.1. The surface obtained after cutting S open along $\gamma_1, \ldots, \gamma_m$ has connected components of various signatures. Removing the Y-pieces among them, we are left with the union S' of all components of signature different from $(0, 3)$. The total length of the boundary $\partial S'$ of S' is less than $4m \log(8\pi(g-1)/m)$.

The inductive procedure is as follows. We look for a suitable simple closed non-boundary geodesic on S', then cut S' open along the new geodesic, let S'' be the union of the components of signature $\neq (0, 3)$ thus obtained and estimate the length of the boundary of S''. After that we look for the next geodesic in the interior of S'', and so on. (It has turned out to be more efficient to work with the total boundary length than with the lengths of the individual boundary geodesics.)

Now assume that after finitely many such steps, the pairwise disjoint simple closed geodesics $\gamma_1, \ldots, \gamma_k$ have been found, and let S^k be the surface which remains after cutting S open along $\gamma_1, \ldots, \gamma_k$ and removing the connected components of signature (0, 3). Assume by induction that

$$\ell(\partial S^k) \le 4k \log \frac{8\pi(g-1)}{k}$$

and that

$$\ell(\gamma_j) \le 4j \log \frac{8\pi(g-1)}{j}, \quad j = 1, \ldots, k.$$

If $k = 3g - 3$, the proof is finished. So assume that $k < 3g - 3$. For $r > 0$ we define

$$Z(r) = \{p \in S^k \mid \text{dist}(p, \partial S^k) \le r\}.$$

For sufficiently small r, $Z(r)$ is a disjoint union of half-collars :

$$Z(r) = [0, r] \times \partial S^k = [0, r] \times \eta_1 \cup \ldots \cup [0, r] \times \eta_n,$$

where $\eta_1, \ldots, \eta_n$ are the boundary geodesics of S^k and each connected component $[0, r] \times \eta_i$ of $Z(r)$ admits Fermi coordinates (ρ, t) with $0 \le \rho \le r$ and $t \in \mathbf{R}/[\tau \mapsto \tau + \ell(\eta_i)]$. The metric tensor in these coordinates is given by $ds^2 = d\rho^2 + \cosh^2\rho \, dt^2$ (cf Theorem 4.1.1), and we obtain

$$\text{area } Z(r) = \ell(\partial S^k) \int_0^r \cosh \rho \, d\rho = \ell(\partial S^k) \sinh r.$$

This formula holds as long as the geodesic arcs of length r emanating perpendicularly from ∂S^k are pairwise disjoint and simple.

Now let r grow continuously until a limiting value $r = r_k$ is reached, where for the first time two perpendicular arcs of length r meet each other. In the universal covering $\tilde{S}^k \subset \mathbf{H}$ (Theorem 1.4.2) the lifts of $\eta_1, \ldots, \eta_n$ (which form the boundary of $\tilde{S}^k$) have pairwise distances greater than or equal to $2r_k$, and there are pairs of lifts whose distance is *equal* to $2r_k$. It follows that the two perpendicular geodesic arcs of length r_k on S^k together form a simple geodesic arc σ of length $2r_k$ which meets ∂S^k orthogonally at both endpoints. We must consider two cases.

Case 1. *σ connects different boundary geodesics of ∂S^k.*

We let η, η' be these boundary geodesics. We then parametrize η, η' and σ in such a way that the closed curve $\eta \sigma \eta' \sigma^{-1}$ is well-defined and such that both η and η' have the same boundary orientation. It follows that $\eta \sigma \eta' \sigma^{-1}$ is freely homotopic on S^k to a *simple* closed curve δ' which together with η and η' cuts out a topological surface of signature (0, 3) as shown in Fig. 5.2.1.

By Theorem 1.6.6, δ' is freely homotopic to a simple closed geodesic δ which together with δ' bounds an annulus. It follows that δ, η and η' bound a pair of pants Y. Since S^k has no components of signature (0, 3), δ is not on

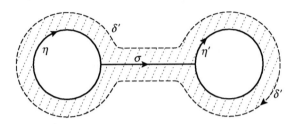

Figure 5.2.1

∂S^k, that is δ is not among the geodesics $\gamma_1, \ldots, \gamma_k$. We therefore let δ be γ_{k+1}. Now we cut S^k open along γ_{k+1} and give away Y to obtain a surface S_*^k. If no component of S_*^k has signature $(0, 3)$, we define S^{k+1} to be S_*^k; otherwise S_*^k has exactly one component of signature $(0, 3)$, say Y_*, and we define $S^{k+1} := S_*^k - Y_*$. In any case we have $\delta \subset \partial S_*^k$ and $\partial S^{k+1} \subset \partial S_*^k$. It suffices therefore to prove that

$$\ell(\partial S_*^k) \leq 4(k+1) \log \frac{8\pi(g-1)}{k+1}.$$

Consider again the domains $Z(r)$ on S^k from above for $r \leq r_k$ and let $\zeta(r) = \partial Z(r) - \partial S^k$. Then

$$\ell(\zeta(r)) = \ell(\partial S^k) \cosh r \text{ for } 0 \leq r \leq r_k.$$

As long as $r < r_k$, $\zeta(r)$ consists of n disjoint simple closed curves which are freely homotopic to the n boundary geodesics of S^k. For $r = r_k$, the two curves which belong to η and η' meet each other and form a closed curve which is homotopic to δ. We conclude that

$$\ell(\partial S_*^k) \leq \ell(\zeta(r_k)).$$

Hence, nothing is to be proved if $\ell(\zeta(r_k)) \leq 4k \log(8\pi(g-1)/k)$, and we may therefore assume that $\zeta(r_k)$ is larger. By the induction hypothesis we then have $r_k' \in [0, r_k[$ satisfying

$$\ell(\zeta(r_k')) = 4k \log(8\pi(g-1)/k).$$

Now, putting $d = r_k - r_k'$ we have

$$\ell(\partial S_*^k) \leq \ell(\zeta(r_k')) + 4d$$

and

$$\text{area}(Z(r_k) - Z(r_k')) \leq \text{area } S = 4\pi(g-1).$$

On the other hand we have

$$\text{area}(Z(r_k) - Z(r_k')) = \ell(\partial S^k) \int_{r_k'}^{r_k'+d} \cosh \rho \, d\rho$$

$$= \ell(\partial S^k)(\sinh r_k' \cosh d + \cosh r_k' \sinh d - \sinh r_k')$$

$$\geq \ell(\partial S^k) \cosh r_k' \sinh d$$

$$= \ell(\zeta(r_k')) \sinh d.$$

The two inequalities together yield

$$k \sinh d \log(8\pi(g-1)/k) \leq \pi(g-1).$$

Using the fact that $k \leq 3(g-1)$ and that $\operatorname{arcsinh} x \leq \log(2x+1)$ for $x \geq 0$, we find the upper bound on d:

$$d \leq \log(1 + 3(g-1)/k).$$

Since

$$\ell(\partial S_*^k) \leq 4k \log(8\pi(g-1)/k) + 4d,$$

it suffices to prove that

$$k \log(8\pi(g-1)/k) + d \leq (k+1) \log(8\pi(g-1)/(k+1))$$

or, equivalently, that

$$(k+1)(\log(k+1) - \log k) + d \leq \log(8\pi(g-1)/k).$$

Since the first term on the left is less than or equal to $\log 4$, this inequality follows from the upper bound of d given above and because $k \leq 3(g-1)$.

Case 2. σ *has its endpoints on one boundary geodesic of* S^k.

The arguments are similar to the preceding ones: Let σ have its endpoints on the boundary geodesic η. The endpoints divide η into two arcs η^1 and η^2 which we parametrize such that $\sigma\eta^1$ and $\sigma(\eta^2)^{-1}$ are defined. Since by the negative curvature, two geodesic arcs in S^k with the same endpoints are non-homotopic, the closed curves $\sigma\eta^1$ and $\sigma(\eta^2)^{-1}$ are homotopically non-trivial. By Theorem 1.6.6, there exist unique simple closed geodesics δ^1, δ^2 in their homotopy classes. Moreover, the pairs $\sigma\eta^1$, δ^1 and $\sigma(\eta^2)^{-1}$, δ^2 both bound an embedded annulus. It follows that η, δ^1 and δ^2 bound a pair of pants Y.

If the component Q of S^k which carries η on the boundary has signature (1, 1), then $\delta^1 = \delta^2$, and Y is immersed, with $Q = Y \mod(\delta^1 = \delta^2)$. In all other cases we have $\delta^1 \neq \delta^2$ and Y is embedded. Since Q does not have signature (0, 3), at least one of the geodesics, say δ^1, is not a boundary geodesic of S^k, and we let δ^1 be γ_{k+1}.

Now we cut S^k open along δ^1 and define S^k_* and S^{k+1} as in case 1. If Q does not have signature $(1, 1)$, the remainder of the proof is exactly as in case 1. If Q has signature $(1, 1)$, then ∂S^{k+1} is a proper subset of ∂S^k, because Y is no longer a component of S^{k+1}, and we need the procedure of case 1 only to estimate the length of δ^1. In this particular case a more direct method yields a better bound. Theorem 5.2.3 is now proved. ◇

We add a few complements. The first concerns the remark at the end of the preceding proof.

5.2.4 Lemma. *Let Q be a hyperbolic surface of signature $(1, 1)$ with boundary geodesic η of length ℓ. Then Q contains a simple closed geodesic δ satisfying $\cosh \frac{1}{2}\ell(\delta) \leq (2\pi/\ell) \sinh \frac{1}{4}\ell$.*

Proof. Define $Z(r) = \{ p \in Q \mid \mathrm{dist}(p, \partial Q) \leq r \}$ as above. Let r' be defined by the equation $\ell \sinh r' = 2\pi = $ area Q. Then there exists a geodesic arc σ of length $\ell(\sigma) \leq 2r'$ standing orthogonally on η. Cut Q open along the geodesic δ which is homotopic to $\sigma\eta^1$ and $\sigma(\eta^2)^{-1}$, where η^1 and η^2 are the arcs into which η is decomposed by the endpoints of σ. The resulting Y-piece consists of two isometric hexagons and each hexagon consists of two right-angled pentagons with sides $\ell/4$, •, $\ell(\delta/2)$, •, $\ell(\sigma)/2$. The pentagon formula (Theorem 2.3.4) yields $\cosh \frac{1}{2}\ell(\delta) \leq \sinh(\ell/4) \sinh r'$. ◇

5.2.5 Remarks. (i) Since a Y-piece is removed at each step, the proof of Theorem 5.1.2 and of Theorem 5.2.3 ends, in fact, after at most $2g - 2$ steps. Taking this into account we obtain a slightly better bound for Bers' constant:

$$L_g \leq 21(g - 1).$$

We may also take into account that the area of S^k becomes smaller at each step and use stronger length estimates based on trigonometry. This would improve the upper bound even further, but at most by a factor of $1/3$. The best bound is probably of order $\sqrt{g}$ (cf. Section 5.3).

(ii) One may ask oneself whether length-controlled pants decompositions exist which are invariant under a group G of isometries. In Buser-Seppälä [1] it is shown that this is possible if G is the cyclic group of order two generated by an orientation reversing isometry. This group occurs when the surface represents a real algebraic curve. Surprisingly, this is the *only* case for which a G-invariant Bers theorem holds. In fact, even if G is the cyclic group of order two, there exists, for arbitrarily large λ and for infinitely many $g \geq 2$, a compact Riemann surface S of genus g on which G acts by orientation-preserving isometries and such that any G-invariant partition of S contains geodesics longer than λ.

We close this section with a look at the non-compact case. In Theorem 4.4.5 we have seen that every unbordered hyperbolic surface is a surface of some signature $(g, 0; q)$ obtained by pasting together Y-pieces (some with cusps) along $3g - 3 + q$ pairs of geodesics. For these surfaces the following is true.

5.2.6 Theorem. (Bers' theorem in the non-compact case). *Let S be a hyperbolic surface of signature $(g, 0; q)$. Then there exists a partition $\gamma_1, \ldots, \gamma_{3g-3+q}$ satisfying*

$$\ell(\gamma_k) \leq 4k \log \frac{4\pi(2g - 2 + q)}{k}, \quad k = 1, \ldots, 3g - 3 + q.$$

Proof. The only difference from the proof in the compact case lies in the search of γ_1 (Lemma 5.2.1). Here the argument is as follows. Take any cusp $\mathscr{E}^*$ of S (Definition 4.4.3) and define

$$Z(r) = \{p \in S - \mathscr{E}^* \mid \text{dist}(p, \partial \mathscr{E}^*) \leq r\}.$$

For small r, $Z(r)$ is isometric to $[-r, 0] \times \boldsymbol{S}^1$ endowed with the metric tensor $d\rho^2 + 4e^{2\rho}dt^2$. As always, we find a geodesic arc σ perpendicular to $\partial \mathscr{E}^*$ at both endpoints which decomposes γ into arcs γ_1, γ_2 such that $\sigma\gamma_1$ and $\gamma_2\sigma^{-1}$ are non-trivial simple closed curves which are either homotopic to a simple closed geodesic or else bound an infinitely long cylinder (cf. Section 4.4). The area of $Z(r_0)$ satisfies

$$\text{area } Z(r_0) = 2e^{r_0} \leq \text{area } S = 2\pi(2g - 2 + q),$$

so that $\sigma\gamma_1$ and $\gamma_2\sigma^{-1}$ have lengths less than or equal to $2\log(4\pi(2g - 2 + q))$. If both curves bound an infinitely long cylinder, then S has signature $(0, 0; 3)$ and nothing has to be proved. In all remaining cases S has a simple closed geodesic γ_1 of length $\ell(\gamma_1) \leq 2\log(4\pi(2g - 2 + q))$. In these cases the induction can be started as in the compact case. ◊

5.3 The Hairy Torus

In this section we construct examples of compact Riemann surfaces of genus $g \geq 2$, where every partition includes considerably large geodesics. The construction is by pasting together geodesic "squares".

We start with a right-angled geodesic pentagon G in the hyperbolic plane with sides a, b, α, c, β, where $\alpha = \beta$ and $a = b$ (Fig. 5.3.1). Such pentagons exist for every $\alpha > 0$. In our examples α is arbitrarily large.

From the relations

$$\sinh b \sinh \alpha = \cosh \beta, \quad \sinh \beta \sinh a = \cosh \alpha$$

(Theorem 2.3.4) and since $\alpha = \beta$, we obtain

$$a = b \operatorname{arcsinh}(\coth \alpha) \geq \operatorname{arcsinh} 1 = 0.88... \ .$$

Observe that $b \to \operatorname{arcsinh} 1$ as $\alpha \to \infty$. Since it is also true that

$$\sinh \alpha \sinh c = \cosh a = \cosh b,$$

it follows that

$$c \to 0 \text{ as } \alpha \to \infty.$$

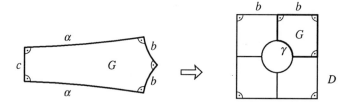

Figure 5.3.1

Now we paste together four copies of G along sides α and β, as indicated in Fig. 5.3.1, to obtain a 1-holed square D, where the hole is a geodesic γ of length

$$\ell(\gamma) = 4c,$$

and the second boundary component is a right-angled polygon with four sides of length $2b$.

Next we let m be a positive integer and paste together m^2 copies of D as in Fig. 5.3.2 to obtain a larger square D_m with m^2 holes of length $4c$ and four sides of length $2mb$. Identifying opposite sides of D_m as in the familiar construction of flat tori, we obtain a surface T_m of signature $(1, m^2)$: the *hairy torus*, so called after the half-collars of the boundary geodesics which are long and thin.

In order to obtain examples for *every* $g \geq 2$, we let

$$k \in \{0, 2, 4, ..., 2m\}, \quad \text{if } m \text{ is even},$$
$$k \in \{1, 3, 5, ..., 2m-1\}, \quad \text{if } m \text{ is odd}.$$

We choose $(m^2 - k)/2$ pairs of boundary geodesics of T_m in an arbitrary way and paste together the two boundary geodesics of each pair. (We do not have to bother about twist parameters here.) Finally, we attach an additional hyperbolic surface of signature $(1, 1)$ with boundary length $4c$ to each of the

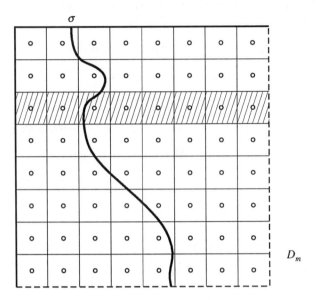

Figure 5.3.2

k remaining holes. The particular geometry of these surfaces is not important. The resulting surface S has genus

$$g = 1 + (m^2 + k)/2.$$

Obviously we obtain examples of every genus $g \geq 2$ in this way.

What are the possible short partitions of S? If one of the geodesics, say η, intersects one of the boundary geodesics γ of T_m transversally, then it follows from Corollary 4.1.2 that

$$\sinh \tfrac{1}{2}\ell(\eta) \sinh \tfrac{1}{2}\ell(\gamma) \geq 1.$$

Since we may take $\ell(\gamma)$ arbitrarily small, $\ell(\eta)$ can be made arbitrarily large, for instance, greater than $\sqrt{6g}$. Hence, we may from now on restrict ourselves to considering partitions of S, where none of the boundary geodesics of T_m are crossed transversally. But then all the boundary geodesics of T_m must belong to the given partition, and we may first cut S open along these to get back T_m. But T_m is a handle, and some geodesic σ of the partition cuts this handle into a surface of signature $(0, m^2 + 2)$. It follows that (possibly after a rotation of Fig. 5.3.2 by $\pi/2$) σ crosses all of the m horizontal m-holed strips of D_m in T_m (one such strip is hatched in Fig. 5.3.2). We claim that each strip has width $2b$.

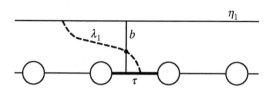

Figure 5.3.3

In fact, let η_1 and η_2 be the two sides of length $2mb$ of a strip. To estimate their distance, let λ be a connecting path from η_1 to η_2. Denote by λ_1 the arc on λ between η_1 and the first intersection point of λ with one of the common perpendiculars τ which connect neighboring holes of the strip (Fig. 5.3.3).

Similarly, let λ_2 denote the arc between the last such intersection point and η_2. Letting the endpoints of λ_1 glide freely on η_1 and τ, we homotope λ_1 into the common perpendicular of length b between η_1 and τ. Hence, $\ell(\lambda_1) \geq b$ and similarly $\ell(\lambda_2) \geq b$. This proves our claim that

$$\mathrm{dist}(\eta_1, \eta_2) = 2b.$$

Since σ crosses all m strips, we have $\ell(\sigma) \geq 2mb$. Thus, every partition of S contains a geodesic of length $\geq 2mb$. Elementary estimates yield $2mb \geq \sqrt{6g} - 2$. This proves Theorem 5.1.3. ◊

5.4 Bers' Constant Without Curvature Bounds

In this section M is a compact orientable surface of genus $g \geq 2$ endowed with an arbitrary Riemannian metric. There may now be infinitely many closed geodesics in a free homotopy class, and we may have homotopically trivial closed geodesics of positive length. In order to extend Bers' theorem to this case we use the following results due to Freedmann-Hass-Scott [1].

5.4.1 Theorem. *Let c be a simple closed non-contractible curve in M and let γ be a shortest closed curve in the free homotopy class of c. Then γ is a simple closed geodesic.*

5.4.2 Theorem. *Let c_1, γ_1 and c_2, γ_2 be as in Theorem 5.4.1. If c_1 and c_2 are disjoint and not homotopic, then γ_1 and γ_2 are disjoint.*

5.4.3 Remark. *Theorems 5.4.1 and 5.4.2 remain valid if M has boundary and is locally convex.*

For the *proofs* we refer the reader to Freedmann-Hass-Scott [1]. ◇

These theorems, together with the theorem of Baer-Zieschang (Theorem A. 3), immediately imply the next corollary. We recall that a 1-homeomorphism is a homeomorphism which is isotopic to the identity (Definition A.1).

5.4.4 Corollary. *Let $c_1, \ldots, c_{3g-3}$ be pairwise disjoint simple closed curves which decompose M topologically into 3-holed spheres. Then there exists a 1-homeomorphism $h : M \to M$ such that $h(c_1), \ldots, h(c_{3g-3})$ is a partition of M, and each $h(c_i)$ is a shortest geodesic in the free homotopy class of c_i, $i = 1, \ldots, 3g-3$.* ◇

We now turn to the proof of Theorem 5.1.4. The strategy is the same as in the case of constant curvature. To estimate areas without involving curvature bounds we need the following technical lemma. N denotes a connected component obtained after cutting M open along a number of pairwise disjoint simple closed geodesics from distinct non-trivial free homotopy classes. For the homotopy classes with gliding endpoints we refer to Definition 1.5.1.

5.4.5 Lemma. *Let η be a boundary geodesic of N. Let $r > 0$ be such that every curve of length $\leq 2r$ with endpoints on η is contractible in the homotopy class with endpoints gliding on η. Assume also that all other boundary geodesics of N are at least distance r from η. Then, for every $\varepsilon > 0$ there exists a family $\{\eta_\sigma\}$, $\sigma \in [0, r]$, of pairwise disjoint simple closed curves η_σ, all homotopic to η, with the following properties:*
  (i) *$\sigma - \varepsilon \leq \mathrm{dist}(p, \eta) \leq \sigma + \varepsilon$ for all $p \in \eta_\sigma$,*
  (ii) *the function $\sigma \mapsto \ell(\eta_\sigma)$ is integrable over $[0, r]$,*
  (iii) *if $0 \leq s < s' \leq r$, then η_s and $\eta_{s'}$ bound an annulus $K_s^{s'}(\eta)$ of area*

$$\mathrm{area}\, K_s^{s'}(\eta) \geq \int_s^{s'} \ell(\eta_\sigma)\, d\sigma - \varepsilon.$$

Proof. To avoid technicalities with curves of infinitely many self-intersections, we approximate the metric of N by a more convenient one (this explains the ε occurring in the lemma). Since no curvature assumptions are made, we can afford a piecewise linear approximation. This is particularly convenient in dimension 2: triangulate N geodesically and replace each triangle with a Euclidean triangle without changing the side lengths. Clearly, we can carry this out in such a way that the new surface is q-quasiisometric to N, with q arbitrarily close to 1. For simplicity we may now assume that N itself is piecewise linear and prove the lemma under this hypothesis without ε. The boundary of N, and in particular the boundary component η, is now piecewise linear instead of geodesic. The distance sets

$$\mu_\sigma = \{p \in N \mid \mathrm{dist}(p, \eta) = \sigma\},$$

for $\sigma \geq 0$, have similarities with the parallel lines arising in the theory of plane isoperimetric problems, as, for instance, in Fejes Tóth [2]. For sufficiently small σ the distance neighborhood $U = \{p \in N \mid \mathrm{dist}(p, \eta) < \sigma\}$ contains only those vertices of the triangulation which lie on η. In this case the metric on U is smooth and Euclidean. Moreover, for $s < \sigma$ (σ still small) μ_s has the following properties:

(1) μ_s *is a simple closed curve homotopic to* η,
(2) μ_s *has a finite number of sides, each side being either a straight line segment or an arc of a Euclidean circle.*

A curve with property (2) will be called *piecewise circular*. A curve satisfying (1) and having constant distance from η will be called a *parallel curve* of η.

Now let σ become larger and assume that μ_σ contain a vertex p of the triangulation. Fig. 5.4.1(a) shows μ_σ as it is seen on N. In Fig. 5.4.1(b) a neighborhood of p has been cut open and developed into the Euclidean plane. We see that μ_σ remains piecewise circular, but the number of sides increases as μ_σ sweeps over p. *Only at such instances can the number of sides increase.* If the angle sum at a vertex is less than 2π (as in the figure), then the number of sides increases by 1; otherwise it increases by 2. Of course, the number of sides may also decrease from time to time.

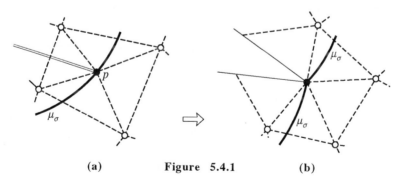

(a) **Figure 5.4.1** (b)

If all μ_σ are parallel curves of η, we set $\eta_\sigma := \mu_\sigma$ for $0 \leq \sigma \leq r$ (where r is as in the lemma). In this case the lemma holds with continuity of the function $\sigma \mapsto \ell(\mu_\sigma)$ and with equality in statement (iii).

Now assume that there exists a $\sigma = \hat{\sigma}$, where for the first time μ_σ fails to be simple. Then $\mu_{\hat{\sigma}}$ is still homotopic to η, but it has a tangential self-intersection at some points p. For ease of description we assume that p is the only self-intersection of $\mu_{\hat{\sigma}}$. Then $\mu_{\hat{\sigma}}$ can be written as a product $\mu_{\hat{\sigma}} = \eta_{\hat{\sigma}} \tau$, where $\eta_{\hat{\sigma}}$

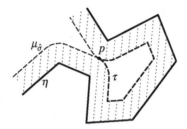

Figure 5.4.2

and τ are simple closed loops at p.

We claim that one of these loops is contractible. Recall that $\mu_{\hat{\sigma}}$ is still homotopic to η. Lift the homotopy to the universal covering $\tilde{N}$ of N to obtain homotopic lifts $\tilde{\mu}_{\hat{\sigma}}$ and $\tilde{\eta}$ in $\tilde{N}$. If $\tilde{\mu}_{\hat{\sigma}}$ has a tangential self-intersection, we are done. So suppose that $\tilde{\mu}_{\hat{\sigma}}$ is tangent to another lift $\mu_{\hat{\sigma}}^*$ which is homotopic to another lift η^* of η. In this case we have dist$(\mu_{\hat{\sigma}}, \eta) = \hat{\sigma} \leq r$, and therefore dist$(\tilde{\eta}, \eta^*) = 2\hat{\sigma} \leq 2r$. There is a connecting curve of length $2\hat{\sigma}$ from $\tilde{\eta}$ to η^* which is not contractible (to a point) with endpoints gliding on $\tilde{\eta}$ and η^*. Its projection in N is a curve of length $\leq 2r$ which has its endpoints on η and is not contractible with endpoints gliding on η. This contradicts the assumption on r in the lemma and proves that one of the loops, say τ, is contractible, as claimed. Now, $\eta_{\hat{\sigma}}$ is a parallel curve of η, and τ is contained in the annulus $K_0^{\hat{\sigma}}(\eta)$ (cf. Proposition A.11). We recall that we assumed, for simplicity, that $\mu_{\hat{\sigma}}$ has only one self-intersection. If $\mu_{\hat{\sigma}}$ has more than one self-intersection, we obtain the same result after removing finitely many contractible loops from $\mu_{\hat{\sigma}}$.

The statements of the lemma are proved thus far for $0 < \sigma, s, s' < \hat{\sigma}$. To continue, we cut off the annulus $K_0^{\hat{\sigma}}(\eta)$ from N. The surface $\hat{N}$ thus obtained satisfies the same hypotheses as N, where $r - \hat{\sigma}$ replaces r, except that the boundary is piecewise circular instead of piecewise linear. We can therefore repeat the arguments on $\hat{N}$ until a second critical value is obtained, and so on. The parallel curves have a uniformly bounded number of sides (at most twice the number of vertices of the triangulation of N) so that only finitely many critical values occur, and we obtain Lemma 5.4.5 in finitely many steps. ◇

For the proof of Theorem 5.1.4 we begin with the following estimate (Gromov-Lafontaine-Pansu [1], Hebda [1]), where $A = $ area M.

5.4.6 Theorem. *Let γ_1 be the globally shortest non-contractible closed geodesic on M. Then*

$$\ell(\gamma_1) \leq \sqrt{2A}.$$

Proof. Cut M open along γ_1 and assume first that the result is two connected components, say N' and N'' with area $N' \leq$ area N''. Then Lemma 5.4.5 applies to all $r < \frac{1}{4}\ell(\gamma_1)$. Since $\ell(\eta_\sigma) \geq \frac{1}{4}\ell(\gamma_1)$, we get

$$\tfrac{1}{4}\ell^2(\gamma_1) \leq \text{area } N' \leq \tfrac{1}{2}A.$$

Next assume that the cutting yields only one component N. Then N has two copies, γ' and γ'', of γ_1 on the boundary with $\text{dist}(\gamma', \gamma'') \geq \ell(\gamma_1)$. Lemma 5.4.5 yields

$$\tfrac{1}{2}\ell^2(\gamma_1) \leq \text{area } N = A. \qquad \diamond$$

The proof of Theorem 5.1.4 is almost identical with the proofs of Theorems 5.1.2 and Theorem 5.2.3; only the numerical estimates are different. Define

$$L_1 = 2\sqrt{2A} \text{ and } L_k = 3\sqrt{kA} \text{ for } k = 2, 3, \ldots .$$

Assume that M has been cut open along geodesics $\gamma_1, \ldots, \gamma_k$ having *minimal length* in their free homotopy classes (cf. Theorems 5.4.1 and 5.4.2), and let M_k be the union of all connected components of signature different from $(0, 3)$. The induction hypothesis is that $\ell(\gamma_j) \leq L_j$, $j = 1, \ldots, k$, and that $\ell(\partial M_k) \leq L_k$.

To find γ_{k+1}, let $\eta^1, \ldots, \eta^n$ be the boundary geodesics of M_k, and let r_k be the supremum of all s' for which the annuli $K_0^{s'}(\eta^1), \ldots, K_0^{s'}(\eta^n)$ of Lemma 5.4.5 exist and are pairwise disjoint. Then we define $s \leq r_k$ to be the supremum of all $\sigma \leq r_k$ for which $\ell(\eta_\sigma^1) + \ldots + \ell(\eta_\sigma^n) \leq L_k$. If $s = r_k$, then either two of the curves $\eta_s^1, \ldots, \eta_s^n$ intersect each other (tangentially), or one of these curves has a double point. In either case there is a geodesic γ_{k+1} of length $\ell(\gamma_{k+1}) < L_k$ which is minimal in its free homotopy class and which gives rise to a decomposition satisfying $\ell(\partial M_{k+1}) < \ell(\partial M_k) \leq L_k$ (Remark 5.4.3). Thus the case in which $s = r_k$ is clear.

Now assume that $s < r_k$. Then we find γ_{k+1} as before, but now the upper bound of $\ell(\gamma_{k+1})$ and $\ell(\partial M_{k+1})$ is $L_k + 4d$, where $d = r_k - s$. It remains to show that $L_k + 4d \leq L_{k+1}$.

The geodesics $\eta^1, \ldots, \eta^n$ have minimal length in their free homotopy classes, and the annuli $K_s^{r_k}(\eta^1), \ldots, K_s^{r_k}(\gamma^n)$ are pairwise disjoint. By the area estimate of Lemma 5.4.5 and by the definition of s we see therefore that

$$L_k d \leq \sum_{v=1}^n \text{area } K_s^{r_k}(\eta^v) \leq A.$$

Hence, $L_k + 4d \leq L_k + 4A/L_k \leq L_{k+1}$. Theorem 5.1.4 is now proved. $\diamond$

Chapter 6

The Teichmüller Space

We introduce the Teichmüller space $\mathcal{T}_g$ based on marked Riemann surfaces. Our main goal is to construct certain rather simple $(6g - 6)$-parameter families of compact Riemann surfaces of genus g and to show that they are models of $\mathcal{T}_g$ in a natural way. The construction has already been outlined in Section 1.7 and gives rise to the Fenchel-Nielsen parameters. In the subsequent chapters we work with these models rather than $\mathcal{T}_g$. The various models are real analytically equivalent and we shall use them to define the real analytic structure of $\mathcal{T}_g$. In Section 6.4 we briefly discuss two distance functions on $\mathcal{T}_g$. In Sections 6.5 and 6.6 we study the Teichmüller modular group. In the final two sections we compare the Fenchel-Nielsen parameters with other parameters which are frequently used in the literature.

6.1 Marked Riemann Surfaces

We introduce the concept of marked Riemann surfaces by distinguishing an isotopy class of a homeomorphism from a base surface to a given surface. In the literature these isotopy classes are usually characterized by their effect on a canonical dissection of the base surface. Here we use a curve system Ω which is better adapted to the construction described in Sections 1.7 and 3.6 and which will be particularly useful in the study of the length spectrum.

For each signature (g, n) with $2g + n \geq 3$ we fix a compact orientable C^∞ surface $F = F_{g, n}$ of genus g with n holes such that the boundary components are smooth closed curves. F is the *base surface* for the homeomorphisms. It need not be a Riemann surface although later on we shall introduce a suitable Riemann surface structure on F for convenience.

6.1.1 Definition. A *marked Riemann surface of signature* (g, n) is a pair (S, φ), where S is a compact Riemann surface of signature (g, n) and $\varphi : F_{g,n} \to S$ is a homeomorphism called the *marking homeomorphism*.

To simplify the notation we often write S instead of (S, φ). It would also be possible to define marked Riemann surfaces as pairs $(S, [\varphi])$, where $[\varphi]$ is the isotopy class of a homeomorphism $\varphi : F_{g,n} \to S$. This would lead to the same notion of Teichmüller space.

6.1.2 Definition. Two marked Riemann surfaces (S, φ) and (S', φ') are *marking equivalent* if there exists an isometry $m : S \to S'$ such that φ' and $m \circ \varphi$ are isotopic. The set of all marking equivalence classes of signature (g, n) is the *Teichmüller space* of signature (g, n) and is denoted by $\mathcal{T}_{g,n}$. If $n = 0$ we write $\mathcal{T}_g$ instead of $\mathcal{T}_{g,0}$.

6.1.3 Remarks. (i) If $\varphi_1, \varphi_2 : F_{g,n} \to S$ are isotopic, then (S, φ_1) and (S, φ_2) are marking equivalent. But the converse does not hold: if $m : S \to S$ is a non-trivial isometry, then, by definition, (S, φ) and $(S, m \circ \varphi)$ are marking equivalent, but φ and $m \circ \varphi$ are not isotopic.

(ii) In every isotopy class there are diffeomorphisms. We could therefore restrict ourselves to marking diffeomorphisms and define the same Teichmüller space.

If φ in (S, φ) is a diffeomorphism, we may pull back the hyperbolic structure from S to $F_{g,n}$. In this way we obtain an equivalent definition of $\mathcal{T}_{g,n}$ in the form of $\mathcal{H}/\text{Diff}^0$, where $\mathcal{H}$ is the set of all smooth hyperbolic structures on $F_{g,n}$ and Diff^0 is the group of diffeomorphisms of $F_{g,n}$ which are isotopic to the identity. Since every hyperbolic structure is the subatlas of a unique conformal structure, and since by the uniformization theorem every conformal structure contains a unique hyperbolic structure, we obtain an equivalent definition of $\mathcal{T}_{g,n}$ in the form of $\mathcal{C}/\text{Diff}^0$, where $\mathcal{C}$ is the set of all conformal structures which are smooth with respect to the C^∞-structure of $F_{g,n}$. There are also other equivalent definitions of $\mathcal{T}_{g,n}$ (see for instance Abikoff [1], Earle-Eells [1], Gardiner [1], Lehto [1], Nag [1], Zieschang-Vogt-Coldewey [1, 2]).

(iii) If the signature is $(0, 3)$, marking homeomorphisms $\varphi_1, \varphi_2 : F_{0,3} \to S$ are isotopic iff $\varphi_1^{-1} \circ \varphi_2$ fixes each boundary component (cf. Proposition A.17). Hence, for signature $(0, 3)$ the preliminary concept of a marked Y-piece given in Section 3.2 is essentially the same as that of Definition 6.1.2.

6.1.4 Definition. Let (S, φ) with $\varphi : F_{g,n} \to S$ be a marked Riemann surface. For every homotopically non-trivial closed curve β on $F_{g,n}$ we denote by $\beta(S)$ the unique closed geodesic in S homotopic to $\varphi \circ \beta$ (Theorem 1.6.6). We recall that the uniqueness is up to a reparametrization of the form $t \mapsto t + \text{const}$, $t \in \boldsymbol{S}^1$. If $\mathcal{B} = \{\beta_1, \beta_2, \dots\}$ is a finite or infinite sequence of such curves on $F_{g,n}$ we denote by

$$\mathcal{B}(S) = \{\beta_1(S), \beta_2(S), \dots\}$$

the corresponding sequence of closed geodesics on S and define

$$\ell\mathcal{B}(S) = \{\ell\beta_1(S), \ell\beta_2(S), \dots\}.$$

From now on we restrict ourselves to surfaces without boundary and let $g \geq 2$ be a fixed genus. Ultimately we want to prove that the surfaces $F(G, L, A)$ which we sketched in Section 1.7 form a model of $\mathcal{T}_g$ in a natural way. For this we define, for each marked cubic graph G, a curve system Ω_G on the base surface $F = F_{g,0}$. This system will be used to characterize the marking equivalence classes.

Let G be given by the list

$$c_k = (c_{i\mu}, c_{j\nu}), \quad k = 1, \dots, 3g - 3$$

as in (3.5.1). It is useful to introduce the following hyperbolic structure on the base surface F. For $i = 1, \dots, 2g - 2$ we let Y_i be a Y-piece with boundary geodesics $\gamma_{i\mu}$, $\mu = 1, 2, 3$, of length 1. The Y-pieces are considered distinct and are pasted together in such a way that whenever $c_k = (c_{i\mu}, c_{j\nu})$, then

$$\gamma_{i\mu}(t) = \gamma_{j\nu}(-t) =: \gamma_k(t), \quad t \in \boldsymbol{S}^1.$$

We then take the resulting surface as our base surface F. (The graph G will be kept fixed in the remainder of this section.) By abuse of notation we also denote by Y_i the image of Y_i under the natural projection $Y_1 \cup \dots \cup Y_{2g-2} \to F$.

Fix k. The two Y-pieces Y_i and Y_j which are pasted together along γ_k form a hyperbolic surface in F which has one of the following four possible signatures: $(1, 1)$, $(1, 2)$, $(0, 4)$ and $(2, 0)$. Signature $(1, 1)$ occurs iff $i = j$; an example with signature $(1, 2)$ is shown in Fig. 6.1.1.

To avoid the distinction of these cases we define a surface X^k together with an isometric immersion $\iota_k : X^k \to F$ as follows. We let Y^i and Y^j be distinct Y-pieces, where Y^i is a copy of Y_i and Y^j is a copy of Y_j. The boundary geodesics of Y^i and Y^j are denoted by γ_r^i and γ_s^j; $r, s = 1, 2, 3$. The surface X^k is the following.

(6.1.5) $\quad X^k = Y^i \cup Y^j \bmod(\gamma_\mu^i(t) = \gamma_\nu^j(-t) =: \gamma^k(t), \, t \in \boldsymbol{S}^1)$.

The natural isometries $Y^i \to Y_i$, $Y^j \to Y_j$ project in an obvious way to an isometric immersion $\iota_k : X^k \to F$. We remark that (6.1.5) corresponds with

(3.3.2) (with $\alpha = 0$) in the following way: Y of (3.3.2) is Y^i of (6.1.5) and Y' of (3.3.2) is Y^j of (6.1.5).

6.1.6 Definition. For $k = 1, \ldots, 3g - 3$ we let δ^k and η^k in X^k be the geodesics as in Definition 3.3.10 and Proposition 3.3.11. We denote their images in F by $\delta_k = \iota_k(\delta^k)$ and $\eta_k = \iota_k(\eta^k)$. The sequence

$$\Omega_G = \{\gamma_1, \ldots, \gamma_{3g-3}, \delta_1, \ldots, \delta_{3g-3}, \eta_1, \ldots, \eta_{3g-3}\}$$

is called the *canonical curve system* with respect to G.

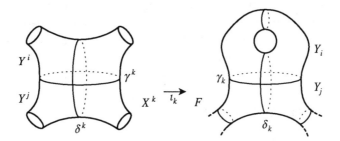

Figure 6.1.1

Ω_G can be used to characterize the isotopy class of a homeomorphism.

6.1.7 Theorem. *Let $\varphi, \varphi' : F \to S$ be marking homeomorphisms such that $\varphi \circ \gamma_k$ is homotopic to $\varphi' \circ \gamma_k$ and such that $\varphi \circ \delta_k$ is homotopic to $\varphi' \circ \delta_k$, $k = 1, \ldots, 3g - 3$. Then φ and φ' are isotopic.*

Proof. Let $\phi = \varphi^{-1} \circ \varphi'$. By Theorem A.3, we may assume that $\phi \circ \gamma_k = \gamma_k$, $k = 1, \ldots, 3g - 3$. We first assume that $g \geq 3$, with the necessary modifications for $g = 2$ added at the end of the proof.

First we prove that each Y_i is fixed under ϕ. For a given Y_i there exists Y_j with $j \neq i$ which is attached to Y_i along some γ_k. Now $Y_i \cup Y_j$ is pasted to another Y-piece along some γ_n. Since ϕ fixes γ_k, it also fixes $Y_i \cup Y_j$ and γ_n. Since γ_n is the boundary geodesic of only one of the surfaces Y_i and Y_j, we conclude that $\phi(Y_i) = Y_i$ and $\phi(Y_j) = Y_j$.

Since ϕ leaves the orientation of γ_k invariant, ϕ is orientation-preserving. (At this point we are using that $\phi \circ \gamma_k = \gamma_k$ and not just $\phi(\gamma_k) = \gamma_k$). By Proposition A.17, the restriction $\phi \mid Y_i$ is isotopic on Y_i to a product of boundary twists (Definition A.14). If Y_i in F has signature $(1, 1)$ as in Fig. 6.1.1, this statement holds, more precisely, for the lift of $\phi \mid Y_i$ in Y^i.

The above statement is valid for all Y_i, $i = 1, \ldots, 2g$, and so ϕ is isotopic to a product of Dehn twists along $\gamma_1, \ldots, \gamma_{3g-3}$ (Definition A.15). Since ϕ fixes

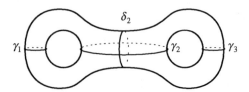

Figure 6.1.2

the homotopy classes of $\delta_1, \ldots, \delta_{3g-3}$, these Dehn twists are trivial. This proves the theorem for $g \geq 3$.

If $g = 2$ and if one of the γ_k is a separating curve, then $\phi(Y_1) = Y_1$, $\phi(Y_2) = Y_2$, and we argue as for $g \geq 3$. If each of γ_1, γ_2, γ_3 is non-separating as in Fig. 6.1.2, then we exchange the roles of γ_2 and δ_2 and argue as before. ◇

An immediate consequence of Theorem 6.1.7 is the following.

6.1.8 Corollary. *Two homeomorphisms φ, $\varphi' : F \to S$ are homotopic iff they are isotopic.* ◇

6.2 Models of Teichmüller Space

In this section we construct a $(6g - 6)$-parameter family of compact Riemann surfaces of genus g, modelled over a given graph, and show that they form a model of Teichmüller space in a natural way (Theorem 6.2.7). In Section 6.3 we shall use this model to define a topology and a real analytic structure on Teichmüller space.

Again let G be a graph with defining relations $c_k = (c_{i\mu}, c_{jv})$, $k = 1, \ldots, 3g - 3$, and consider the various Riemann surfaces $F^{LA} = F(G, L, A)$, where

(6.2.1) $(L, A) = (\ell_1, \ldots, \ell_{3g-3}, \alpha_1, \ldots, \alpha_{3g-3}) \in \mathbf{R}_+^{3g-3} \times \mathbf{R}^{3g-3} =: \mathcal{R}^{6g-6}$.

We turn this set into a model of $\mathcal{T}_g$ by introducing suitable marking homeomorphisms.

To simplify the notation we write ω instead of (L, A) and F^ω, instead of F^{LA}, etc. Let us recall the definition of F^ω. For given $\omega = (L, A) \in \mathcal{R}^{6g-6}$ we have a unique sequence of distinct Y-pieces Y_i^ω, $i = 1, \ldots, 2g-2$, with boundary geodesics γ_{i1}^ω, γ_{i2}^ω, γ_{i3}^ω such that whenever $c_k = (c_{i\mu}, c_{jv})$, then

$$\ell(\gamma_{i\mu}^\omega) = \ell(\gamma_{jv}^\omega) = \ell_k.$$

F^ω is defined as the quotient

$$F^\omega = Y_1^\omega \cup \ldots \cup Y_{2g-2}^\omega \mod(\wp^\omega),$$

where ($\wp^\omega$) is the *pasting scheme*

$$(\wp^\omega) \quad \gamma_{i\mu}^\omega(t) = \gamma_{j\nu}^\omega(\alpha_k - t) =: \gamma_k^\omega(t), \quad t \in \mathbf{S}^1, \quad k = 1, \ldots, 3g-3.$$

All Y_i^ω are assumed to be given in the standard form of Section 3.1.

We denote by

$$\pi^\omega: Y_1^\omega \cup \ldots \cup Y_{2g-2}^\omega \to F^\omega$$

the natural projection. In the special case where $A = (0, \ldots, 0) = A_0$, we shall use superscript L instead of LA_0, and for the base surface itself (where $L = (1, \ldots, 1) = L_0$, and $A = A_0$) we omit the superscript. Thus, the base surface F has pasting scheme ($\wp$) together with the natural projection π and closed geodesics $\gamma_1, \ldots, \gamma_{3g-3}$, etc; the intermediate surface F^L has pasting scheme ($\wp^L$) together with the natural projection π^L and the closed geodesics $\gamma_1^L, \ldots, \gamma_{3g-3}^L$, etc.

The marking homeomorphisms will be introduced in two parts as a product $\varphi^\omega = \tau^\omega \circ \sigma^L$, where $\sigma^L : F \to F^L$ is some sort of "stretch" and $\tau^\omega : F^L \to F^\omega$ is a product of "twists".

For $i = 1, \ldots, 2g-2$ we let $\sigma_i^L : Y_i \to Y_i^L$ be the homeomorphism introduced in Section 3.2 which preserves the standard parametrization of the boundary geodesics (Definition 3.2.4 and Lemma 3.2.5). Since all twist parameters are zero, the following is a well-defined mapping from F to F^L:

(6.2.2) $\quad \sigma^L = \{\pi^L \circ \sigma_i^L \circ \pi^{-1} \text{ on } \pi(Y_i), \quad i = 1, \ldots, 2g-2.$

$\sigma^L : F \to F^L$ is a homeomorphism with the property that

$$\sigma^L \circ \gamma_k(t) = \gamma_k^L(t), \quad t \in \mathbf{S}^1, k = 1, \ldots, 3g-3.$$

To define the twist homeomorphisms τ^ω we proceed as in Section 3.3. Let

$$\mathscr{C}_k^\omega = \{p \in F^\omega \mid \operatorname{dist}(p, \gamma_k^\omega) \le w_k^\omega\},$$

where $w_k^\omega = \operatorname{arcsinh}\{1/\sinh \tfrac{1}{2}\ell_k\}$. By the collar theorem, the $\mathscr{C}_k^\omega$ are pairwise disjoint annuli which admit Fermi coordinates (ρ, t) with $-w_k^\omega \le \rho \le w_k^\omega$ and $t \in \mathbf{S}^1$. The sign convention is such that ρ is negative on the left-hand side of γ_k and positive on the right-hand side (with respect to the orientation of γ_k). This is the same convention as that in Definition 3.3.5 Note that for given L and k the collars $\mathscr{C}_k^L$ and $\mathscr{C}_k^\omega$ are isometric for any A in $\omega = (L, A)$. We define smooth mappings

$$T_k^\omega : \mathscr{C}_k^L \to \mathscr{C}_k^\omega$$

via Fermi coordinates as follows

$$(6.2.3) \qquad T_k^\omega(\rho, t) = (\rho, t + \alpha_k \frac{w_k^\omega + \rho}{2w_k^\omega}).$$

The mapping is shown in Fig. 6.2.1 for small positive α_k; the "\" indicate $\gamma_k^L(0)$ and $\gamma_k^\omega(0)$. The figure is in accordance with Fig. 3.3.1 and Fig. 3.3.2.

If $I^\omega : Y_1^L \cup \ldots \cup Y_{2g-2}^L \to Y_1^\omega \cup \ldots \cup Y_{2g-2}^\omega$ (disjoint unions) is the natural identification, then the following is a well-defined homeomorphism from F^L to F^ω

$$(6.2.4) \qquad \tau^\omega = \begin{cases} T_k^\omega & \text{on } \mathscr{C}_k^L, \ k = 1, \ldots, 3g - 3 \\ \pi^\omega \circ I^\omega \circ (\pi^L)^{-1} & \text{elsewhere.} \end{cases}$$

This is the analog of the twist homeomorphisms for X-pieces in (3.3.7) and for 1-holed tori in (3.4.4).

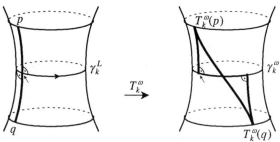

Figure 6.2.1

6.2.5 Definition. Let G be the same graph as above. Then S^ω denotes the marked Riemann surface $S^\omega := (F^\omega, \varphi^\omega)$, where

$$\varphi^\omega := \tau^\omega \circ \sigma^L.$$

The set of all marked Riemann surfaces S^ω based on the graph G and with $\omega \in \mathcal{R}^{6g-6}$ is denoted by $\mathcal{T}_G$.

From (6.2.3) and (6.2.4) we note that

$$(6.2.6) \qquad \varphi^\omega \circ \gamma_k(t) = \gamma_k^\omega(t + \frac{\alpha_k}{2}), \qquad t \in \mathbf{S}^1, \qquad k = 1, \ldots, 3g - 3,$$

where, with the terminology of Definition 6.1.4,

$$\gamma_k^\omega = \gamma_k(S^\omega).$$

The next theorem is the goal of this section. Among other things, it shows that Teichmüller space is a cell (with respect to the topology defined in the next section). For an approach to the cell structure from the point of view of minimal surfaces we refer the reader to Fischer-Tromba [1, 2] and Jost [1].

6.2.7 Theorem. *Let G be given. Then for every marked Riemann surface (S, φ) (with marking homeomorphism $\varphi : F \to S$), there exists a unique $S^\omega \in \mathcal{T}_G$ which is marking equivalent to (S, φ).*

Proof. We have to determine the appropriate value of $\omega = (L, A)$ in S. By Theorem A.3, we may assume that $\varphi(\gamma_k)$ is the closed geodesic $\gamma_k(S)$, $k = 1, \ldots, 3g - 3$ (Definition 6.1.4). The candidate for L is, of course,

$$L(S) := (\ell\gamma_1(S), \ldots, \ell\gamma_{3g-3}(S)).$$

Fix k and consider again the X-piece X^k as in (6.1.5) with isometric immersion $\iota_k : X^k \to F$. This immersion is a local isometry. We can therefore pull back the structure of the hyperbolic surface $\varphi(\iota_k(X^k))$ onto X^k, that is, there exists an X-piece $X^k(S)$ (not contained in S) together with an isometric immersion $\iota'_k : X^k(S) \to S$ and a homeomorphism $\varphi^k : X^k \to X^k(S)$ such that the following diagram is commutative

$$\begin{array}{ccc} X^k & \xrightarrow{\varphi^k} & X^k(S) \\ \downarrow \iota_k & & \downarrow \iota'_k \\ F & \xrightarrow{\cdot\varphi} & S \end{array}$$

Note that φ^k maps the Y-pieces Y^i and Y^j of X^k onto Y-pieces in $X^k(S)$. In the next two lines we use Definition 6.1.6. On $X^k(S)$ we let $\gamma^k(S) = \varphi^k(\gamma^k)$, $\delta^k(S) = \varphi^k(\delta^k)$ and $\eta^k(S) = \varphi^k(\eta^k)$. Then

$$\iota'_k(\gamma^k(S)) = \gamma_k(S), \quad \iota'_k(\delta^k(S)) = \delta_k(S), \quad \iota'_k(\eta^k(S)) = \eta_k(S).$$

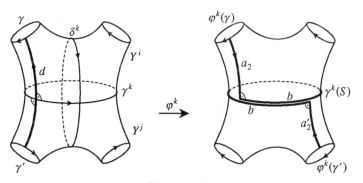

Figure 6.2.2

Let γ and γ' with $\gamma \subset \partial Y^i$ and $\gamma' \subset \partial Y^j$ be boundary geodesics of X^k as in Fig. 6.2.2 and which are not separated by δ^k, and let d be the shortest connection in $X^k - \delta^k$ from γ' to γ. These curves are temporarily parametrized in such a way that

δ^k is homotopic to the curve $d\gamma d^{-1}\gamma'$

(first along d, then along γ, etc.). This corresponds with the remark preceding Definition 3.3.10.

$\varphi^k(d)$ in $X^k(S)$ is homotopic with endpoints gliding on the boundary to a unique curve $a_2' b a_2^{-1}$ with the following properties (Fig. 6.2.2): a_2' is the shortest connection from $\varphi^k(\gamma')$ to $\gamma^k(S)$; b is a geodesic arc on $\gamma^k(S)$ (not simple, in general); and a_2 is the shortest connection from $\varphi^k(\gamma)$ to $\gamma^k(S)$.

We let $\beta_k(S)$ be the directed length of b: positive if b leads to the left hand side of a_2' (as in Fig. 6.2.2), negative otherwise. Setting

$$\alpha_k(S) := \beta_k(S)/\ell\gamma_k(S), \quad k = 1, \ldots, 3g - 3,$$

we have the candidate for A:

$$A(S) := (\alpha_1(S), \ldots, \alpha_{3g-3}(S)).$$

Together with the above candidate $L(S)$ we set $\omega := (L(S), A(S))$ and $\hat{S} := S^\omega$. We shall prove that $S = (S, \varphi)$ and $\hat{S}$ are marking equivalent. Again fix k.

As for S we consider the isometric immersion $\hat{\iota}_k : X^k(\hat{S}) \to \hat{S}$ with the commutative diagram

$$\begin{array}{ccc} X^k & \xrightarrow{\hat{\vartheta}^k} & X^k(\hat{S}) \\ \downarrow{\iota_k} & & \downarrow{\hat{\iota}_k} \\ F & \xrightarrow{\hat{\vartheta}} & \hat{S} \end{array}$$

where $\hat{\vartheta} = \varphi^\omega = \tau^\omega \circ \sigma^L$ with $L = L(S)$. The analog of $a_2' b a_2^{-1}$ is $\hat{a}_2' \hat{b} \hat{a}_2^{-1}$. We check with the definition of τ^ω in (6.2.3) and (6.2.4) that $\hat{b}$ and b have the same directed length (all surfaces are assumed oriented in such a way that the marking homeomorphisms preserve orientation).

We have therefore an orientation-preserving isometry $m^k : X^k(S) \to X^k(\hat{S})$ which maps a_2' to $\hat{a}_2'$, b to $\hat{b}$ and a_2 to $\hat{a}_2$. Since δ^k is homotopic to $d\gamma d^{-1}\gamma'$, it follows that the geodesics $m^k(\delta^k(S))$ and $\delta^k(\hat{S})$ are homotopic. By the uniqueness of a geodesic in its homotopy class we get

$$m^k \circ \delta^k(S) = \delta^k(\hat{S}).$$

We now observe that the restriction $m^k \mid \varphi^k(Y^i)$ is the *unique* orientation-preserving isometry from $\varphi^k(Y^i)$ to $\hat{\vartheta}^k(Y^i)$ with the property that $m^k(c) = \hat{\vartheta}^k \circ (\varphi^k)^{-1}(c)$ for each boundary geodesic c of $\varphi^k(Y^i)$. This implies that the local isometries

$$\hat{\iota}_k \circ m^k \circ (\iota_k')^{-1} : \iota_k' (\text{int } X^k(S)) \to \hat{\iota}_k(\text{int } X^k(\hat{S}))$$

for $k = 1, \ldots, 3g - 3$ together define a single isometry $m : S \to \hat{S}$ satisfying

$$m \circ \delta_k(S) = \delta_k(\hat{S}), \quad k = 1, \ldots, 3g - 3.$$

By Theorem 6.1.7, S and $\hat{S} = S^\omega$ are marking equivalent as claimed.

Finally, if S^ω and $S^{\omega'} \in \mathcal{T}_G$ are marking equivalent, then we have an isometry $m : S^\omega \to S^{\omega'}$ such that $m \circ \varphi^\omega$ is isotopic to $\varphi^{\omega'}$. Hence, m maps $\Omega_G(S^\omega)$ onto $\Omega_G(S^{\omega'})$ so that $\ell\Omega_G(S^\omega) = \ell\Omega_G(S^{\omega'})$. The last equation implies $L = L'$ and $A = A'$: for L this is clear, for A we use Proposition 3.3.12 or formula (2) in the proof of Lemma 3.3.14 (replace α, δ^α and η^α in Proposition 3.3.12 respectively by α_i, δ_i and $\eta_i \in \Omega_G$, $i = 1, \ldots, 3g - 3$). ◊

The following consequence of Theorem 6.2.7 results from the last remark of the proof.

6.2.8 Corollary. *Two marked Riemann surfaces S and S' are marking equivalent if and only if $\ell\Omega_G(S) = \ell\Omega_G(S')$.* ◊

6.3 The Real Analytic Structure of $\mathcal{T}_g$

Again we fix $g \geq 2$. For every marked cubic graph G with $2g - 2$ vertices, the set

$$\mathcal{T}_G = \{ S^\omega = S_G^\omega \mid \omega \in \mathcal{R}^{6g-6} \}$$

is a model of $\mathcal{T}_g$, where

$$\mathcal{R}^{6g-6} = \mathbf{R}_+^{3g-3} \times \mathbf{R}^{3g-3}$$

(Theorem 6.2.7). We use this fact to define a *topology* and at the same time a *real analytic structure* in $\mathcal{T}_g$. (A different, more intrinsic definition of the topology of $\mathcal{T}_g$ will follow in Section 6.4.) The topology will be defined by means of coordinate maps $\omega_G : \mathcal{T}_g \to \mathcal{R}^{6g-6}$.

By abuse of notation we also denote by S the marking equivalence class of a marked Riemann surface $S = (S, \varphi)$. Thus, if $S \in \mathcal{T}_g$, we identify S with any of its representatives and deal with S as with a Riemann surface. The definition which follows is based on Theorem 6.2.7.

6.3.1 Definition. Let G be as above. For every $S \in \mathcal{T}_g$ we let $\omega(S) = \omega_G(S)$ denote the unique $\omega \in \mathcal{R}^{6g-6}$ such that S is marking equivalent to S_G^ω. The components of $\omega_G(S)$ are called the *Fenchel-Nielsen coordinates* of S.

The main result of this section is the following.

6.3.2 Theorem. *If two graphs G and G' are given with coordinate maps ω_G and $\omega_{G'}$, respectively, then the overlap map $\omega_G \circ \omega_{G'}^{-1} : \mathcal{R}^{6g-6} \to \mathcal{R}^{6g-6}$ is a real analytic diffeomorphism.*

6.3.3 Definition. On $\mathcal{T}_g$ we introduce the unique real analytic structure such that the mappings $\omega_G : \mathcal{T}_g \to \mathcal{R}^{6g-6}$ are real analytic diffeomorphisms.

For the proof of Theorem 6.3.2 we proceed in two steps. First we show that the Fenchel-Nielsen parameters are analytic functions of $\ell\Omega_G(S)$. Then we prove, more generally, that all lengths of closed geodesics are analytic functions of the Fenchel-Nielsen parameters.

6.3.4 Lemma. *Let G be given and let $\mathcal{L} = \{\ell\Omega_G(S) \mid S \in \mathcal{T}_g\}$. There exists an open neighborhood $\mathcal{D}$ of $\mathcal{L}$ in $\mathbf{R}^{9g-9}$ together with a real analytic mapping $\mathcal{A} = \mathcal{A}_G : \mathcal{D} \to \mathbf{R}^{6g-6}$ such that $\omega_G(S) = \mathcal{A}_G(\ell\Omega_G(S))$ for all $S \in \mathcal{T}_g$.*

Proof. We prove this in the model $\mathcal{T}_G$. The first $3g - 3$ coordinates, forming a vector $L = (\ell_1, \ldots, \ell_{3g-3})$, are part of $\ell\Omega(S^\omega)$. It suffices therefore to prove that each twist parameter α_k is a real analytic function of $\ell\Omega_G(S^\omega)$. For this we consider, as before (see the lines next to Fig. 6.2.2), the isometric immersion $\iota'_k : X^k(S^\omega) \to S^\omega$ and the lifts $\gamma^k(S^\omega)$, $\delta^k(S^\omega)$ and $\eta^k(S^\omega)$ of $\gamma_k(S^\omega)$, $\delta_k(S^\omega)$ and $\eta_k(S^\omega)$, together with the boundary geodesics of $X^k(S^\omega)$ which form the curve system on $X^k(S^\omega)$ as in (3.3.13). The present curves $\gamma^k(S^\omega)$, $\delta^k(S^\omega)$ and $\eta^k(S^\omega)$ are γ^α, δ^α and η^α of (3.3.13) respectively with $\alpha = \alpha_k$, and the boundary geodesics of $X^k(S^\omega)$ are γ_2, γ_3, γ'_2, γ'_3. Now α_k is determined by the analytic function $\mathcal{A}$ of Lemma 3.3.14. ◇

In the following, the graph G is again fixed.

6.3.5 Theorem. *For any closed curve β on the base surface F, the function $\omega \mapsto \ell\beta(S^\omega)$, $\omega \in \mathcal{R}^{6g-6}$ is real analytic.*

Proof. Fix ω_0. We compute $\ell\beta(S^\omega)$ for ω in a neighborhood of ω_0. This will be achieved in 3 steps. First we replace $\beta(S^\omega)$ by a right-angled geodesic polygon which we consider as a piecewise geodesic loop at some base point $p(S^\omega)$. Then we use this polygon to compute the geodesic loop $\beta^*(S^\omega)$ in its homotopy class with fixed base point, and finally we compute $\beta(S^\omega)$ in terms of $\beta^*(S^\omega)$.

In view of a later application (the proof of Lemma 6.7.9), we choose the base point as follows:

(1) $$p(S^\omega) = \gamma_1^\omega(\tfrac{1}{2}\alpha_1 + \tfrac{1}{4}),$$

where α_1 is the twist parameter at $\gamma_1^\omega = \gamma_1(S^\omega)$. Next we choose a geodesic loop $\beta^*(S^{\omega_0})$ on S^{ω_0} with base point $p(S^{\omega_0})$ which is freely homotopic to $\beta(S^{\omega_0})$. This loop is determined up to conjugacy in the fundamental group of S^{ω_0} at $p(S^{\omega_0})$, and we take it in such a way that it forms an angle $\vartheta < \pi$ at $p(S^{\omega_0})$.

The right-angled piecewise geodesic loop is defined as follows. The right-angled hexagons of $Y_1, \ldots, Y_{2g}$ in S^{ω_0} tessellate S^{ω_0}. We let

$$\bar{\beta}^{\omega_0} = b_1^{\omega_0}, \ldots, b_{2n+1}^{\omega_0}$$

be a piecewise geodesic loop with base point $p(S^{\omega_0})$ which runs along the sides of the hexagons of the tessellation such that the following hold:

(2) *$\bar{\beta}^{\omega_0}$ is homotopic, with fixed base point, to $\beta^*(S^{\omega_0})$;*

(3) *each side $b_\nu^{\omega_0}$ with ν odd runs along some $\gamma_k^{\omega_0}$ (with k depending on ν);*

(4) *each side $b_\nu^{\omega_0}$ with ν even is one of the three common perpendiculars between the boundary geodesics of some Y_i.*

In (3) $b_\nu^{\omega_0}$ is allowed to have length zero (if the twist parameter at $\gamma_k^{\omega_0}$ is an integer multiple of $\tfrac{1}{2}$). We also allow $b_\nu^{\omega_0}$ to run several times around $\gamma_k^{\omega_0}$. The reader will convince himself that such loops exist. They are, of course, not unique.

The angle of $\bar{\beta}^{\omega_0}$ at $p(S^{\omega_0})$ is either 0 or π. All other angles of $\bar{\beta}^{\omega_0}$ are right angles.

The sides with an even number have positive length. For the remaining sides, i.e. for those on $\gamma_1^{\omega_0}, \ldots, \gamma_{3g-3}^{\omega_0}$ we must introduce *directed lengths*: if $b_\nu^{\omega_0}$ on $\gamma_k^{\omega_0}$ has the same orientation as $\gamma_k^{\omega_0}$ (with respect to the orientation of the loop $\bar{\beta}^{\omega_0}$), then $\ell(b_\nu^{\omega_0}) \geq 0$; otherwise $\ell(b_\nu^{\omega_0}) \leq 0$.

Now let ω be near ω_0 and consider the corresponding tessellation of S^ω. It differs only little from that of S^{ω_0} and we have a corresponding right-angled piecewise geodesic loop

$$\bar{\beta}^\omega = b_1^\omega, \ldots, b_{2n+1}^\omega$$

with base point $p(S^\omega)$ which differs only little from $\bar{\beta}^{\omega_0}$. By the above sign convention, *the lengths of $b_1^\omega, \ldots, b_{2n+1}^\omega$ are continuous and therefore analytic functions of ω*.

The marking homeomorphisms $\varphi^\omega : F \to S^\omega$ in the model $\mathcal{T}_G$ are defined in such a way that

(5) $$\varphi^\omega \circ (\varphi^{\omega_0})^{-1}(p(S^{\omega_0})) = p(S^\omega)$$

(cf. (1)) and (6.2.6). We check with the definition of φ^ω in Section 6.2 that

(6) $\bar{\beta}^\omega$ *is homotopic with fixed base point to* $\varphi^\omega \circ (\varphi^{\omega_0})^{-1} (\bar{\beta}^{\omega_0})$.

Hence, the closed geodesic $\beta = \beta(S^\omega)$ is in the free homotopy class of $\bar{\beta} = \bar{\beta}^\omega$. It suffices therefore to find an analytic function which computes the length of β in terms of $b_1 = b_1^\omega, \ldots, b_{2n+1} = b_{2n+1}^\omega$. For ease of notation we do this without writing the superscript ω.

Without changing the notation we lift $\bar{\beta}$ into the unit disk $\mathbf{D}$ in such a way that b_1 lies on the positive real axis with initial point p_0 at the origin. For $\nu = 1, \ldots, 2n+1$ we denote by p_ν the endpoint of b_ν and let B_ν^- and B_ν^+ be the endpoints at infinity ($\partial \mathbf{D}$) of the geodesic extension of the arc b_ν. If on S^ω, b_ν lies on γ_k^ω, then in $\mathbf{D}$, b_ν lies on a lift of γ_k^ω, and B_ν^- and B_ν^+ are the initial and endpoints at infinity of this lift.

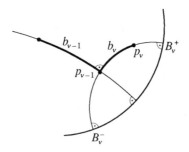

Figure 6.3.1

An induction over ν shows that the Euclidean coordinates of the points p_ν, B_ν^- and B_ν^+ are computable in terms of $\ell(b_1), \ldots, \ell(b_{2n+1})$ by means of analytic functions.

Let β^* be the geodesic arc in $\mathbf{D}$ from p_0 to p_{2n+1}. Then $\ell(\beta^*)$ and the two angles between β^* and b_1, and β^* and b_{2n+1}, are computable by analytic functions of $\ell(b_1), \ldots, \ell(b_{2n+1})$.

Again, on S^ω the projection $\beta^*(S^\omega)$ of β^* is a geodesic loop at $p(S^\omega)$ which is homotopic with fixed base point to $\bar{\beta} = \bar{\beta}^\omega$. Together with (6) this gives the following:

(7) $\beta^*(S^\omega)$ *is homotopic with fixed base point to* $\varphi^\omega \circ (\varphi^{\omega_0})^{-1}(\beta^*(S^{\omega_0}))$.

From the above remarks we also have the next statement which we shall use again in the proof of Lemma 6.7.9.

(8) *The length of* $\beta^*(S^\omega)$ *and the angles at the base point between* $\beta^*(S^\omega)$ *and* γ_1^ω *are analytic functions of* ω.

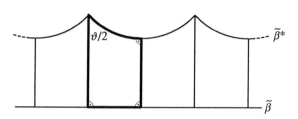

Figure 6.3.2

Once again we lift the (periodically extended) freely homotopic closed curves β^* and β into the universal covering $\mathbf{D}$ to obtain homotopic curves $\tilde{\beta}^*$ and $\tilde{\beta}$ with the same endpoints at infinity. For ω sufficiently close to ω_0 the angle ϑ of the loop β^* at the base point is positive. Therefore $\tilde{\beta}^*$ and $\tilde{\beta}$ are disjoint and bound a strip which is paved by isometric trirectangles as shown in Fig. 6.3.2. The acute angle of the trirectangle is $\vartheta/2$ and the sides are $\bullet$, $\frac{1}{2}\ell(\beta^*)$, $\bullet$, $\frac{1}{2}\ell(\beta)$. From formula (iii) in Theorem 2.3.1 we obtain

(9) $$\cosh \tfrac{1}{2}\ell(\beta) = \cosh \tfrac{1}{2}\ell(\beta^*) \sin \tfrac{1}{2}\vartheta.$$

Together with (8) this proves Theorem 6.3.5, and together with Lemma 6.3.4 it also proves Theorem 6.3.2. ◇

The following corollary will be needed in Section 6.7 when we compare the Fenchel-Nielsen coordinates with those of Zieschang-Vogt-Coldewey. F is again the base surface for the marking homeomorphisms, G is a fixed graph and $\omega = (\ell_1, \ldots, \ell_{3g-3}, \alpha_1, \ldots, \alpha_{3g-3})$.

6.3.6 Corollary. *Let $p_0 = \gamma_1(\tfrac{1}{4}) \in F$. Then for any $S^\omega = (S^\omega, \varphi^\omega) \in \mathcal{T}_G$ we have $\varphi^\omega(p_0) = \gamma_1^\omega(\tfrac{1}{4} + \tfrac{1}{2}\alpha_1)$.*

Let β^* be a homotopically non-trivial loop in F with base point p_0, and denote by $\beta^*(S^\omega)$ the geodesic loop which is homotopic with fixed endpoint to $\varphi^\omega \circ \beta^*$. Then the length of $\beta^*(S^\omega)$ and the angle at $\varphi^\omega(p_0)$ between $\beta^*(S^\omega)$ and γ_1^ω are analytic functions of ω.

Proof. The corollary is a resumé of (7), (8) and (6.2.6). ◇

The corollary holds for each model $\mathcal{T}_G$ of $\mathcal{T}_g$. It would take some extra effort to formulate and prove it for $\mathcal{T}_g$ itself since we have no control over the position of the image of p_0 under the various marking homeomorphisms for a given marking equivalence class.

6.4 Distances in $\mathcal{T}_g$

It is sometimes useful to have at hand a distance function which is compatible with the topology of $\mathcal{T}_g$ induced by the coordinate maps of the preceding sections. Various such distances functions have been studied in the literature (we refer to Nag [1] pp. 403-405 for a brief overview). For our needs the distance function based on quasi isometries will do. Since the distance function based on quasiconformal mappings is more common in the literature, we briefly compare the two in the second part of this section. For the geometry of Teichmüller space induced by these, and related, distance functions, we refer the reader to Wolpert [4 - 6]. We again let $g \geq 2$ be a fixed genus.

6.4.1 Definition. For $S = (S, \varphi)$ and $S' = (S', \varphi') \in \mathcal{T}_g$ the *distance δ* is defined as

$$\delta(S, S') := \inf \log q[\phi],$$

where ϕ runs through the quasi isometries, $\phi : S \to S'$ in the isotopy class of $\varphi' \circ \varphi^{-1}$ and $q[\phi]$ is the maximal length distortion of ϕ (cf. Definition 3.2.3).

6.4.2 Theorem. *The distance function δ is compatible with the topology given by the real analytic structure on $\mathcal{T}_g$.*

Proof. Since both topologies are metric topologies it suffices to prove that a sequence in $\mathcal{T}_g$ converges with respect to one topology if and only if it converges with respect to the other. We let G be a marked cubic graph and work in the model $\mathcal{T}_G$ with the corresponding curve system $\Omega = \Omega_G$

Assume that $(S_n, \varphi_n) \to (S, \varphi)$ with respect to the topology of $\mathcal{T}_G$. Then by the definition of the marking homeomorphisms φ_n and φ (Definition 6.2.5) and by virtue of Lemma 3.2.6 (convergence of Y-pieces) and Lemma 3.3.8 (convergence of X-pieces), it follows that $q[\varphi_n \circ \varphi^{-1}] \to 1$. Consequently, the distance converges to 0.

Now assume that $\delta(S_n, S) \to 0$. A q-quasi isometry in the isotopy class of $\varphi_n \circ \varphi^{-1}$ maps each geodesic $\beta(S)$, $\beta \in \Omega$, to a curve in the homotopy class of $\beta(S_n)$ which is not longer than $q\ell\beta(S)$. Hence, $\ell\beta(S_n) \leq \ell\beta(S) \exp \delta(S_n, S)$. Similarly, $\ell\beta(S) \leq \ell\beta(S_n) \exp(\delta(S_n, S))$. This proves that $\ell\Omega(S_n) \to \ell\Omega(S)$, and therefore that $S_n \to S$ in the topology of $\mathcal{T}_G$. ◇

We now compare δ to another distance function on $\mathcal{T}_g$ which is based on quasiconformal mappings. Since quasiconformal mappings are not used in the book, the proofs will only be outlined.

A homeomorphism $\phi : S \to S'$ is called *k-quasiconformal* if for almost all $p \in S$

$$\limsup_{t \downarrow 0} \frac{\max}{\min} \{ \text{dist}(\phi(x), \phi(y)) \mid \text{dist}(x, y) = t \} \leq k.$$

(Here (max/min){} means max{}/min{}.) We define $k[\phi] \geq 1$ to be the greatest lower bound of all κ such that ϕ is κ-quasiconformal.

Any q-quasi isometry is q^2-quasiconformal, but the converse does not hold. However, we have the following theorem (Buser [3], Wolpert [3]).

6.4.3 Theorem. *Let S, S' be compact Riemann surfaces of genus $g \geq 2$. If a homeomorphism $\phi : S \to S'$ is k-quasiconformal, then for each closed geodesic β on S, the closed geodesic $\phi[\beta]$ in the homotopy class of $\phi \circ \beta$ satisfies*

$$\frac{1}{k}\ell(\beta) \leq \ell\phi[\beta] \leq k\ell(\beta).$$

As a consequence, the distance function

(6.4.4) $$d(S, S') := \inf \log k[\phi],$$

with ϕ running through the quasiconformal mapping in the isotopy class of $\varphi' \circ \varphi^{-1}$, is compatible with the above distance function δ.

We outline the proof of Theorem 6.4.3 (cf. also Wolpert [3]). The basic tool is the following generalization of Schwarz's lemma provided by Hersch [1], (cf. also Lehto-Virtanen [1] section II.3 and Seppälä-Sorvali [4] section 4.9). Let $w : \mathbf{D} \to \mathbf{D}$ be a k-quasiconformal homeomorphism from the unit disk $\mathbf{D}$ into itself such that $w(0) = 0$. Then

$$|w(z)| \leq \mu^{-1}(\frac{1}{k}\mu(|z|)), \quad z \in \mathbf{D},$$

where $\mu :]0, 1[\to]0, \infty[$ is a continuous monotone decreasing function with the asymptotic behavior

$$\mu(r) = \frac{1}{2\pi} \log \frac{4}{r} + O(r^2) \quad \text{if } r \to 0,$$

$$\mu(r) = \frac{\pi}{\log(\frac{8}{1-r})} + o(1-r) \quad \text{if } r \to 1.$$

In terms of the *hyperbolic distance* the second relation implies the inequality $\text{dist}(w(z), 0) \leq k \, \text{dist}(z, 0) + O(1)$. Using the isometry group we conclude that the following inequality

$$\text{dist}(w(x), w(y)) \leq k \, \text{dist}(x, y) + O(1)$$

holds for all $x, y \in \mathbf{D}$ and for any k-quasiconformal homeomorphism $w : \mathbf{D} \to w(\mathbf{D}) \subset \mathbf{D}$. Now let $\pi : \mathbf{D} \to S$ and $\pi' : \mathbf{D} \to S'$ be universal coverings. Let $\phi : S \to S'$ be a k-quasiconformal homeomorphism in the isotopy class of $\varphi' \circ \varphi^{-1}$ and let $\tilde{\phi} : \mathbf{D} \to \mathbf{D}$ be a lift of ϕ. If β is a closed curve on S, then the above inequality applied to $w = \tilde{\phi}$ implies that $\ell(\phi(\beta)) \leq k\ell(\beta)$. In particular, $\ell(\phi[\beta]) \leq \ell(\phi(\beta)) \leq k\ell(\beta)$ for every closed geodesic β on S. This is the second inequality of Theorem 6.4.3. Applied to ϕ^{-1} it yields the first. ◇

6.5 The Teichmüller Modular Group

Every homeomorphism $h : F \to F$ of the base surface onto itself defines an action on $\mathcal{T}_g$ via the rule $(S, \varphi) \mapsto (S, \varphi \circ h)$. Homeomorphisms which are isotopic define the same action. In this way the so-called *mapping class group* operates on $\mathcal{T}_g$ and, as we shall prove, the space of isometry classes of compact Riemann surfaces is $\mathcal{T}_g$ modulo this action.

Two homeomorphisms $F \to F$ are said to be *equivalent* if they are isotopic. The equivalence class of a homeomorphism is sometimes called a *mapping class*.

6.5.1 Definition. The *mapping class group* $\mathfrak{M}_g$ is the group of all equivalence classes of homeomorphisms $F \to F$ modulo isotopy.

In contrast to complex analysis, we admit also *orientation-reversing* homeomorphisms. The subgroup $\mathfrak{M}_g'$ of $\mathfrak{M}_g$ corresponding to the orientation-preserving homeomorphisms has index 2 in $\mathfrak{M}_g$. All theorems below can easily be adjusted to $\mathfrak{M}_g'$.

For the algebraic properties of $\mathfrak{M}_g'$ and $\mathfrak{M}_g$ we refer to Birman [1]. A detailed account from a geometric point of view may be found in Zieschang [1]. Here we are primarily concerned with the action of $\mathfrak{M}_g$ (or $\mathfrak{M}_g'$) on $\mathcal{T}_g$.

As usual, we identify a homeomorphism with its equivalence class. For every $h \in \mathfrak{M}_g$ we define an action $m[h] : \mathcal{T}_g \to \mathcal{T}_g$ via the rule

$$m[h](S, \varphi) = (S, \varphi \circ h).$$

6.5.2 Definition. The *Teichmüller modular group*, denoted by $\mathcal{M}_g$, is the group of transformations

$$\mathcal{M}_g = \{ m[h] \mid h \in \mathfrak{M}_g \}.$$

The elements of $\mathcal{M}_g$ are called *Teichmüller mappings*.

Here too, one may restrict oneself to the subgroup $\mathcal{M}_g'$ which corresponds to $\mathfrak{M}_g'$. The reader may check that the following theorem (and its proof) remains valid if we replace $\mathfrak{M}_g$ and $\mathcal{M}_g$ by $\mathfrak{M}_g'$ and $\mathcal{M}_g'$.

6.5.3 Theorem. *For $g \geq 3$, $m : \mathfrak{M}_g \to \mathcal{M}_g$ is an isomorphism. For $g = 2$, m has kernel $\mathbf{Z}_2$.*

Proof. First let $g \geq 3$. If (S, φ) and $(S, \varphi \circ h)$ represent the same point in $\mathcal{T}_g$, then $\varphi \circ h \circ \varphi^{-1}$ is isotopic to an isometry of S onto itself. Hence, to prove that m has trivial kernel, it suffices to find an example of a compact Riemann surface with trivial isometry group.

Take a pair of pants Y with boundary geodesics γ_i of length $0 < \ell(\gamma_1) < \ell(\gamma_2) < \ell(\gamma_3) < 1$ and extend it to a surface S^t of genus g such that all other closed geodesics on S^t have length greater than $\ell(\gamma_3)$ and such that the twist parameter at γ_1 is in the interval $]0, \frac{1}{4}[$. By virtue of the collar theorem, this is easy to achieve. Since $g \neq 2$, there exists no other Y-piece in S^t with the same boundary lengths as Y. The only non-trivial isometry of Y onto itself which fixes the boundary geodesics of Y is the symmetry which exchanges the two right-angled hexagons of Y. The twist parameter at γ_1 makes it impossible to extend this isometry from Y to S^t, and hence S^t has trivial isometry group.

Now let $g = 2$ and take the model $\mathcal{T}_G$ based on the underlying graph G in the example of Fig. 6.1.2.

There exists an orientation-preserving isometry $\rho : F \to F$ of order two with 6 fixed points which exchanges Y_1, Y_2 and fixes $\gamma_1, \gamma_2, \gamma_3$. For arbitrary $(S, \varphi) \in \mathcal{T}_G$, the mapping $\varphi \circ \rho \circ \varphi^{-1}$ is an isometry (check) and thus, the surfaces $(S, \varphi \circ \rho)$ and (S, φ) are marking equivalent (cf. Remark 6.1.3(i)). Therefore, ρ is in the kernel of m. In order to prove that ρ is the only non-trivial element in the kernel, it suffices to find an example of a compact Riemann surface of genus 2 whose isometry group is $\{id, \varphi \circ \rho \circ \varphi^{-1}\}$. This is done in the same way as for $g \geq 3$. ◊

6.5.4 Theorem. *$S, S' \in \mathcal{T}_g$ are isometric if and only if there exists $\mu \in \mathcal{M}_g$ such that $S' = \mu(S)$. If $g \geq 3$, the isotropy group*

$$I_S = \{\mu \in \mathcal{M}_g \mid \mu(S) = S\}$$

is canonically isomorphic to the isometry group $\mathrm{Is}(S)$. If $g = 2$ then I_S is canonically isomorphic to $\mathrm{Is}(S)/\mathbf{Z}_2$.

Proof. Since $m[h]$ only changes the marking, $m[h](S)$ is trivially isometric to S. Conversely, let (S, φ) and (S', φ') be isometric with isometry $j : S' \to S$. Setting $h = \varphi^{-1} \circ j \circ \varphi'$, we obtain the marking equivalent surfaces (S', φ') and $(S, j \circ \varphi') = (S, \varphi \circ h) = m[h](S, \varphi)$. This proves the first part.

For the second part we let I'_S denote the set

$$I'_S = \{ h \in \mathfrak{M}_g \mid m[h](S) = S \}.$$

For $h \in I'_S$, (S, φ) and $(S, \varphi \circ h)$ are marking equivalent. By the definition of marking equivalence there exists an isometry $j = J[h]$ in the isotopy class of $\varphi \circ h \circ \varphi^{-1}$. By Lemma 6.5.5 (below), this isometry is uniquely determined by h. Hence, we have a homomorphism

$$J : I'_S \to \mathrm{Is}(S)$$

which sends h to $j = J[h]$. Trivially, J is injective, and the same arguments as in the proof of the first part show that J is surjective. Together with Theorem 6.5.3 this proves the theorem. ◇

6.5.5 Lemma. *Let $j : S \to S$ be an isometry. If j is isotopic to the identity, then j is the identity.*

Proof. We may assume that $S \in \mathcal{T}_G$. If j is isotopic to the identity, then $j \circ \gamma_k(S)$ is homotopic to $\gamma_k(S)$ for $k = 1, \ldots, 3g - 3$, where the γ_k are from Ω_G. By the uniqueness of a closed geodesic in its homotopy class (Theorem 1.6.6), $j \circ \gamma_k(S) = \gamma_k(S)$ (up to a parameter change of type $t \mapsto t + \mathrm{const}$, $t \in \mathbf{S}^1$). Furthermore, j is an orientation-preserving mapping. Hence, j fixes each Y_i, $i = 1, \ldots, 2g - 2$.

The only orientation-preserving isometry of a Y-piece which fixes the boundary geodesics is the identity. Hence $j = \mathrm{id}$. ◇

If in Theorem 6.5.4 the group $\mathcal{M}_g$ is replaced by $\mathcal{M}'_g$, then "isometric" must be replaced by "direct isometric" and $\mathrm{Is}(S)$ by $\mathrm{Is}^+(S)$. (Direct = orientation-preserving; by definition, the orientation of S in (S, φ) is the one for which φ is direct.) In this case, $\mathcal{T}_g/\mathcal{M}'_g$ is the space of all hyperbolic structures modulo orientation-preserving isometries. By the uniformization theorem, every conformal structure contains a hyperbolic atlas. Conversely, every hyperbolic atlas extends to a unique conformal structure. Thus $\mathcal{T}_g/\mathcal{M}'_g$ can be interpreted as the so-called *Riemann space* of all conformal equivalence classes. The natural projection $\mathcal{T}_g/\mathcal{M}'_g \to \mathcal{T}_g/\mathcal{M}_g$ is a branched covering of order 2.

6.5.6 Theorem. *$\mathcal{M}_g$ acts properly discontinuously on $\mathcal{T}_g$ by real analytic diffeomorphims which leave the distance functions δ and d of Section 6.4 invariant.*

Proof. By the definition of marking equivalence each $\mu = m[h] \in \mathcal{M}_g$ is a bijection of $\mathcal{T}_g$ and leaves the distances d and δ of any pair (S, φ), (S', φ') invariant (or any other extremum in the isotopy class of $\varphi' \circ \varphi^{-1}$). To prove analyticity, we fix a graph G with corresponding base surface F and consider the canonical curve system $\Omega = \Omega_G$. For each homeomorphism $h : F \to F$, we let $h \circ \Omega$ be the sequence of all closed curves $h \circ \beta$, $\beta \in \Omega$. For S^ω and $S^{\omega'} := m[h](S^\omega)$ we then have

$$\ell(h \circ \Omega(S^\omega)) = \ell(\Omega(S^{\omega'})).$$

By Lemma 6.3.4, there exists an analytic function $\mathcal{A}_G$ with the property that $\mathcal{A}_G(\ell\Omega(S^{\omega'})) = \omega'$. By Theorem 6.3.5, $\omega' = \mathcal{A}_G(\ell(h \Omega (S^\omega)))$ is an analytic function of ω. This proves that the mappings $m[h] \in \mathcal{M}_g$ act by analytic diffeomorphisms.

Since $\mathcal{M}_g$ acts by isometries (with respect to δ), discontinuity and proper discontinuity are equivalent properties; we prove discontinuity. Thus, let $S_0 = (S_0, \varphi_0) \in \mathcal{T}_g$ and set

$$\mathcal{U} = \{ S \in \mathcal{T}_g \mid \delta(S_0, S) < \log 2 \},$$

where δ is the distance function of Definition 6.4.1. By Theorem 6.4.2, $\mathcal{U}$ is an open neighborhood of S_0 in $\mathcal{T}_g$. Let $\lambda = \max\{ \ell\beta(S_0) \mid \beta \in \Omega \}$, and consider the set W of all closed geodesics β' on F satisfying $\ell\beta'(S_0) \leq 2\lambda$. Then W is finite, and by Theorem 6.1.7, there exist only finitely many isotopy classes of homeomorphisms $h : F \to F$ satisfying $h \circ \Omega(S_0) \subset W(S_0)$. If $m[h](S_0) = (S_0, \varphi_0 \circ h)$ lies in $\mathcal{U}$, then $h\Omega(S_0) \subset W(S_0)$ because the lengths of $\beta(S_0)$ and $\beta(m[h](S_0))$ differ at most by a factor 2. Hence, there are only finitely many $\mu = m[h] \in \mathcal{M}_g$ satisfying $\mu(S_0) \in \mathcal{U}$. ◇

In connection with Theorem 6.5.4 the proper discontinuity of $\mathcal{M}_g$ implies that each compact Riemann surface has a finite isometry group. Below we give a direct geometric proof of Hurwitz' theorem that the order of the isometry group is at most $168(g - 1)$.

In the next theorem, $I_g \subset \mathcal{T}_g$ is defined as follows. If $g \geq 3$, then I_g is the set of all surfaces in $\mathcal{T}_g$ which have a non-trivial isometry group. If $g = 2$, then every surface $S = (S, \varphi)$ has an isometry of order 2 (cf. the proof of Theorem 6.5.3), and we let I_2 be the set of all $S = (S, \varphi) \in \mathcal{T}_2$ having an isometry group of order ≥ 3.

6.5.7 Theorem. *I_g is a proper closed real analytic subvariety of $\mathcal{T}_g$. For any $S \in \mathcal{T}_g - I_g$ there exists an open neighborhood of S consisting of pairwise non-isometric surfaces.*

Proof. By Theorem 6.5.4 we have

$S \in I_g$ *if and only if S is a fixed point of* $\mathcal{M}_g$.

Since $\mathcal{M}_g$ acts properly discontinuously, I_g is locally the union of the fixed points of finitely many $\mu \in \mathcal{M}_g$. Since each μ is an analytic diffeomorphism, the fixed point set of μ is a closed real analytic subvariety. The surface S^t in the proof of Theorem 6.5.3 shows that it is proper, and therefore I_g is proper.

Now let $S = (S, \varphi) \in \mathcal{T}_g$ be such that there exists a sequence $\{(S_n, \varphi_n), (S_n, \varphi_n')\}_{n=1}^{\infty}$ of pairs of isometric but not marking equivalent surfaces with $(S_n, \varphi_n) \to (S, \varphi)$ and with $(S_n, \varphi_n') \to (S, \varphi)$. Since the lengths of the closed geodesics of S form a discrete set, it follows again from Theorem 6.1.7 that in the sequence $\{\varphi_n^{-1} \circ \varphi_n'\}_{n=1}^{\infty}$ of homeomorphisms of F only finitely many isotopy classes occur. We may therefore select the sequence such that all $\varphi_n^{-1} \circ \varphi_n'$ are in the same isotopy class, say in the class of some h, where $m[h] \neq id$. By continuity we now conclude that $\ell\Omega(S) = \ell(h \circ \Omega(S))$. By Corollary 6.2.8, (S, φ) and $(S, \varphi \circ h)$ are marking equivalent. This proves that $S \in I_g$. ◇

It is interesting to observe that the second statement in Theorem 6.5.7 holds also for analytic subvarieties in the following form.

6.5.8 Lemma. *Let* $g \geq 2$, *let* $\mathcal{W} \subset \mathcal{T}_g$ *be an open neighborhood and let* $\mathcal{T} \subset \mathcal{W}$ *be a smooth real analytic subvariety of* $\mathcal{W}$. *Then there exists a proper real analytic subvariety* $\mathcal{I}$ *of* $\mathcal{T}$ *with the following property. For any* $S \in \mathcal{T} - \mathcal{I}$ *there exists an open neighborhood of* S *in* $\mathcal{T}$ *consisting of pairwise non-isometric surfaces.*

Proof. It suffices to prove this for the case that $\mathcal{W}$ has compact closure in $\mathcal{T}_g$. The proof is similar to the preceding one. We call $S \in \mathcal{T}$ a *particular point* of $\mathcal{T}$, if in any neighborhood $\mathcal{U}_S \subset \mathcal{T}$ of S there exists $S' \neq S''$ such that S' and S'' are isometric surfaces. We then have $\tau \in \mathcal{M}_g$ such that $\tau(S') = S''$. Since $\mathcal{M}_g$ acts properly discontinuously on $\mathcal{T}_g$, τ fixes S if we assume that $\mathcal{U}_S$ is sufficiently small. Since τ does not fix all point of $\mathcal{U}_S$ ($\tau(S') = S''$), the set

$$\mathcal{I}_\tau = \{ x \in \mathcal{T} \mid \tau(x) = x \}$$

is a proper real analytic subvariety of $\mathcal{T}$. Since $\mathcal{W}$ has compact closure and since $\mathcal{M}_g$ is properly discontinuous, only finitely many elements of $\mathcal{M}_g$ have fixed points in $\mathcal{W}$, and we let $\tau_1, \ldots, \tau_r \in \mathcal{M}_g$ be those which fix some but not all points of $\mathcal{T}$. The set

Ch.6, §5]   The Teichmüller Modular Group   159

$$\mathcal{I} = \mathcal{I}_{\tau_1} \cup \ldots \cup \mathcal{I}_{\tau_r} = \{\, x \in \mathcal{T} \mid \exists\, \tau \in \mathcal{M}_g,\ \tau(x) = x,\ \tau \mid \mathcal{T} \ne id_{\mathcal{T}}\,\}$$

is a proper real analytic subvariety of $\mathcal{T}$ which contains all particular points of $\mathcal{T}$.   ◇

We now prove Hurwitz' theorem.

6.5.9 Theorem. *The isometry group of any compact Riemann surface of genus $g \ge 2$ has order $\le 168(g - 1)$.*

Proof. The proof is similar to Greenberg's proof in Greenberg [3], p. 219, but is in terms of surfaces instead of Fuchsian groups.

The proof is in two steps. First we prove that $\mathrm{Is}(S)$ is finite, and then we use the finiteness to estimate the order.

For the finiteness we consider a figure eight geodesic on S, i.e. a closed geodesic γ with exactly one self-intersection. At most four isometries fix γ as point set. Hence, at most four isometries φ have the same image $\varphi(\gamma)$. Since only finitely many closed geodesics on S have the same length as γ (Theorem 1.6.11), this proves the finiteness of $\mathrm{Is}(S)$.

Now we estimate the order. Since $\mathrm{Is}^+(S)$ is a normal subgroup of $\mathrm{Is}(S)$ of index ≤ 2, it suffices to prove that $\mathrm{Is}^+(S)$ has order $\#\mathrm{Is}^+(S) \le 84(g - 1)$.

To explain the idea of proof, we first assume that $\mathrm{Is}^+(S)$ acts without fixed points. Here the quotient $M = \mathrm{Is}^+(S)\backslash S$ is again a compact Riemann surface of some genus $h \ge 2$, and S is tessellated with $\#\mathrm{Is}^+(S)$ fundamental domains of area $4\pi(h - 1)$. Hence, $\#\mathrm{Is}^+(S) = (g - 1)/(h - 1) \le g - 1$.

In the general case, the finiteness of $\mathrm{Is}^+(S)$ implies that the quotient M is a *compact Riemann surface with cone-like singularities of integer orders.* This structure is defined as follows (cf. also Definition 12.1.4). Consider the quotient metric on M. If $p \in S$ is not a fixed point of $\mathrm{Is}^+(S)$, then the image $\bar{p}$ of p in M has an ε-neighborhood which is isometric to a disk of radius ε in the hyperbolic plane. If p is a fixed point of order v, then a sufficiently small ε-neighborhood of $\bar{p}$ is isometric to

$$\bar{p} \cup\,]0,\, \varepsilon[\, \times \mathbf{S}^1 \text{ with metric } ds^2 = d\rho^2 + \frac{1}{v^2} \sinh^2\!\rho\, d\sigma^2.$$

We say that $\bar{p}$ is a *cone-like singularity of order* v. Since $\mathrm{Is}^+(S)$ has only finitely many fixed points, M has only finitely many singularities. In the next lemma we show that area $M \ge \pi/21$. Admitting this here, we see that S is tessellated with $\#\mathrm{Is}^+(S)$ fundamental domains of area $\ge \pi/21$ and the theorem follows.   ◇

A typical example of a surface with cone-like singularities is obtained by pasting together two isometric geodesic triangles of the hyperbolic plane, say with interior angles π/v_1, π/v_2 and π/v_3, where v_1, v_2, v_3 are integers. The resulting surface has genus 0 and three singularities of orders v_1, v_2 and v_3.

6.5.10 Lemma. *Let M be a compact hyperbolic surface of genus $h \geq 0$ with k cone-like singularities of orders $v_1, \ldots, v_k$. Then*

$$\text{area } M = 4\pi(h-1) + 2\pi(k - (\frac{1}{v_1} + \ldots + \frac{1}{v_k})) \geq \pi/21.$$

Proof. Triangulate M geodesically such that every singularity of M is a vertex of the triangulation. If n_0, n_1 and n_2 denote the number of vertices, edges and triangles respectively, then $3n_2 = 2n_1$, and by Euler's formula, $2 - 2h = n_0 - n_1 + n_2$. From the area formula of a hyperbolic geodesic triangle (Theorem 1.1.7) we obtain area $M = \pi n_2 - \theta$, where θ denotes the sum over the interior angles of the triangles. Grouping these angles together vertexwise we obtain the above area formula.

The lower bound follows from a straightforward checking of cases. The minimum is obtained for $h = 0$ and $k = 3$ with $v_1 = 2$, $v_2 = 3$ and $v_3 = 7$. ◊

We remark without proof that the upper bound of the cardinality of Is(S) in Theorem 6.5.9 is sharp for infinitely many but not all g (Macbeath [1, 2], Maclachlan [1]). For this and for related questions we refer to Accola [1], Beardon [1] chapter 10, Magnus [1], and, in particular, the review article by Conder [1].

6.6 A Rough Fundamental Domain

The moduli problem for compact Riemann surfaces of a fixed genus $g \geq 2$ or, equivalently, the classification problem, consists of describing explicitly a domain W of moduli such that each modulus is a set of geometric invariants with the following properties.

- Every modulus $w \in W$ determines a unique compact Riemann surface (up to isometry).
- Every compact Riemann surface is represented by a unique modulus $w \in W$.

If Riemann surfaces were rectangles with sides a and b then the pairs $w = (a, b) \in \mathbf{R}^2$ satisfying $a \leq b$ would be such moduli.

A modulus may have finitely many components, each of which should have a geometric meaning on the corresponding surface. (In the literature it is usually the components which are called the moduli.)

The moduli problem may also be understood as that of describing the so-called *Riemann* or *moduli space*

(6.6.1) $$\mathcal{R}_g = \mathcal{T}_g / \mathcal{M}_g,$$

where $\mathcal{M}_g$ is the Teichmüller modular group. A possible way of approaching this problem is to give an explicit fundamental domain for the action of $\mathcal{M}_g$. This has been carried out by Keen [5] for the compact Riemann surfaces of signature (1, 1) and by Semmler [1] for the compact Riemann surfaces of genus 2. In both cases the fundamental domain is given as the intersection of a finite number of topological half-spaces. For higher genus, the problem is unsolved and, in any case, is extremely complicated. In Keen [6, 7] Linda Keen has therefore introduced the concept of a *rough fundamental domain* in which any compact Riemann surface of genus g is represented at least once and at most finitely many times. A variant of this domain will now be described.

Let $\mathcal{G}$ be a set of marked cubic graphs with $2g - 2$ vertices such that the elements in $\mathcal{G}$ are pairwise non-isomorphic and such that every isomorphism class is represented in $\mathcal{G}$. To each $\mathcal{G} \in G$ and to each

$$\omega = (L, A) = (\ell_1, \ldots, \ell_{3g-3}, \alpha_1, \ldots, \alpha_{3g-3}) \in \mathcal{R}^{6g-6}$$

corresponds a marked Riemann surface $S^\omega = S_G^\omega$ with some marking homeomorphism $\varphi^\omega : F \to S^\omega$. The base surface F is independent of G, but for every G a fixed hyperbolic structure has been defined on F and on the basis of which the curve system Ω_G has been defined. In this section we view each S_G^ω as an element of the abstract Teichmüller space $\mathcal{T}_g$ by identifying each model $\mathcal{T}_G$ (cf. Definition 6.2.5) with $\mathcal{T}_g$ in a fixed way (for instance via the mappings ω_G as in Definition 6.3.1). Now let

$$W^0 = \{ S_G^\omega \in \mathcal{T}_g \mid 0 < \ell_1 < \ldots < \ell_{3g-3} < \operatorname{arccosh} 3;$$
$$0 < \alpha_1, \ldots, \alpha_{3g-3} < \tfrac{1}{4} ; G \in \mathcal{G} \}$$

(6.6.2)

$$W^1 = \{ S_G^\omega \in \mathcal{T}_g \mid 0 < \ell_1, \ldots, \ell_{3g-3} \leq L_g ;$$
$$0 \leq \alpha_1, \ldots, \alpha_{3g-3} < 1 ; G \in \mathcal{G} \},$$

where $L_g \leq 26(g - 1)$ is Bers' constant (Section 5.1).

6.6.3 Theorem. *W^1 is a rough fundamental domain for the Teichmüller modular group in $\mathcal{T}_g$. The number of occurrences of a given surface in W^1 is at most a_g, where a_g depends only on g. In W^0 the surfaces are pairwise non-isometric.*

Proof. By Bers' theorem (Theorem 5.1.2), every $S \in \mathcal{T}_g$ has at least one occurrence in W^1. The statement about W^0 was proved at the end of Section 3.6 (W^0 coincides with $\mathcal{F}_g$ of (3.6.6)). It remains to estimate how many times a given $S \in \mathcal{T}_g$ may occur in W^1.

For every $S_G^\omega \in \mathcal{T}_g$ which is isometric to S, we fix an isometry $j : S_G^\omega \to S$. The geodesics $\gamma_1(S_G^\omega), \ldots, \gamma_{3g-3}(S_G^\omega)$ with the γ_k from the curve system Ω_G, form a partition of S_G^ω. Similarly, the images $j \circ \gamma_1(S_G^\omega), \ldots, j \circ \gamma_{3g-3}(S_G^\omega)$ are oriented, closed geodesics of length $\leq L_g \leq 26(g-1)$ which form a Bers partition of S. By Lemma 6.6.4, below, S has at most $\exp(27g)$ oriented, closed geodesics of length $\leq L_g$ and therefore at most $\exp(27g(3g-3))$ such partitions. We show that the $S_G^\omega \in W^1$ isometric to S are uniquely determined by the sequence $j \circ \gamma_1(S_G^\omega), \ldots, j \circ \gamma_{3g-3}(S_G^\omega)$, except for $g = 2$, where the number of surfaces with the same sequence may be 2. This will prove the theorem with $a_g \leq \exp(81g^2)$.

Thus, assume without loss of generality that $S = S_G^\omega$ with $j = id$, and let $S' = S_{G'}^{\omega'} \in W^1$ with isometry $j' : S_{G'}^{\omega'} \to S_G^\omega$ be such that, up to a parameter change of type $t \mapsto t + \text{const}$, $t \in \mathbf{S}^1$, we have

$$j' \circ \gamma_k'(S_{G'}^{\omega'}) = \gamma_k(S_G^\omega), \quad k = 1, \ldots, 3g-3,$$

where $\gamma_1', \ldots, \gamma_{3g-3}'$ are from the curve system $\Omega_{G'}$. We first observe that G and G' are isomorphic graphs and therefore $G = G'$ by our choice of $\mathcal{G}$.

Let $g \geq 3$. Then j' maps each pair of pants Y_i' in the partition of S' onto the corresponding Y_i in the partition of S. Moreover, since j' preserves the orientation of $\gamma_1', \ldots, \gamma_{3g-3}'$, we see that $j' : Y_i' \to Y_i$ is the canonical identification which preserves the standard presentation, $i = 1, \ldots, 2g-2$. It follows that for $k = 1, \ldots, 3g-3$ we have $\alpha_k = \alpha_k' \mod(\mathbf{Z})$ and therefore $\alpha_k = \alpha_k'$ since $S, S' \in W^1$. This shows that $S = S'$ and proves the theorem for $g \geq 3$. A similar argument shows that for $g = 2$ a given partition occurs at most twice in W^1 (if $g = 2$ the above j' may map Y_1' to Y_2 and Y_2' to Y_1). ◊

The lemma which follows is stated in a form which is adapted to the proof of Theorem 13.1.1. In Section 9.4 we use a completely different approach to show that for fixed S and for $L \to \infty$ the number of closed geodesics of length $\leq L$ is asymptotically equal to $\frac{1}{L}\exp L$, independently of the genus.

6.6.4 Lemma. *Let S be a compact Riemann surface of genus $g \geq 2$ and let $L > 0$. There are at most $(g-1)\exp(L+6)$ oriented closed geodesics of length $\leq L$ on S which are not iterates of closed geodesics of length $\leq 2 \operatorname{arcsinh} 1$.*

Proof. Let $\beta_1, \ldots, \beta_k$, $k \leq 3g-3$ be the simple closed geodesics of length $\ell \leq 2 \operatorname{arcsinh} 1$ on S (Theorem 4.1.6), and let $p_1, \ldots, p_s$ be a maximal set of points on S such that

(1) at each p_σ the injectivity radius $r_{p_\sigma}(S)$ of S satisfies $r_{p_\sigma}(S) \geq r := \operatorname{arcsinh} 1$,

(2) $\operatorname{dist}(p_\sigma, p_\tau) \geq r$ for $\sigma, \tau = 1, \ldots, s$, $\sigma \neq \tau$.

Since disks of radius $r/2$ around $p_1, \ldots, p_s$ are pairwise disjoint and have area equal to $2\pi(\cosh \frac{r}{2} - 1)$, it follows that

(3) $$s \leq 2(g-1)/(\cosh \tfrac{r}{2} - 1).$$

A closed geodesic c of length $> 2r$ on S which is not an iterate of β_κ for $\kappa \in \{1, \ldots, k\}$, cannot be entirely contained in the collar $\mathscr{C}(\beta_\kappa)$. By Theorem 4.1.6, c passes through a point where the injectivity radius is $\geq r$. By the maximality of the set $p_1, \ldots, p_s$, this point is a distance $d \leq r$ from some p_σ. It follows that c is homotopic to a geodesic loop at p_σ of length $\leq \ell(c) + 2r$.

It remains to estimate the number of geodesic loops at p_σ of length $\leq L + 2r$. We use a universal covering $\mathbf{H} \to S$ and lift the loops onto geodesic arcs in $\mathbf{H}$ with a common initial point. Since the injectivity radius of S at p_σ is $r_{p_\sigma}(S) \geq r$, the endpoints of these arcs have pairwise distances $\geq 2r$. Comparison of the areas now shows that there are at most

$$\frac{\cosh(L + 3r) - 1}{\cosh r - 1} \leq \frac{e^{L+3r}}{2(\cosh r - 1)}$$

such loops. Together with (3) the theorem follows. ◇

As another application of Bers' theorem we note here the following so called Mahler compactness theorem (Bers [2], Mumford [1]).

6.6.5 Theorem. *Let $\varepsilon > 0$. There exists a compact subset $Q(\varepsilon) \subset \mathscr{T}_g$ such that for any compact Riemann surface S of genus g and injectivity radius bounded below by ε there exists $S' \in Q(\varepsilon)$ isometric to S.*

Proof. Again let $\mathscr{G}$ be a set of marked cubic graphs with $2g - 2$ vertices such that the elements in $\mathscr{G}$ are pairwise non-isomorphic and such that every isomorphism class is represented in $\mathscr{G}$. As in the definition of W^0 and W^1 (cf. (6.6.2)) we identify each model $\mathscr{T}_G$ with $\mathscr{T}_g$ in a fixed way. The set

$$Q(\varepsilon) = \{ S_G^\omega \in \mathscr{T}_g \mid 2\varepsilon \leq \ell_1, \ldots, \ell_{3g-3} \leq L_g ; 0 \leq \alpha_1, \ldots, \alpha_{3g-3} < 1 ; G \in \mathscr{G}\}$$

is compact, and by Bers' theorem (Theorem 5.2.3), every $S \in \mathscr{T}_g$ is isometric to some $S' \in Q(\varepsilon)$. ◇

6.7 The Coordinates of Zieschang-Vogt-Coldewey

Besides the Fenchel-Nielsen coordinates, many other parameters are frequently used in the literature. We describe two of them. In this section we consider the coordinates of Zieschang-Vogt-Coldewey which are based on canonical polygons. In the next section we shall consider Bers' coordinates which are adapted to Fuchsian groups. We shall show that both coordinates are analytically equivalent with the Fenchel-Nielsen coordinates. For complex coordinates we refer the reader to Kra [2, 3], Maskit [1], and Nag [1].

We recall some facts about canonical generators of the fundamental group and canonical polygons. References for this are for instance Massey [1], Stillwell [1], Zieschang-Vogt-Coldewey [1, 2].

Let P be a simply connected geodesic polygon or, more generally, any compact topological disk in $\mathbf{R}^2$ whose boundary is a Jordan curve consisting of the consecutive sides $b_1, b_2, \bar{b}_1, \bar{b}_2, \ldots, b_{2g-1}, b_{2g}, \bar{b}_{2g-1}, \bar{b}_{2g}$, where $g \geq 2$. We assume that each side is parametrized on the interval $[0, 1]$ in such a way that the endpoint of b_1 is the initial point of b_2, the endpoint of b_2 is the initial point of $\bar{b}_1$, etc. If we paste the sides together according to the pasting scheme

(6.7.1) $\quad b_n(t) = \bar{b}_n(1 - t) =: \beta_n(t), \, t \in [0, 1], \quad n = 1, \ldots, 2g,$

we obtain a compact orientable surface $F = P \bmod (6.7.1)$ of genus g. The presentation of F by means of (6.7.1) is the so-called *normal form*. On F the loops $\beta_1, \ldots, \beta_{2g}$ have the following properties.

(1) $\beta_1, \ldots, \beta_{2g}$ *are simple loops with a common base point p. For $i \neq j$, β_i intersects β_j only at p.*
(2) *The surface obtained by cutting F open along $\beta_1, \ldots, \beta_{2g}$ is a topological disk.*
(3) *The loops $\beta_1, \ldots, \beta_{2g}$ generate the fundamental group $\pi_1(F)$ at p with the single defining relation*

$$\prod_{v=1}^{g} \beta_{2v-1} \beta_{2v} \beta_{2v-1}^{-1} \beta_{2v}^{-1} = 1.$$

The first two properties follow from the construction. For the last property we refer to the book of Massey [1], or Stillwell [1]. The fact that $\beta_1, \ldots, \beta_{2g}$ together with the single defining relation (3) yield a presentation of the fundamental group will not be used in this book.

Figure 6.7.1 shows an example for $g = 3$ where the polygon P is a geodesic polygon in the hyperbolic plane.

Ch.6, §7]   The Coordinates of Zieschang-Vogt-Coldewey   165

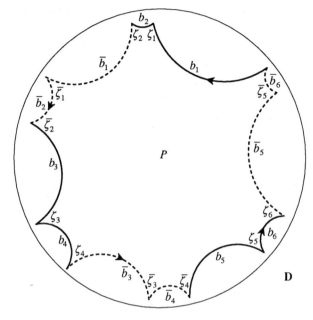

Figure 6.7.1

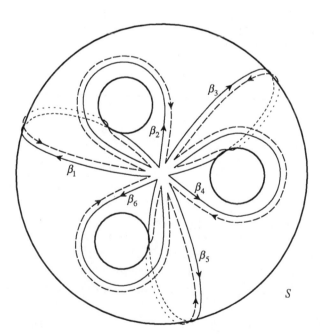

Figure 6.7.2

6.7.2 Definition. Let S be a compact orientable surface of genus g. A sequence $\beta_1', \ldots, \beta_{2g}'$ of closed loops on S with a common base point is called a *canonical dissection* if there exists a homeomorphism $\phi : F \to S$ such that $\phi \circ \beta_n = \beta_n'$, $n = 1, \ldots, 2g$.

Now let F as above be the base surface for the marking homeomorphisms and consider a marked Riemann surface $(S, \varphi) \in \mathcal{T}_g$, where $\varphi : F \to S$. Then the sequence $\varphi \circ \beta_1, \ldots, \varphi \circ \beta_{2g}$ is a canonical dissection and we wish to modify φ such that the $\varphi \circ \beta_n$ are *geodesic loops*.

As shown in Example 1.6.10, this cannot be achieved simply by replacing each $\varphi \circ \beta_n$ by the geodesic loop in its homotopy class. We shall see, however, that things work smoothly if the base point is chosen properly.

6.7.3 Definition. Two canonical dissections $\beta_1', \ldots, \beta_{2g}'$ and $\beta_1'', \ldots, \beta_{2g}''$ of S are called *equivalent* if there exists a 1-homeomorphism $h : S \to S$ such that $h \circ \beta_n' = \beta_n''$ for $n = 1, \ldots, 2g$.

We emphasize that equivalent dissections are not required to have the same base point. The following theorem has been proved in Zieschang-Vogt-Coldewey [1, 2]. For simplicity we abbreviate this reference by Z-V-C.

6.7.4 Theorem. *Let $S = (S, \varphi) \in \mathcal{T}_g$. Then the canonical dissection $\varphi \circ \beta_1, \ldots, \varphi \circ \beta_{2g}$ is equivalent to a unique canonical dissection $\beta_1(S), \ldots, \beta_{2g}(S)$ with the following properties*

(i) $\beta_1(S)$ and $\beta_2(S)$ are smooth closed geodesics with a unique intersection point $p(S)$,

(ii) $\beta_3(S), \ldots, \beta_{2g}(S)$ are geodesic loops at $p(S)$.

A dissection with the properties as stated in Theorem 6.7.4 will be called a *normal canonical dissection*.

Proof. We shall successively modify φ by adding 1-homeomorphisms $S \to S$ until the curves $\varphi \circ \beta_1, \ldots, \varphi \circ \beta_{2g}$ have the required properties. The proof is based on Theorems A.3 - A.5 on isotopies.

Let $\beta_1(S), \beta_2(S)$ be the smooth closed geodesics in the free homotopy classes of $\varphi \circ \beta_1, \varphi \circ \beta_2$. By Theorem 1.6.7, they intersect each other in a unique point $p(S)$. By Theorem A.3, we can modify φ with a first 1-homeomorphism such that $\varphi \circ \beta_1 = \beta_1(S)$ and such that $\varphi(p_0) = p(S)$, where p_0 is the base point of $\beta_1, \ldots, \beta_{2g}$.

Next, since $\varphi \circ \beta_2$ and $\beta_2(S)$ are freely homotopic and since both curves intersect $\beta_1(S)$ only once, a standard covering argument shows that for some integer m, $\varphi \circ \beta_2$ and $(\beta_1(S))^m \circ \beta_2(S) \circ (\beta_1(S))^{-m}$ are homotopic with fixed base point $p(S)$. We add a 1-homeomorphism which fixes $\beta_1(S)$ (as set) and which is the identity outside a tubular neighborhood of $\beta_1(S)$ such that this m becomes zero. Theorem A.5 applied to S after having been cut open along $\beta_1(S)$ allows the further modification of φ such that now $\varphi \circ \beta_1 = \beta_1(S)$ and $\varphi \circ \beta_2 = \beta_2(S)$.

The rest is easier. Assume by induction that $\varphi \circ \beta_i = \beta_i(S)$ is a geodesic loop at $p(S)$ for $i = 1, \ldots, n$. Cut S open temporarily along $\beta_1(S), \ldots, \beta_n(S)$. Since the curves belong to a canonical dissection, the resulting surface S' is connected and $\varphi \circ \beta_{n+1}$ connects two copies of $p(S)$ on the boundary of S'. By Theorem A.5, there exists a 1-homeomorphism $S' \to S'$ which fixes the boundary pointwise and which maps the arc $\varphi \circ \beta_{n+1}$ onto a geodesic arc $\beta_{n+1}(S)$. On S again, $\beta_{n+1}(S)$ is a geodesic loop and, modified with this new 1-homeomorphism, φ now satisfies $\varphi \circ \beta_i = \beta_i(S)$ for $i = 1, \ldots, n + 1$. This proves the existence of a normal canonical dissection in the given equivalence class.

To prove its uniqueness, assume that $J : [0, 1] \times S \to S$ with $J(0, \ ) = id$ and $J(1, \ ) = h$ is an isotopy (as in Definition A.1) such that h maps $\beta_1(S), \ldots, \beta_{2g}(S)$ onto another normal canonical system. By the uniqueness of the smooth closed geodesic in a homotopy class we have $h \circ \beta_i(S) = \beta_i(S)$ for $i = 1, 2$ (except possibly for a parameter change of the form $t \mapsto t + \text{const}, t \in \mathbf{S}^1$). In particular, $h(p(S)) = p(S)$. Now consider the closed curve $\tau \mapsto \eta(\tau) := J(\tau, p(S))$, $\tau \in [0, 1]$.

For $n = 1, \ldots, 2g$ the curves $h \circ \beta_n(S)$ and $\eta \beta_n(S) \eta^{-1}$ are homotopic with fixed base point $p(S)$. Since $\beta_1(S)$ and $h \circ \beta_1(S)$ are smooth closed geodesics this is only possible if η is homotopic (with $p(S)$ fixed) to β_1^m for some m. For the same reason η is homotopic to β_2^k for some k. We conclude that $m = k = 0$ and that η is a trivial loop. Hence, $h \circ \beta_n(S)$ and $\beta_n(S)$ are homotopic with fixed base point. By the uniqueness of geodesic loops in their homotopy classes, we have $h \circ \beta_n(S) = \beta_n(S)$ for $n = 1, \ldots, 2g$. ◇

We investigate the geodesic polygon P_S which results if S is cut open along the normal canonical dissection. Both the sides and the side lengths of P_S are denoted by $b_n = b_n(S)$, $\bar{b}_n = \bar{b}_n(S)$ (cf. Fig. 6.7.1). The lengths are

$$b_n = \bar{b}_n = \ell(\beta_n(S)), \quad n = 1, \ldots, 2g.$$

The interior angle between side b_n (respectively $\bar{b}_n$) and the subsequent side is denoted by $\zeta_n = \zeta_n(S)$, (respectively by $\bar{\zeta}_n = \bar{\zeta}_n(S)$). Figure 6.7.3 shows how these angles come together at $p(S)$ in S.

Every P_S has the property that the interior angles sum up to 2π and that $\bar{\zeta}_1 + \zeta_2 = \zeta_2 + \bar{\zeta}_1 = \pi$. By Corollary 1.4.3, P_S may be considered a geodesic

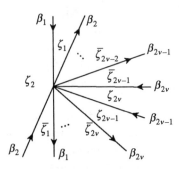

Figure 6.7.3

polygon in the hyperbolic plane. This leads to the following.

6.7.5 Definition. Let $g \geq 2$ and let P be a geodesic polygon in the hyperbolic plane with sides $b_1, \ldots, \bar{b}_{2g}$ and angles $\zeta_1, \ldots, \bar{\zeta}_{2g}$ labelled as above. P is called a *canonical polygon* if
  (i) $b_n = \bar{b}_n$, $n = 1, \ldots, 2g$,
  (ii) $\zeta_1 + \ldots + \bar{\zeta}_{2g} = 2\pi$.
A canonical polygon is called *normal* if in addition
  (iii) $\zeta_1 + \zeta_2 = \bar{\zeta}_1 + \bar{\zeta}_2 = \pi$.
Two normal canonical polygons P with sides $b_1, \ldots, \bar{b}_{2g}$ and P' with sides $b'_1, \ldots, \bar{b}'_{2g}$ are considered *equivalent* if there exists an isometry $P \to P'$ which sends b_1 to b'_1 and b_2 to b'_2.

Note that any normal canonical polygon is equivalent to one for which the cycle of sides $b_1, \ldots, \bar{b}_{2g}$ along the boundary is positively oriented. In the figures, the polygons will always be drawn in this way.

6.7.6 Definition. (i) For every $g \geq 2$, $\mathcal{P}_g$ denotes the set of all equivalence classes of normal canonical polygons with $4g$ sides.
 (ii) For every $S = (S, \varphi) \in \mathcal{T}_g$ we have the normal canonical polygon P_S obtained by cutting S open along the normal canonical dissection $\beta_1(S), \ldots, \beta_{2g}(S)$ given by Theorem 6.7.4. Denoting this polygon by $\pi(S)$ we obtain a mapping $\pi : \mathcal{T}_g \to \mathcal{P}_g$.

6.7.7 Theorem. $\pi : \mathcal{T}_g \to \mathcal{P}_g$ *is one-to-one and onto.*

Proof. If we paste the sides of $P \in \mathcal{P}_g$ together as in (6.7.1), we obtain a compact Riemann surface, and from this we easily see that π is onto.
 Now let $S = (S, \varphi)$, $S' = (S', \varphi') \in \mathcal{T}_g$ with $\pi(S) = \pi(S')$. By Theorem

6.7.4, we may choose φ and φ' within their isotopy classes in such a way that $\varphi \circ \beta_n = \beta_n(S)$ and $\varphi' \circ \beta_n = \beta_n(S')$ for $n = 1, \ldots, 2g$, where $\beta_1, \ldots, \beta_{2g}$ is the canonical dissection of the base surface F. By hypothesis there exists an isometry $j : S \to S'$ satisfying $j(\beta_n(S)) = \beta_n(S')$, $n = 1, \ldots, 2g$. It follows that $\varphi^{-1} \circ j^{-1} \circ \varphi'$ is a homeomorphism of F which fixes $\beta_1, \ldots, \beta_{2g}$ pointwise. Since $F - (\beta_1 \cup \ldots \cup \beta_{2g})$ is a disk, Alexander's theorem (Theorem A.2) implies that $\varphi^{-1} \circ j^{-1} \circ \varphi'$ is isotopic to the identity. Hence, φ' and $j \circ \varphi$ are isotopic and thus (S, φ) and (S', φ') are marking equivalent. This proves that π is one-to-one. ◇

In Z-V-C [1, 2] it is shown by a direct investigation of the geometry of the polygons that $\mathcal{P}_g$ has a natural topology in which $\mathcal{P}_g$ is homeomorphic to $\mathbf{R}^{6g-6}$. We shall obtain this result indirectly via Theorem 6.7.7 and via the following coordinates which are essentially those used in Z-V-C [1, 2].

6.7.8 Definition. For any $P \in \mathcal{P}_g$ with sides $b_1, \ldots, \bar{b}_{2g}$ and interior angles $\zeta_1, \ldots, \bar{\zeta}_{2g}$ as above, we define

$$Z(P) = (b_3, \ldots, b_{2g}, \zeta_3, \ldots, \zeta_{2g}, \bar{\zeta}_3, \ldots, \bar{\zeta}_{2g}).$$

For $S \in \mathcal{T}_g$ we define $\Pi(S) = Z(\pi(S))$. Finally we set

$$\mathcal{Z}_g = \{ Z(P) \mid P \in \mathcal{P}_g \} = \{ \Pi(S) \mid S \in \mathcal{T}_g \}.$$

$Z(P)$ and $\Pi(S)$ are the sequences of *Zieschang-Vogt-Coldewey (Z-V-C) coordinates* of P and S.

The proof that the mapping Z is one-to-one will be given together with the proof of Theorem 6.8.13.

6.7.9 Lemma. *The mapping* $\Pi : \mathcal{T}_g \to \mathcal{Z}_g \subset \mathbf{R}^{6g-6}$ *is real analytic.*

Proof. We prove this in a particular model $\mathcal{T}_G$. We take G such that on the base surface F the geodesic γ_2 cuts off a surface Q of signature $(1, 1)$ carrying γ_1 in the interior. Then we set $\beta_1 = \gamma_1$ and let β_2 be the shortest simple closed geodesic in Q which intersects β_1 in exactly one point p_0. (Fig. 6.7.4). Since the twist parameter at γ_1 is 0, we have $p_0 = \gamma_1(\frac{1}{4})$ (cf. the discussion at the end of Section 3.4).

Now extend β_1, β_2 to a normal canonical dissection $\beta_1, \beta_2, \ldots, \beta_{2g}$ of F. By Corollary 6.3.6, the marking homeomorphism $\varphi^\omega : F \to S^\omega$, $S^\omega \in \mathcal{T}_G$, has the property that $\varphi^\omega(p_0) = \gamma_1(S^\omega)(\frac{1}{4} + \alpha_1)$, where $\omega = (\ell_1, \ldots, \alpha_{3g-3})$. We use the definition of φ^ω (Definition 6.2.5) to check that $\beta_2(S^\omega)$ (the closed geodesic in the free homotopy class of $\varphi^\omega(\beta_2)$) intersects the geodesic

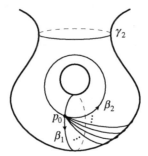

Figure 6.7.4

$\beta_1(S^\omega) = \gamma_1^\omega$ at the point $\gamma_1^\omega(\frac{1}{4} + \alpha_1) = \varphi^\omega(p_0)$ and, moreover, that $\varphi^\omega(\beta_2)$ is homotopic *with fixed base point* to $\beta_2(S^\omega)$. In the proof of Theorem 6.7.4 we have shown that under these conditions there exists a 1-homeomorphism $h : S^\omega \to S^\omega$ fixing $\beta_1(S^\omega)$ pointwise and transforming $\varphi^\omega(\beta_1), \ldots, \varphi^\omega(\beta_{2g})$ into the normal canonical dissection $\beta_1(S^\omega), \ldots, \beta_{2g}(S^\omega)$. By (1) and (8) in the proof of Theorem 6.3.5, the lengths of the geodesic loops $\beta_3(S^\omega), \ldots, \beta_{2g}(S^\omega)$ and the angles they form with $\gamma_1(S^\omega)$ are analytic functions of ω. Since $\zeta_3, \ldots, \bar{\zeta}_{2g}$ are sums and differences of these angles, this proves the lemma. ◇

The mapping $\Pi : \mathcal{T}_g \to \mathcal{L}_g$ is in fact a bianalytic diffeomorphism. This will be shown in the next section via Bers' coordinates.

6.8 Fuchsian Groups and Bers' Coordinates

The coordinates of Bers [1] are adapted to the description of compact Riemann surfaces in the form $S = \Gamma\backslash \mathbf{H}$, where Γ is a Fuchsian group. The aim of this section is to prove the real analytic equivalence of Bers' coordinates with those of Zieschang-Vogt-Coldewey (Z-V-C) and Fenchel-Nielsen. For this purpose we only consider a very restricted type of Fuchsian group. We work in the upper half-plane $\mathbf{H}$. Recall that $\mathrm{Is}^+(\mathbf{H}) = \mathrm{PSL}(2, \mathbf{R})$.

6.8.1 Definition. A *Fuchsian group of genus g* is a discrete subgroup $\Gamma \subset \mathrm{PSL}(2, \mathbf{R})$ for which the quotient $\Gamma\backslash \mathbf{H}$ is a compact Riemann surface of genus g.

We fix $g \geq 2$. "Fuchsian group" is always taken to mean Fuchsian group of genus g.

For any Fuchsian group Γ we define *markings* as follows. Since $\Gamma\backslash \mathbf{H}$

is a compact Riemann surface, it has normal canonical dissections. Let β_1, $\beta_2, \ldots, \beta_{2g}$ be one. These loops are generators of the fundamental group and satisfy the relation

$$\prod_{v=1}^{g} \beta_{2v-1} \beta_{2v} \beta_{2v-1}^{-1} \beta_{2v}^{-1} = 1.$$

Consider the canonical projection $\mathbf{H} \to \Gamma\backslash\mathbf{H}$ (a universal covering) and let $p \in \mathbf{H}$ be an inverse image of the base point p_0 of the dissection $\beta_1, \ldots, \beta_{2g}$. For $n = 1, \ldots, 2g$ we have geodesic arcs b'_n of length $\ell(b'_n) = \ell(\beta_n)$, which are lifts of β_n and emanating from p. Denote by $B_n \in \Gamma$ the unique element which maps p to the endpoint of b'_n. Then B_n is the image of β_n under the canonical isomorphism of the fundamental group of $\Gamma\backslash\mathbf{H}$ and Γ. In particular, $B_1, \ldots, B_{2g}$ are generators of Γ and satisfy the relation

(6.8.2) $$\prod_{v=1}^{g} B_{2v-1} B_{2v} B_{2v-1}^{-1} B_{2v}^{-1} = 1.$$

We say that an ordered set $B_1, \ldots, B_{2g}$ is a *canonical* set of generators of Γ if *it arises from a construction as above*. This definition is convenient for our purposes but less satisfactory from the point of view of Fuchsian groups. We mention therefore without proof that any ordered set of generators of a Fuchsian group of genus g satisfying (6.8.2) is canonical. This follows from the Dehn-Nielsen theorem which states that every automorphism of the fundamental group of $\Gamma\backslash\mathbf{H}$ is induced by a homeomorphism $\Gamma\backslash\mathbf{H} \to \Gamma\backslash\mathbf{H}$ (for a proof of this theorem refer to Z-V-C [2] Sections 3.3 and 5.6).

The following corresponds to the marking of surfaces.

6.8.3 Definition. A Fuchsian group of genus g together with an ordered set $(B_1, \ldots, B_{2g})$ of canonical generators is called a *marked Fuchsian group*. Two such groups Γ and Γ' marked with $(B_1, \ldots, B_{2g})$ and $(B'_1, \ldots, B'_{2g})$ are called *marking equivalent* if they are conjugate in Is($\mathbf{H}$) by an element A satisfying $AB_nA^{-1} = B'_n$, $n = 1, \ldots, 2g$. The set of all marking equivalence classes of Fuchsian groups of genus g is denoted by $\mathscr{F}_g$.

A variant of this definition which amounts to the same would be to call marked groups *marking equivalent* if they are conjugate in Is$^+$($\mathbf{H}$) = PSL(2, $\mathbf{R}$) instead of Is($\mathbf{H}$), and then to define $\mathscr{F}_g$ as the identity component of the set of all marking equivalence classes (with respect to the natural topology).

We are approaching Bers' coordinates. Recall that each $B \in \Gamma$ is a hyperbolic transformation which is uniquely determined by its axis (the oriented invariant geodesic) and the *displacement length* (the distance dist(z, Bz) for any point z on the axis). The axis will always be considered oriented with the orientation "from z to Bz" for any z on the axis.

The following is clear (observe that we allow conjugation in Is(**H**) and not just in Is$^+$(**H**)).

6.8.4 Lemma. *In any marking equivalence class of a Fuchsian group of genus g there exists a unique representative Γ with canonically generators $B_1, \ldots, B_{2g}$ such that the following hold.*

  (i) *The axis of B_1 is the imaginary axis oriented from ∞ to 0.*
  (ii) *The axis of B_2 intersects the imaginary axis at $z = i$.*
  (iii) *The orientation of the axis of B_2 is such that it crosses the imaginary axis from $\mathcal{R}\wp z < 0$ to $\mathcal{R}\wp z > 0$.* ◊

Canonical generators which satisfy (i) - (iii) of Lemma 6.8.4 will be called *normal canonical generators*. In the following the elements of $\mathcal{F}_g$ will always be represented by a group Γ and a set of normal canonical generators $B_1, \ldots, B_{2g}$. By abuse of notation we shall say that $\Gamma \in \mathcal{F}_g$.

With every generator B_n are associated three real numbers: σ_n, τ_n and λ_n. The numbers σ_n and τ_n are on the extended real line and are the endpoints at infinity of the axis of B_n. The notation is such that the axis is oriented from σ_n to τ_n. The number λ_n is defined as $\lambda_n := e^{\ell_n}$, where ℓ_n is the displacement length of B_n. Observe that $\sigma_1 = \infty$, $\tau_1 = 0$, $\sigma_2 < 0$ and $\tau_2 > 0$.

Note that for any $z \in \mathbf{H}$, $B_n^k(z) \to \tau_n$ (convergence in **C**) as $k \to \infty$ and $B_n^k(z) \to \sigma_n$ as $k \to -\infty$. The points σ_n and τ_n are called the *repulsive* and the *attractive* fixed points (at infinity) of B_n, $n = 1, \ldots, 2g$.

6.8.5 Definition. We let $B : \mathcal{F}_g \to \mathbf{R}^{6g-6}$ be the mapping which to every $\Gamma \in \mathcal{F}_g$ (together with the choice of normal canonical generators as above) associates the ordered set $B(\Gamma) := (\sigma_3, \ldots, \sigma_{2g}, \tau_3, \ldots, \tau_{2g}, \lambda_3, \ldots, \lambda_{2g})$. The range of B is denoted by $\mathcal{B}_g$:

$$\mathcal{B}_g = \{B(\Gamma) \mid \Gamma \in \mathcal{F}_g\}.$$

$B(\Gamma)$ is the sequence of *Bers coordinates* of Γ.

The proof that the mapping B is one-to-one will be given together with the proof of Theorem 6.8.13.

For $P \in \mathcal{P}_g$, a marked Fuchsian group Γ_P which has the form as described in Lemma 6.8.4 can be constructed in an obvious way: paste the sides of each pair b_n, $\bar{b}_n$ together as in (6.7.1) to obtain the compact Riemann surface S with the corresponding normal canonical generators $\beta_1(S), \ldots, \beta_{2g}(S)$ of the fundamental group. Then take the unique universal covering

(6.8.6) $$h : \mathbf{H} \to S$$

for which the closed geodesics $\beta_1(S), \beta_2(S)$ have lifts $\tilde{\beta}_1, \tilde{\beta}_2$ with the following properties.

(i) $\tilde{\beta}_1(S)$ *is the imaginary axis oriented from ∞ to 0,*
(ii) $\tilde{\beta}_2(S)$ *intersects $\tilde{\beta}_1(S)$ at $z = i$,*
(iii) *The orientation of $\tilde{\beta}_2(S)$ is such that it crosses the imaginary axis from $\mathcal{R}ez < 0$ to $\mathcal{R}ez > 0$.*

We let Γ_P be the covering transformation group with respect to h and let $B_1, \ldots, B_{2g}$ be the generators of Γ_P which correspond to $\beta_1(S), \ldots, \beta_{2g}(S)$ under h.

We then have $S = \Gamma \backslash \mathbf{H}$, and the $B_1, \ldots, B_{2g}$ are normal canonical generators arising from $\beta_1(S), \ldots, \beta_{2g}(S)$ as described a few lines prior to (6.8.2).

6.8.7 Definition. For $P \in \mathcal{P}_g$ we denote by $\psi(P) \in \mathcal{F}_g$ the marking equivalence class represented by the above Γ_P with the generators $B_1, \ldots, B_{2g}$.

6.8.8 Lemma. $\psi : \mathcal{P}_g \to \mathcal{F}_g$ *is one-to-one and onto.*

Proof. Let $\Gamma \in \mathcal{F}_g$ with generators $B_1, \ldots, B_{2g}$ be as in Lemma 6.8.4. For $n = 1, \ldots, 2g$ we let b'_n be the geodesic arc in $\mathbf{H}$ from $z = i$ to $B_n(i)$. On $S = \Gamma \backslash \mathbf{H}$ the images of $b'_1, \ldots, b'_{2g}$ form a canonical dissection. Cut S open along this dissection to obtain a polygon $P \in \mathcal{P}_g$ (here we are using Corollary 1.4.3) and check that Γ is a conjugate in $\mathrm{Is}(\mathbf{H})$ of the above representative Γ_P of $\psi(P)$. Hence, we see that ψ is onto. Since the sides and angles of P can be read off from the geodesic arcs which connect i to $B_n(i)$, $n = 1, \ldots, 2g$, ψ is one-to-one. ◇

In conjunction with Theorem 6.7.7 we have the following bijective mappings, where τ denotes $(\psi \circ \pi)^{-1}$

(6.8.9) $$\mathcal{T}_g \xrightarrow{\pi} \mathcal{P}_g \xrightarrow{\psi} \mathcal{F}_g \xrightarrow{\tau} \mathcal{T}_g.$$

The mapping τ may also be described as follows. Let $S = \Gamma \backslash \mathbf{H}$ and let $\beta_1(S), \ldots, \beta_{2g}(S)$ be the canonical dissection of S, where $\beta_n(S)$ is the image of b'_n as described in the proof of Lemma 6.8.8. The surface S after having been cut open along $\beta_1(S), \ldots, \beta_{2g}(S)$ is a topological disk as is the base surface F cut open along the canonical dissection $\beta_1, \ldots, \beta_{2g}$ introduced at the beginning of Section 6.7. We have therefore a homeomorphism $\varphi : F \to S$ satisfying $\varphi \circ \beta_n = \beta_n(S)$ for $n = 1, \ldots, 2g$. By Alexander's theorem (Theorem A.2), the isotopy class of φ is uniquely determined by this condition, and we can check that indeed $(S, \varphi) = \tau(\Gamma)$.

For better reference we note the following properties of τ resulting from this description.

6.8.10 Remark. Let $\Gamma \in \mathcal{F}_g$ be given with normal canonical generators $B_1, \ldots, B_{2g}$ (properties (i) - (iii) of Lemma 6.8.4), and let $S = (S, \varphi) = \tau(\Gamma)$, where the choice of the marking homeomorphism φ in its isotopy class is such that $\varphi \circ \beta_1, \ldots, \varphi \circ \beta_{2g}$ is a normal canonical dissection (Theorem 6.7.4), say with base point p. Then $S = \Gamma \backslash \mathbf{H}$. The universal covering map $\mathbf{H} \to \Gamma \backslash \mathbf{H}$ maps the axes of B_1 and B_2 onto $\varphi \circ \beta_1$ and $\varphi \circ \beta_2$ respectively. The natural isomorphism between the fundamental group $\pi_1(S, p)$ and Γ sends $\varphi \circ \beta_n$ to B_n, $n = 1, \ldots, 2g$.

6.8.11 Definition. With (6.8.9) and with the mapping $B : \mathcal{F}_g \to \mathcal{B}_g \subset \mathbf{R}^{6g-6}$ as in Definition 6.8.5, we define $B \circ \tau^{-1}(S)$ to be the *Bers coordinates* of $S \in \mathcal{T}_g$.

In the next lemma $Z : \mathcal{P}_g \to \mathcal{L}_g \subset \mathbf{R}^{6g-6}$ is the Z-V-C coordinate mapping as in Definition 6.7.8.

6.8.12 Lemma. *There exists a real analytic mapping* $\Psi : \mathcal{L}_g \to \mathcal{B}_g$ *such that* $B \circ \psi = \Psi \circ Z$.

Proof. Take $P \in \mathcal{P}_g$ with sides $b_1, b_2, \bar{b}_1, \bar{b}_2, \ldots, \bar{b}_{2g}$ and angles $\zeta_1, \zeta_2, \bar{\zeta}_1, \bar{\zeta}_2, \ldots, \bar{\zeta}_{2g}$. Consider the marked Fuchsian group $\Gamma = \psi(P) \in \mathcal{F}_g$ marked by the generators $B_1, \ldots, B_{2g}$ and let $S = \Gamma \backslash \mathbf{H}$ be the corresponding quotient surface as above. We want to compute $B_1, \ldots, B_{2g}$ in terms of the coordinates $Z(P) = (b_3, \ldots, b_{2g}, \zeta_3, \ldots, \zeta_{2g}, \bar{\zeta}_3, \ldots, \bar{\zeta}_{2g})$.

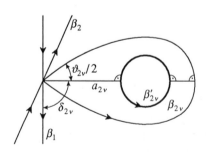

Figure 6.8.1

For $v = 2, \ldots, g$ the geodesic loops β_{2v-1}, β_{2v} on S which correspond to the generators B_{2v-1}, B_{2v} and to the sides b_{2v-1}, b_{2v} have angles $\vartheta_{2v-1}, \vartheta_{2v}$ given by

(1) $$\vartheta_{2v-1} = \zeta_{2v} + \bar\zeta_{2v-1}, \quad \vartheta_{2v} = \zeta_{2v-1} + \zeta_{2v}.$$

This can be seen in Fig. 6.7.3. For $n = 3, \ldots, 2g$ we let β'_n be the closed geodesic in the free homotopy class of β_n. Then β_n and β'_n bound an annulus which is the union of two isometric trirectangles with sides $a_n, \tfrac{1}{2}\beta_n, \bullet, \tfrac{1}{2}\beta'_n$ and acute angle $\tfrac{1}{2}\vartheta_n$ between a_n and $\tfrac{1}{2}\beta_n$ (lift the curves to $\mathbf{H}$ to obtain a configuration as in Fig. 6.3.2 for the proof). In Fig. 6.7.3 we also see that the angle δ_n between β_1 and a_n is given as follows.

(2) $$\begin{aligned}\delta_{2v-1} &= -\tfrac{1}{2}\vartheta_{2v-1} + \sum_{k=2v-1}^{2g}(\zeta_k + \bar\zeta_k), \\ \delta_{2v} &= \delta_{2v-1} - \tfrac{1}{2}\zeta_{2v-1} - \tfrac{1}{2}\bar\zeta_{2v-1}.\end{aligned} \quad v = 2, \ldots, g,$$

Formulae (iii) and (ii) of Theorem 2.3.1 applied to the trirectangles yield

(3) $$\begin{aligned}\cosh\tfrac{1}{2}\ell_n &= \cosh\tfrac{1}{2}b_n \sin\tfrac{1}{2}\vartheta_n, \\ \tanh a_n &= \coth\tfrac{1}{2}b_n \cos\tfrac{1}{2}\vartheta_n,\end{aligned} \quad n = 3, \ldots, 2g,$$

where $\ell_n = \ell(\beta'_n)$ is the displacement length of B_n, i.e.

(4) $$\lambda_n = e^{\ell_n}.$$

Consider now the universal covering $h : \mathbf{H} \to S$ as in (6.8.6) and let $\tilde\beta_n$ be the axis of B_n. Then $h(\tilde\beta_n) = \beta'_n$, $n = 1, \ldots, 2g$. We recall that $\tilde\beta_1$ is the imaginary axis oriented from ∞ to 0 and $\tilde\beta_2$ intersects $\tilde\beta_1$ at $z = i$. The perpendicular $\tilde a_n$ from i to $\tilde\beta_n$ is a lift of a_n, $n = 3, \ldots, 2g$ and forms an angle δ_n with $\tilde\beta_1$ (Fig. 6.8.2). Since an arbitrarily small homotopy can remove the intersec-

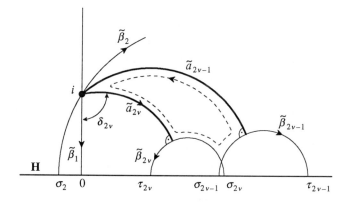

Figure 6.8.2

tion of β_1 and β_n on S ($n \geq 3$), Theorem 1.6.7 implies that β_1 and β'_n are disjoint and therefore $\tilde{\beta}_1$ and $\tilde{\beta}_n$ are disjoint. It follows that the endpoints σ_n, τ_n at infinity of $\tilde{\beta}_n$ satisfy $\sigma_n > 0$, $\tau_n > 0$. Note that $\tilde{\beta}_n$ is oriented from σ_n to τ_n so that $\sigma_n > \tau_n$ if n is even and $\sigma_n < \tau_n$ if n is odd.

We are ready to compute σ_n and τ_n, for $n = 3, \ldots, 2g$. Recall that for $\delta \in \mathbf{R}$ the element $R_\delta \in \mathrm{SL}(2, \mathbf{R})$ given by

$$R_\delta = \begin{pmatrix} \cos \delta/2 & \sin \delta/2 \\ -\sin \delta/2 & \cos \delta/2 \end{pmatrix}$$

acts on $\mathbf{H}$ as a rotation with fixed point i and with angle of rotation δ. If $a > 0$, then the geodesic, with endpoints at infinity $-e^{-a}$ and e^{-a}, intersects $\tilde{\beta}_1$ orthogonally at $z = i\,e^{-a}$ and consequently for $n = 3, \ldots, 2g$

(5) $\qquad \sigma_n = R_{\delta_n}(\varepsilon_n\, e^{-a_n}), \qquad \tau_n = R_{\delta_n}(-\varepsilon_n\, e^{-a_n}),$

where $\varepsilon_n = 1$ if n is even and $\varepsilon_n = -1$ if n is odd. Formulae (1) - (5) define a real analytic mapping $\Psi: \mathcal{B}'_g \to \mathbf{R}^{6g-6}$ for some neighborhood $\mathcal{B}'_g \subset \mathbf{R}^{6g-6}$ which contains $\mathcal{B}_g$, and where Ψ restricted to $\mathcal{B}_g$ satisfies the requirement of Lemma 6.8.12. ◊

Finally we prove the following theorem in which θ is defined by the statement of the theorem.

6.8.13 Theorem. *The diagram*

$$\begin{array}{ccccccc} \mathcal{T}_g & \xrightarrow{\pi} & \mathcal{P}_g & \xrightarrow{\psi} & \mathcal{F}_g & \xrightarrow{\tau} & \mathcal{T}_g \\ & \searrow{\Pi} & \downarrow{Z} & & \downarrow{B} & \nearrow{\theta} & \\ & & \mathcal{L}_g & \xrightarrow{\Psi} & \mathcal{B}_g & & \end{array}$$

is commutative. $\mathcal{L}_g$ and $\mathcal{B}_g$ are open subsets of $\mathbf{R}^{6g-6}$ and are homeomorphic to $\mathbf{R}^{6g-6}$. All mappings are one-to-one and onto. The mappings Π, Ψ and θ are real analytic diffeomorphisms, and $\theta \circ \Psi \circ \Pi = \mathrm{id}$.

Proof. We first show that B is one-to-one by computing the generators $B_1, \ldots, B_{2g}$ of $\Gamma \in \mathcal{F}_g$ (in the normal form of Lemma 6.8.4). We compute these generators in terms of the Bers coordinates. For this we write each B_n as a matrix in $\mathrm{SL}(2, \mathbf{R})$. This matrix is determined up to a factor ± 1. To make it unique we add the *additional condition* that the trace of the matrix be non-negative.

For $n = 3, \ldots, 2g$ we easily check that

$$(1) \quad B_n = \frac{1}{(\tau_n - \sigma_n)\sqrt{\lambda_n}} \begin{pmatrix} \tau_n \lambda_n - \sigma_n & -\sigma_n \tau_n (\lambda_n - 1) \\ \lambda_n - 1 & -\sigma_n \lambda_n + \tau_n \end{pmatrix}.$$

For instance, to see that the axis of B_n is indeed oriented from σ_n to τ_n, we compute the derivative f' of the holomorphic function $z \to f(z) = B_n(z)$. We obtain $f'(\sigma_n) = \lambda_n$, $f'(\tau_n) = 1/\lambda_n$. Since $\lambda_n = e^{\ell_n} > 1$, σ_n is the repulsive and τ_n the attractive fixed point of B_n.

For $n = 1, 2$ the B_n assume the more special forms

$$(2) \quad B_1 = \begin{pmatrix} \lambda_1^{-1/2} & 0 \\ 0 & \lambda_1^{1/2} \end{pmatrix}, \quad B_2 = \frac{1}{(1 + \tau_2^2)\sqrt{\lambda_2}} \begin{pmatrix} \tau_2^2 \lambda_2 + 1 & \tau_2(\lambda_2 - 1) \\ \tau_2(\lambda_2 - 1) & \lambda_2 + \tau_2^2 \end{pmatrix}.$$

For $n \geq 3$ the components of B_n are in terms of the Bers coordinates. It suffices therefore to find an expression for B_1 and B_2 in terms of $B_3, \ldots, B_{2g}$. Now Γ is a subgroup of PSL(2, **R**) whose generators satisfy the relation (6.8.2). However, in the present case we interpret $B_1, \ldots, B_{2g}$ as elements of SL(2, **R**) and can only say that

$$\prod_{v=1}^{g} [B_{2v-1}, B_{2v}] = J,$$

where $J = \varepsilon \begin{pmatrix} 1 & 0 \\ 0 & 1 \end{pmatrix}$ with $\varepsilon = \pm 1$. Here [,] again is the commutator

$$[A, B] = ABA^{-1}B^{-1}.$$

As S varies through $\mathcal{T}_g$, the sides and angles of the polygon $\pi(S)$ and hence the components of the generators of $\psi \circ \pi(S)$ are continuous functions. The same is true for the components of J. It follows that $\varepsilon = \varepsilon_g \in \{1, -1\}$ is a constant which depends only on the genus. Below we shall prove that $\varepsilon_g = 1$. For the present proof the value of ε_g is not important. We abbreviate

$$(3) \quad [B_1, B_2] = \varepsilon_g \Big(\prod_{v=2}^{g} [B_{2v-1}, B_{2v}]\Big)^{-1} =: \begin{pmatrix} \omega_1 & \omega_2 \\ \omega_3 & \omega_4 \end{pmatrix}$$

and compute the components of B_1 and B_2 in terms of $\omega_1, \ldots, \omega_4$. For this we set

$$(4) \quad B_2 = \begin{pmatrix} a & b \\ b & d \end{pmatrix},$$

where we know from above that $b \neq 0$. The commutator becomes

$$(5) \quad [B_1, B_2] = \begin{pmatrix} 1 + b^2(1 - \frac{1}{\lambda_1}) & ab(\frac{1}{\lambda_1} - 1) \\ bd(\lambda_1 - 1) & 1 + b^2(1 - \lambda_1) \end{pmatrix}.$$

We now find

(6)
$$\lambda_1 = \frac{1-\omega_4}{\omega_1-1}, \qquad b = \left(\frac{1-\omega_4}{\lambda_1-1}\right)^{1/2}$$
$$a = \frac{\lambda_1 \omega_2}{b(1-\lambda_1)}, \qquad d = \frac{\omega_3}{b(\lambda_1-1)}.$$

The relations (1) - (6) define a real analytic mapping $A : \mathcal{B}'_g \to (SL(2,\mathbf{R}))^{2g}$ on a neighborhood $\mathcal{B}'_g$ containing $\mathcal{B}_g$, such that the restriction of A to $\mathcal{B}_g$ is an inverse of B (B is as in Theorem 6.8.13 and in Definition 6.8.5).

Setting $\theta := \tau \circ B^{-1}$, it follows that all mappings in the diagram of Theorem 6.8.13 are bijections. By the definition of τ we have $\tau \circ \psi \circ \pi = id$ and therefore $\theta \circ \Psi \circ \Pi = id$.

If Γ is a marked Fuchsian group with generators $B_1, \ldots, B_{2g}$ in the normal form of Lemma 6.8.4, and if C is any product (= word) of the generators, then the closed geodesic γ on the surface $S = \tau(\Gamma) = \Gamma\backslash\mathbf{H}$ corresponding to C has length

$$\ell(\gamma) = 2 \operatorname{arccosh}\{\tfrac{1}{2}|\operatorname{tr} C|\}.$$

If the word C is fixed and the parameters vary through $\mathcal{B}_g$, then either everywhere tr $C > 2$ or everywhere tr $C < -2$. Using the mapping A we see that $\ell(\gamma)$ is given by an analytic function which is defined on an open neighborhood containing $\mathcal{B}_g$. Now we take a canonical curve system Ω_G, identify $\mathcal{T}_g$ with $\mathcal{T}_G$ and apply this remark to the words which define the elements of Ω_G. Thus, there exists an open neighborhood $\mathcal{B}''_g$ with $\mathcal{B}_g \subset \mathcal{B}''_g \subset \mathcal{B}'_g$ and a real analytic mapping $L_G : \mathcal{B}''_g \to \mathbf{R}^{9g-9}$ such that $L_G(y) = \ell\Omega_G(\tau \circ B^{-1}(y))$ for $y \in \mathcal{B}_g$. For $S = \tau \circ B^{-1}(y)$, $y \in \mathcal{B}_g$, the Fenchel-Nielsen parameters $\omega_G(S)$ are now given by $\omega_G(\tau \circ B^{-1}(y)) = \mathcal{A}_G(\ell\Omega_G(S)) = \mathcal{A}_G(L_G(y))$, where $\mathcal{A}_G$ is the analytic mapping as in Lemma 6.3.4. Hence, $\omega_G \circ \tau \circ B^{-1} : \mathcal{B}_g \to \mathbf{R}^{6g-6}$ is the restriction to $\mathcal{B}_g$ of an analytic mapping defined on $\mathcal{B}''_g$. Since $\omega_G \circ \tau \circ B^{-1} \circ \Psi \circ \Pi = \omega_G$, the analytic mappings Ψ (Lemma 6.8.12) and Π (Lemma 6.7.9) are everywhere non-degenerate. Therefore, $\mathcal{T}_g \subset \mathbf{R}^{6g-6}$ and $\mathcal{B}_g \subset \mathbf{R}^{6g-6}$ are open and Ψ and Π are analytic diffeomorphisms. Theorem 6.8.13 is now proved. ◊

If we write the above generators B_n in the form

$$B_n = \begin{pmatrix} a_n & b_n \\ c_n & d_n \end{pmatrix},$$

then, for $n = 3, \ldots, 2g$, the components a_n, c_n and d_n are analytic functions of λ_n, σ_n and τ_n given by (1), and we have $c_n \neq 0$. Conversely, the Bers coordinates are determined by these components via the following equations, where we recall that the matrices B_n have positive traces for $n = 1, \ldots, 2g$.

(6.8.14)
$$\lambda_n^{1/2} = \tfrac{1}{2}(a_n + d_n + \sqrt{(a_n + d_n)^2 - 4})$$
$$\tau_n = \frac{1}{c_n}(\lambda_n^{1/2} - d_n), \quad \sigma_n = \frac{1}{c_n}(\lambda_n^{-1/2} - d_n).$$
$n = 3, \ldots, 2g,$

The coordinates $(a_3, c_3, d_3, \ldots, a_{2g}, c_{2g}, d_{2g})$ are therefore real analytically equivalent to the Bers coordinates. They are called the *Fricke coordinates* of the Fuchsian group with generators $B_1, \ldots, B_{2g}$.

We conclude the section with a remark about the defining relation (6.8.2) for the case that the B_n are elements of SL(2, **R**) and not of PSL(2, **R**) as in (6.8.2). For this remark we also refer to Seppälä-Sorvali [3] and Kra [1].

6.8.15 Lemma. *Let $B_1, \ldots, B_{2g} \in$ SL(2, **R**) represent canonical generators of a Fuchsian group of genus g. Then*

$$\prod_{v=1}^{g} [B_{2v-1}, B_{2v}] = \begin{pmatrix} 1 & 0 \\ 0 & 1 \end{pmatrix}$$

Proof. (Sketch). Let Γ be the Fuchsian group, let $S = \Gamma \backslash \mathbf{H}$ and let $\beta_1, \ldots, \beta_{2g}$ be a canonical dissection of S corresponding to the generators of Γ which are represented by $B_1, \ldots, B_{2g}$. From the intersection properties of the curves in the dissection (cf. Fig. 6.7.2) we have the following. If we let the closed curves $\beta_1, \ldots, \beta_{2g}$ and $[\beta_1, \beta_2], \ldots, [\beta_{2g-1}, \beta_{2g}]$ vary through their free homotopy classes, we get simple closed geodesics $\beta_1', \ldots, \beta_{2g}', \eta_1, \eta_3, \eta_5, \ldots, \eta_{2g-1}$ with the following properties (cf. Fig. 6.8.3 in which k stands for v).

The g geodesics $\eta_1, \eta_3, \ldots, \eta_{2g-1}$ cut away hyperbolic surfaces $Q_1, Q_3, \ldots, Q_{2g-1}$ of signature $(1, 1)$ which are pairwise disjoint except for $g = 2$, where Q_1 and Q_3 meet each other along their boundaries. For $v = 1, \ldots, g$,

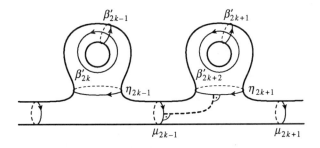

Figure 6.8.3

the closed geodesics β'_{2v-1} and β'_{2v} are contained in Q_{2v-1} and intersect each other in exactly one point (Fig. 6.8.3). We make two claims.

(a) $\quad\quad\quad\quad \text{tr}\,[B_{2v-1}, B_{2v}] < -2, \quad v = 1, \ldots, g,$

(b) $\quad\quad\quad\quad \text{tr}\,\prod_{v=1}^{k}[B_{2v-1}, B_{2v}] < -2, \quad k = 1, \ldots, g-1.$

Since $B_1, \ldots, B_{2g}$ represent canonical generators we have

$$\prod_{v=1}^{g-1}[B_{2v-1}, B_{2v}] = \varepsilon([B_{2g-1}, B_{2g}])^{-1},$$

where $\varepsilon = \pm 1$. By (a) and (b), the matrices on either side have negative trace, hence, $\varepsilon = 1$. The lemma is thus reduced to the proof of (a) and (b).

To prove (a) we may restrict ourselves to $v = 1$ (the generators need not be normal so that we may use cyclic permutation). Conjugate the group in SL(2, **R**) such that up to a factor ± 1 (which will drop out), B_1 and B_2 assume the form (6.8.13(2)) with $\lambda_1 > 1$ and $\lambda_2 > 1$. In contrast to Lemma 6.8.4(iii), the axis of B_2 may now cross the imaginary axis from $\mathcal{R}ez > 0$ to $\mathcal{R}ez < 0$ because presently we only allow conjugation in $\text{Is}^+(\mathbf{H})$; this will, however, not affect the arguments. By (6.8.13(5)), $\text{tr}\,[B_1, B_2] < 2$. But $[B_1, B_2]$ represents a hyperbolic transformation, i.e. one with $|\text{trace}| \geq 2$. Hence, $\text{tr}\,[B_1, B_2] < -2$.

Statement (b) is proved by induction. Assume that it holds for $k \leq g - 2$. The closed geodesic μ_{2k-1} in the free homotopy class of

$$m_{2k-1} := \prod_{v=1}^{k}[\beta_{2v-1}, \beta_{2v}]$$

corresponds to

$$M_{2k-1} := \prod_{v=1}^{k}[B_{2v-1}, B_{2v}].$$

The geodesics μ_{2k-1} and η_{2k+1} are boundary geodesics with positive bound-

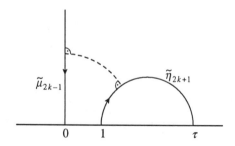

Figure 6.8.4

ary orientation of an embedded pair of pants Y.

This time we conjugate the groups in SL(2, **R**) such that the axes $\tilde{\mu}_{2k-1}$ and $\tilde{\eta}_{2k+1}$ of M_{2k-1} and $[B_{2k+1}, B_{2k+2}]$ are oriented from ∞ to 0 and from $\sigma = 1$ to $\tau > 1$ or from $\sigma = -1$ to $\tau < -1$, as shown in Fig. 6.8.4. Let λ and λ' denote the displacement lengths of M_{2k-1} and $[B_{2k+1}, B_{2k+2}]$ or, equivalently, the lengths of the closed geodesics μ_{2k-1} and η_{2k+1}. Since the matrices M_{2k-1} and $[B_{2k+1}, B_{2k+2}]$ have traces of the same sign, we compute the product $M_{2k+1} = M_{2k-1}[B_{2k+1}, B_{2k+2}]$ with (6.8.13(2)) and (6.8.13(1)) and obtain

(c) $$\operatorname{tr} M_{2k+1} = \frac{1}{(|\tau|-1)\sqrt{\lambda\lambda'}} (|\tau|(\lambda'+\lambda) - 1 - \lambda\lambda').$$

To avoid lengthy computations we determine the sign of this trace by a continuity argument. Deform the Y-piece Y continuously such that λ and λ' are increasing and such that $\operatorname{dist}(\mu_{2k-1}, \eta_{2k+1})$ is kept fixed. Then $|\tau|$ in (c) is fixed and M_{2k+1} represents a hyperbolic transformation for all λ, λ' in this deformation. Hence, either $\operatorname{tr} M_{2k+1} > 2$ is always true or $\operatorname{tr} M_{2k+1} < -2$ is always true. But for large λ and λ' the trace is negative. This proves (b). ◇

Chapter 7

The Spectrum of the Laplacian

This chapter gives a self-contained introduction into the Laplacian of compact Riemann surfaces. We prove the spectral theorem using the heat kernel which is given explicitly in Section 7.4 for the hyperbolic plane and in Section 7.5 for the compact quotients. As a tool we use the Abel transform which is introduced in Section 7.3. This transform will again show up in Chapter 9 in connection with Selberg's trace formula.

7.1 Introduction

We give a brief overview, with some history, of those topics in the spectral geometry of Riemann surfaces which will be covered in this book. Introductory texts for the Laplacian on Riemannian manifolds are Bérard [1], Berger-Gauduchon-Mazet [1] and Chavel [1]. A general overview of the Laplacian may be found in Simon-Wissner [1]. The articles by Elstrodt [2] and Venkov [1] give a detailed overview of the Laplacian on Riemann surfaces in connection with Selberg's trace formula. The article by Bérard [2] collects the developments in isospectrality up to 1989.

Originally, the Laplacian has been studied in mathematical physics in connection with the wave and the heat equations. See, for instance, Lagrange [1], Laplace [1], and Rayleigh [1] for the forerunners. One of the earliest spectral results which shows the interplay between the eigenvalues of the Laplacian and the geometry of the underlying domain is Weyl's asymptotic law [1, 2] of 1911:

(7.1.1) $$\lambda_k^{m/2}(M) \sim k \frac{c_m}{\text{vol } M}.$$

Introduction

Here $\lambda_k(M)$ is the k-th eigenvalue of the Laplacian on a compact domain $M \subset \mathbf{R}^m$ with respect to Dirichlet boundary conditions; c_m is a dimension constant and $\sim$ denotes asymptotic equality as $k \to \infty$. Up to the year 1949 the spectrum of the Laplacian was studied mainly on domains in Euclidean space. In 1949 Minakshisundaram and Pleijel [1] published a fundamental article in which they gave a proof of the spectral theorem (Theorem 7.2.6) for an arbitrary compact Riemannian manifold M. Using the zeta function

$$(7.1.2) \qquad Z(s) = \sum_{k=1}^{\infty} \lambda_k^{-s}(M), \qquad s \in \mathbf{C}$$

(with real part $\mathcal{R}\!e\, s$ sufficiently large), they proved Weyl's law (7.1.1) for compact manifolds. They also showed that $Z(s)$ has an analytic continuation in the entire complex plane, thereby introducing methods from analytic number theory into Riemannian geometry. In the same year Maaß [1] introduced certain automorphic eigenfunctions of the Laplacian for Fuchsian groups of the first kind in connection with Dirichlet series and Siegel modular forms. The articles of Minakshisundaram-Pleijel and Maaß may be considered as the initiators of the "modern" spectral geometry of manifolds.

In 1954 Roelcke [1] proved general existence theorems for eigenfunctions and eigenpackets for Fuchsian groups of the first kind (this work was later continued in Elstrodt [1]). In the same year Huber [1] used the eigenfunction expansion of a new kind of Dirichlet series in order to study the asymptotic distribution of lattice points in the hyperbolic plane. At the same time Selberg [1, 2] undertook his investigations in harmonic analysis and found the celebrated Selberg trace formula (Section 9.5).

Huber [2] introduced a new geometric quantity, the *length spectrum* which is the sequence of the lengths of the closed geodesics listed in ascending order (cf. Definitions 9.2.8 and 10.1.1). He proved that for compact Riemann surfaces the length spectrum and the eigenvalue spectrum are equivalent geometric quantities. This result is also a consequence of Selberg's trace formula, and we shall give a proof in Section 9.2 which lies somewhere between Huber's and Selberg's methods.

Since a finite set of the lengths (for instance those belonging to a canonical curve system) determine the Riemann surface up to isometry, it looks plausible that compact Riemann surfaces which are *isospectral*, i.e. which have the same spectrum, are isometric. This was conjectured by Gel'fand [1] in 1962. As a first step toward the conjecture, Gel'fand proved in [1] that continuous isospectral deformations cannot occur. On the other hand, Milnor [1] in 1964 gave examples of 16-dimensional isospectral non-isometric flat tori. Beyond this, little progress was made on behalf of the Laplacian on manifolds until the lecture notes of Berger-Gauduchon-Mazet [1] appeared in

1971, greatly influencing the interest in the Laplacian. (The comprehensive reference list of Bérard-Berger [1] contains about 40 titles before 1970 and about 750 titles between 1970 and 1982.)

In 1972 McKean [1] showed that the cardinality of a set of isospectral non-isometric compact Riemann surfaces is always finite. We shall give an upper bound on this cardinality in Chapter 13. This bound depends only on the genus. In 1977 Wolpert [1, 3] proved that for a generic surface in the moduli space of genus g Gel'fand's conjecture is true, the possible exceptional surfaces being contained in a proper real analytic subvariety $\mathcal{V}_g$. At about the same time Marie France Vignéras [1, 2, 3] found examples showing that $\mathcal{V}_g \neq \emptyset$ for infinitely many values of g. Vignéras' examples are modelled over quaternion algebras and are not easy to describe geometrically.

In 1985 Sunada [3] discovered a general way of constructing pairs of isospectral non-isometric Riemannian manifolds. His method gave access to a more systematic investigation of the phenomenon of isospectrality. We shall discuss Sunada's method in detail in Chapter 11. In Chapter 12 we shall follow Brooks [1], Brooks-Tse [1] and some of our own work (Buser [8]) to show that $\mathcal{V}_g$ has positive dimension for any $g \geq 4$.

In the next sections we shall introduce the Laplacian in a fairly self-contained manner. The trace formula techniques and the connection with the length spectrum will follow in Chapter 9.

7.2 The Spectrum and the Heat Equation

Let M be an arbitrary complete m-dimensional Riemannian manifold. The Laplace-Beltrami operator $\Delta = -\text{div grad}$ has the coordinate expression

$$(7.2.1) \qquad \Delta u = -\frac{1}{\sqrt{g}} \sum_{j,k=1}^{m} \partial_j (g^{jk} \sqrt{g}\, \partial_k u), \quad u \in C^{\infty}(M),$$

where ∂_j denotes the partial differentiation with respect to the j-th coordinate; $g = \det(g_{ij})$ is the determinant of the metric tensor, where the g_{ij} are the components of the metric tensor with respect to the local coordinates and the g^{jk} are the components of the inverse tensor. A direct though tedious computation shows that (7.2.1) is invariant under a parameter change so that we may consider (7.2.1) as a definition of the Laplacian. At the same time this shows that Δ is an isometry invariant.

From Berger-Gauduchon-Mazet [1], p. 126/127 we obtain the following geometric definition of Δ. Let $p \in M$ and consider m unit speed geodesics $x_i \mapsto \gamma_i(x_i)$, $x_i \in\,]-\varepsilon, \varepsilon[$ satisfying $\gamma_i(0) = p$, $i = 1, \ldots, m$, which are pairwise orthogonal at p. Abbreviate $u(\gamma_i(x_i)) =: u(x_i)$. Then Δ has the form

$$\Delta u(p) = -\sum_{i=1}^{m} \frac{\partial^2 u}{\partial x_i^2}(0).$$

Our goal is the spectral theorem Theorem 7.2.6 below which we shall approach following Dodziuk [1]. The particular advantage of this approach is that it yields at the same time part of Huber's theorem (Section 9.2).

We first give an overview. The proof of the spectral theorem is based on the heat equation. In the case of a Riemann surface the fundamental solution of the heat equation, the so called *heat kernel*, will be given explicitly in Sections 7.4 and 7.5. For the existence proof for general Riemannian manifolds we refer to either Bérard [1], Berger-Gauduchon-Mazet [1] or Chavel [1]. With the heat kernel

$$p = p(x, y, t)$$

a family of integral operators $\mathcal{P}_t$, $t > 0$, is defined which has the semi-group property $\mathcal{P}_s \mathcal{P}_t = \mathcal{P}_{s+t}$. The Hilbert-Schmidt theorem provides a complete orthonormal sequence of eigenfunctions in the L^2-space for each $\mathcal{P}_t$, and the semi-group property then implies that all $\mathcal{P}_t$ have the same eigenspaces. Finally we shall prove that these eigenspaces are also the eigenspaces of the Laplacian. An additional advantage of this approach is that the eigenfunctions are easily seen to be of class C^∞, and we do not have to refer to the regularity theorems for elliptic operators.

The function space in question is the Hilbert space $L^2(M)$ of square integrable functions $f: M \to \mathbf{R}$ on the compact connected unbordered Riemannian manifold M with inner product

$$f, g \mapsto \int_M fg \, dM, \quad f, g \in L^2(M),$$

where dM is the volume element. The amount of Riemannian geometry needed here is small and is essentially reduced to the notion of the gradient, the definition of the Laplacian and *Green's formula*. Greens's formula is

(7.2.2) $$\int_M f \Delta g \, dM = \int_M \langle \operatorname{grad} f, \operatorname{grad} g \rangle \, dM,$$

where grad is the gradient vectorfield and $\langle \,.\,,\,.\, \rangle$ is the metric tensor.

7.2.3 Definition. Let A and B be smooth manifolds and let $\mathbf{F} = \mathbf{R}$ or $\mathbf{F} = \mathbf{C}$. We denote by $C^{\ell,k}(A, B; \mathbf{F})$ the set of all functions $f: A \times B \to \mathbf{F}$ whose mixed partial derivatives up to ℓ times with respect to A and up to k times with respect to B exist and are continuous on $A \times B$.

7.2.4 Definition. Let M be any connected unbordered Riemannian manifold (M need not be complete in this definition). A continuous function $p = p(x, y, t) : M \times M \times \,]0, \infty[\, \to \mathbf{R}$ is called a *fundamental solution to the heat equation* on M if it belongs to $C^{2,1}(M \times M, \,]0, \infty[\,; \mathbf{R})$ and satisfies the following conditions.

(i) $\dfrac{\partial p}{\partial t} = -\Delta_x p$

(ii) $p(x, y, t) = p(y, x, t)$

(iii) $\lim\limits_{t \downarrow 0} \displaystyle\int_M p(x, y, t) f(y) \, dM(y) = f(x)$.

Here Δ_x is the Laplacian with respect to the first argument of p, and the convergence in (iii) is locally uniform in x for every continuous function f with compact support on M.

We remark that condition (ii) can be concluded from the remaining conditions by Duhamel's principle (Chavel [1], p. 137). We also remark that for noncompact M, a fundamental solution, should it exist, need not be unique unless suitable boundary conditions are given.

The following theorem will be proved for compact Riemann surfaces by giving an explicit formula in Theorem 7.4.1 and Theorem 7.5.11. For the proof in the general case we refer to Bérard [1] or Chavel [1].

7.2.5 Theorem. *Let M be any m-dimensional compact connected Riemannian manifold without boundary. Then M has a unique fundamental solution p_M of the heat equation. The function p_M belongs to $C^\infty(M \times M \times \,]0, \infty[\,)$. For $0 < t < 1$, p_M has the following bounds, where the constant c_M depends on M.*

$$0 \le p_M(x, y, t) \le c_M \, t^{-m/2}.$$

The unique solution $p_M = p_M(x, y, t)$ is called the *heat kernel* of M. Although we do not need this in this book, we remark that the inequality in the preceding theorem has a stronger version. In fact, p_M has the asymptotic expansion of Minakshisundaram-Pleijel

$$p_M(x, y, t) \sim (4\pi t)^{-m/2} e^{-r^2/4t}(u_0 + u_1 t + \dots)$$

($t \downarrow 0$), where $r = \text{dist}(x, y)$ and the u_k are smooth functions on $M \times M$. (Minakshisundaram-Pleijel [1] or Berger-Gauduchon-Mazet [1, p. 204].)

7.2.6 Theorem. (Spectral theorem). *Let M be a compact connected Riemannian manifold without boundary. The eigenvalue problem*

$$\Delta \varphi = \lambda \varphi$$

has a complete orthonormal system of C^∞-eigenfunctions $\varphi_0, \varphi_1, \ldots$ in $L^2(M)$ with corresponding eigenvalues $\lambda_0, \lambda_1, \ldots$. These have the following properties.

(i) $0 = \lambda_0 < \lambda_1 \leq \lambda_2 \leq \ldots, \lambda_n \to \infty$ as $n \to \infty$.

(ii) $p_M(x, y, t) = \sum_{n=0}^{\infty} e^{-\lambda_n t} \varphi_n(x) \varphi_n(y)$,

where the series converges uniformly on $M \times M$ for each $t > 0$.

There is, of course, also a version for manifolds with boundary. For the proof of Theorem 7.2.6 we use two theorems from the analysis of integral operators: the Hilbert-Schmidt theorem and Mercer's theorem. For the proofs of these theorems we refer to Jörgens [1], p. 78 and p. 123. We remark that the proofs in Riesz-Nagy [1], paragraph 97, given for domains in $\mathbf{R}^n$ are easily translated to manifolds. The theorems are the following.

7.2.7 Theorem. (Hilbert-Schmidt theorem). *Let M be a compact connected Riemannian manifold and let $\mathcal{K}$ be the integral operator defined by*

$$\mathcal{K}[f](x) = \int_M K(x, y) f(y) \, dM(y), \quad f \in L^2(M),$$

where $K : M \times M \to \mathbf{R}$ is a continuous function which is symmetric: $K(x, y) = K(y, x)$. Then the eigenvalue problem $\mathcal{K}[\varphi] = \eta \varphi$ has a complete orthonormal system of eigenfunctions $\varphi_0, \varphi_1, \ldots$ in $L^2(M)$ with corresponding eigenvalues $\eta_0, \eta_1, \ldots$, where $\eta_n \to 0$ as $n \to \infty$. The kernel K has the following expansion in the L^2-sense:

$$K(x, y) = \sum_{n=0}^{\infty} \eta_n \varphi_n(x) \varphi_n(y).$$

7.2.8 Theorem. (Mercer's theorem). *Let M, K and $\mathcal{K}$ be as in Theorem 7.2.7. Assume in addition that almost all eigenvalues $\eta_0, \eta_1, \ldots$ are nonnegative. Then K has the expansion*

$$K(x, y) = \sum_{n=0}^{\infty} \eta_n \varphi_n(x) \varphi_n(y),$$

where the convergence of the series is uniform on $M \times M$.

We now start with the proof of Theorem 7.2.6 following Chavel [1] and Dodziuk [1]. In the remainder of this section M will always be a compact connected Riemannian manifold without boundary.

7.2.9 Definition. Let $f: M \to \mathbf{R}$ be a continuous function. A continuous function $u = u(x, t): M \times [0, \infty[\to \mathbf{R}$ is called a *solution of the heat equation for the initial condition* $u(x, 0) = f(x)$, if $u \in C^{2,1}(M,]0, \infty[; \mathbf{R})$ (cf. Definition 7.2.3), and if u satisfies the *heat equation*

(i) $\quad \dfrac{\partial u}{\partial t} = -\Delta u$

with respect to the *initial conditions*

(ii) $\quad u(x, 0) = f(x)$, for all $x \in M$.

(Δ is with respect to the first variable.)

Note that in contrast to the fundamental solution to the heat equation, $u(x, t)$ must be defined for $t = 0$.

7.2.10 Lemma. *Let $p = p(x, y, t)$ be a fundamental solution to the heat equation on M and let $f: M \to \mathbf{R}$ be a continuous function. The function*

$$u(x, t) := \int_M p(x, y, t) f(y) \, dM(y), \quad t > 0$$

has a continuous extension to $t = 0$, and this extended function is a solution to the heat equation for the initial condition $u(x, 0) = f(x)$.

Proof. The continuous extension to $t = 0$ is due to the continuity of f and to condition (iii) in Definition 7.2.4, where the convergence is uniform on compact sets. Since M is compact and since the partial derivatives are continuous functions on $M \times M \times]0, \infty[$, the operators Δ_x and $\partial/\partial t$ pass under the integration sign. The lemma follows. ◊

7.2.11 Lemma. *The solution to the heat equation given by Lemma 7.2.10 is unique and has the following properties*

$$\frac{d}{dt} \int_M u \, dM = 0, \quad \frac{d}{dt} \int_M u^2 \, dM \leq 0, \quad t > 0.$$

Proof. Let u be any solution. Then for $t > 0$ by Green's formula

$$\frac{d}{dt} \int_M u \, dM = \int_M \frac{\partial u}{\partial t} \, dM = -\int_M \Delta u \, dM = 0.$$

Similarly we have

$$\frac{d}{dt}\int_M u^2\, dM = -2\int_M u\, \Delta u\, dM = -2\int_M \langle \operatorname{grad} u, \operatorname{grad} u \rangle\, dM \le 0.$$

Now let u_1, u_2 be two solutions and set $v = u_1 - u_2$. Then v is a solution of the heat equation with initial condition $v(x, 0) = 0$. Since $\int_M v^2\, dM$ is continuous on $[0, \infty[$ and monotone decreasing on $]0, \infty[$, we have $v \equiv 0$. ◇

7.2.12 Remark. *The preceding lemmata yield the uniqueness of the fundamental solution of the heat equation on a compact manifold.*

Proof. Let $q = q(x, y, t)$ be another solution. Let f be any continuous function on M. Then by Lemma 7.2.10 and Lemma 7.2.11

$$\int_M p(x, y, t) f(y)\, dM(y) = \int_M q(x, y, t) f(y)\, dM(y)$$

for all $t > 0$. Therefore

$$\int_M (p - q)(x, y, t) f(y)\, dM(y) \equiv 0 \text{ on } M \times]0, \infty[$$

for any f. Hence, $p = q$. ◇

In the following, $p = p(x, y, t) = p_M(x, y, t)$ will always denote the heat kernel of M.

7.2.13 Lemma. $\displaystyle\int_M p(x, y, t)\, dM(y) = 1.$

Proof. Apply Lemma 7.2.10 to the constant function $f(y) = 1$. Observe that f is a solution of the heat equation and apply the uniqueness given by Lemma 7.2.11. ◇

7.2.14 Definition. For every $f \in L^2(M)$ and for every $t > 0$, we define

$$\mathcal{P}_t[f](x) = \int_M p(x, y, t) f(y)\, dM(y), \quad x \in M.$$

7.2.15 Lemma. *Each $\mathcal{P}_t$ is a compact positive self-adjoint operator. For any $f \in L^2(M)$, we have $\mathcal{P}_t[f] \in C^\infty(M)$.*

Proof. Since M is compact and since the partial derivatives are continuous in y, the derivations pass under the integration sign. Together with the smoothness of p, this proves the second statement. For the first statement we note that the operator $\mathcal{P}_t$ has a continuous kernel on the compact manifold and is therefore compact. Since p is a symmetric kernel, $\mathcal{P}_t$ is self-adjoint. That $\mathcal{P}_t$ is positive will be proved after the next lemma.

7.2.16 Lemma. *For $s > 0$ and $t > 0$ we have the semi group property*

$$\mathcal{P}_s \circ \mathcal{P}_t = \mathcal{P}_{s+t}.$$

In particular, $\mathcal{P}_s \circ \mathcal{P}_t = \mathcal{P}_t \circ \mathcal{P}_s$.

Proof. $\mathcal{P}_s \circ \mathcal{P}_t$ and $\mathcal{P}_{s+t}$ are continuous operators, and $C^0(M)$ is dense in $L^2(M)$. It suffices therefore to prove that

$$\mathcal{P}_s[\mathcal{P}_t[f]] = \mathcal{P}_{s+t}[f]$$

for any *continuous* function f. Now, for fixed $t > 0$ the function

$$x, s \mapsto v(x, s) := \mathcal{P}_{s+t}[f](x)$$

is a solution of the heat equation with initial condition $v(x, 0) = \mathcal{P}_t[f](x)$ (Lemma 7.2.10), and the same is true for the function

$$x, s \mapsto w(x, s) := \mathcal{P}_s[\mathcal{P}_t[f]](x).$$

Since $\mathcal{P}_t[f]$ is a continuous function (first part of Lemma 7.2.15), it follows from Lemma 7.2.11 that $v = w$. ◇

We now prove that $\mathcal{P}_t$ is a positive operator. By the above, $\mathcal{P}_t = \mathcal{P}_{t/2} \circ \mathcal{P}_{t/2}$ and therefore, since $\mathcal{P}_{t/2}$ is self-adjoint, we have that

$$\int_M \mathcal{P}_t[f] f \, dM = \int_M (\mathcal{P}_{t/2}[f])^2 \, dM \geq 0$$

for all $f \in L^2(M)$. ◇

7.2.17 Lemma. *For any $f \in L^2(M)$ we have, in the L^2-sense,*

$$\lim_{t \downarrow 0} \mathcal{P}_t[f] = f.$$

Proof. First let f be a continuous function. The function $\mathcal{P}_t[f]$ is a solution of the heat equation so that by Lemma 7.2.11, $\dfrac{d}{dt} \|\mathcal{P}_t[f]\| \leq 0$, where

$$\|u\| = \left(\int_M u^2\, dM\right)^{1/2}$$

for $u \in L^2(M)$. Since f is continuous, $\lim_{t\downarrow 0} \mathcal{P}_t[f] = f$ uniformly on M (cf. Definition 7.2.4). Hence, $\|\mathcal{P}_t[f]\| \leq \|f\|$. Since $C^0(M)$ is dense in $L^2(M)$ and since $p(x, y, t)$ is continuous, we have

$$\|\mathcal{P}_t[f]\| \leq \|f\| \quad \text{for all } f \in L^2(M).$$

Since $C^0(M)$ is dense in $L^2(M)$, the lemma follows. ◇

After these preparations we apply the Hilbert-Schmidt theorem to each of the operators $\mathcal{P}_t$. We begin with $t = 1$ and let

$$\varphi_0, \varphi_1, \varphi_2, \ldots$$

be a complete orthonormal system in $L^2(M)$ consisting of eigenfunctions of $\mathcal{P}_1$ with corresponding eigenvalues

$$\eta_0, \eta_1, \eta_2, \ldots \geq 0; \quad \eta_j \to 0 \text{ as } j \to \infty.$$

We claim that the φ_j are the eigenfunctions of any other $\mathcal{P}_t$. For this we first let $t = 1/k$ with $k \in \mathbf{N} - \{0\}$. If φ is an eigenfunction of $\mathcal{P}_{1/k}$ with eigenvalue μ, then, since $\mathcal{P}_1 = (\mathcal{P}_{1/k})^k$, φ is an eigenfunction of $\mathcal{P}_1$ with eigenvalue μ^k. It follows from the completeness of the system of eigenfunctions of $\mathcal{P}_{1/k}$ that $\mathcal{P}_{1/k}$ has the same orthogonal system of eigenspaces as $\mathcal{P}_1$ so that the sequence $\varphi_0, \varphi_1, \varphi_2, \ldots$ is a complete orthonormal system of eigenfunctions of $\mathcal{P}_{1/k}$. The eigenvalues of $\mathcal{P}_{1/k}$ are $\eta_0^{1/k}, \eta_1^{1/k}, \eta_2^{1/k}, \ldots$. By Lemma 7.2.16, the φ_j are eigenfunctions of $\mathcal{P}_t$ with eigenvalues η_j^t for all positive rational t. By the continuity of the function $p(x, y, t)$ we have

$$\mathcal{P}_t[\varphi_j] = \eta_j^t \varphi_j, \quad j = 0, 1, 2, \ldots,$$

for *all* $t > 0$. By Lemma 7.2.17, $\lim_{t\downarrow 0} \mathcal{P}_t[\varphi_j] = \varphi_j$, so that $\eta_j^t \to 1$ as $t \downarrow 0$. We find that

$$\eta_j > 0 \text{ for all } j.$$

Lemma 7.2.15 implies that

$$\varphi_j \in C^\infty(M) \text{ for all } j.$$

By the compactness of $\mathcal{P}_1$ all eigenspaces are now finite dimensional. We may therefore assume that the eigenvalues are arranged in decreasing order. We claim that

$$1 = \eta_0 > \eta_1 \geq \eta_2 \geq \ldots > 0.$$

By Lemma 7.2.13 we know that 1 is an eigenvalue for the constant function. Now let φ_j be a non-constant eigenfunction of $\mathcal{P}_1$. The argument in the proof of Lemma 7.2.11 can be improved as follows.

$$\frac{d}{dt}\|\mathcal{P}_t[\varphi_j]\|^2 = -2\int_M \langle \text{ grad } \mathcal{P}_t[\varphi_j], \text{ grad } \mathcal{P}_t[\varphi_j] \rangle dM$$

$$= -2\eta_j^{2t}\int_M \langle \text{ grad } \varphi_j, \text{ grad } \varphi_j \rangle dM < 0.$$

Here we have used that $\eta_j > 0$. It follows that $\|\mathcal{P}_t[\varphi_j]\| < \|[\varphi_j]\|$ and therefore $\eta_j < 1$. This proves the above claim.

To prove Theorem 7.2.6 we let

$$\lambda_j = -\log \eta_j, \quad j = 0, 1, 2, \ldots.$$

From the equation

$$0 = \Delta \mathcal{P}_t[\varphi_j] + \frac{\partial}{\partial t}\mathcal{P}_t[\varphi_j]$$
$$= e^{-t\lambda_j}(\Delta \varphi_j - \lambda_j \varphi_j)$$

we see that the φ_j are eigenfunctions of the Laplacian, and the properties of the η_j correspond to the properties (i) in the theorem. Statement (ii) follows from Mercer's theorem applied to the compact positive operator $\mathcal{P}_t$. ◊

In the proof of the prime number theorem in Section 9.4 we need the following improvement of (ii) in Theorem 7.2.6, where $\{\varphi_n\}_{n=0}^{\infty}$ is again a complete orthonormal sequence of eigenfunctions of the Laplacian on the compact Riemannian manifold M and $0 = \lambda_0 < \lambda_1 \leq \lambda_2 \leq \ldots$ is the corresponding sequence of eigenvalues.

7.2.18 Theorem. *Let m be the dimension of M. For any $\sigma > m/2$, the series*

$$\sum_{n=1}^{\infty} \lambda_n^{-\sigma} \varphi_n(x) \varphi_n(y)$$

converges absolutely and uniformly on $M \times M$.

Proof. We follow Minakshisundaram-Pleijel [1], p. 322. In order to go from $e^{-\lambda_n t}$ to $\lambda_n^{-\sigma}$ we multiply $e^{-\lambda_n t}$ with $t^{\sigma-1}$ and integrate. From the integral representation of the Gamma function, then, for $\sigma > 1$

$$\int_0^\infty t^{\sigma-1} e^{-\lambda_n t} dt = \Gamma(\sigma) \lambda_n^{-\sigma}.$$

Thus we consider the φ_n as the eigenfunctions of a new integral operator with kernel G_σ

$$G_\sigma(x, y) = \int_0^\infty (p(x, y, t) - \mathrm{vol}^{-1}(M)) t^{\sigma-1} dt.$$

To see that G_σ is well defined and continuous on $M \times M$, we first observe that for any $\sigma > 1$, we find an integer $n(\sigma)$ such that for all $n \geq n(\sigma)$ and for all $\varepsilon \in \,]0, \tfrac{1}{2}[$

$$\int_\varepsilon^\infty e^{-\lambda_n t} t^{\sigma-1} dt \leq e^{-\lambda_n \varepsilon}.$$

From the expansion

$$p(x, y, t) - \mathrm{vol}^{-1}(M) = \sum_{n=1}^\infty e^{-\lambda_n t} \varphi_n(x) \varphi_n(y),$$

we see therefore that for all $\varepsilon > 0$ the function

$$G_{\sigma,\varepsilon}(x, y) := \int_\varepsilon^\infty (p(x, y, t) - \mathrm{vol}^{-1}(M)) t^{\sigma-1} dt$$

is well defined, continuous on $M \times M$ and satisfies

$$\lim_{\varepsilon \downarrow 0} \int_M G_{\sigma,\varepsilon}(x, y) \varphi_n(y) \, dM(y) = \Gamma(\sigma) \lambda_n^{-\sigma} \varphi_n(x)$$

pointwise in $x \in M$, for any $n \geq 1$.

On the other hand we have $p(x, y, t) \leq c_M t^{-m/2}$ for $t < 1$ (cf. Theorem 7.2.5) so that for all $\delta \in \,]0, \varepsilon[$

$$\int_\delta^\varepsilon p(x, y, t) t^{\sigma-1} dt \leq \frac{c_M}{\sigma - m/2} \varepsilon^{\sigma - m/2}.$$

This shows that $G_\sigma(x, y)$ is well defined on $M \times M \times \,]1, \infty[$ and that $G_{\sigma,\varepsilon}$ converges to G_σ uniformly on $M \times M$ as $\varepsilon \downarrow 0$. Hence,

$$\int_M G_\sigma(x, y) \varphi_n(y) \, dM(y) = \Gamma(\sigma) \lambda_n^{-\sigma} \varphi_n(x).$$

That is, the $\varphi_1, \varphi_2, \ldots$ are the eigenfunctions with the positive eigenvalues $\Gamma(\sigma) \lambda_n^{-\sigma}$ of the compact integral operator whose kernel is G_σ. Mercer's theorem now implies that

$$G_\sigma(x, y) = \Gamma(\sigma) \sum_{n=1}^{\infty} \lambda_n^{-\sigma} \varphi_n(x) \varphi_n(y)$$

with *uniform* convergence of the series on $M \times M$. This proves Theorem 7.2.18. ◇

7.3 The Abel Transform

The next two sections are from private lectures by Philippe Anker in Lausanne. In the present section we give a self-contained introduction to the Abel transformation. In Section 7.4 this transformation will be used to compute an explicit fundamental solution of the heat equation in **H**. The Abel transform will show up again in Section 9.2 in connection with Huber's theorem and in Section 9.5 in connection with Selberg's trace formula. A geometric motivation for this transformation will be given at the beginning of the next section, where we shall transform radial functions into functions which are constant along horocycles. For a more detailed account of the Abel transformation and its applications to harmonic analysis on symmetric spaces we refer to Anker [1] and Koornwinder [1].

7.3.1 Definition. (Abel transform I). For any continuous function $f : [1, \infty[\to \mathbf{R}$ with the property that $|f(t)| \leq c t^{-\varepsilon - 1/2}$ for some $\varepsilon > 0$ and some constant c, the *Abel transform* $A_1[f]$ is defined by

$$A_1[f](x) = \frac{1}{\sqrt{\pi}} \int_x^{\infty} \frac{f(y)}{\sqrt{y-x}} dy, \qquad x \in [1, \infty[.$$

The Abel transform is more interesting if more regularity is required for f. For instance, if f is of class C^1 and if the derivative of f is bounded above by $|f'(t)| \leq c' t^{-\delta - 3/2}$ for some $\delta > 0$ and some constant c', then we may substitute

$$t = y - x$$

and integrate by parts to obtain

(7.3.2) $$A_1[f](x) = \frac{-2}{\sqrt{\pi}} \int_0^{\infty} \sqrt{t} f'(x+t) dt.$$

In this form the Abel transform looks less singular. We now restrict ourselves to the Schwartz space.

7.3.3 Definition. The *Schwartz space* $\mathscr{S}_1 = \mathscr{S}_1([1, \infty[)$ is defined as the set of all functions $f \in C^\infty([1, \infty[)$ such that for any $m, n \in \mathbf{N} \cup \{0\}$ the function $t \mapsto t^m D^n f(t)$ is bounded on $[1, \infty[$.

In this definition D^0 denotes the identity operator and D^n is the n-th derivative for $n = 1, 2, \ldots$.

7.3.4 Theorem. $A_1 : \mathscr{S}_1 \to \mathscr{S}_1$ *is one-to-one and onto. The inverse of* A_1 *is given by*

$$A_1^{-1}[g](y) = \frac{-1}{\sqrt{\pi}} \int_y^\infty \frac{Dg(x)}{\sqrt{x-y}} \, dx.$$

Proof. We work with (7.3.2). We first observe that for all $m \in \mathbf{N}$ the function $x \mapsto x^m A_1[f](x)$ is bounded. We also see that $A_1[f]$ has a derivative and that

(7.3.5) $$DA_1[f] = A_1[Df].$$

It now follows by induction that $A_1[f] \in C^\infty([1, \infty[)$. Hence, $A_1 : \mathscr{S}_1 \to \mathscr{S}_1$.
Since $D : \mathscr{S}_1 \to \mathscr{S}_1$, the operator $\tilde{A}_1 := -A_1 D$ is well defined and maps $\mathscr{S}_1$ to $\mathscr{S}_1$. For $g \in \mathscr{S}_1$ we compute

$$A_1 \tilde{A}_1[g](x) = \frac{-1}{\pi} \int_x^\infty \int_y^\infty Dg(z)(z-y)^{-1/2}(y-x)^{-1/2} \, dz \, dy$$

$$= \frac{-1}{\pi} \int_x^\infty \int_x^z Dg(z)(z-y)^{-1/2}(y-x)^{-1/2} \, dy \, dz.$$

The inner integral of the second equation equals

$$Dg(z) \int_x^z (z-y)^{-1/2}(y-x)^{-1/2} dy = Dg(z) \int_0^1 (1-t)^{-1/2} t^{-1/2} dt = \pi Dg(z).$$

Since $g(z) \to 0$ as $z \to \infty$, $A_1 \tilde{A}_1[g](x) = g(x)$ so that $A_1 \tilde{A}_1 = id$. By (7.3.5) we also have $\tilde{A}_1 A_1 = -A_1 DA_1 = -A_1 A_1 D = A_1 \tilde{A}_1 = id$. This proves the theorem. ◇

Later, the Abel transform will show up under a parameter change of type $x = \cosh r$. We shall therefore rewrite the transform in terms of the new variables. In the next definition, a smooth function defined on $[0, \infty[$ is called *even* if it is the restriction to $[0, \infty[$ of a smooth even function.

7.3.6 Definition. The *modified Schwartz space* $\mathscr{S}$ is defined as the set of all even functions $F \in C^\infty([0, \infty[)$ such that for all $m, n \in \mathbf{N} \cup \{0\}$ the function $r \mapsto e^{mr} D^n F(r)$ is bounded.

Since the hyperbolic cosine is an even function which grows as $\frac{1}{2}e^r$, we obtain the following.

7.3.7 Lemma. *The mapping* $M : \mathscr{S}_1 \to \mathscr{S}$ *defined by* $M[f](r) = f(\cosh r)$ *is one-to-one and onto.*

Proof. This follows from elementary considerations. For instance, to prove the smoothness of $M^{-1}[F]$ at $t = 1$, observe that at $r = 0$ we may write $F(r) = G(r^2)$, where G is a smooth function in a neighborhood of 0. Similarly, we may write $\cosh r = h(r^2)$, where h has a smooth inverse h^{-1}. Now $F(r) = G(h^{-1}(\cosh r))$, etc. ◇

7.3.8 Definition. (Abel transform II). For $F \in \mathscr{S}$ or, more generally, for any F with growth of order $O(e^{-\rho(1+\delta)})$ for some $\delta > 0$ as $\rho \to \infty$, we set

$$A[F](r) = \frac{1}{\sqrt{\pi}} \int_r^\infty \frac{F(\rho) \sinh \rho}{\sqrt{\cosh \rho - \cosh r}} \, d\rho.$$

It is easy to check that $MA_1 = AM$ or, since M is one-to-one and onto,

(7.3.9) $\qquad\qquad A = MA_1 M^{-1}.$

From Theorem 7.3.4 we obtain therefore the following.

7.3.10 Theorem. $A : \mathscr{S} \to \mathscr{S}$ *is one-to-one and onto. The inverse of* A *is given by*

$$A^{-1}[G](\rho) = \frac{-1}{\sqrt{\pi}} \int_\rho^\infty \frac{DG(r)}{\sqrt{\cosh r - \cosh \rho}} \, dr.$$

◇

To demonstrate the smoothness of the heat kernel on **H**, we shall have to consider two variables. For this we let I be any open interval in **R** (or, more generally, an open subset of some $\mathbf{R}^k$) and define

$$\mathscr{S}_1^I = \{f = f(x, \tau) \in C^\infty([1, \infty[\times I) \mid f(\,, \tau) \in \mathscr{S}_1 \text{ for any } \tau \in I\}$$

(7.3.11)

$$\mathscr{S}^I = \{F = F(r, \tau) \in C^\infty([0, \infty[\times I) \mid F(\,, \tau) \in \mathscr{S} \text{ for any } \tau \in I\}.$$

7.3.12 Lemma. *The mapping* $M : \mathscr{S}_1^I \to \mathscr{S}^I$ *defined by the equation* $M[f](r, \tau) = f(\cosh r, \tau)$, *is one-to-one and onto.*

Proof. Clearly, M is one-to-one and $M(\mathscr{S}_1^I) \subset \mathscr{S}^I$. The inverse mapping is given by $M^{-1}[F](x, \tau) = F(\operatorname{arccosh} x, \tau)$. This function has the required decay conditions and is smooth on $]1, \infty[\times I$. The only point to check is therefore the smoothness at $x = 1$. For this we first see from the power series expansion of the hyperbolic cosine function that the function

$$z \mapsto w = \cosh(\sqrt{z})$$

is holomorphic for $z \in \mathbf{C}$ with non-vanishing derivative at $z = 0$. There exists therefore a holomorphic inverse in a neighborhood U of $w = 1$. On $U \cap [1, \infty[$ this inverse coincides with $(\operatorname{arccosh} w)^2$. Hence, the function $x \mapsto (\operatorname{arccosh} x)^2$ is C^∞ at $x = 1$. Since $F(\ , \tau)$ is an even function for each τ, we see from the Taylor formula that there exists a function $G \in C^\infty(\mathbf{R} \times I)$ satisfying $F(r, \tau) = G(r^2, \tau)$. Hence, $M^{-1}[F](x, \tau) = G((\operatorname{arccosh} x)^2, \tau)$ so that $M^{-1}[F]$ is smooth at $x = 1$. ◇

The Abel transformation may, without modification, be applied to functions of $\mathscr{S}_1^I$ and $\mathscr{S}^I$ respectively. From (7.3.2) and from the rapid decay which allows us to pass all derivations under the integration sign, we see that $A_1 : \mathscr{S}_1^I \to \mathscr{S}_1^I$ is one-to-one and onto. In conjunction with (7.3.9) and Lemma 7.3.12 we have therefore the following theorem.

7.3.13 Theorem. *$A : \mathscr{S}^I \to \mathscr{S}^I$ is one-to-one and onto.* ◇

7.4 The Heat Kernel of the Hyperbolic Plane

Our next goal is an explicit formula for the heat kernel in the hyperbolic plane. This formula will yield half of the trace formula (9.2.11) needed to prove Huber's theorem in Section 9.2. It will also allow us to obtain the heat kernel of a compact quotient $\Gamma \backslash \mathbf{H}$ simply by summing over the action of Γ (cf. Section 7.5). For an account of the heat kernel of hyperbolic spaces we also refer to the book of Davies [1]. The formula is as follows.

7.4.1 Theorem. *The function $p_\mathbf{H} : \mathbf{H} \times \mathbf{H} \times]0, \infty[\to \mathbf{R}$ given by*

$$p_\mathbf{H}(z, w, t) = \frac{\sqrt{2}}{(4\pi t)^{3/2}} e^{-t/4} \int_{d(z, w)}^{\infty} \frac{r e^{-r^2/4t}}{\sqrt{\cosh r - \cosh d(z, w)}} \, dr$$

is a fundamental solution of the heat equation.

Here $d(z, w) := \operatorname{dist}(z, w)$ is the hyperbolic distance of z and w in the upper half-plane **H**. We recall (1.1.2):

(7.4.2) $\quad \cosh \operatorname{dist}(z, w) = \cosh d(z, w) = 1 + \dfrac{|z-w|^2}{2\, \operatorname{Im} z\, \operatorname{Im} w}.$

For comparison we mention the formula for the heat kernel h in $\mathbf{R}^1$:

(7.4.3) $\quad h(x, y, t) = (4\pi t)^{-1/2}\, e^{-(y-x)^2/4t}, \quad x, y \in \mathbf{R},\ t > 0.$

While it is easy to check that $\partial^2 h/\partial x^2 - \partial h/\partial t = 0$, a direct verification of Theorem 7.4.1 is quite tedious. Also, a direct check does not reveal how an integral like that in Theorem 7.4.1 arises. We prefer therefore to give a deductive proof. For this we apply the Abel transform and first solve the transformed equation. A similar deductive approach based on the Mehler transform is given in Chavel [1] pp. 242 - 246. Yet other transform techniques are described in Terras [1].

For the proof of Theorem 7.4.1 we proceed in three steps. First we give a simple geometric construction in which the Abel transformation shows up in a natural way. This will make the idea of using the Abel transform less cumbersome. In the second step we shall guess a fundamental solution of the heat equation by looking at the transformed equation. In the final step we shall prove that $p_{\mathbf{H}}$ as given in Theorem 7.4.1 has the required initial value property for $t \downarrow 0$.

The Abel transformation shows up when we transform radial eigenfunctions of the Laplacian on **H** into eigenfunctions which are constant along horocycles. To make this precise we introduce horocyclic coordinates centered at infinity. They are defined as follows (cf. Chapter 1). For any complex number $z = x + iy \in \mathbf{H}$ we set

$$r = \log y$$

and define (r, x) to be the horocyclic coordinates of z. The metric tensor with respect to these coordinates looks as follows,

$$ds^2 = dr^2 + e^{-2r}\, dx^2$$

(cf. (1.1.10) for $t = x$ and $\rho = -r$). For any function $v = v(r, x)$ which is constant along the horocycles $r = \text{const}$, the Laplacian applied to v is given by the expression

(7.4.4) $\quad \Delta v = -\dfrac{\partial^2 v}{\partial r^2} + \dfrac{\partial v}{\partial r}.$

For any function u which is radial, say which is constant along distance circles centered at the point $p_0 = i$, the Laplacian of u becomes

(7.4.5) $$\Delta u = -\frac{\partial^2 u}{\partial \rho^2} - \coth \rho \frac{\partial u}{\partial \rho},$$

where ρ denotes the hyperbolic distance to i. An advantage of using functions which are constant along horocycles is that (7.4.4) is simpler to work with than (7.4.5). Thus, assume that a *radial* function u is given in the first place. How can we transform it into a horocyclic function? Since certain decay conditions will show up, let us assume that u has the form

(7.4.6) $\qquad u = u(r, x) = H(\rho(r, x))$ with $H \in \mathscr{S}$,

where $\rho = \rho(r, x)$ is the hyperbolic distance from i to the point $x + ie^r$. By (7.4.2) we have

(7.4.7) $$\cosh \rho(r, x) = \cosh r + \tfrac{1}{2} x^2 e^{-r}.$$

Now we integrate u over the horocycles or, equivalently, over the action of the subgroup

$$N = \{ n_a \in \mathrm{Is}^+(\mathbf{H}) \mid n_a(r, x) = (r, x + a), a \in \mathbf{R} \}.$$

By the rapid decay of H, the following integrals are well defined.

(7.4.8)
$$\begin{aligned} v(r, x) &:= \int_N u \circ n_a(r, x) \, da \\ &= \int_{-\infty}^{+\infty} H(\rho(r, x + a)) \, da \\ &= 2 \int_0^{\infty} H(\rho(r, a)) \, da. \end{aligned}$$

This function is constant along the horocycles $r = \mathrm{const}$. Let us consider the Laplacian. From (7.4.5) we have

(7.4.9) $\qquad \Delta u(r, x) = -LH(\rho(r, x)),$

where L is the differential operator

$$L = \frac{d^2}{d\rho^2} + \coth \rho \, \frac{d}{d\rho}.$$

Clearly, $H \in \mathscr{S}$ implies $LH \in \mathscr{S}$ (observe that $\coth$ is an odd function). The computations in (7.4.10) will therefore be admissible. From the fact that the Laplacian commutes with the isometries, we have in conjunction with (7.4.9)

$$\Delta v(r, x) = \Delta \int_N u \circ n_a(r, x)\, da$$

$$= \int_N \Delta(u \circ n_a)(r, x)\, da$$

(7.4.10)

$$= \int_N (\Delta u) \circ n_a(r, x)\, da$$

$$= -2 \int_0^\infty LH(\rho(r, a))\, da.$$

The transformation (7.4.8) is not yet the one we are looking for; a slight modification will replace (7.4.4) by an even more manageable expression. Let

$$v^*(r, x) = e^{-r/2} v(r, x).$$

Then

(7.4.11) $$\left(\frac{\partial^2 v}{\partial r^2} - \frac{\partial v}{\partial r} \right)^* = \frac{\partial^2 v^*}{\partial r^2} - \frac{1}{4} v^*.$$

Hence, we finally define, with a normalization constant for later use,

$$w(r, x) = \frac{1}{\sqrt{2\pi}} e^{-r/2} \int_N u \circ n_a(r, x)\, da$$

(7.4.12)

$$= \frac{\sqrt{2}}{\sqrt{\pi}} e^{-r/2} \int_0^\infty H(\rho(r, a))\, da.$$

In the second integral, ρ is given by (7.4.7). Substituting ρ for a we have therefore

(7.4.13) $$w(r, x) = \frac{1}{\sqrt{\pi}} \int_r^\infty \frac{H(\rho) \sinh \rho}{\sqrt{\cosh \rho - \cosh r}}\, d\rho = A[H](r).$$

This is where the Abel transform shows up.

As mentioned before, we have $LH \in \mathcal{S}$. Applying (7.4.13) and (7.4.12) we obtain

$$A[LH](r) = \frac{\sqrt{2}}{\sqrt{\pi}} e^{-r/2} \int_0^\infty LH(\rho(r, a))\, da.$$

Using (7.4.10), (7.4.4), (7.4.11) and (7.4.13) in this order we get

(7.4.14) $$A[LH](r) = \left(\frac{d^2}{dr^2} - \frac{1}{4}\right)A[H](r).$$

In the second step we now show a way of guessing a fundamental solution of the heat equation. This is a function of two spaces variables and a time variable, say t, where $t \in]0, \infty[$. Since **H** is a two-point homogeneous space, the heat kernel should only depend on the *distance* of the two space variables. We therefore let one of the space variables be the point i and represent the other variable in the form $z = x + ie^r$. We then look for a function $p_t = p_t(r, x)$ which is a solution of the equation

(7.4.15) $$\Delta p_t + \frac{\partial p_t}{\partial t} = 0$$

subject to the initial condition

(7.4.16) $$\lim_{t \downarrow 0} \int_H p_t(r, x)\varphi(r, x)\,d\mathbf{H} = \varphi(0, 0)$$

for any continuous function φ with compact support. ((0, 0) is the point i in horocyclic coordinates.) Since p_t depends only on the distance to i, p_t has the form

$$p_t(r, x) = P_t(\rho(r, x)),$$

where P_t is a solution to the equation

$$LP_t - \frac{\partial P_t}{\partial t} = 0$$

(cf. (7.4.9)). In view of the rapid decay of the heat kernel in Euclidean space (cf. (7.4.3)), it looks reasonable to continue our search under the additional assumption that $P_t \in \mathcal{S}$. We may then look for its transform

$$Q_t := A[P_t]$$

which should be a solution to the equation $A[LP_t] - A[\partial P_t/\partial t] = 0$ or, by virtue of (7.4.14) (and hoping that A and $\partial/\partial t$ commute) a solution to

(7.4.17) $$\frac{\partial^2 Q_t}{\partial r^2} - \frac{1}{4}Q_t - \frac{\partial Q_t}{\partial t} = 0.$$

This is almost the heat equation in $\mathbf{R}^1$ whose solution is given in (7.4.3) (with $x = 0$ and $y = r$). Thus, we try in (7.4.17) with

$$Q_t(r) = \alpha(t)\,e^{-\beta r^2/t}$$

and find

$$\alpha(t) = \text{const } t^{-1/2} e^{-t/4}, \qquad \beta = \tfrac{1}{4}.$$

Anticipating the constant we finally define

(7.4.18) $$Q_t(r) := \frac{\sqrt{2}}{4\pi} t^{-1/2} e^{-t/4} e^{-r^2/4t}.$$

We then have the following.

The function $(r, t) \mapsto Q_t(r)$ belongs to $\mathscr{S}^I$ with $I =]0, \infty[$ (cf. (7.3.11)) and satisfies the equation (7.4.17).

We may therefore apply A^{-1} to obtain the candidate

(7.4.19) $$p_t(r, x) = P_t(\rho(r, x)) \text{ with } P_t = A^{-1}[Q_t].$$

By virtue of Theorem 7.3.10 this is

(7.4.20)
$$P_t(\rho) = \frac{-1}{\sqrt{\pi}} \int_\rho^\infty \frac{\partial Q_t(r)/\partial r}{\sqrt{\cosh r - \cosh \rho}} \, dr$$

$$= \frac{\sqrt{2}}{(4\pi t)^{3/2}} e^{-t/4} \int_\rho^\infty \frac{r e^{-r^2/4t}}{\sqrt{\cosh r - \cosh \rho}} \, dr.$$

In the third step we go through all the preceding arguments and prove that p_t is indeed a solution to (7.4.15) and (7.4.16).

First we observe using Theorem 7.3.13 that the function $(\rho, t) \mapsto P_t(\rho)$ belongs to $\mathscr{S}^I$ with $I =]0, \infty[$ (cf. (7.3.11)). Next we see in (7.4.20) that in order to differentiate P_t with respect to t we may pass the differentiation under the integration sign. Hence, by Theorem 7.3.13

(7.4.21)
$$\frac{\partial^n P_t}{\partial t^n} = A^{-1}\left[\frac{\partial^n Q_t}{\partial t^n}\right] \in \mathscr{S}^I \text{ and}$$

$$\frac{\partial^n Q_t}{\partial t^n} = A\left[\frac{\partial^n P_t}{\partial t^n}\right] \text{ for } n = 1, 2, \dots.$$

With (7.4.6) and (7.4.9) we have

$$\Delta p_t(r, x) + \frac{\partial p_t}{\partial t}(r, x) = -LP_t(\rho(r, x)) + \frac{\partial P_t}{\partial t}(\rho(r, x)).$$

Using that $A[P_t] = Q_t$, we see from (7.4.14) that

$$A\left[LP_t - \frac{\partial P_t}{\partial t}\right] = \left(\frac{\partial^2}{\partial r^2} - \frac{1}{4}\right)Q_t - \frac{\partial Q_t}{\partial t} = 0.$$

Using that A is invertible we obtain $LP_t - \partial P_t/\partial t = 0$. Thus p_t is indeed a solution to (7.4.15).

Condition (7.4.16) is tested by integrating p_t over $\mathbf{H}$. (Recall the rapid decay of p_t.) In the horocyclic coordinates (r, x) the area element is

$$d\mathbf{H} = e^{-r} dr\, dx.$$

Integrating over the action of N we compute, using (7.4.12) and (7.4.13)

$$\int_{\mathbf{H}} p_t\, d\mathbf{H} = \int_{-\infty}^{+\infty} \left[\int_N p_t(n_a(r, x))\, da \right] e^{-r}\, dr$$

$$= (2\pi)^{1/2} \int_{-\infty}^{+\infty} A[P_t](r)\, e^{-r/2}\, dr,$$

where $A[P_t] = Q_t$ is given by (7.4.18). Replacing $r/(2\sqrt{t})$ by s and using that

$$\int_{\mathbf{R}} e^{-s^2}\, ds = \pi^{1/2},$$

we obtain

(7.4.22) $$\lim_{t \downarrow 0} \int_{\mathbf{H}} p_t\, d\mathbf{H} = 1.$$

Since p_t decays rapidly, the standard arguments now imply (7.4.16).

The function in Theorem 7.4.1 is the following.

(7.4.23) $$p_{\mathbf{H}}(z, w, t) = P_t(\text{dist}(z, w)).$$

By (7.4.21) we have

(7.4.24) $$p_{\mathbf{H}} \in C^{\infty}(\mathbf{H} \times \mathbf{H} \times \,]0, \infty[),$$

and by (7.4.15)

$$(\Delta_z + \frac{\partial}{\partial t}) p_{\mathbf{H}}(z, w, t) = 0.$$

From the rapid decay of p_t we conclude, using Green's formula (7.2.2), that

$$\int_{\mathbf{H}} \Delta p_t\, d\mathbf{H} = 0.$$

Since $\Delta p_t = -\partial p_t/\partial t$, it follows from (7.4.22) that

$$\int_{\mathbf{H}} p_t\, d\mathbf{H} = 1$$

for all $t > 0$. Hence we have the following.

7.4.25 Lemma. *For any $z \in \mathbf{H}$,*

$$\int_{\mathbf{H}} p_{\mathbf{H}}(z, w, t) \, d\mathbf{H}(w) = 1.$$ ◇

In the literature there exist quite precise estimates of $p_{\mathbf{H}}$ and of the heat kernel of other symmetric spaces (Anker [1], Davies [1], Davies-Mandouvalos [1]). For the inequality in Theorem 7.2.5 the following simple estimate will be sufficient. The symbol "const" after the inequality sign means that there exists a constant for which the inequality holds.

7.4.26 Lemma.

$$p_{\mathbf{H}}(z, w, t) \leq \text{const } t^{-1} e^{-d^2(z, w)/8t}.$$

Proof. For $\rho > 0$ we have

$$\int_{\rho}^{2\rho} \frac{r \, e^{-r^2/4t}}{\sqrt{\cosh r - \cosh \rho}} \, dr \leq 2\rho \, e^{-\rho^2/4t} \int_{\rho}^{2\rho} \frac{dr}{\sqrt{(r-\rho)\sinh\rho}} \leq 4\rho \, e^{-\rho^2/4t}.$$

From the inequality $x \, e^{-x^2} \leq e^{-x^2/2}$,

$$4\rho \, e^{-\rho^2/4t} \leq \text{const } \sqrt{t} \, e^{-\rho^2/8t}.$$

Using that $\cosh r - \cosh \rho \geq \frac{1}{2}(r-\rho)^2$ we obtain, setting $s = r - 2\rho$,

$$\int_{2\rho}^{\infty} \frac{r \, e^{-r^2/4t}}{\sqrt{\cosh r - \cosh \rho}} \, dr \leq 3 \int_{2\rho}^{\infty} e^{-r^2/4t} \, dr$$

$$\leq 3 \, e^{-\rho^2/t} \int_{0}^{\infty} e^{-s^2/4t} \, ds$$

$$\leq \text{const } \sqrt{t} \, e^{-\rho^2/t}.$$

Together with (7.4.23) and (7.4.20) this proves the lemma. ◇

From Lemma 7.4.25 and Lemma 7.4.26 we have immediately

7.4.27 Lemma. *Let $f: \mathbf{H} \to \mathbf{C}$ be a continuous bounded function (f need not have compact support). Then*

$$\lim_{t \downarrow 0} \int_{\mathbf{H}} p_{\mathbf{H}}(z, w, t) f(w) \, d\mathbf{H}(w) = f(z)$$

uniformly on compact sets in $\mathbf{H}$. ◇

Theorem 7.4.1 is now proved. ◇

7.5 The Heat Kernel of $\Gamma \backslash \mathbf{H}$

In this section Γ is a subgroup of PSL(2, **R**) which operates freely and properly discontinuously on the upper half-plane **H** with compact fundamental domain. The goal is to show that the heat kernel p_M of the compact Riemann surface $M = \Gamma \backslash \mathbf{H}$ is obtained by summation of $p_\mathbf{H}$ over the action of Γ (Theorem 7.5.11). In view of the relationship between the lengths of the closed geodesics and the eigenvalues of the Laplacian we consider, more generally, any complex valued function

(7.5.1) $\qquad k(z, w) = K(\text{dist}(z, w)), \qquad z, w \in \mathbf{H},$

where $K : [0, \infty[\to \mathbf{C}$ is a so called *generating function* which satisfies the decay condition

(7.5.2) $\qquad |K(\rho)| \leq \text{const } e^{-\rho(1+\delta)}$

for some $\delta > 0$. Typical examples of generating functions are $K(\rho) = P_t(\rho)$ for fixed t (they generate the heat kernel $p_\mathbf{H}$) or $K(\rho) = (\cosh \rho)^{-s}$ for fixed $s \in \mathbf{C}$ with $\mathcal{R}e\, s > 1$.

The first step is to define a Γ-bi-invariant kernel. For this we need that the summations over Γ are well defined. In the proof of the next lemma we use *Dirichlet fundamental domains*. They are defined as follows: one fixes any point $p \in \mathbf{H}$, and sets $D = \{ z \in \mathbf{H} \mid \text{dist}(z, p) \leq \text{dist}(Tz, p) \text{ for any } T \in \Gamma \}$. D is a convex geodesic polygon domain (see Beardon [1] for properties of Dirichlet domains).

7.5.3 Lemma. *Let $p \in \mathbf{H}$ and define for $m = 0, 1, \ldots,$*

$$\Gamma(m) = \{ T \in \Gamma \mid m \leq \text{dist}(p, Tp) < m+1 \}.$$

Then $\Gamma(m)$ has cardinality $\#\Gamma(m) \leq \text{const } e^m$.

Proof. Let D be a Dirichlet fundamental domain of Γ which contains p, and let a be its diameter. The domains $T(D), T \in \Gamma(m)$ are contained in the disk of radius $r = m + 1 + a$ around p. This disk has area $= 2\pi(-1 + \cosh r)$, and the domains do not overlap. The lemma follows. ◊

As an immediate consequence we have the following.

7.5.4 Lemma. *Let k be given as in (7.5.1) and (7.5.2). For any compact subset $A \subset \mathbf{H}$ and for any $\varepsilon > 0$ there exists a finite subset $\Delta \subset \Gamma$ such that*

$$\sum_{T \in \Gamma - \Delta} |k(z, Tw)| < \varepsilon \text{ for all } z, w \in A. \qquad ◊$$

In view of Lemmata 7.5.3 and 7.5.4 the following definition makes sense.

7.5.5 Definition and Lemma. *The function*

$$\mathcal{K}_\Gamma(z, w) := \sum_{T \in \Gamma} k(z, Tw) = \sum_{T \in \Gamma} K(\mathrm{dist}(z, Tw)), \quad z, w \in \mathbf{H}$$

is Γ-bi-invariant,

$$\mathcal{K}_\Gamma(Rz, Sw) = \mathcal{K}_\Gamma(z, w) \text{ for all } R, S \in \Gamma.$$

$\mathcal{K}_\Gamma$ induces a kernel $\mathcal{K}_M$ on the quotient $M = \Gamma \backslash \mathbf{H}$ defined by

$$\mathcal{K}_M(x, y) := \mathcal{K}_\Gamma(\tilde{x}, \tilde{y}), \quad x, y \in M,$$

where $\tilde{x}$ and $\tilde{y}$ are any inverse images of x and y under the projection $\mathbf{H} \to \Gamma \backslash \mathbf{H} = M$. The kernel $\mathcal{K}_M$ is symmetric,

$$\mathcal{K}_M(x, y) = \mathcal{K}_M(y, x), \quad x, y \in M.$$

Proof. Since $k(z, w)$ depends only on the distance, we have $k(Rz, TSw) = k(z, R^{-1}TSw)$. As T runs through Γ so does $R^{-1}TS$, and the bi-invariance of $\mathcal{K}_\Gamma$ follows. The symmetry follows from the relation

$$\sum_{T \in \Gamma} k(\tilde{x}, T\tilde{y}) = \sum_{T \in \Gamma} k(T\tilde{y}, \tilde{x}) = \sum_{T \in \Gamma} k(\tilde{y}, T^{-1}\tilde{x}). \quad \diamond$$

We next consider the differentiability properties of $\mathcal{K}_M$. In the applications, the generating function K usually depends on additional real or complex parameters. We include the differentiability properties with respect to these parameters.

Let B be a real or complex parameter space. In our later applications, B will be either an interval on the real line or an open subset of the complex plane. At present we assume that B is an open subset of $\mathbf{R}^d$ or of $\mathbf{C}^d$.

Let A be a manifold. In Definition 7.2.3 we introduced the class $C^{\nu, \lambda}(A, B; \mathbf{C})$ of all continuous functions $F : A \times B \to \mathbf{C}$ whose partial derivatives (with respect to local coordinates) exist and are continuous on $A \times B$ up to the ν-th derivatives with respect to A and up to the λ-th derivatives with respect to B. We denote these derivatives by $F^{(n, \ell)}$, $n = 0, \ldots, \nu$, $\ell = 0, \ldots, \lambda$. When $B \subset \mathbf{C}^d$, the derivatives with respect to B are understood to be *complex derivatives*.

In what follows, a function $F : [0, \infty[\times B \to \mathbf{C}$ of a given differentiability class is *even* if it is the restriction to $[0, \infty[\times B$ of a function $\mathbf{R} \times B \to \mathbf{C}$ of the same differentiability class which is even with respect to the first variable.

7.5.6 Lemma. *Assume that the generating function*

$$K = K(\rho, b) : [0, \infty[\times B \to \mathbf{C}$$

is an even function belonging to $C^{\nu, \lambda}([0, \infty[, B; \mathbf{C})$ and satisfying the inequality

$$|K^{(n, \ell)}(\rho, b)| \le c\, e^{-\rho(1+\delta)} \text{ on } [0, \infty[\times B,$$

$n = 0, \ldots, \nu; \ell = 0, \ldots, \lambda$, *where c and δ are positive constants. Then $\mathcal{K}_M$ belongs to $C^{\nu, \lambda}(M \times M, B; \mathbf{C})$.*

Proof. It suffices to prove that $\mathcal{K}_\Gamma$ belongs to $C^{\nu, \lambda}(\mathbf{H} \times \mathbf{H}, B; \mathbf{C})$. We first observe that

$$k \in C^{\nu, \lambda}(\mathbf{H} \times \mathbf{H}, B; \mathbf{C}).$$

This is true because K is an even function and because the function $(z, w) \mapsto \text{dist}^2(z, w)$ is of class $C^\infty(\mathbf{H} \times \mathbf{H})$. (The point here is that k is differentiable on the diagonal of $\mathbf{H} \times \mathbf{H}$).

Let $U \subset \mathbf{H}$ be an open disk and let $z_0 \in \mathbf{H}$ be a point at some positive distance from U. Then

$$(7.5.7) \qquad \Gamma = \bigcup_{m=0}^{\infty} \Gamma(m)$$

with $\Gamma(m) = \{T \in \Gamma \mid m \le \text{dist}(z_0, Tz_0) < m + 1\}$. By Lemma 7.5.3 we have the estimate

$$(7.5.8) \qquad \# \Gamma(m) \le \text{const } e^m.$$

It suffices therefore to show that any of our partial derivations, say $\mathcal{D}$, satisfies an inequality of the following type, where the constant may depend on U.

$(7.5.9) \quad |\mathcal{D}[k(z, Tw)]| \le \text{const } e^{-m(1+\delta)}$ for all $z, w \in U, T \in \Gamma(m)$.

$\mathcal{D}[.]$ is $\mathcal{D}$ applied to the function $(z, w) \mapsto k(z, Tw)$. This derivative has the form

$$\mathcal{D}[k(z, Tw)] = \sum_{j=1}^{\nu} (\mathcal{D}_B K^{(j)})(\text{dist}(z, Tw)) D_j[\text{dist}(z, Tw)],$$

where $\mathcal{D}_B$ is a partial derivation with respect to B; $K^{(j)} = \partial^j K / \partial \rho^j$, and D_j is a partial derivation with respect to $\mathbf{H} \times \mathbf{H}$. We introduce polar coordinates for $\mathbf{H}$ centered at z_0 and let (ρ, σ) and (r, s) be the polar coordinates of $z \in U$ and $w \in U$ respectively. In this way (ρ, σ, r, s) are coordinates for $U \times U$ (recall that z_0 has positive distance from U).

We shall write $z = z(\rho, \sigma)$ and $w = w(r, s)$. The proof of (7.5.9) is now reduced to show that for any partial derivation $\mathscr{D}$ with respect to (ρ, σ, r, s) on $U \times U$ there exists a positive constant c satisfying

(7.5.10) $\qquad |\mathscr{D}[\mathrm{dist}(z(\rho, \sigma), Tw(r, s))]| \leq c,$

for all $z, w \in U$, and all $T \in \Gamma$. Here we use a geometric argument. Let γ be the geodesic joining z_0 and Tz_0. Without loss of generality we may assume that σ is the angle between γ and the geodesic arc joining z_0 and z. Similarly we may assume that s is the angle between γ and the geodesic arc joining Tz_0 and Tw (otherwise add a constant to s). This is shown in Fig. 7.5.1.

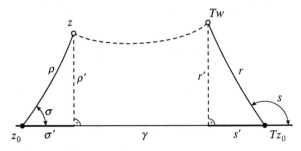

Figure 7.5.1

We drop the perpendiculars ρ' from z to γ and r' from Tw to γ to obtain right-angled geodesic triangles with sides σ', ρ', ρ and r, r', s'. The partial derivatives of ρ', σ' and r', s' with respect to ρ, σ, r, s are bounded independently of T, and (7.5.10) follows from formula (2.3.2) which in the present case looks as follows:

$$\cosh(\mathrm{dist}(z, Tw)) = \cosh\rho' \cosh r' \cosh(\mathrm{dist}(z_0, Tz_0) - \sigma' + s')$$
$$- \sinh\rho' \sinh r'$$

(σ' and s' are oriented lengths). This proves Lemma 7.5.6. $\qquad \diamond$

7.5.11 Theorem. *Let* $p_{\mathbf{H}}(z, w, t) = P_t(\mathrm{dist}(z, w))$ *be the heat kernel of the hyperbolic plane and define*

$$p_\Gamma(z, w, t) = \sum_{T \in \Gamma} p_{\mathbf{H}}(z, Tw, t), \qquad p_M(x, y, t) = p_\Gamma(\tilde{x}, \tilde{y}, t)$$

as in Definition 7.5.5, where $\tilde{x}$ and $\tilde{y}$ are inverse images under the natural projection $\mathbf{H} \to M = \Gamma\backslash\mathbf{H}$. Then p_M is a fundamental solution of the heat equation on M. Moreover, $p_M \in C^\infty(M \times M \times \,]0, \infty[)$ and

$$p_M \leq \mathrm{const}\, t^{-1} \text{ for } 0 < t < 1$$

with a constant which depends on M.

Proof. By (7.4.21), the generating function P_t satisfies the hypothesis of Lemma 7.5.6 so that $p_M \in C^\infty(M \times M \times \,]0, \infty[)$. By (7.5.7) - (7.5.9), the equation $\Delta_z p_H = -\partial p_H / \partial t$ yields $\Delta_z p_\Gamma = -\partial p_\Gamma / \partial t$, and therefore $\Delta_x p_M = -\partial p_M / \partial t$. The symmetry of p_M is given by Lemma 7.5.5.

The initial condition is checked as follows. Let $f: M \to \mathbf{R}$ be a continuous function and let F be a Γ-automorphic lift of f in $\mathbf{H}$. If M^* denotes a compact fundamental domain of Γ with boundary measure 0 (for instance a Dirichlet fundamental domain), then

$$\int_{M^*} p_\Gamma(z, w, t) F(w) \, d\mathbf{H}(w) = \sum_{T \in \Gamma} \int_{M^*} p_\mathbf{H}(z, Tw, t) F(w) \, d\mathbf{H}(w)$$

$$= \sum_{T \in \Gamma} \int_{T(M^*)} p_\mathbf{H}(z, w, t) F(w) \, d\mathbf{H}(w)$$

$$= \int_\mathbf{H} p_\mathbf{H}(z, w, t) F(w) \, d\mathbf{H}(w),$$

and the initial condition (iii) of Definition 7.2.4 is checked with Lemma 7.4.27. Finally, the inequality $p_M \leq \text{const } t^{-1}$ or, equivalently, the inequality $p_\Gamma \leq \text{const } t^{-1}$ for $t \in \,]0, 1[$ follows from Lemma 7.4.26 and Lemma 7.5.3. $\diamond$

Chapter 8

Small Eigenvalues

The asymptotic length distribution of the closed geodesics on a compact Riemann surface S is strongly influenced by the eigenvalues of the Laplacian in the interval $[0, \frac{1}{4}]$. In the prime number theorem (Theorem 9.6.1) and in many applications of Selberg's trace formula, the distinction between eigenvalues above and eigenvalues below $\frac{1}{4}$ is essentially for technical reasons and has to do with the fact that for $x \to \infty$ the function $x \mapsto \sin(x(\lambda_n - \frac{1}{4})^{1/2})$ is bounded if $\lambda_n \geq \frac{1}{4}$ but grows exponentially if $\lambda_n < \frac{1}{4}$. Eigenvalues below $\frac{1}{4}$ are called *small* (Huber [2], II, p. 386). In this chapter we show that there are also geometric reasons why the eigenvalues below $\frac{1}{4}$ should be distinguished from those above $\frac{1}{4}$.

In Section 8.1 we state the theorems and show a connection between the eigenvalue problem and the isoperimetric problem. In Sections 8.2 and 8.3 we review our main tools, the minimax principles and Cheeger's inequality. The proofs then follow in Section 8.4, together with a brief history and a discussion of related results.

8.1 The Interval $[0, \frac{1}{4}]$

Historically, the number $\frac{1}{4}$ shows up for technical reasons. In Maaß [1], p.142 the eigenvalues of the Laplacian are written in the form $\lambda = r^2 + \frac{1}{4}$ in order to explain the Γ-factors

$$\Gamma(\tfrac{1}{2}(s+ir))\Gamma(\tfrac{1}{2}(s-ir)) \quad \text{and} \quad \Gamma(\tfrac{1}{2}(s+1+ir))\Gamma(\tfrac{1}{2}(s+1-ir))$$

($s, r \in \mathbb{C}$; Γ is the gamma function) occurring in the functional equation of certain Dirichlet series. The notation $\lambda = r^2 + \frac{1}{4}$ was later adopted by Selberg [2] and Huber [1] and lead to various number theoretic and geometric results, such as Theorems 9.6.1 and 9.7.1 in the next chapter, which depend on the eigenvalues in the interval $[0, \frac{1}{4}]$. It is therefore of interest to know whether there is some a priori information about the distribution of the eigenvalues in the interval $[0, \frac{1}{4}]$.

The first three theorems which follow are from Buser [1]. Theorem 8.1.4 is new. $\mathcal{R}_g$ denotes the set (or moduli space) of all compact Riemann surfaces of genus g, $g \geq 2$.

8.1.1 Theorem. *For any* $S \in \mathcal{R}_g$, *we have* $\lambda_{4g-2}(S) > \frac{1}{4}$.

8.1.2 Theorem. *For any* $n \in \mathbf{N}$ *and for arbitrarily small* $\varepsilon > 0$, *there exists* $S' \in \mathcal{R}_g$ *with* $\lambda_n(S') \leq \frac{1}{4} + \varepsilon$.

At first sight it is quite surprising that there is a universal bound for the number of eigenvalues in $[0, \frac{1}{4}]$ but not for any interval $[0, \frac{1}{4} + \varepsilon]$, $\varepsilon > 0$. We shall therefore give a heuristic argument which makes this phenomenon more plausible.

From Cheeger's inequality combined with the minimax principles, it will follow that for a compact Riemann surface to have small eigenvalues, the surface must be decomposable into domains whose isoperimetric quotient is smaller than 1 (the isoperimetric quotient is the boundary length divided by the enclosed surface area; see below). The number of such domains must be of the same order of magnitude as the number of small eigenvalues. We are thus led to isoperimetric considerations on a Riemann surface. There we have the following analogous phenomenon.

Let D be a compact domain with piecewise smooth boundary ∂D on a Riemann surface, as shown in Fig. 8.1.1, which is either simply connected or an annulus. Then, independently of the size of D,

$$\frac{\ell(\partial D)}{\text{area } D} > 1.$$

Figure 8.1.1

This may be seen by using polar coordinates in the first case and Fermi coordinates in the second. (The details will follow in Section 8.4 in the proof of Theorem 8.1.1.) The *isoperimetric quotient* on the left-hand side of the above inequality will be denoted by $J(D)$. If the surface S has a sufficiently long collar, we may cut out arbitrarily many disjoint annuli D with $J(D)$ as close to 1 as we wish. We point out that the number of these annuli is not restricted by the genus because their union is again an annulus.

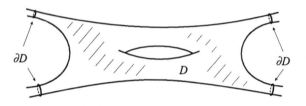

Figure 8.1.2

Now suppose that we want to subdivide S into domains with $J(D) < 1$. Then any D must have some signature (h, m), where either $m \geq 3$ or $h \geq 1$, a condition which "consumes" topology. Fig. 8.1.2 shows such a domain for comparison.

Hence, the number of domains in a decomposition with $J \leq 1$ is bounded by the genus, but no such bound exists for decompositions with $J \leq 1 + \varepsilon$. We convert these arguments into a correct proof in Section 8.4.

We state two more theorems. In view of the prime number theorem with error terms (Theorem 9.6.1), it is of interest to know whether there are eigenvalues in $]0, \frac{1}{4}[$ at all. This question already arises in a seemingly unnoticed paper of Delsarte [1, 2]. The first examples of compact Riemann surfaces with small eigenvalues were given by Randol [1]. The following holds.

8.1.3 Theorem. *For any $\delta > 0$, there exists $S' \in \mathcal{R}_g$ with $\lambda_{2g-3}(S') \leq \delta$.*

This does not fully match with Theorem 8.1.1. However, we shall prove the following result.

8.1.4 Theorem. *There exists a universal constant $c > 0$ which is independent of g and such that $\lambda_{2g-2}(S) \geq c$ for any $S \in \mathcal{R}_g$ and any $g \geq 2$.*

The proof in Section 8.4 will yield the estimate $c > 10^{-12}$ but the conjecture is that $c = \frac{1}{4}$.

8.2 The Minimax Principles

The relation between the eigenvalues and the isoperimetric quotients mentioned in the preceding section is obtained via the minimax principles and Cheeger's inequality. We reprove them here and in the next section for convenience.

Let M be an arbitrary compact Riemannian manifold without boundary. For any smooth function f on M we set

$$\operatorname{supp} f = \text{closure of } \{x \in M \mid f(x) \neq 0\}.$$

The domain $\operatorname{supp} f$ is called the *support* of f. For $k = 0, 1, \ldots$ we let $\lambda_k(M)$ be the k-th eigenvalue of the Laplacian of M.

8.2.1 Theorem. (Minimax principles). (i) *Let* $f_0, \ldots, f_k \in C^\infty(M)$ *with norm*

$$\int_M f_j^2 \, dM = 1$$

be such that $\operatorname{vol}(\operatorname{supp} f_i \cap \operatorname{supp} f_j) = 0$ *for* $0 \leq i < j \leq k$. *Then*

$$\lambda_k(M) \leq \max_{0 \leq \kappa \leq k} \int_M \|\operatorname{grad} f_\kappa\|^2 \, dM.$$

(ii) *Let* $N_1, \ldots, N_k$ *be compact domains with positive measure such that* $M = N_1 \cup \ldots \cup N_k$, $\operatorname{vol}(N_i \cap N_j) = 0$ *for* $1 \leq i < j \leq k$, *and let*

$$v(N_\kappa) = \inf \int_{N_\kappa} \|\operatorname{grad} f\|^2 \, dM,$$

where f ranges over all smooth functions with

$$\int_{N_\kappa} f^2 \, dM = 1 \quad \text{and} \quad \int_{N_\kappa} f \, dM = 0.$$

Then

$$\lambda_k(M) \geq \min_{1 \leq \kappa \leq k} v(N_\kappa).$$

Proof. Let $\varphi_0, \varphi_1, \ldots$ be the complete orthonormal system of eigenfunctions in $C^\infty(M)$ with eigenvalues $\lambda_0(M), \lambda_1(M), \ldots$. Since the f_κ in (i) are linearly independent, we find a linear combination $f = \beta_0 f_0 + \ldots + \beta_k f_k$ with norm

$$\int_M f^2 \, dM = 1$$

and orthogonal to $\varphi_0, \ldots, \varphi_{k-1}$. Now let α_j be the Fourier coefficient

$$\alpha_j = \int_M f \varphi_j \, dM.$$

By Green's formula we then have for all $m > k$,

$$0 \leq \int_M \| \operatorname{grad}(f - \sum_{j=k}^m \alpha_j \varphi_j) \|^2 \, dM$$

$$= \int_M \| \operatorname{grad} f \|^2 \, dM - \sum_{j=k}^m \alpha_j^2 \lambda_j.$$

Here $\lambda_j(M) \geq \lambda_k(M)$, and by Parseval's relation

$$\sum_{j=k}^\infty \alpha_j^2 = 1.$$

Recalling that $\operatorname{vol}(\operatorname{supp} f_i \cap \operatorname{supp} f_j) = 0$ for $i \neq j$ and recalling that f has norm

$$\int_M f^2 \, dM = \sum_{i=0}^k \beta_i^2 = 1,$$

we obtain from this

$$\lambda_k(M) \leq \int_M \| \operatorname{grad} f \|^2 \, dM = \sum_{i=0}^k \beta_i^2 \int_M \| \operatorname{grad} f_i \|^2 \, dM$$

$$\leq \max_{0 \leq \kappa \leq k} \int_M \| \operatorname{grad} f_\kappa \|^2 \, dM.$$

This proves (i).

For the proof of (ii) we work in $L^2(M)$ instead of $C^\infty(M)$. Let χ_κ be the characteristic function defined by $\chi_\kappa(x) = 1$ if $x \in N_\kappa$ and $\chi_\kappa(x) = 0$ if $x \notin N_\kappa$, $\kappa = 1, \ldots, k$. Consider the linear combination $\varphi = \alpha_0 \varphi_0 + \ldots + \alpha_k \varphi_k$ with norm

$$\int_M \varphi^2 \, dM = 1$$

and orthogonal to $\chi_1, \ldots, \chi_k$, so that on each N_κ, φ has the mean value

$$\int_{N_\kappa} \varphi \, dM = 0.$$

By the definition of $v(N_\kappa)$, we then have

$$\int_{N_\kappa} \|\operatorname{grad}\varphi\|^2\, dM \geq \nu(N_\kappa)\int_{N_\kappa} \varphi^2\, dM$$

and therefore

$$\int_M \|\operatorname{grad}\varphi\|^2\, dM \geq \min_{1\leq \kappa \leq k} \nu(N_\kappa).$$

On the other hand, since

$$\int_M \varphi^2\, dM = \alpha_0^2 + \ldots + \alpha_k^2 = 1,$$

we have, by Green's formula,

$$\int_M \|\operatorname{grad}\varphi\|^2\, dM = \int_M \varphi\Delta\varphi\, dM = \sum_{j=0}^k \lambda_j(M)\alpha_j^2 \leq \lambda_k(M). \quad \diamond$$

For a more general form of the minimax principles we refer to Chavel [1] Section I. 5.

8.3 Cheeger's Inequality

For the estimates of $\nu(N_\kappa)$ we use isoperimetric quotients. Cheeger [1] and Yau [1] introduced the following constant. For our purposes we assume that M is two dimensional. For a general account of Cheeger's inequality we refer the reader to Chavel [1] and Buser [6]. See also Huxley [1] and Schmutz [2] for Cheeger's inequality for mixed boundary value problems.

8.3.1 Definition. Let M be a compact connected two-dimensional Riemannian manifold without boundary. Let N be a closed subset of M with positive area and with piecewise smooth boundary (possibly $N = M$). Then *Cheeger's isoperimetric constant* (of Neumann type) is the quantity

$$h(N) = \inf \frac{\ell(A)}{\min\{\operatorname{area} B, \operatorname{area} B'\}},$$

where A ranges over the set of all finite unions of piecewise smooth curves on N which separate N into disjoint relatively open (not necessarily connected) subsets B and B' such that $N - A = B \cup B'$ and $A \subset \partial B \cap \partial B'$.

We point out that $h(N)$ is not a scaling invariant, but $h(N)/\nu(N)$ is. Here $\nu(N)$ is the first eigenvalue of the Neumann problem

$$\text{(8.3.2)} \qquad v(N) := \inf \int_N \|\operatorname{grad} f\|^2 \, dM,$$

where f ranges over the smooth functions on M satisfying

$$\int_N f^2 \, dM = 1 \quad \text{and} \quad \int_N f \, dM = 0.$$

8.3.3 Theorem. (Cheeger's inequality). *Let N be a closed domain in M with positive area and with piecewise smooth boundary (possibly $N = M$). Then*

$$v(N) \geq \tfrac{1}{4} h^2(N).$$

Proof. Let f be as in (8.3.2). Then for any constant $c \in \mathbf{R}$

$$\int_N (f+c)^2 \, dM = \int_N (f^2 + c^2) \, dM \geq \int_N f^2 \, dM.$$

Now take c such that each of

$$N_1 := \{x \in N \mid (f+c)(x) < 0\} \quad \text{and} \quad N_2 := \{x \in N \mid (f+c)(x) > 0\}$$

has area $\leq \tfrac{1}{2} \operatorname{area} N$. It suffices therefore to show that on either domain the function $F := f + c$ satisfies

$$\int_{N_i} \|\operatorname{grad} F\|^2 \, dM \geq \tfrac{1}{4} h^2(N) \int_{N_i} F^2 \, dM.$$

The relation $\operatorname{grad}(F^2) = 2F \operatorname{grad} F$ and the Cauchy-Schwarz inequality yield

$$4 \int_{N_i} F^2 \, dM \int_{N_i} \|\operatorname{grad} F\|^2 \, dM \geq \left(\int_{N_i} \|\operatorname{grad} F^2\| \, dM \right)^2.$$

It remains to prove that the function $u := F^2$ satisfies the following inequality for $i = 1, 2$

$$\text{(8.3.4)} \qquad \int_{N_i} \|\operatorname{grad} u\| \, dM \geq h(N) \int_{N_i} u \, dM.$$

The inequality will be proved by means of the co-area formula (Chavel [1], p. 85). In order to bypass the technical difficulties in the proof of the co-area formula, we first prove (8.3.4) under the assumption that in an open neighborhood of N in M the function u has only finitely many critical values and that the level lines

$$A_i(t) = \{x \in N_i \mid u(x) = t\}$$

intersect ∂N transversally except for at most finitely many t. Then the co-area formula, in its simplest form, states that

(8.3.5) $$\int_{N_i} \|\operatorname{grad} u\| \, dM = \int_0^\infty \ell(A_i(t)) \, dt.$$

Roughly speaking, this formula is true because the level lines $A_i(t)$ are orthogonal to the gradient vectors, and because the infinitesimal strip between $A_i(t)$ and $A_i(t + dt)$ has width $dt/\|\operatorname{grad} u\|$ at each point. Let $\beta_i(t)$ be the area of

$$B_i(t) = \{x \in N_i \mid u(x) > t\}.$$

Then $\beta_i(t)$ is a piecewise differentiable monotone decreasing function of t, and the infinitesimal strip between $A(t)$ and $A(t + dt)$ has area $= -\beta_i'(t)$, where β_i' is the derivative of β_i. Finally, since area $B_i(t) \leq \operatorname{area} N_i \leq \frac{1}{2}\operatorname{area} N$, we have

$$\ell(A_i(t)) \geq h(N) \beta_i(t).$$

Then with (8.3.5) we obtain

$$\int_{N_i} \|\operatorname{grad} u\| \, dM \geq h(N) \int_0^\infty \beta_i(t) \, dt$$

$$= -h(N) \int_0^\infty t \beta_i'(t) \, dt$$

$$= h(N) \int_{N_i} u \, dM.$$

This proves (8.3.4) under the assumed regularity conditions. The general case is obtained by a standard approximation argument. It is also possible to give a completely elementary proof of (8.3.4) along the above lines by approximating both N and u piecewise linearly with the same method as in Section 5.4 (see the proof of Lemma 5.4.5). ◊

The following helpful fact was first observed by Yau [1].

8.3.6 Lemma. *Let N be as in Definition 8.3.1 and assume that N is connected (otherwise $h(N) = 0$). In the definition of $h(N)$ we then may restrict A to curve families for which B and B' are connected.*

Proof. The claim is that $\ell(A)/\min\{\text{area } B, \text{area } B'\} \geq h^*(N)$, where $h^*(N)$ is the infimum obtained under the restriction of the lemma.

For the proof of the claim we may restrict ourselves to curve families A which are union sets of pairwise disjoint closed Jordan curves and of Jordan arcs, where the latter have both endpoints on the boundary of N.

We proceed by induction over the number m of connected components of A. If $m = 1$, then B and B' are connected and nothing has to be proved. Now let the claim be proved for all $\tilde{A}$ with at most m components and let A have $m + 1$ components, say $A = \sigma_1 \cup \ldots \cup \sigma_{m+1}$. Nothing has to be proved if both B and B' are connected. We assume therefore that B' is the disjoint union $B' = P \cup B''$, where P is a connected component. We enumerate the components of A in such a way that for some $k < m + 1$, the components $\sigma_1, \ldots, \sigma_k$ are on the boundary of P and the $\sigma_{k+1}, \ldots, \sigma_{m+1}$ are on the boundary of B''. By the induction hypothesis we have

$$\ell(\sigma_1) + \ldots + \ell(\sigma_k) \geq h^*(N) \min\{\text{area } P, \text{area } B + \text{area } B''\},$$

$$\ell(\sigma_{k+1}) + \ldots + \ell(\sigma_{m+1}) \geq h^*(N) \min\{\text{area } B'', \text{area } B + \text{area } P\}.$$

This implies

$$\ell(\sigma_1) + \ldots + \ell(\sigma_{m+1}) \geq h^*(N) \min\{\text{area } P \cup B'', \text{area } B\}.$$

The lemma follows. ◇

In the next section we shall give an example showing that Cheeger's inequality is sharp.

8.4 Eigenvalue Estimates

We now prove Theorems 8.1.1 - 8.1.4. We first construct the necessary examples. For S' as in Theorem 8.1.3 we paste together $2g - 2$ pairwise isometric pairs of pants $Y_1, \ldots, Y_{2g-2}$ with boundary geodesics of length $\varepsilon/6$. For each such Y_κ we define a function f_κ as follows.

$$f_\kappa(x) = \begin{cases} \text{dist}(x, \partial Y_\kappa) & \text{if } x \in Y_\kappa \text{ and } \text{dist}(x, \partial Y_\kappa) \leq 1 \\ 1 & \text{if } x \in Y_\kappa \text{ and } \text{dist}(x, \partial Y_\kappa) \geq 1 \\ 0 & \text{if } x \notin Y_\kappa. \end{cases}$$

The area of the set $\{x \in Y_\kappa \mid \text{dist}(x, \partial Y_\kappa) \leq 1\}$ is equal to $\frac{\varepsilon}{2}\sinh(1)$ (use Proposition 3.1.8 and introduce Fermi coordinates as in Example 1.3.2). Since $\|\text{grad} f_\kappa\| = 1$ on this set and $\|\text{grad} f_\kappa\| = 0$ elsewhere, we can smooth the functions $f_1, \ldots, f_{2g-2}$ and then normalize them in such a way that they satisfy the hypothesis of Theorem 8.2.1(i) and such that they also satisfy

$$\int_{S'} \|\operatorname{grad} f_\kappa\|^2 \, dM < \varepsilon.$$

This proves Theorem 8.1.3. ◇

To prove Theorem 8.1.2, it suffices to take S' with a sufficiently large collar. This collar is isometric to $[-w, w] \times \mathbf{R}/[\tau \mapsto \tau + \ell]$ with metric $ds^2 = d\rho^2 + \cosh^2\rho \, d\tau^2$ (Proposition 3.1.8 and Fermi coordinates), where w may be chosen arbitrarily large. Now let $0 < a < b \leq w$ and consider the function f on the annulus

$$C_{ab} = [a, b] \times \mathbf{R}/[\tau \mapsto \tau + \ell]$$

defined by

$$f(\rho, t) = f(\rho) = e^{-\rho/2} \sin \frac{\pi(\rho - a)}{b - a}, \quad a \leq \rho \leq b.$$

Then, with $f' = df/d\rho$,

$$\int_{C_{ab}} \|\operatorname{grad} f\|^2 \, dM = \ell \int_a^b (f'(\rho))^2 \cosh \rho \, d\rho.$$

Using that $\cosh \rho \geq \frac{1}{2} e^\rho$ and that, for $\rho \geq a$, $\cosh \rho \leq \frac{1}{2}(1 + e^{-2a}) e^\rho$, we obtain by direct computation

$$\int_{C_{ab}} \|\operatorname{grad} f\|^2 \, dM \leq \left(\frac{1}{4} + \frac{\pi^2}{(b-a)^2}\right) (1 + e^{-2a}) \int_{C_{ab}} f^2 \, dM.$$

We now split the collar into n disjoint annuli of the above kind and let the values of a and $(b - a)$ be sufficiently large. Then we smooth the functions f as before and obtain Theorem 8.1.2 via the minimax principle (Theorem 8.2.1(i)). ◇

For the proof of Theorem 8.1.1 we let $S \in \mathcal{R}_g$ and consider a canonical fundamental polygon with $4g$ sides, as in Section 6.7. Drawing the diagonals from a fixed vertex p to the remaining vertices of the polygon, we obtain a decomposition of S into $4g - 2$ geodesic triangles $T_1, \ldots, T_{4g-2}$, thus satisfying the hypotheses of the minimax principle (ii) in Theorem 8.2.1. It suffices to prove that

(8.4.1) $$v(T) > \tfrac{1}{4}$$

for any geodesic triangle T. For this we use Cheeger's inequality (Theorem

8.3.3) together with Lemma 8.3.6. We have to find a number $\varepsilon(T) > 0$ such that any one dimensional submanifold A of T which decomposes T into connected relatively open subsets B and B' satisfies the inequality

$$\ell(A)/\min\{\text{area } B, \text{area } B'\} \geq 1 + \varepsilon(T).$$

It is easy to see that A is a *connected* curve and that there are three cases : i) A is a closed curve, ii) A is an arc having both endpoints on the same side of T, iii) A is an arc having endpoints on different sides of T. We consider only case iii), the remaining cases being similar.

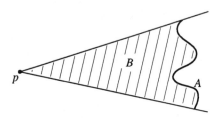

Figure 8.4.1

We introduce polar coordinates (ρ, σ) centered at the common vertex p of the two sides. The metric tensor with respect to these coordinates becomes $ds^2 = d\rho^2 + \sinh^2\rho \, d\sigma^2$ (cf. (1.1.8)). We let B be the domain which has p on its boundary. (Fig. 8.4.1). We represent A as a parametrized curve $t \mapsto \alpha(t) \in T, 0 \leq t \leq 1$, where $\alpha(t) = (\rho(t), \sigma(t))$. Let R be the diameter of T. With the quantity $(1 + \varepsilon(T)) := (\sinh R)/(-1 + \cosh R)$ we obtain

$$\ell(\alpha(t)) = \int_0^1 (\dot{\rho}^2(t) + (\sinh^2\rho(t)) \, \dot{\sigma}^2(t))^{1/2} dt$$

$$\geq \int_0^1 \sinh \rho(t) \, |\dot{\sigma}(t)| \, dt$$

$$\geq (1 + \varepsilon(T)) \int_0^1 (\cosh \rho(t) - 1) |\dot{\sigma}(t)| \, dt$$

$$\geq (1 + \varepsilon(T)) \text{ area } B.$$

This proves (8.4.1), and Theorem 8.1.1 follows. ◊

8.4.2 Example. (Cheeger's inequality is sharp). If in Fig. 8.4.1 the sides of the triangle at p become large, then the sides cut out an angular domain c_{ab} which is isometric to the sector $\{(\rho, \sigma) \mid a \leq \rho \leq b, 0 \leq \sigma \leq \vartheta\}$ with metric $ds^2 = d\rho^2 + \sinh^2\rho \, d\sigma^2$, where ϑ is the interior angle of T at p, and where a

and $b - a$ may become arbitrarily large. Consider the function f which is zero on $T - c_{ab}$ and defined on c_{ab} by

$$f(\rho, \sigma) = e^{-\rho/2} \sin\frac{2\pi(\rho - a)}{b - a}, \, a \le \rho \le b.$$

As in the proof of Theorem 8.1.3, this function can be used to show that $v(T) < \frac{1}{4} + \varepsilon$ with arbitrarily small ε, provided T is sufficiently large. Together with (8.4.1) this shows that Cheeger's inequality cannot be improved unless additional geometric invariants are used. Further examples may be found in Buser [6]. ◇

Finally we prove Theorem 8.1.4 using the results of Section 4.5.

We cut S open along the pairwise disjoint simple closed geodesics of length $\le \log 4$, provided there are any. Then we triangulate each connected component of the resulting surface with trigons as in Theorem 4.5.2.

Let S_1 be one of these components. The triangulation of S_1 may be interpreted as a cubic graph G_1, each trigon X of which is interpreted as a vertex whose emanating half-edges are the sides of X. The vertex set of G_1 is a metric space with the distance dist(X, Y) between vertices X and Y defined as the length of the shortest edge path from X to Y, the length of an edge path being defined as the number of edges in the path.

In G_1 we choose a largest possible set $\mathcal{P}_1$ of vertices with pairwise distances greater than c, where c is a constant which will be determined below. If $\mathcal{P}_1$ is largest, then for any vertex $X \in G_1$, there exists a vertex $P \in \mathcal{P}_1$ satisfying dist$(X, P) \le c$. Thus, the metric balls of radius c around the vertices of $\mathcal{P}_1$ cover G_1, and the balls of radius $c/2$ around these vertices are pairwise disjoint. It is now easy to find subsets (Dirichlet domains in the metric space G_1) $D(P)$, $P \in \mathcal{P}_1$, with the following properties:

(i) *the union $\bigcup_{P \in \mathcal{P}_1} D(P)$ is the vertex set of G_1;*

(ii) *the domains $D(P)$ are pairwise disjoint;*

(iii) *each $D(P)$, $P \in \mathcal{P}_1$, is contained in the metric ball of radius c around P, and $D(P)$ contains the ball of radius $c/2$ around P;*

(iv) *each $D(P)$ is connected.*

In (iii), the metric ball of radius $c/2$ may be the entire graph G_1: in this case $\mathcal{P}_1$ consists of exactly one vertex. Statement (iv) means that if $X, Y \in D(P)$, then there exists an edge path from X to Y contained in $D(P)$.

For each $P \in \mathcal{P}_1$ we define $D_S(P) \subset S_1$ to be the union set of all trigons of $D(P)$. Then $D_S(P)$ is connected. Moreover, if $X, Y \subset D_S(P)$ are two trigons, then there exists a chain $X_1, \ldots, X_n$ of trigons of $D(P)$ such that

$X_1 = X$, $X_n = Y$ and such that X_i and X_{i+1} are adjacent along a common side for $i = 1, \ldots, n-1$. By Theorem 4.5.2, any of our trigons X satisfies the inequality

$$0.19 \leq \text{area } X \leq 1.36.$$

Since $D(P)$ has at most 3^c vertices, we conclude from the bounds of the interior angles of the trigons (Section 4.5), that the isoperimetric constant $h(D_S(P))$ has a positive lower bound which depends only on c. We now fix $c = 68$. This implies the inequality

$$\text{area } D_S(P) \geq 2\pi.$$

In fact, if $\mathcal{P}_1$ consists only of one point, say P, then $D_S(P)$ is the connected component S_1 and has area $2\pi k$ for some integer $k \geq 1$. If $\mathcal{P}_1$ consists of more than one point, then any $D(P)$ contains at least $c/2 = 34$ vertices, and we obtain the inequality area $D_S(P) \geq 34 \cdot 0.18 \geq 2\pi$.

This shows that the total number of domains $D_S(P)$ is at most $2g - 2$. Theorem 8.1.4 is now proved ◊

Notes. We complete the list of references about the small eigenvalues. The question about the possibility of small eigenvalues appears first in an article of Delsarte, [1, 2] on hyperbolic lattice point problems; this paper, however, seems to have been unnoticed for a long time.

The advantage of writing eigenvalues in the form $\lambda = \frac{1}{4} + r^2$ and working in terms of r instead of λ has been recognized by Maaß [1] in connection with Hecke theory, and by Selberg [1-3] in connection with the zeroes of the Selberg zeta function. Huber [2] and, some years later, Patterson [1, 2, 3] showed the influence of the small eigenvalues on the asymptotic distribution of the lengths of the closed geodesics and on the distribution of lattice points.

An early version of Theorem 8.1.1 is contained in a footnote on p. 74 of Selberg [2]. The same footnote gives a hint for the possible existence of Riemann surfaces with small eigenvalues. After an erroneous statement of McKean [1] which greatly influenced the interest in small eigenvalues, Randol [1] gave the first explicit construction of examples with small eigenvalues via Selberg's trace formula. More precisely, he proved that every compact Riemann surface has finite coverings with arbitrarily many eigenvalues smaller than $\frac{1}{4}$. At the same time Huber [3] gave an upper bound for λ_1 which approaches $\frac{1}{4}$ as the genus goes to infinity.

Huber's bound is probably sharp. From work of Langlands, Rankin and Selberg it follows that there are sequences of compact Riemann surfaces with genus approaching infinity and with λ_1 bounded below by 3/16. The conjecture is that in these examples $\frac{3}{16}$ can be replaced by $\frac{1}{4}$. Various authors give elementary constructions of surfaces with $g \to \infty$ and with λ_1 bounded away from zero. The best elementary bound, so far, is 5/32 and is given in Sarnak-Xue [1].

Theorems 8.1.1-3 are from Buser [1]. Zograf [1, 2, 3] generalizes Theorem 8.1.1 to Fuchsian groups of arbitrary signature. Huxley [1, 2] gives a refinement of Cheeger's inequality to prove that some of the classical examples have no eigenvalues in $]0, \frac{1}{4}[$. Zograf [1, 2, 3] also gives such examples. A construction of compact Riemann surfaces with particularly large λ_1 (for $g = 2$) and a discussion about the best possible such example is given in Jenni [1, 2] and in Huber [7].

Refinements of the estimates of the small eigenvalues are given in Schoen-Wolpert-Yau [1] and in Dodziuk-Pignataro-Randol-Sullivan [1]. Colbois [1] gives precise asymptotic formulae for small eigenvalues in sequences of compact Riemann surfaces which suitably collapse onto a graph. This was used by Colbois and Colin de Verdière [1] to construct compact Riemann surfaces whose first eigenvalue has a large multiplicity. Burger [2] gives a version which estimates the difference between the small eigenvalues of a Riemann surface and the corresponding eigenvalues of the nearby graph. He then uses this result to give upper bounds on the multiplicity of λ_1.

Results about the multiplicity of the higher eigenvalues may be found in Besson [1] and Huber [5, 6].

In Colbois-Courtois [1, 2] it is shown that if a sequence of compact Riemann surfaces converges to a non-compact Riemann surface S with finite area, then the eigenfunctions with eigenvalues below $\frac{1}{4} - \delta$ converge to L^2-eigenfunctions on S, for any given $\delta > 0$. In this way they obtain criteria for the existence of small eigenvalues on S. For eigenvalues $\geq \frac{1}{4}$ their method does not apply. In fact, it is a still unsolved question, known as the Roelcke problem, whether any non-compact Riemann surface of finite area has L^2-eigenfunctions with eigenvalues $> \frac{1}{4}$. Colin de Verdière [2] makes the difficulties clear. In fact, he constructs non-compact surfaces of finite area with variable negative curvature arbitrarily close to -1 and having no L^2-eigenfunctions of the Laplacian.

We also mention that the problem of whether or not the inequality $\lambda_{2g-2} \geq \frac{1}{4}$ is true for any compact Riemann surface of genus g is still unsolved. Randol [8] shows that if $\lambda_{2g-3} < \varepsilon$ for sufficiently small ε, then no eigenvalue occurs in $[\varepsilon, \frac{1}{4}]$. Schmutz [1, 2] uses mixed boundary value problems together with an interesting monotonicity argument which yields better bounds on Y-pieces. In [3] Schmutz proves the inequality $\lambda_{2g-2} \geq \frac{1}{4}$ for all compact Riemann surfaces of genus $g = 2$.

Chapter 9

Closed Geodesics and Huber's Theorem

There is no general procedure to be found in the literature which allows us to write down the spectrum of the Laplacian of a compact Riemann surface in closed form, say as a function of the moduli. Nevertheless, isospectrality and related problems can be solved by quite explicit geometric constructions. This is possible through Huber's theorem, which states that the spectra of the Laplacian for two compact Riemann surfaces are the same if and only if the surfaces have the same length spectrum, i.e. the same sequence of lengths of closed geodesics.

Huber's theorem is the main goal of this chapter and will be proved in Section 9.2 so that we might have closed the chapter with this section. However, there are many interesting additional relations between the length spectrum and the eigenvalue spectrum, in particular those concerning the small eigenvalues. In Sections 9.3 - 9.7 we deal with some of these relations, although they will not be used elsewhere in the book.

In this development we partly follow Huber's original arguments and partly use Selberg's transform pair technique, which we approach in three steps, with an application after each step. The first step is the same in Huber's and Selberg's approach and consists of a fundamental domain argument which leads to a length trace formula (Theorem 9.2.10). This step is the most important one and we first give an outline in Section 1.

As an immediate consequence of the length trace formula, we obtain Huber's theorem by looking at the heat kernel. This argument follows McKean [1]. We take advantage of the fact that by Theorem 7.2.6 the eigenfunction expansion of the heat kernel is already known. In Section 9.3 a similar eigenfunction expansion will then be given for more general kernels (Theorem 9.3.7). As an application of this, in Section 9.4 we prove the prime number theorem for Riemann surfaces after a brief sketch of the corresponding theorem in number theory. In Section 9.5 the length trace formula and the eigenfunction expansion will merge into Selberg's trace formula. In Sections 9.6 and 9.7 we return to the small eigenvalues.

9.1 The Origin of the Length Spectrum

It is interesting to see how the relationship between the eigenvalues and the closed geodesics is "born" in Huber [1, 2]. The starting point is a hyperbolic lattice point problem. Let $\Gamma \in \text{Is}^+(\mathbf{H})$ be a discrete subgroup such that $\Gamma \backslash \mathbf{H}$ is a compact Riemann surface, and consider two given points $z, w \in \mathbf{H}$. How many lattice points $T(w)$, $T \in \Gamma$, are there inside a circle of radius t around w? More precisely, let N be the counting function

$$N(t; z, w) = \#\{T \in \Gamma \mid \text{dist}(z, Tw) \leq t\},$$

where # denotes the cardinality. What is the growth of $N(t; z, w)$ as $t \to \infty$? To solve the problem Huber introduces the Dirichlet series

$$G(s; z, w) = \sum_{T \in \Gamma} (\cosh \text{dist}(z, Tw))^{-s},$$

which for each $(z, w) \in \mathbf{H} \times \mathbf{H}$ is a holomorphic function of s in the half-plane $\{s \in \mathbf{C} \mid \mathcal{R}es > 1\}$. This is analogous to a familiar procedure in analytic number theory. The function $(z, w) \mapsto G(s; z, w)$ is Γ-bi-automorphic on $\mathbf{H} \times \mathbf{H}$ and has an eigenfunction expansion

$$G(s; z, w) = \sum_{n=0}^{\infty} h(s; \lambda_n) \phi_n(z) \phi_n(w),$$

in which $\phi_0, \phi_1, \ldots$ are the lifts in $\mathbf{H}$ of the eigenfunctions $\varphi_0, \varphi_1, \ldots$ with eigenvalues $\lambda_0, \lambda_1, \ldots$ of the Laplacian of $\Gamma \backslash \mathbf{H}$. The coefficients $h(s; \lambda_n)$ can be computed explicitly in terms of special functions and are meromorphic on $\mathbf{C}$. Moreover the uniform absolute convergence of $\sum \lambda_n^{-2} \phi_n(z) \phi_n(w)$ on $\mathbf{H} \times \mathbf{H}$ (cf. Theorem 7.2.18) is strong enough to show that the expansion itself is meromorphic on $\mathbf{C}$. A Tauberian argument then yields the following asymptotic equality (cf. Definition 9.2.15 for the asymptotic equality $\sim$)

$$N(t; z, w) \sim \frac{1}{4(g-1)} e^t, \quad t \to \infty.$$

We remark that up to this point the Laplacian enters only *as a tool*. Now let $F \subset \mathbf{H}$ be a compact fundamental domain of Γ and integrate the function $z \mapsto G(s; z, z)$ over F. From the above expansion we then get

$$\sum_{T\in\Gamma} \int_F (\cosh \text{dist}(z, Tz))^{-s} \, d\mathbf{H}(z) = \sum_{n=0}^{\infty} h(s; \lambda_n),$$

in which the left-hand side is a purely geometric quantity.

The length spectrum now enters through the following argument (Huber [1]; this argument is also inherent in Selberg's approach). Namely, let P be a given element in Γ. How much does the conjugacy class

$$\{P\} = \{S^{-1}PS \mid S \in \Gamma\}$$

contribute to the above sum? One first observes that

$$S^{-1}PS = R^{-1}PR \quad \textit{iff} \quad SR^{-1} \textit{ commutes with } P.$$

Hence, if Z denotes the centralizer of P in Γ and if $S_1, S_2, \ldots$ is a system of representatives of the right cosets of Z such that Γ is the disjoint union

$$\Gamma = Z S_1 \cup Z S_2 \cup \ldots,$$

then the sequence $S_1^{-1}PS_1, S_2^{-1}PS_2, \ldots$ runs exactly once through each conjugate of P. Setting $F_p^* = S_1(F) \cup S_2(F) \cup \ldots$, we get

$$\sum_{T\in\{P\}} \int_F (\cosh \text{dist}(z, Tz))^{-s} \, d\mathbf{H}(z) = \int_{F_p^*} (\cosh \text{dist}(z, Pz))^{-s} \, d\mathbf{H}(z).$$

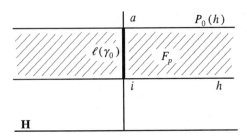

Figure 9.1.1

The *observation* is now that F_p^* is a fundamental domain of the group Z and that the integrand is Z-automorphic. F_p^* can therefore be replaced by any other sufficiently regular fundamental domain. A more convenient one is the strip between h and $P_0(h)$, where P_0 is a generator of Z (with $P = P_0^\nu$ for some ν) and h is a horocycle orthogonal to the common axis a of P_0 and P (cf. Fig. 9.1.1 where a is the positive imaginary axis).

In the integral over F_p the *width* of F_p and the exponent ν in $P = P_0^\nu$ become dominating. But the width is the same as the length of the closed geodesic in $\Gamma\backslash\mathbf{H}$ belonging to the conjugacy class of P_0. Hence, we obtain a

9.2 Summation over the Lengths

We let $M = \Gamma\backslash\mathbf{H}$ be a fixed compact Riemann surface. In Lemma 7.5.5 we have seen that every L^2-function $K : [0, \infty[\to \mathbf{C}$ which satisfies the decay condition

(9.2.1) $$|K(\rho)| \leq \text{const } e^{-\rho(1+\delta)}$$

for some $\delta > 0$, generates a Γ-bi-invariant function

(9.2.2) $$\mathcal{K}_\Gamma(z, w) = \sum_{T \in \Gamma} k(z, Tw), \quad z, w \in \mathbf{H},$$

with $k(\,,\,) = K(\text{dist}(\,,\,))$. The function $\mathcal{K}_\Gamma$ induces a symmetric kernel $\mathcal{K}_M : M \times M \to \mathbf{C}$. The goal is to relate the trace

$$\text{tr } \mathcal{K}_M := \int_M \mathcal{K}_M(x, x)\, dM(x)$$

to the lengths of the closed geodesics of M and to prove Huber's theorem. When working in connection with the trace formulae we shall consider the closed geodesics *oriented* (as in Definition 1.6.3) so that we have a one-to-one correspondence between the closed geodesics of $M = \Gamma\backslash\mathbf{H}$ and the conjugacy classes in $\Gamma - \{id\}$.

We use the following notation. Every $T \in \Gamma - \{id\}$ has a unique invariant geodesic a_T, the *axis* of T. We denote by $\ell(T)$ the *displacement length* of T, i.e. the real number ℓ satisfying

$$\text{dist}(z, Tz) = \ell \text{ for all } z \in a_T, \quad \text{dist}(z, Tz) > \ell \text{ for all } z \in \mathbf{H} - a_T.$$

If a_T is parametrized with unit speed and if the orientation is such that the tangent vectors point from z to Tz for $z \in a_T$, then we have

$$T(a_T(s)) = a_T(s + \ell(T)), \quad s \in \mathbf{R}.$$

The displacement length is given as follows.

(9.2.3) $$\ell(T) = 2 \operatorname{arccosh} \tfrac{1}{2} |\operatorname{tr} T|.$$

Conjugate elements have the same displacement length so that we may consider $\ell(T)$ as an attribute of the conjugacy class of T. The associated closed geodesic γ_T on $\Gamma\backslash\mathbf{H}$ has length

(9.2.4) $$\ell(\gamma_T) = \ell(T).$$

If $S = T^m$, $m \in \mathbf{Z} - \{0\}$, we shall write $\gamma_S = \gamma_T^m$ and say that γ_S is the *m-fold iterate* of γ_T. This coincides with the notation in Definition 1.6.4. Part of the next definition is a recall of Definition 1.6.4.

9.2.5 Definition. $S \in \Gamma - \{0\}$ is *primitive* if it cannot be written in the form $S = R^m$ with $R \in \Gamma$ and $m \geq 2$. Similarly, a closed geodesic γ_S on $M = \Gamma \backslash \mathbf{H}$ is *primitive* if it is not the *m*-fold iterate with $m \geq 2$ of another closed geodesic γ_R on M.

A primitive closed geodesic will also be called a *prime geodesic*.

9.2.6 Lemma. *For every $T \in \Gamma - \{id\}$ there exists a unique primitive $S \in \Gamma$ such that $T = S^m$ for some $m \geq 1$. Similarly, for every non-trivial closed geodesic γ on $M = \Gamma \backslash \mathbf{H}$, there exists a unique primitive closed geodesic γ_0 such that $\gamma = \gamma_0^m$ for some $m \geq 1$. The elements S^n, $n \in \mathbf{Z}$, are pairwise non-conjugate in Γ and the centralizer of T in Γ is*

$$Z_T = \{S^n \mid n \in \mathbf{Z}\}.$$

Proof. Let a_T be the axis of T. The subgroup Z_T of Γ which fixes a_T acts properly discontinuously and without fixed points on a_T. The restriction $Z_T \mid a_T$ is therefore a Euclidean lattice in $\mathbf{R}^1$ and there exists $S \in Z_T$ such that $\ell(S) > 0$ and such that $\ell(S) \leq \ell(R)$ for all $R \in Z_T$. It follows that for any $R \in Z_T$ there exists a unique $n \in \mathbf{Z}$ such that $R \mid a_T = (S \mid a_T)^n$. Since $(S \mid a_T)^n = S^n \mid a_T$ and since Γ acts without fixed points, this implies $R = S^n$. In particular, we have a unique m such that $T = S^m$, replacing S by S^{-1}, if necessary, so that m is always positive. If $T = \Sigma^\mu$ for some $\Sigma \in \Gamma$ and $\mu \geq 1$, then T leaves the axis of Σ invariant. From the uniqueness of the axis of T we conclude that Σ and T have the same axis. Hence, $\Sigma \in Z_T$ and therefore $\Sigma = S^r$ for some r. If Σ is primitive this is only possible for $r = \pm 1$, and since $S^m = T = \Sigma^\mu = S^{r\mu}$ with m and μ positive, we see that $r = 1$. This proves the uniqueness of S. The corresponding statement about the closed geodesics γ and γ_0 is clear.

It remains to prove that Z_T is the centralizer of T. If $R \in \Gamma$ and if $TR = RT$, then $T(R(a_T)) = R(T(a_T)) = R(a_T)$. By the uniqueness of the axis of T we have $R(a_T) = a_T$ and thus $R \in Z_T$. This shows that Z_T contains the centralizer of T, and since Z_T is abelian, Z_T is the centralizer of T. ◇

We need an estimate of the growth rate of the number of geodesics up to a given length. For the purpose of the trace formulae the estimate of the function Φ in the next lemma will be sufficient. In Theorem 9.4.14 we shall prove the asymptotic formula $\Phi(t) \sim t^{-1}e^t$ as $t \to \infty$.

9.2.7 Lemma. *Let $\Phi(L)$ be the number of closed geodesics of length $\ell \leq L$ on a compact Riemann surface of genus $g \geq 2$. Then Φ has growth rate $\Phi(L) = O(e^L)$ as $L \to \infty$.*

Proof. This follows from Lemma 6.6.4 or from Lemma 7.5.3. For convenience we repeat the simple argument. Fix any base point $x \in M$ and let $\tilde{x}$ be a lift of x in $\mathbf{H}$ with respect to the universal covering $\mathbf{H} \to M = \Gamma \backslash \mathbf{H}$. For each closed geodesic γ on M there exists a freely homotopic geodesic loop γ_x at x of length $\ell(\gamma_x) \leq \ell(\gamma) + 2d$, where d is the diameter of M. Lifting these loops to the universal covering, we see that $\Phi(L)$ is bounded above by the number of Dirichlet fundamental domains with respect to the point set $\Gamma(\tilde{x})$ which intersect the hyperbolic disk of radius $L + 2d$ centered at $\tilde{x}$. Since the area of this disk has growth rate $O(e^L)$, we are done. ◇

We are now in a position to introduce the length spectrum. We recall (Definition 1.6.3) that two parametrized geodesics

$$\gamma, \delta : \mathbf{S}^1 = \mathbf{R}^1/[t \mapsto t+1] \to M$$

(parametrized with constant speed) are considered *equivalent* if and only if there exists a constant c such that $\gamma(t) = \delta(t + c)$, $t \in \mathbf{S}^1$. To avoid confusion, we shall from now on call the corresponding equivalence classes *oriented closed geodesics*.

9.2.8 Definition. Let $M = \Gamma \backslash \mathbf{H}$ be a compact Riemann surface. We denote by $\mathscr{C}(M)$ the set of all oriented closed geodesics on M and by $\mathscr{P}(M)$ the subset of all primitive oriented closed geodesics.

The sequence of all lengths $\ell(\gamma)$, $\gamma \in \mathscr{C}(M)$ arranged in ascending order is called the *length spectrum* of M. The subsequence of all lengths $\ell(\gamma)$, $\gamma \in \mathscr{P}(M)$, also arranged in ascending order, is called the *primitive length spectrum* of M.

Huber's theorem [2] is as follows.

9.2.9. Theorem. *Two compact Riemann surfaces of genus $g \geq 2$ have the same spectrum of the Laplacian if and only if they have the same length spectrum.*

We point out that compact Riemann surfaces have the same spectrum only if they have the same genus. This will follow from Weyl's asymptotic law (Theorem 9.2.14). For generalizations of Huber's theorem we refer to the discussion at the end of this section.

The proof is based on the following theorem. Note that the summation is over all $n \in \mathbf{N} - \{0\}$. If we replace $\mathcal{P}(M)$ by the set of all *non-oriented* primitive closed geodesics, then the summation is over all $n \in \mathbf{Z} - \{0\}$.

9.2.10 Theorem. (Length trace formula). *Consider the compact Riemann surface $M = \Gamma\backslash\mathbf{H}$ and let $\mathcal{K}_M : M \times M \to \mathbf{C}$ be a kernel defined by the generating function K satisfying the inequality as in (9.2.1). Then*

$$\int_M \mathcal{K}_M(x, x) \, dM(x) = K(0) \text{ area } M$$

$$+ \sum_{n=1}^{\infty} \sum_{\gamma \in \mathcal{P}(M)} \frac{\ell(\gamma)}{\sqrt{\cosh \ell(\gamma^n) - 1}} \int_{\ell(\gamma^n)}^{\infty} \frac{K(\rho) \sinh \rho}{\sqrt{\cosh \rho - \cosh \ell(\gamma^n)}} \, d\rho.$$

Proof. Let Π be a set of primitive elements of Γ such that each $T \in \Gamma - \{id\}$ is conjugate to S^n for some unique $S \in \Pi$ and some unique $n \in \mathbf{Z} - \{0\}$ (use Lemma 9.2.6). Thus, $\Pi \cup \Pi^{-1}$ is a minimal set of representatives of the conjugacy classes of the primitive elements of $\Gamma - \{id\}$, and $\Pi \cap \Pi^{-1} = \emptyset$. For each $S \in \Pi$ we have the cyclic subgroup $Z_S \subset \Gamma$,

$$Z_S = \{S^n \mid n \in \mathbf{Z}\}.$$

We let $\mathcal{R}_S = \{R_1, R_2, \dots\}$ be a minimal set of representatives of the right cosets of Z_S so that Γ is the disjoint union

(1) $\quad \Gamma = Z_S R_1 \cup Z_S R_2 \cup \dots$.

Since $R_j R_i^{-1}$ commutes with S if and only if $R_j R_i^{-1} \in Z_S$ (Lemma 9.2.6), it follows that

$$R_i^{-1} S R_i = R_j^{-1} S R_j \text{ if and only if } i = j.$$

Each $T \in \Gamma - \{id\}$ therefore has the unique representation $T = R_i^{-1} S^n R_i$, yielding the decomposition

(2) $\quad \Gamma - \{id\} = \bigcup_{n \neq 0} \bigcup_{S \in \Pi} \bigcup_{R_i \in \mathcal{R}_S} \{R_i^{-1} S^n R_i\}$ (disjoint union).

In what follows a *fundamental domain* of Z_S will be understood to be the closure $\mathcal{F}$ of an open domain $\mathring{\mathcal{F}} \subset \mathbf{H}$ such that

(i) the images $T(\mathscr{F})$, $T \in Z_S$, cover $\mathbf{H}$,
(ii) $T(\mathring{\mathscr{F}}) \cap T'(\mathring{\mathscr{F}}) = \emptyset$ for $T \neq T'$,
(iii) $\mathscr{F} - \mathring{\mathscr{F}}$ is a set of measure 0.

The length trace formula will result from the comparison of two different fundamental domains. First we let F be a compact Dirichlet fundamental domain of Γ with interior $\mathring{F}$ and set

$$\mathscr{F}_S = \bigcup_{R_i \in \mathscr{R}_S} R_i(F), \qquad \mathring{\mathscr{F}}_S = \bigcup_{R_i \in \mathscr{R}_S} R_i(\mathring{F}).$$

Under Z_S the images of $\mathscr{F}_S$ cover $\mathbf{H}$ by (1), and the images of $\mathring{\mathscr{F}}_S$ are pairwise disjoint by (2). Hence, $\mathscr{F}_S$ is a fundamental domain of Z_S.

For every $T = R_i^{-1} S^n R_i$ in (2) we have

$$(3) \quad \int_F k(z, R_i^{-1} S^n R_i z) \, d\mathbf{H}(z) = \int_F k(R_i z, S^n R_i z) \, d\mathbf{H}(z)$$

$$= \int_{R_i(F)} k(z, S^n z) \, d\mathbf{H}(z).$$

Hence, by (2)

$$(4) \quad \int_F \mathscr{K}_\Gamma(z, z) \, d\mathbf{H}(z) = K(0) \int_F d\mathbf{H}(z)$$

$$+ \sum_{n \neq 0} \sum_{S \in \Pi} \sum_{R_i \in \mathscr{R}_S} \int_{R_i(F)} k(z, S^n z) \, d\mathbf{H}(z).$$

By Lemma 7.5.3 (or by the argument in the proof of Lemma 9.2.7), we have

$$\sum_{T \in \Gamma} \int_F |k(z, Tz)| \, d\mathbf{H}(z) < \infty$$

so that the above changes in the order of summation are allowed. Moreover, Lemma 7.5.3 together with (3) shows that the function $z \mapsto k(z, S^n z)$ is absolutely integrable over the domain $\mathscr{F}_S$. Since this function is Z_S-automorphic and since $\mathscr{F}_S$ is a fundamental domain of Z_S, it follows that

$$(5) \quad \sum_{R_i \in \mathscr{R}_S} \int_{R_i(F)} k(z, S^n z) \, d\mathbf{H}(z) = \int_{\mathscr{F}_S} k(z, S^n z) \, d\mathbf{H}(z)$$

$$= \int_{\mathscr{F}} k(z, S^n z) \, d\mathbf{H}(z)$$

for *any other* fundamental domain $\mathcal{F}$ of Z_S. To compute this integral we temporarily conjugate Γ in $PSL(2, \mathbf{R})$ such that S becomes the transformation

$$z \mapsto S(z) = \lambda z, \quad z \in \mathbf{H}, \quad \text{where } \lambda = e^{\ell(S)},$$

(cf. (9.2.3)). We use the following fundamental domain of Z_S

$$\mathcal{F} = \{z \in \mathbf{H} \mid 1 \leq \operatorname{Im} z \leq \lambda\}.$$

Writing $z = x + iy$, we find for (5)

$$\int_{\mathcal{F}} k(z, S^n z) \, d\mathbf{H}(z) = 2 \int_1^\lambda \int_0^\infty K(\operatorname{dist}(z, \lambda^n z)) \frac{dx}{y} \frac{dy}{y}.$$

The inner integral is almost that of (7.4.12), which led us to the Abel transform. In fact, if we apply the isometry $w \mapsto (w - x)/y$, $w \in \mathbf{H}$, $(x, y$ fixed), to the pair z, $\lambda^n z$, then z moves to i and $\lambda^n z$ moves to $a + i\lambda^n$, where a is given as $a = (\lambda^n - 1)x/y$. Substituting a for x, we obtain the expression

$$\int_{\mathcal{F}} k(z, S^n z) \, d\mathbf{H}(z) = \frac{2 \log \lambda}{\lambda^n - 1} \int_0^\infty K(\operatorname{dist}(i, a + i\lambda^n)) \, da.$$

Except for the coefficient, this is now the same as the integral in (7.4.12), in which $\rho(r, a)$ is the distance from i to the point $a + ie^r$. We repeat the argument which led to (7.4.13) and obtain

$$\int_{\mathcal{F}} k(z, S^n z) \, d\mathbf{H}(z) = \frac{\ell(S)}{\sqrt{\cosh \ell(S^n) - 1}} \int_{\ell(S^n)}^\infty \frac{K(\rho) \sinh \rho}{\sqrt{\cosh \rho - \cosh \ell(S^n)}} \, d\rho.$$

(Replace a by $\rho = \operatorname{dist}(i, a + i\lambda^n)$, where $\cosh \rho = 1 + (a^2 + (\lambda^n - 1)^2)/(2\lambda^n)$ and replace λ by $e^{\ell(S)}$.) Combined with (4) and (5), this yields

(6)
$$\int_F \mathcal{K}_\Gamma(z, z) \, d\mathbf{H}(z) = K(0) \operatorname{area} F + \pi^{1/2} \sum_{n \neq 0} \sum_{S \in \Pi} \frac{\ell(S)}{\sqrt{\cosh \ell(S^n) - 1}} A[K](\ell(S^n)),$$

where A is the Abel transform (given in Definition 7.3.8). Since the mapping $S \mapsto \gamma_S$ sets up a one-to-one correspondence between $\Pi \cup \Pi^{-1}$ and $\mathcal{P}(M)$, this proves the trace formula. ◇

Following McKean [1], we plug the heat kernel p_M of M into the trace formula to prove Huber's theorem. We first take a look at how p_M fits into the general theory. We keep $t > 0$ fixed. The heat kernel $p_\mathbf{H}$ of the hyperbolic plane is defined in (7.4.23) in the form $p_\mathbf{H}(z, w, t) = P_t(\operatorname{dist}(z, w))$, where P_t is defined in (7.4.19) and (7.4.18). The function P_t is the generating function

K as in (7.5.2), and $p_{\mathbf{H}} = p_{\mathbf{H}}(z, w, t)$ is the corresponding symmetric function $k = k(z, w) = K(\text{dist}(z, w))$ on $\mathbf{H} \times \mathbf{H}$ as in (7.5.1). The functions $p_\Gamma = p_\Gamma(z, w, t)$ and $p_M = p_M(x, y, t)$ (defined in Theorem 7.5.11) are $\mathcal{K}_\Gamma = \mathcal{K}_\Gamma(z, w)$ and $\mathcal{K}_M = \mathcal{K}_M(x, y)$ (defined in Lemma 7.5.5). By Lemma 7.4.26, the generating function $K(\rho) = P_t(\rho)$ decays more rapidly than $\exp(-\rho(1 + \delta))$ for arbitrarily large δ, in fact. Hence, the hypotheses of the trace formula are all satisfied.

By Theorem 7.2.6, the trace of p_M is given as follows

$$\operatorname{tr} p_M = \sum_{n=0}^{\infty} e^{-\lambda_n t},$$

where $\lambda_0, \lambda_1, \ldots$ are the eigenvalues of the Laplacian of $M = \Gamma \backslash \mathbf{H}$. As a consequence, we obtain the following formula of McKean [1]

$$\sum_{n=0}^{\infty} e^{-\lambda_n t} = \operatorname{area} M \, (4\pi t)^{-3/2} \, e^{-t/4} \int_0^\infty \frac{r e^{-r^2/4t}}{\sinh r/2} \, dr$$

(9.2.11)

$$+ \tfrac{1}{2}(4\pi t)^{-1/2} \, e^{-t/4} \sum_{n=1}^{\infty} \sum_{\gamma \in \mathcal{P}(M)} \frac{\ell(\gamma)}{\sinh \tfrac{1}{2} \ell(\gamma^n)} \, e^{-\ell^2(\gamma^n)/4t}.$$

Proof. All the necessary computations have been carried out earlier: the trace is

$$\operatorname{tr} p_M = \int_M p_M(x, x, t) \, dM(x) = \int_F p_\Gamma(z, z, t) \, d\mathbf{H}(z).$$

The integral is given in (6) above with $p_\Gamma = \mathcal{K}_\Gamma$ and $K = P_t$. The expression for $K(0) = P_t(0)$ is given in (7.4.20) and the expression for $A[K] = A[P_t]$ is given in (7.4.19) and (7.4.18). Finally we note that $\operatorname{area}(F) = \operatorname{area}(M)$ since F is a fundamental domain. ◊

With this we now prove Huber's theorem.

Proof. We must show that the eigenvalues determine the lengths of the closed geodesics and vice versa.

Assume first that the eigenvalues are given. Then the function of t in (9.2.11) is known to us, and we want to extract from it the area and each individual length, together with its multiplicity. We begin with the area. By Lemma 9.2.7, the second term on the right-hand side of (9.2.11) is bounded above by $k \times t^{-1/2} \exp(-1/8t)$, k some constant, and therefore converges to 0 as $t \to 0$. Using the equation

$$\int_0^\infty e^{-x^2/4t}\, dx = (4\pi t)^{1/2},$$

we obtain

(9.2.12) $$\sum_{n=0}^{\infty} e^{-\lambda_n t} = \frac{\text{area } M}{4t}(1 + o(1))$$

for $t \downarrow 0$. This determines the area of M. It follows that the function

$$f(t) = \sum_{n=1}^{\infty} \sum_{\gamma \in \mathcal{P}(M)} \frac{\ell(\gamma)}{\sinh \frac{1}{2}\ell(\gamma^n)} e^{-\ell^2(\gamma^n)/4t}$$

is determined by the eigenvalue spectrum. The length $\ell(\gamma_1)$ of the shortest element in $\mathcal{P}(M)$ is characterized as the unique $\omega > 0$ for which $f(t)e^{\omega^2/4t}$ has finite positive limit as $t \to 0$. Hence, $\ell(\gamma_1)$ *and the multiplicity* of this length are determined by the eigenvalue spectrum. Knowing $\ell(\gamma_1)$, we remove all powers of all primitive geodesics of length $\ell(\gamma_1)$ from $f(t)$ and then determine the second primitive length and so on.

Now assume that it is the length spectrum which is given. Here we consider the case $t \to \infty$ instead of $t \to 0$. Since the length spectrum is known to us, the function

$$F(t) = \sum_{0 < \lambda_n \le 1/4} e^{-\lambda_n t} - \sigma(t)e^{-t/4}\text{area}\, M + \sum_{\lambda_n > 1/4} e^{-\lambda_n t}$$

is known to us also, where $\sigma(t)$ is the factor on the right-hand side of the term area M in (9.2.11). Its growth rate is given by the expression $\sigma(t) = O(t^{-3/2})$ as $t \to \infty$. From $F(t)$ we first determine the small eigenvalues (if there are any) by taking the unique ω for which

$$\lim_{t \to \infty} e^{\omega t} F(t) =: m_\omega$$

is finite and positive. Then m_ω is the multiplicity of λ_1 and we remove $m_\omega e^{-\lambda_1 t}$ from F. In this way we proceed until the sum over all $\lambda_n \le 1/4$ is removed. Then the area is determined by multiplication with $\sigma^{-1}(t)e^{t/4}$, and we continue as before. Huber's theorem is now proved. ◇

We conclude this section with a few remarks.

Huber's theorem can also be stated for the primitive length spectrum by the following simple observation, whose proof we leave as an exercise.

9.2.13 Remark. *$\Gamma\backslash\mathbf{H}$ and $\Gamma'\backslash\mathbf{H}$ have the same length spectrum if and only if they have the same primitive length spectrum.* ◇

In the statement of Huber's theorem we did not assume that the two surfaces have the same genus. However, isospectral compact Riemann surfaces have the same genus. This follows from formula (9.2.12). We also mention that with Karamata's Tauberian theorem (see e.g. Widder [1], p. 192 Theorem 4.3), we may translate (9.2.12) into the following.

9.2.14 Theorem. (Weyl's asymptotic law).

$$\lambda_n(M) \sim \frac{4\pi n}{\text{area } M}, \qquad n \to \infty.$$

The symbol $\sim$ has the following meaning.

9.2.15 Definition. Let f and g be functions of a real (or integer) variable and let $a \in \mathbf{R}$ or $a = \infty$. Then f and g are *asymptotically equal* for $t \to a$, and we write

$$f(t) \sim g(t), \qquad t \to a,$$

if and only if $\lim_{t \to a} f(t)/g(t) = 1$.

Huber's theorem holds in a wider range. Riggenbach [1] and Bérard-Bergery [1] have generalized the theorem to compact hyperbolic manifolds of arbitrary dimension with the length spectrum replaced by a modified length spectrum (called the *geometric spectrum*) in which the lengths are multiplied by a factor depending on the holonomy of the geodesic. A similar theorem has been proved by Gangolli [1] for compact quotients of rank 1 symmetric spaces. The result of Bérard-Bergery and Riggenbach shows that the lengths are not, in general, sufficient to determine the eigenvalues. But the converse holds, at least in a generic sense. This has been shown by Colin de Verdière [1] and by Duistermaat Guillemin [1, 2]. Examples where the length and eigenvalue spectra are not equivalent are given by Gordon [2]. The length spectrum on compact Riemann surfaces of signature (g, n) is studied in Guillopé [1].

9.3 Summation over the Eigenvalues

In the preceding section we expressed the trace of a kernel $\mathcal{K}_M$ in terms of the length spectrum. We now prove a similar formula in terms of the eigenvalue spectrum (Theorems 9.3.7 and 9.3.11). In the particular case that $\mathcal{K}_M$ is the heat kernel p_M, this formula is

$$\operatorname{tr} p_M = \sum_{n=0}^{\infty} e^{-\lambda_n t}.$$

We first give an outline of the approach (cf. Hejhal [2], vol I, pp. 8 - 10 and, in general form, Selberg [2] pp. 53 - 55). Let $\phi_0, \phi_1, \ldots$ be the Γ-invariant lifts in $\mathbf{H}$ of the eigenfunctions $\varphi_0, \varphi_1, \ldots$ of the Laplacians on $\Gamma\backslash\mathbf{H}$. Fixing an arbitrary point $w \in \mathbf{H}$ we construct *radial* eigenfunctions, that is, eigenfunctions ψ_n, $n = 0, 1, \ldots$, which depend on $\operatorname{dist}(\,, w)$. The function ψ_n is obtained by integrating ϕ_n over the action of $\mathbf{S}_w^1$, where $\mathbf{S}_w^1$ is the subgroup of $\operatorname{Is}^+(\mathbf{H})$ which fixes point w. Since ψ_n is radial, it is essentially a function of *one* variable and is the solution to an ordinary differential equation of second order with initial values $\psi_n(0) = \phi_n(w)$, $\psi_n'(0) = 0$. *This solution is unique.* Hence, if Ω_n is another eigenfunction with the same eigenvalue and, for instance, such that $\Omega_n(w) = 1$, then, momentarily leaving aside questions of convergence,

$$\int_{\mathbf{S}_w^1} \phi_n\, d\sigma = \phi_n(w) \int_{\mathbf{S}_w^1} \Omega_n\, d\sigma,$$

where $d\sigma$ is the invariant measure on $\mathbf{S}_w^1$ with volume 2π. Using the decomposition

$$\int_{\mathbf{H}} = \int_0^{\infty} \int_{\mathbf{S}_w^1},$$

we obtain, with $k(z, w) = K(\operatorname{dist}(z, w))$,

$$\int_{\mathbf{H}} k(z, w)\phi_n(z)\, d\mathbf{H}(z) = \phi_n(w) \int_{\mathbf{H}} k(z, w)\Omega_n(z)\, d\mathbf{H}(z).$$

Ignoring momentarily the details which will follow below, we translate this into

$$\int_M \mathcal{K}_M(x, y)\varphi_n(x)\, dM(x) = \varphi_n(y) h_n,$$

where

$$h_n = \int_{\mathbf{H}} k(z, w)\Omega_n(z)\, d\mathbf{H}(z).$$

This number depends on k and on the eigenvalue λ_n of φ_n. We learn from this that the *eigenfunctions of the Laplacian are also the eigenfunctions of the integral operator with kernel $\mathcal{K}_M$, and the eigenvalues h_n are certain transforms of the λ_n.* The important fact will be that these transforms can be written in a closed form. The trace formula will finally be obtained from the double expansion

$$\mathcal{K}_M(x, y) = \sum_{n=0}^{\infty} h_n \varphi_n(x) \varphi_n(y).$$

The required regularity of the generating function K is as follows. In the first steps it will be sufficient to assume that $K : [0, \infty[\to \mathbb{C}$ is an L^2-function satisfying the decay condition

$$|K(\rho)| \leq \text{const } e^{-\rho(1+\delta)}$$

for some $\delta > 0$. This is the same decay condition as in Section 9.2. It will be strong enough to prove the above double expansion in the L^2- sense. Later we shall restrict ourselves to more regular K in order to obtain the pointwise expansion.

We now give the details of the above development. We first need some preparations. It is sometimes more convenient to rewrite K in the form

(9.3.1) $$K(\rho) = L(\cosh \rho),$$

so that by (1.1.2) the function $k(z, w) = K(\text{dist}(z, w))$ becomes

(9.3.2) $$k(z, w) = L(1 + \frac{|z-w|^2}{2 \, \text{Im} z \, \text{Im} w}).$$

L is then a function $L : [1, \infty[\to \mathbb{C}$ with growth rate $O(t^{-(1+\delta)})$ as $t \to \infty$. Instead of the eigenfunction Ω_n mentioned above, we take the following

(9.3.3) $$\Omega(z) = y^{1/2 + ir}, \quad y := \text{Im} z,$$

where $r \in \mathbb{C}$ is a complex parameter ranging in the strip

$$\mathcal{S}_{\delta'} = \{r \in \mathbb{C} \mid |\text{Im} r| < \tfrac{1}{2} + \delta'\}$$

and δ' is a constant satisfying $0 < \delta' < \delta$. We check that

(9.3.4) $$\Delta \Omega - (r^2 + \tfrac{1}{4}) \Omega = 0.$$

To simplify the notation we now fix $w = i$.

9.3.5 Lemma. *The function $k(z, i)\Omega(z)$ is absolutely integrable over $\mathbf{H}$ and the following equations hold.*

$$h(r) := \int_{\mathbf{H}} k(z, i)\Omega(z) \, d\mathbf{H}(z) = \int_{-\infty}^{+\infty} \int_0^{\infty} y^{1/2+ir} L(\frac{1+x^2+y^2}{2y}) \frac{dy}{y^2} dx$$

$$= \sqrt{2} \int_{-\infty}^{+\infty} e^{iru} \int_{|u|}^{\infty} \frac{K(\rho) \sinh \rho}{\sqrt{\cosh \rho - \cosh u}} \, d\rho \, du.$$

Proof. On the distance circle $\{z \in \mathbf{H} \mid \operatorname{dist}(z, i) = \rho\}$ we have $\operatorname{Im} z \leq e^\rho$. This is easily seen if we observe that the distance circle in the upper half-plane $\mathbf{H}$ is also a Euclidean circle with its uppermost point on the imaginary axis. Since $\operatorname{Im} z \leq e^\rho$, it follows that

$$|\Omega(z)| \leq e^{\rho(1/2 + \operatorname{Im} r)} \leq e^{\rho(1+\delta')},$$

and the absolute integrability follows from the rapid decay of K together with the choice of the strip $\mathscr{S}_{\delta'}$.

Now we substitute successively, first $(1 + x^2 + y^2)/2y = t(x) = t$ (with $dx = y\, dt/x$), then e^u, and finally $t = \cosh \rho$. ◇

The function h provides the transformation of the eigenvalues of the Laplacian into those of the integral operator with kernel $\mathscr{K}_M$. For this we set

(9.3.6) $\qquad r_n = \begin{cases} i\sqrt[+]{\tfrac{1}{4} - \lambda_n} & \text{if } 0 \leq \lambda_n \leq \tfrac{1}{4} \\ \sqrt[+]{\lambda_n - \tfrac{1}{4}} & \text{if } \lambda_n \geq \tfrac{1}{4}. \end{cases}$

Observe that all r_n are contained in the closure of the strip $\mathscr{S}_0$, where $\mathscr{S}_0 = \{r \in \mathbf{C} \mid |\operatorname{Im} r| \leq \tfrac{1}{2}\}$.

9.3.7 Theorem. (Pre-trace formula). *If $K : [0, \infty[\to \mathbf{C}$ is an L^2-function satisfying $|K(\rho)| \leq \operatorname{const} \exp(-\rho(1 + \delta))$ for some $\delta > 0$, then $\mathscr{K}_M$ has the following expansion in the L^2-sense, where h is as in Lemma 9.3.5*

$$\mathscr{K}_M(x, y) = \sum_{n=0}^{\infty} h(r_n)\varphi_n(x)\varphi_n(y).$$

(For the pointwise version see Theorem 9.3.11.)

Proof. Let ϕ_n be a Γ-automorphic lift of the eigenfunction φ_n with respect to the covering $\mathbf{H} \to M = \Gamma\backslash\mathbf{H}$. We show that

(9.3.8) $\qquad \displaystyle\int_\mathbf{H} k(z, w)\phi_n(z)\, d\mathbf{H}(z) = h(r_n)\phi_n(w).$

Observe that ϕ_n is a bounded function on $\mathbf{H}$, so that the integrand is absolutely integrable. For simplicity we conjugate Γ in $\operatorname{PSL}(2, \mathbf{R})$ in such a way that we may assume $w = i$. Let $z = (\rho, \sigma)$ be the representation of z in hyperbolic polar coordinates, where $\rho = \operatorname{dist}(i, z)$ and $\sigma \in \mathbf{S}^1$, and write $\phi_n = \phi_n(\rho, \sigma)$. Since the Laplacian is an isometry invariant and since $\mathbf{S}^1$ acts on $\mathbf{H}$ by rotations with center i, the averaged function

$$(9.3.9) \qquad \psi(\rho, \sigma) = \psi(\rho) = \frac{1}{2\pi} \int_{S^1} \phi_n(\rho, \sigma) \, d\sigma$$

is again a solution (not Γ-automorphic) to the eigenvalue equation $\Delta \psi = \lambda_n \psi$, where $\lambda_n = \frac{1}{4} + r_n^2$. From the expression (7.4.5) of the Laplacian for radial functions, we see that $\psi = \psi(\rho)$ is a solution to

$$\psi''(\rho) + (\coth \rho) \, \psi'(\rho) + \lambda_n \psi(\rho) = 0, \qquad \rho \in [0, \infty[$$

with respect to the initial conditions $\psi(0) = \phi_n(i)$, $\psi'(0) = 0$. Except for initial conditions, the same is true, if we average the function Ω of (9.3.3), with $r = r_n$. By the uniqueness of the solution of the initial value problem we have

$$(9.3.10) \qquad \psi(\rho) = \phi_n(i) \frac{1}{2\pi} \int_{S^1} \Omega(\rho, \sigma) \, d\sigma.$$

Using the decomposition $\int_{\mathbf{H}} = \int_0^\infty \int_{S^1}$ together with (9.3.9), we obtain

$$\int_{\mathbf{H}} k(z, i) \phi_n(z) \, d\mathbf{H}(z) = \phi_n(i) \int_{\mathbf{H}} k(z, i) \Omega(z) \, d\mathbf{H}(z)$$
$$= h(r_n) \phi_n(i).$$

This proves (9.3.8). Now let F be a compact fundamental domain of Γ. Then

$$\int_M \mathcal{K}_M(x, y) \varphi_n(x) \, dM(x) = \sum_{T \in \Gamma} \int_F k(w, Tz) \phi_n(z) \, d\mathbf{H}(z)$$
$$= \sum_{T \in \Gamma} \int_F k(w, Tz) \phi_n(Tz) \, d\mathbf{H}(z)$$
$$= \int_{\mathbf{H}} k(z, w) \phi_n(z) \, d\mathbf{H}(z).$$

Here w is a lift in $\mathbf{H}$ of y, and the interchanging of integration and summation is admissible because ϕ_n (as lift of φ_n) is a bounded function. With the notaion of (9.3.6) we obtain from (9.3.8)

$$\int_M \mathcal{K}_M(x, y) \varphi_n(x) \, dM(x) = h(r_n) \varphi_n(y).$$

The pre-trace formula follows now from the Hilbert-Schmidt theorem (Theorem 7.2.7) and from the earlier remark that the eigenfunctions of the Laplacian are also the eigenfunctions of the integral operator with kernel $\mathcal{K}_M$. ◇

9.3.11 Theorem. *Assume that the generating function K occurring in Theorem 9.3.7 and its transform h as defined in Lemma 9.3.5 satisfy at least one of the following additional two conditions.*

(i) *K is continuous and h has growth rate $h(r) = O((1 + |r|^2)^{-(1+\varepsilon)})$ for some $\varepsilon > 0$ as $r \to \infty$,*

(ii) $\mathcal{K}_M \in C^2(M \times M, \mathbf{C})$.

Then

$$\mathcal{K}_M(x, y) = \sum_{n=0}^{\infty} h(r_n)\varphi_n(x)\varphi_n(y)$$

with absolute uniform convergence on $M \times M$. In particular

$$\operatorname{tr} \mathcal{K}_M = \sum_{n=0}^{\infty} h(r_n).$$

(A sufficient criterion for $\mathcal{K}_M$ to be of class C^2 is given in Lemma 7.5.6.)

Proof. Since equality holds in the L^2-sense, we need only prove the absolute uniform convergence.

If condition (i) holds, then $h(r_n) = O(\lambda_n^{-(1+\varepsilon)})$, and the series has a majorant, given by Theorem 7.2.18.

If condition (ii) holds, then the statement of the theorem is an immediate consequence of Proposition 9.3.12 below. In the particular case that K is of class C^∞ and has compact support, we have a simpler argument: h is the Fourier transform of a smooth function. Repeated integration by parts yields that for $k = 1, 2, \dots$ the function h has growth rate

$$h(r) = O((1 + |r|)^{-k}), \text{ as } |r| \to \infty,$$

uniformly in any strip $S = \{r \in \mathbf{C} \mid |\operatorname{Im} r| \leq \text{const}\}$. Now use (i). ◇

9.3.12 Proposition. *Let $f = f(x, y) \in C^2(M \times M, \mathbf{C})$ and let $c_n(x)$ denote the Fourier coefficients of the function $y \mapsto f(x, y)$ for each fixed x. Then the Fourier series*

$$f(x, y) = \sum_{n=0}^{\infty} c_n(x)\varphi_n(y)$$

converges absolutely and uniformly on $M \times M$.

Proof. We consider x as a parameter and understand the operators Δ and $\int dM$ to be with respect to the variable y. It suffices to prove the proposition under the hypothesis that f has the first Fourier coefficient $c_0 \equiv 0$. Observe first that

$$\Delta f = \sum_{n=1}^{\infty} \lambda_n c_n(x) \varphi_n.$$

This holds because Δf (as a function of y for fixed x) has a unique expansion

$$\Delta f = \sum_{n=1}^{\infty} b_n(x) \varphi_n,$$

where by Green's formula,

$$b_n(x) = \int_M \Delta f\, \varphi_n\, dM = \int_M f\, \Delta \varphi_n\, dM = \lambda_n c_n(x).$$

By Parseval's relation,

$$\int_M |\Delta f|^2\, dM = \sum_{n=1}^{\infty} \lambda_n^2 |c_n(x)|^2.$$

The left-hand side depends continuously on x, as do the individual terms on the right-hand side. Moreover these terms are positive functions. Dini's theorem (see for instance Jörgens [1] p. 95) states that under these circumstances the right-hand side converges uniformly on M. The inequality

$$\left(\sum_{n=a}^{b} |c_n(x)\varphi_n|\right)^2 = \left(\sum_{n=a}^{b} |\lambda_n c_n(x)| \frac{|\varphi_n|}{\lambda_n}\right)^2$$

$$\leq \sum_{n=a}^{b} |\lambda_n c_n(x)|^2 \cdot \sum_{n=a}^{b} \frac{1}{\lambda_n^2} |\varphi_n|^2$$

together with the uniform convergence of $\sum_{n=1}^{\infty} |\varphi_n|^2 \lambda_n^{-2}$ (cf. Theorem 7.2.18) now yields the proposition. ◇

9.4 The Prime Number Theorem

In number theory the prime number theorem states that the cardinality $\pi(x)$ of the set of all prime numbers in the interval $[0, x]$ satisfies the asymptotic equality

(9.4.1) $$\pi(x) \sim \frac{x}{\log x}, \quad x \to \infty.$$

As Selberg points out in [3], there is a strong analogy between the prime numbers and the norms of the elements in a discrete subgroup of $SL(2, \mathbf{R})$ or, in our case, the lengths of the closed geodesics. This analogy is most strikingly reflected in Selberg's zeta function We refer to Hejhal [2] and

Selberg [2] as a reference for the Selberg zeta function; an overview may be found in Elstrodt [2] and Venkov [1].

Although the Selberg zeta function is not a subject of this book, we shall consider one of the connections between the prime numbers and the closed geodesics: the prime number theorem for the compact Riemann surfaces. This analogy was discovered independently by Huber [1, 2] in connection with the distribution of lattice points and by Selberg through the study of the zeta function. In the present section we shall prove the prime number theorem for Riemann surfaces as an application of the trace formulae given in Theorems 9.2.10 and 9.3.11. We shall follow Huber's original approach [2] via the generating function $K(\rho) = (\cosh \rho)^{-s}$ ($\mathcal{R}es > 1$). Huber also proved a prime number theorem with error terms in which the small eigenvalues of the Laplacian occur. This result will be proved in Section 9.6 via Selberg's trace formula. For a proof via Selberg's zeta function we refer to Hejhal [1] and Randol [2, 3].

To point out some of the similarities between the distributions of the prime numbers and the lengths of the closed geodesics, we make a small digression and look at some of the steps which led to the prime number theorem for prime numbers. For the proofs and some history we refer to Apostol [1], Patterson [4] and Edwards [1].

In 1737 Euler [1] found the following theorem (not stated here in full generality). Let $f : \mathbf{N} - \{0\} \to \mathbf{C}$ be a *completely multiplicative function*, that is, by definition, a function f satisfying

$$f(mn) = f(m)f(n) \text{ for all } m, n \in \mathbf{N} - \{0\}.$$

A typical example is the function given by $f(m) = m^s$ for fixed $s \in \mathbf{C}$. If the series $\sum_{n=1}^{\infty} f(n)$ is absolutely convergent, then

$$(9.4.2) \qquad \sum_{n=1}^{\infty} f(n) = \prod_p \frac{1}{1-f(p)},$$

where the product is extended over all *prime numbers p*. The right-hand side of (9.4.2) is called the *Euler product* of the series (see for instance Apostol [1] for a proof). Using this identity, Euler gave a new proof that there exist infinitely many prime numbers. In fact, he showed that the distribution of the prime numbers among the positive integers is so dense that the series $\sum 1/p$ extended over all prime numbers is divergent. Identity (9.4.2) is an early form of an analytic relationship expressed in terms of the prime numbers.

Dirichlet later applied analytic methods to number theory in a systematic way. The power of these methods can be seen in the following simple proof of the divergence of the series $\sum 1/p$. For *real* $s > 1$, let

(9.4.3) $$\zeta(s) = \sum_{n=1}^{\infty} n^{-s}.$$

This series converges uniformly on each interval $[1 + \delta, \infty[$, $\delta > 0$, and is an analytic function satisfying $\zeta(s) \to \infty$ as $s \to 1$. The function was later studied by Riemann for complex s also and is called the *Riemann ζ-function*. Since for fixed s the function $n \mapsto n^{-s}$, $n \in \mathbf{N}$, is completely multiplicative, we may express $\zeta(s)$ using the Euler product

(9.4.4) $$\zeta(s) = \prod_{p} \left(\frac{1}{1-p^{-s}}\right).$$

Now take the logarithm:

$$\log \zeta(s) = \sum_{p} \log\left(\frac{1}{1-p^{-s}}\right) \leq 2 \sum_{p} p^{-s},$$

and divergence follows from the fact that $\log \zeta(s) \to \infty$ as $s \to 1$.

The asymptotic law (9.4.1) was conjectured from inspection of tables of primes by Gauss in 1792 and by Legendre in 1798. The first proofs were given independently by Hadamard [1] and de la Vallée Poussin [1] in 1896.

To understand the behavior of $\pi(x)$, other enumerating functions have proven to be useful, such as the Chebyshev functions

(9.4.5) $$\vartheta(x) = \sum_{p \leq x} \log p, \qquad \psi(x) = \sum_{n \leq x} \Lambda(n),$$

where the first summation is over all prime numbers smaller than or equal to x and the second summation is over all positive integers smaller than or equal to x. In the second sum, Λ is the von Mangoldt function

(9.4.6) $$\Lambda(n) = \begin{cases} \log p & \text{if } n = p^m \text{ for some prime } p \text{ and } m \geq 1, \\ 0 & \text{otherwise.} \end{cases}$$

We mention these functions here because they will appear in connection with the length spectrum.

Wiener [2, Ch. IV] gave a proof of the prime number theorem using a Tauberian theorem which is as follows. (This theorem will later be needed in the proof of Theorem 9.4.14.)

9.4.7 Theorem. (Wiener-Ikehara). *Let $t \mapsto g(t) \in \mathbf{R}$, $t \in [0, \infty[$ be a non-negative non-decreasing function and assume that on the open half-plane $\{s \in \mathbf{C} \mid \mathcal{R}\!e\, s > 1\}$ we have*

$$\int_0^\infty e^{-st} g(t)\, dt = \frac{1}{s-1} + f(s)$$

for some continuous function f which extends continuously to the closed half-plane $\{s \in \mathbf{C} \mid \mathcal{R}es \geq 1\}$. Then

$$g(t) \sim e^t, \quad t \to \infty.$$

For a proof we refer to Widder [1], p. 233. The major steps in the proof of the prime number theorem are then as follows. In a first step the Euler product (9.4.4) is used to prove the following analytic property of the logarithmic derivative of the ζ-function

(9.4.8) $$-\frac{\zeta'(s)}{\zeta(s)} = f(s) + \frac{1}{s-1},$$

where f is holomorphic in an open domain Ω which contains the half-plane $\mathcal{R}es \geq 1$. Here the prime number aspect is not yet important. This will come in the next step, where the logarithmic derivative is expressed by means of the Cheybyshev function.

Now the following equations hold, where the proof of the first line is the difficult part (we do not give the proof of this equation).

(9.4.9)
$$\begin{aligned}
-\frac{\zeta'(s)}{\zeta(s)} &= \sum_p \sum_{m=1}^\infty p^{-ms} \log p \\
&= \sum_{n=1}^\infty \Lambda(n)\, n^{-s} \\
&= \int_1^\infty x^{-s}\, d\psi(x) \\
&= \int_0^\infty e^{-st}\, d\psi(e^t) \\
&= s \int_0^\infty e^{-st}\, \psi(e^t)\, dt.
\end{aligned}$$

Here $\int d\alpha(x)$ (with $\alpha = \psi$) is the *Stieltjes integral*

$$\int_a^b h(x)\, d\alpha(x) := \lim_{\delta \to 0} \sum_{i=0}^{n-1} h(\xi_i)(\alpha(x_{i+1}) - \alpha(x_i))$$

with subdivisions of the interval $[a, b]$ of width $\max_i |x_{i+1} - x_i| \leq \delta$ and with $\xi_i \in [x_i, x_{i+1}]$.

For the Stieltjes integral we refer to Widder [1] chapter I. Among its elementary properties we mention the frequently used *integration by parts* (already used in (9.4.9))

$$\int_a^b h(x)\,d\alpha(x) = h(b)\alpha(b) - h(a)\alpha(a) - \int_a^b \alpha(x)\,dh(x).$$

If h is differentiable, then $dh(x) = h'(x)\,dx$.

The Wiener-Ikehara theorem yields $\psi(e^t) \sim e^t$ or, equivalently,

(9.4.10) $$\psi(x) \sim x, \quad x \to \infty.$$

The remaining arguments are elementary and will also show up in connection with the closed geodesics. The first observation is that ψ does not grow much more rapidly than ϑ, the difference is

$$\psi(x) - \vartheta(x) = \sum_{m=2}^{m_x} \vartheta(x^{1/m}),$$

where m_x is the largest integer m satisfying $2^m \leq x$. Since $m_x \leq \log_2 x$ and since $\vartheta(x^{1/m}) \leq \vartheta(x^{1/2}) \leq \psi(x^{1/2}) = O(x^{1/2})$, we obtain

(9.4.11) $$\vartheta(x) \sim x, \quad x \to \infty.$$

Integration by parts yields

$$\pi(x) = \sum_{p \leq x} 1 = \int_{3/2}^x \frac{d\vartheta(\tau)}{\log \tau} = \frac{\vartheta(x)}{\log x} + \int_{3/2}^x \frac{\vartheta(\tau)}{\tau(\log \tau)^2}\,d\tau.$$

Hence, $\pi(x) \sim x/\log x$.

We return to compact Riemann surfaces. Products of closed geodesics are not defined, but powers γ^m of such geodesics make sense. A closed geodesic is called *prime* if it is not the power of another closed geodesic for some exponent $m \geq 2$ (cf. Definition 9.2.5). Rather than lengths, it is sometimes more natural to use the *norms* of the geodesics

(9.4.12) $$N_\gamma := e^{\ell(\gamma)}.$$

Since $N_{\gamma^k} = (N_\gamma)^k$, the behavior of the norms is closer to that of the integers as far as "multiplication" is concerned.

In analogy with the classical number theoretic functions the following are defined for compact (and more general) Riemann surfaces.

9.4.13 Definition. Let M be a compact Riemann surface of genus $g \geq 2$. For every closed geodesic $\gamma \in \mathscr{C}(M)$ there exists a unique prime geodesic $\gamma_0 \in \mathscr{P}(M)$ and a unique exponent $m \geq 1$ such that $\gamma = \gamma_0^m$. With this notation

we set
$$\Lambda(\gamma) = \ell(\gamma_0) = \log N_{\gamma_0}.$$
The counting functions on $\mathscr{C}(M)$ in terms of the lengths are defined in the following way.

$$\phi(t) = \#\{\gamma \in \mathscr{C}(M) \mid \ell(\gamma) \leq t\}$$
$$\Pi(t) = \#\{\gamma \in \mathscr{P}(M) \mid \ell(\gamma) \leq t\}$$
$$\Psi(t) = \sum_{\ell(\gamma) \leq t} \Lambda(\gamma)$$
$$\theta(t) = {\sum_{\ell(\gamma) \leq t}}' \ell(\gamma) = \int_0^t \tau \, d\Pi(\tau).$$

The symbol # denotes cardinality and the prime in the definition of θ means that the summation is restricted to the prime geodesics. In terms of the norms, letting $x = e^t$, we define correspondingly

$$\varphi(x) = \#\{\gamma \in \mathscr{C}(M) \mid N_\gamma \leq x\} = \sum_{N_\gamma \leq x} 1$$
$$\pi(x) = \#\{\gamma \in \mathscr{P}(M) \mid N_\gamma \leq x\} = {\sum_{N_\gamma \leq x}}' 1$$
$$\psi(x) = \sum_{N_\gamma \leq x} \Lambda(\gamma)$$
$$\vartheta(x) = {\sum_{N_\gamma \leq x}}' \log N_\gamma = \int_1^x \log \xi \, d\pi(\xi).$$

9.4.14 Theorem. (Prime number theorem for compact Riemann surfaces). *For any compact Riemann surface of genus $g \geq 2$,*

$$\pi(x) \sim x/\log x, \qquad \varphi(x) \sim x/\log x, \qquad x \to \infty.$$

(The proof will follow below.) For a stronger version of the theorem and for some references we refer to Theorem 9.6.1 and the comments thereafter.

In order to round out the above sketch of the prime number theorem, we mention that Euler's formula (9.4.2) and the Riemann ζ-function (9.4.4) have as a descendant Selberg's zeta function

$$Z(s) = \prod_{\gamma \in \mathscr{P}(M)} \prod_{v=0}^{\infty} (1 - N_\gamma^{-s-v}), \qquad \mathcal{R}\!e\, s > 1.$$

Ch.9, §4]   The Prime Number Theorem   247

There are various ways in which the prime number theorem for Riemann surfaces can be obtained from Selberg's zeta function. This is carried out in detail in Hejhal [2]. We also refer to Randol [2]. An overview of the Selberg zeta function can be found in Elstrodt [2] and in Venkov [1]. In Elstrodt [2] we also find an interesting interpretation of $Z(s)$ in terms of the resolvent of the Laplacian.

Proof of Theorem 9.4.14. We now give Huber's proof of the prime number theorem (in a form which is adapted to Theorem 9.3.11). The proof also yields the asymptotic formula for the distribution of the lattice points in Theorem 9.4.23. An independent proof based on Selberg's trace formula will follow in Section 9.6.

Let M be given in the form $M = \Gamma \backslash \mathbf{H}$, where Γ is a subgroup of $\mathrm{PSL}(2, \mathbf{R})$ acting freely and properly discontinuously on $\mathbf{H}$. For $s \in \mathbf{C}$ with $\mathcal{R}\!e\, s > 1$ the generating function

$$K(\rho) = (\cosh \rho)^{-s}$$

satisfies the decay conditions of Lemma 7.5.6 and therefore generates a smooth kernel $\mathcal{K}_M$ on M. Its Γ- bi-invariant lift $\mathcal{K}_\Gamma$ on $\mathbf{H}$ is the Dirichlet series

$$(9.4.15) \qquad G(s; z, w) = \mathcal{K}_\Gamma(z, w) = \sum_{T \in \Gamma} (\cosh \mathrm{dist}(z, Tw))^{-s}$$

as introduced in Section 9.1. By Theorem 9.3.11, the following expansions are absolutely and uniformly convergent

$$(9.4.16) \qquad \begin{aligned} \mathcal{K}_M(x, y) &= \sum_{n=0}^{\infty} h_s(r_n) \varphi_n(x) \varphi_n(y), \quad x, y \in M \\ G(s; z, w) &= \sum_{n=0}^{\infty} h_s(r_n) \phi_n(z) \phi_n(w), \quad z, w \in \mathbf{H}, \end{aligned}$$

where for $n = 0, 1, \ldots$ the function ϕ_n is the Γ-invariant lift of the eigenfunction φ_n of the Laplacian on M. The numbers r_n are as in (9.3.6):

$$r_n = \begin{cases} i\sqrt[+]{\tfrac{1}{4} - \lambda_n} & \text{if } 0 \le \lambda_n \le \tfrac{1}{4} \\ \sqrt[+]{\lambda_n - \tfrac{1}{4}} & \text{if } \lambda_n \ge \tfrac{1}{4}. \end{cases}$$

The function $h = h_s$ is the transform of $K = \cosh^{-s}$ as in Lemma 9.3.5.

In (9.4.15), $G(s; z, w)$ is defined and proved to be a holomorphic function of s for $\mathcal{R}\!e\, s > 1$. We shall now prove that the series on the right-hand side of

(9.4.16) converges in a larger region. (This will give access to the Wiener-Ikehara theorem.) To this end we compute $h_s(r_n)$ using properties of the gamma function Γ. For this and related functions we shall use Magnus-Oberhettinger-Soni (M-O-S) [1] as a reference.

Under the conditions

$$\mathcal{R}e\,\zeta > -1, \quad \mathcal{R}e\,\omega > \tfrac{1}{2} + \tfrac{1}{2}\mathcal{R}e\,\zeta; \quad a > 0, \ b > 0,$$

the following identity holds (M-O-S, p. 6):

(9.4.17) $$\int_0^\infty \frac{t^\zeta\, dt}{(a+bt^2)^\omega} = \frac{a^{(\zeta+1-2\omega)/2}}{2b^{(\zeta+1)/2}} \frac{1}{\Gamma(\omega)} \Gamma(\tfrac{1}{2}(\zeta+1))\Gamma(\omega - \tfrac{1}{2}(\zeta+1)).$$

In particular, for $|Im\, r| \leq \tfrac{1}{2}$ and $\mathcal{R}e\, s > 1$, if $z = x + iy \in \mathbf{H}$, then

$$\int_0^\infty y^{ir-3/2} \left(\frac{1+x^2+y^2}{2y}\right)^{-s} dy = $$
$$2^{s-1}(1+x^2)^{(-1/2+ir-s)/2} \frac{1}{\Gamma(s)} \Gamma(\tfrac{1}{2}(s+ir-\tfrac{1}{2}))\Gamma(\tfrac{1}{2}(s-ir+\tfrac{1}{2})).$$

This equation together with another application of (9.4.17) shows that the function $h = h_s$ in Lemma 9.3.5 (with $K(\rho) = (\cosh \rho)^{-s}$ and $L(\tau) = \tau^{-s}$) may be defined as follows.

(9.4.18) $$h_s(r_n) = 2^{s-1} \pi^{1/2} \frac{1}{\Gamma(s)} \Gamma(\tfrac{1}{2}(s-\tfrac{1}{2}+ir_n))\Gamma(\tfrac{1}{2}(s-\tfrac{1}{2}-ir_n)).$$

For $n = 0$, we have $r_0 = \tfrac{i}{2}$. Using the functional equation

$$\Gamma(2z) = \pi^{-1/2}\, 2^{2z-1}\, \Gamma(z)\Gamma(\tfrac{1}{2}+z)$$

(M-O-S, p. 3), we obtain

(9.4.19) $$h_s(r_0) = \frac{2\pi}{s-1}.$$

For $n \geq 1$, we have $\lambda_n \geq \lambda_1 > 0$ and therefore $\mathcal{R}e(\tfrac{1}{2}-ir_n) \leq 1-\delta$ for some $\delta > 0$ which depends on the surface M. The gamma function is holomorphic in $\mathbf{C}$ except at the points $z = -n$, $n = 0, 1, 2, \ldots$, where it has simple poles with residues $(-1)^n/n!$. Since $\Gamma(z) \neq 0$ for all $z \in \mathbf{C}$, it follows that for each $n \geq 1$ the function $s \to h_s(r_n)$ is meromorphic in $\mathbf{C}$ with poles contained in the set $\{\tfrac{1}{2} \pm ir_n - 2k \mid k = 0, 1, 2, \ldots\}$. In particular, $h_s(r_n)$ is holomorphic in the half-plane $\mathcal{R}e\, s \geq 1 - \delta$, $n = 1, 2, \ldots$.

We next consider the growth rate of $h_s(r_n)$ as $n \to \infty$ for $s \in \mathcal{B}$, where $\mathcal{B} \subset \{s \in \mathbf{C} \mid \mathcal{R}e\, s \geq 1 - \delta\}$ is an arbitrary fixed compact domain. By Stirling's asymptotic formula there exists, for any compact subset $\mathcal{B}' \subset \mathbf{C}$, a constant c' such that for any $\zeta \in \mathcal{B}'$ and for any real number $\rho \geq 1$

$$|\Gamma(\zeta + i\rho)\Gamma(\zeta - i\rho)| \le c'\rho^{2\mathcal{R}e\zeta - 1}e^{-\pi\rho}.$$

In particular, there exists a constant c such that

$$|h_s(r_n)| \le c\lambda_n^{-2}$$

for any $s \in \mathcal{B}$ and any $n \ge 1$. The uniform absolute convergence of the series $\sum \lambda_n^{-2}\phi_n(z)\phi_n(w)$ on $\mathbf{H} \times \mathbf{H}$ (cf. Theorem 7.2.18) implies that the following functions of s are holomorphic in $\mathcal{R}e\, s > 1 - \delta$:

(9.4.20) $\quad G_1(s; z, w) := \sum_{n=1}^{\infty} h_s(r_n)\phi_n(z)\phi_n(w), \quad z, w \in \mathbf{H},$

and

(9.4.21) $\quad H_1(s) := \sum_{n=1}^{\infty} h_s(r_n).$

Since $\phi_0(z) = \text{const} = (\text{area } M)^{-1/2}$, where area $M = 4\pi(g - 1)$ and g is the genus of $M = \Gamma\backslash\mathbf{H}$, we obtain from (9.4.19)

(9.4.22) $\quad G(s; z, w) = \dfrac{1}{2(g-1)} \dfrac{1}{s-1} + G_1(s; z, w).$

This allows us to prove the asymptotic formula for the counting function

$$N(t) = N(t; z, w) := \#\{T \in \Gamma \mid \text{dist}(z, Tw) \le t\}.$$

From (9.4.15),

$$G(s; z, w) = m_0 + \int_0^{\infty} (\cosh t)^{-s} dN(t),$$

where m_0 is the number of elements $T \in \Gamma$ satisfying $\text{dist}(z, Tw) = 0$. To use the notation as in the Wiener-Ikehara theorem (Theorem 9.4.7), we make the change of variable $\cosh t = e^{\tau}$ and integrate by parts:

$$G(s; z, w) = m_0 + \int_0^{\infty} e^{-s\tau} dN(t(\tau))$$

$$= s \int_0^{\infty} e^{-s\tau} N(t(\tau))\, d\tau.$$

The Wiener-Ikehara theorem together with (9.4.22) implies the asymptotic equation $2(g-1)N(t(\tau)) \sim e^{\tau} = \cosh t(\tau)$, $\tau \to \infty$. From this we obtain the following theorem (Delsarte [1, 2], Huber [2]).

9.4.23 Theorem. *The counting function $N(t; z, w)$ satisfies the asymptotic equality*

$$N(t; z, w) \sim \frac{1}{4(g-1)} e^t, \quad t \to \infty.$$ ◊

For further references and remarks we refer to the discussion at the end of Section 9.7 which gives an asymptotic formula for $N(t; z, w)$, with error terms.

To continue with the proof of the prime number theorem, we combine Theorems 9.3.11 and 9.2.10 to obtain the formula

$$\sum_{n=0}^{\infty} h_s(r_n) = 4\pi(g-1) + \sum_{\gamma \in \mathscr{C}(M)} \frac{\ell(\gamma_0)}{\sqrt{\cosh \ell(\gamma) - 1}} \int_{\ell(\gamma)}^{\infty} \frac{(\cosh \rho)^{-s} \sinh \rho}{\sqrt{\cosh \rho - \cosh \ell(\gamma)}} \, d\rho$$

in which γ_0 is the unique primitive closed geodesic satisfying $\gamma = \gamma_0^m$ for some $m \geq 1$ (cf. Lemma 9.2.6). In the integral we substitute as follows,

$$\cosh \rho = \cosh \ell(\gamma) + \tfrac{1}{2} x^2 e^{-\ell(\gamma)}.$$

This is the same substitution as in (7.4.7) and (7.4.12/13). Another application of (9.4.17) then yields

$$(9.4.24) \quad \sum_{n=0}^{\infty} h_s(r_n) = 4\pi(g-1) + \pi^{1/2} \frac{\Gamma(s - \tfrac{1}{2})}{\Gamma(s)} \sum_{\gamma \in \mathscr{C}(M)} \Lambda^*(\gamma)(\cosh \ell(\gamma))^{-s},$$

where we use the abbreviation

$$\Lambda^*(\gamma) := \Lambda(\gamma)(\cosh \ell(\gamma))^{1/2} (\cosh \ell(\gamma) - 1)^{-1/2}.$$

Since the function $H_1(s)$ in (9.4.21) is holomorphic for $\mathcal{R}\!es > 1 - \delta$ and since $h_s(r_0) = 2\pi/(s-1)$, we have, using that $\Gamma(\tfrac{1}{2}) = \pi^{1/2}$,

$$\sum_{\gamma \in \mathscr{C}(M)} \Lambda^*(\gamma)(\cosh \ell(\gamma))^{-s} = \frac{2}{s-1} + H_2(s),$$

where H_2 is a holomorphic function in the domain $\mathcal{R}\!es > 1 - \delta$.

The remainder of the proof is now the same as in the case of the classical prime number theorem, and we proceed as in (9.4.9), setting

$$\Psi^*(t) = \sum_{\ell(\gamma) \leq t} \Lambda^*(\gamma).$$

By Lemma 9.2.7, $\Psi^*(t) = O(te^t)$ as $t \to \infty$. Therefore, since $\mathcal{R}\!es > 1$, we obtain, with $t(\tau) = \operatorname{arccosh}(e^\tau)$,

$$\sum_{\gamma \in \mathcal{C}(M)} \Lambda^*(\gamma)(\cosh \ell(\gamma))^{-s} = \int_0^\infty (\cosh t)^{-s} d\Psi^*(t)$$

$$= \int_0^\infty e^{-s\tau} d\Psi^*(t(\tau))$$

$$= s \int_0^\infty e^{-s\tau} \Psi^*(t(\tau)) \, d\tau.$$

The Wiener-Ikehara theorem yields $\Psi^*(t(\tau)) \sim 2e^\tau = 2\cosh t(\tau)$, $\tau \to \infty$. Therefore the Chebyshev function Ψ (cf. Definition 9.4.13) satisfies

$$\Psi(t) = \sum_{\ell(\gamma) \le t} \Lambda(\gamma) \sim e^t, \qquad t \to \infty.$$

Denote by $\mu > 0$ a positive lower bound for the lengths of the closed geodesics of $\Gamma \backslash \mathbf{H}$ and let

$$m(t) = \text{int}(t/\mu)$$

be the integer part of t/μ for $t > 0$. The Chebyshev functions Ψ and θ are related by

(9.4.25) $$\Psi(t) = \theta(t) + \sum_{m=2}^{m(t)} \theta(t/m).$$

The sum on the right-hand side is bounded above by $m(t)\theta(t/2)$, where

$$m(t)\theta(t/2) \le \frac{t}{\mu} \Psi(t/2) = O(te^{t/2}), \qquad t \to \infty,$$

and therefore

(9.4.26) $$\theta(t) \sim e^t, \qquad t \to \infty.$$

Since

$$\Pi(t) = \int_{\mu/2}^t \frac{d\theta(\tau)}{\tau} = \frac{\theta(t)}{t} + \int_{\mu/2}^t \frac{\theta(\tau)}{\tau^2} d\tau,$$

we obtain $\Pi(t) \sim e^t/t$ and $\pi(x) \sim x/\log x$. Finally, ϕ and Π have the same asymptotic growth because

(9.4.27) $$\phi(t) = \Pi(t) + \sum_{m=2}^{m(t)} \Pi(t/m).$$

This proves Theorem 9.4.14. ◇

9.5 Selberg's Trace Formula

Theorem 9.2.10 and Theorem 9.3.11 together yield a relation between the length and the eigenvalue spectrum of a compact Riemann surface. This relation is in terms of functions g and h which are transforms of the generating function K. In many applications, either h or g is given but not K. It is therefore desirable to state the relation independently of K and under fairly general conditions. This is obtained in Selberg's trace formula. The following condition has proved to be useful although it is not the most general one.

9.5.1 Definition. Let $\varepsilon > 0$, set
$$\mathscr{S}_\varepsilon = \{r \in \mathbf{C} \mid |Imr| < \tfrac{1}{2} + \varepsilon\}$$
and assume that $h : \mathscr{S}_\varepsilon \to \mathbf{C}$ is a holomorphic even function with the decay property
$$h(r) = O((1 + |r|^2)^{-1-\varepsilon}) \text{ uniformly on } \mathscr{S}_\varepsilon.$$
Then h has the Fourier transform (up to a constant)
$$g(u) := \frac{1}{2\pi} \int_{-\infty}^{+\infty} e^{-iru} h(r)\, dr = \frac{1}{\pi} \int_0^\infty (\cos ru)\, h(r)\, dr, \quad u \in \mathbf{R}.$$
The pair h, g will be called an *admissible transform pair*.

9.5.2 Example. Let $g : \,]-\infty, +\infty[\, \to \mathbf{C}$ be an even C^∞-function with compact support and let $h = \hat{g}$ be its Fourier transform. Then h is holomorphic in the entire complex plane. Repeated integration by parts yields, for all $k = 1, 2, \ldots$ and for any $c > 0$,
$$h(r) = O(1 + |r|)^{-k}),$$
uniformly in the strip $S = \{r \in \mathbf{C} \mid |Imr| \leq c\}$. Hence, h and g form an admissible transform pair.

For every compact Riemann surface $M = \Gamma \backslash \mathbf{H}$, we have the set $\mathscr{C}(M)$ of all oriented closed geodesics and the set $\mathscr{P}(M)$ of all primitive oriented closed geodesics (Definition 9.2.8). We also recall the r_n from (9.3.6),
$$r_n = \begin{cases} i\sqrt[+]{\tfrac{1}{4} - \lambda_n} & \text{if } 0 \leq \lambda_n \leq \tfrac{1}{4} \\ \sqrt[+]{\lambda_n - \tfrac{1}{4}} & \text{if } \lambda_n \geq \tfrac{1}{4}, \end{cases}$$
where $\lambda_0 < \lambda_1 \leq \lambda_2 \leq \ldots$ are the eigenvalues of the Laplacian of M.

We now state Selberg's trace formula [2]. In this formula, $N_\gamma = e^{\ell(\gamma)}$ is the *norm* of γ and $\Lambda(\gamma)$ is the length $\Lambda(\gamma) = \ell(\gamma_0)$, where γ_0 is the unique primitive oriented closed geodesic satisfying $\gamma = \gamma_0^m$ for some $m \geq 1$ (cf. Lemma 9.2.6).

9.5.3 Theorem. *Let M be a compact Riemann surface. For any admissible transform pair g, h, we have*

$$\sum_{n=0}^{\infty} h(r_n) = \frac{\text{area } M}{4\pi} \int_{-\infty}^{+\infty} rh(r)\tanh(\pi r)\, dr + \sum_{\gamma \in \mathscr{C}(M)} \frac{\Lambda(\gamma)}{N_\gamma^{1/2} - N_\gamma^{-1/2}} g(\log N_\gamma).$$

On both sides the series are absolutely convergent.

We point out that the sum on the right-hand side can also be written as

$$\sum_{\gamma \in \mathscr{P}(M)} \sum_{k=1}^{\infty} \frac{\ell(\gamma)}{2\sinh \ell(\gamma^k)/2} g(\ell(\gamma^k)).$$

Proof of Selberg's trace formula. We follow Hejhal [2], chapter I. First we prove that both sides of the trace formula are well defined and absolutely convergent.

Assume more explicitly that

(1) $$|h(r)| \leq c(1 + |r|^2)^{-1-\varepsilon}$$

for some constant $c > 0$. The standard Fourier inversion argument implies that

(2) $$|g(u)| \leq \frac{2c}{\pi} e^{-(1/2 + \varepsilon)|u|}.$$

Indeed, let $\varepsilon' \in\,]0, \varepsilon[$ and assume w.l.o.g. that $u < 0$. The Cauchy integral theorem yields

(3) $$g(u) = \frac{1}{2\pi} \int_{E'} h(r) e^{-iru}\, dr,$$

where $E' = \mathbf{R} + i(\frac{1}{2} + \varepsilon')$. Hence,

$$|g(u)| \leq \frac{c}{2\pi} e^{-(1/2 + \varepsilon')|u|} \int_{-\infty}^{+\infty} (1 + x^2)^{-1}dx.$$

This proves (2).

By Weyl's asymptotic law (Theorem 9.2.14) or, equivalently by formula (9.2.12), we have

$$\sum_{n=1}^{\infty} \lambda_n^{-1-\sigma} < \infty \text{ for any } \sigma > 0.$$

This implies that

(4) $$\sum_{n=0}^{\infty}(1+|r_n|^2)^{-1-\varepsilon} < \infty.$$

This yields the absolute convergence of the left-hand side of Selberg's trace formula. By Lemma 9.2.7, the number of closed geodesics of length $\leq t$ is of order $O(e^t)$. Together with (2) this yields the absolute convergence of the right-hand side.

To prove the formula, we approximate g and h by functions which are the transforms of an admissible generating function K and apply the trace formulae of Theorems 9.2.10 and 9.3.11. We proceed in three steps.

We point out that in many applications (such as e.g. in Section 9.6) the Selberg trace formula is applied under the stronger hypothesis of step 1, in which case the proof as given in step 1 is sufficient.

Step 1. We assume that g is a smooth even function with compact support.

Since g is even, the restriction $g\,|\,[0,\infty[$ belongs to the Schwartz space $\mathscr{S}$ (cf Definition 7.3.6). By Theorem 7.3.10, its inverse Abel transform

$$K := (2\pi)^{-1/2}A^{-1}[g]$$

is defined and belongs to $\mathscr{S}$. This yields g and h in terms of K:

(5)
$$g(u) = \sqrt{2}\int_{|u|}^{\infty}\frac{K(\rho)\sinh\rho}{\sqrt{\cosh\rho-\cosh u}}\,d\rho = (2\pi)^{1/2}A[K](|u|), \quad u \in \mathbf{R},$$

$$h(r) = \int_{-\infty}^{+\infty}e^{iru}g(u)\,du, \quad r \in \mathbf{C}.$$

Comparing this with Lemma 9.3.5, we see that g and h coincide with the transforms of K occurring in Theorems 9.2.10 and 9.3.11. As noted in Example 9.5.2, h and g form an admissible transform pair. To prove the trace formula, it therefore only remains to express $K(0)$ in terms of h. Applying A^{-1} as given in Theorem 7.3.4, we see that

$$K(0) = \frac{-1}{2\pi}\int_0^{\infty}\frac{g'(u)}{\sinh u/2}\,du.$$

In the defining equation of g (Definition 9.5.1), $r \in \mathbf{R}$ and h has growth rate $h(r) = O((1+r^2)^{-1})$, so that we may differentiate under the integration sign to get

$$g'(u) = \frac{-1}{\pi}\int_0^{\infty}rh(r)\sin ru\,dr.$$

The function $(r, u) \mapsto (\sinh(r/2))^{-1} r h(r) \sin ru$ is absolutely integrable over $[0, \infty[\times [0, \infty[$ and from M-O-S, p. 422, we obtain

$$K(0) = \frac{1}{2\pi^2} \int_0^\infty rh(r) \int_0^\infty \frac{\sin ru}{\sinh u/2} \, du \, dr = \frac{1}{2\pi} \int_0^\infty rh(r) \tanh(\pi r) \, dr.$$

This proves the trace formula under the hypothesis of step 1.

For simplicity we abbreviate the formula as

$$\mathscr{L}(h) = \mathscr{R}(g)$$

with

$$\mathscr{L}(h) := -\frac{\text{area } M}{4\pi} \int_{-\infty}^{+\infty} rh(r) \tanh(\pi r) \, dr + \sum_{n=0}^\infty h(r_n),$$

$$\mathscr{R}(g) := \sum_{k=1}^\infty \sum_{\gamma \in \mathscr{P}(M)} \frac{\ell(\gamma)}{2 \sinh \frac{1}{2}\ell(\gamma^k)} g(\ell(\gamma^k)).$$

Step 2. We assume that h decays rapidly, say, $h(r) = O(e^{-\delta|\mathscr{R}er|^2})$, on the strip $\mathscr{S}_\varepsilon$ for some $\delta > 0$. With the argument used in (3) but now applied to the ν-th derivative $g^{(\nu)}$, then

(6) $$g^{(\nu)}(u) = O(e^{-(1/2 + \varepsilon)|u|})$$

for each fixed ν-th derivative. We approximate g using cut-off functions. Let $\varphi : [0, \infty[\to \mathbf{R}$ be a smooth function with the following properties: $\varphi = 1$ on $[0, 1]$; $1 \geq \varphi \geq 0$ everywhere; φ is monotone decreasing on $[1, 2]$; $\varphi = 0$ on $[2, \infty[$. For $m \in \mathbf{N}$ we define

$$\varphi_m(u) = \begin{cases} 1 & \text{if } |u| \leq m \\ \varphi(|u| - m) & \text{if } |u| \geq m \end{cases}$$

and, for $u \in \mathbf{R}$ and $r \in \mathbf{C}$,

$$g_m(u) = g(u) \varphi_m(u), \quad h_m(r) = \int_{-\infty}^{+\infty} e^{iru} g_m(u) \, du.$$

It follows that g_m is an even function satisfying $|g_m| \leq |g|$ and $g_m \to g$ as $m \to \infty$, uniformly on compact sets. Thus, by the absolute convergence of $\mathscr{R}(g)$, $\mathscr{R}(g_m) \to \mathscr{R}(g)$ as $m \to \infty$. To prove the analog for $\mathscr{L}$, we observe by (6) that for each ν-th derivative

$$g_m^{(\nu)}(u) = O(e^{-(1/2 + \varepsilon)|u|})$$

uniformly in u and m. Repeated integration by parts in the defining integral of h_m shows that for each ν

$$h_m(r) = O((1 + |r|)^{-\nu})$$

uniformly in m and r for $|\mathrm{Im}\,r| \leq \frac{1}{2}$. Together with (4) this shows that $\mathscr{L}(h_m) \to \mathscr{L}(h)$ as $m \to \infty$. Then using Step 1 we get $\mathscr{L}(h) = \mathscr{R}(g)$.

Step 3. The general case. For arbitrarily small $\delta > 0$ we set

$$h_\delta(r) = h(r) e^{-\delta r^2}, \quad r \in \mathscr{S}_\varepsilon,$$

$$g_\delta(u) = \frac{1}{2\pi} \int_{-\infty}^{+\infty} e^{-iru} h_\delta(r)\, dr.$$

Then h_δ and g_δ form a transform pair which satisfies the hypothesis of Step 2. As before, $\mathscr{L}(h_\delta) \to \mathscr{L}(h)$ as $\delta \to 0$. By (1),

$$|h_\delta(r)| \leq |h(r)| e^{\delta |\mathrm{Im}\, r|^2} \leq 2c(1 + |r|^2)^{-1-\varepsilon}$$

so that by (2)

$$|g_\delta(u)| \leq \frac{4c}{\pi} e^{-(1/2 + \varepsilon)|u|}.$$

Since $g_\delta \to g$ as $\delta \to 0$ uniformly on $\mathbf{R}$, then $\mathscr{R}(g_\delta) \to \mathscr{R}(g)$ as $\delta \to 0$ and again we obtain $\mathscr{L}(h) = \mathscr{R}(g)$. ◇

9.6 The Prime Number Theorem with Error Terms

A somewhat surprising fact about the asymptotic growth of $\pi(x)$ (or $\Pi(t)$) in the prime number theorem is its independence of the genus of the surface. To see that this is not evident, let us look at the following example. For $k = 1, 2, \ldots$ we let M_k be a compact Riemann surface which consists of k copies of a surface of signature $(1, 2)$ glued together along the small boundary geodesics γ, γ' as shown in Fig. 9.6.1. M_k has genus $k + 1$ and is the k-fold cyclic covering of M_1.

Then $\Pi(t) \sim e^t/t$, independently of k. If, on the other hand, M_k^* is the non-connected disjoint union of k copies of M_1, then the corresponding counting function Π^* satisfies $\Pi^*(t) \sim ke^t/t$. Now let $\ell(\gamma) = \ell(\gamma_1) = \ldots = \ell(\gamma_k) \to 0$, where the γ_i are the closed geodesics on M_k along which the building blocks of signature $(1, 2)$ were pasted together. Then, in a sense which can be made precise (cf. e.g. Colbois-Courtois [1]), both M_k and M_k^* converge to the same limiting surface, and we expect to see a corresponding effect in the behavior of Π and Π^*. In the asymptotic formula of Theorem 9.4.14 this is not the case. It is, however, possible to see a corre-

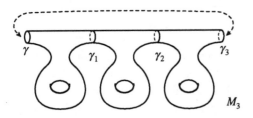 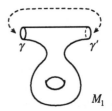

Figure 9.6.1

sponding effect if we consider error terms. In Theorem 9.6.1 we shall see that each small eigenvalue λ contributes to the asymptotic expansion of Π an error term of order $O(e^{\lambda' t}/t)$, where $\lambda' = \frac{1}{2} + (\frac{1}{4} - \lambda)^{1/2}$. As $\ell(\gamma) \to 0$, the first k eigenvalues $\lambda_0, \ldots, \lambda_{k-1}$ of M_k converge to 0, whereas $\lambda_k(M_k) \geq c$ and $\lambda_1(M_1) \geq c$ for some positive constant c. Hence, if we add the error terms, then the asymptotic expansion of Π converges to that of Π^*.

Instead of using $\Pi(t)$, we work with the function

$$\pi(x) = \#\{\gamma \in \mathscr{P}(M) \mid N_\gamma \leq x\}$$

(cf. Definition 9.4.13). The relation between π and Π is

$$\pi(x) = \Pi(t), \quad \text{where } x = e^t.$$

For the statement of the result we abbreviate

$$s_k = \tfrac{1}{2} + (\tfrac{1}{4} - \lambda_k)^{1/2} = \tfrac{1}{2} - ir_k, \quad \text{for } \lambda_k \leq \tfrac{1}{4},$$

where again $\lambda_0 = 0 < \lambda_1 \leq \lambda_2 \leq \ldots$ are the eigenvalues of a given compact Riemann surface $M = \Gamma\backslash\mathbf{H}$.

9.6.1 Theorem. (Prime number theorem with error terms). *For every compact Riemann surface of genus $g \geq 2$ the following asymptotic equation holds, as $x \to \infty$,*

$$\pi(x) = \mathrm{li}(x) + \sum_{1 > s_k > 3/4} \mathrm{li}(x^{s_k}) + O(x^{3/4}/\log x).$$

Here li is the *integral logarithm*:

$$\mathrm{li}(x) = \int_2^x \frac{d\tau}{\log \tau} \sim x/\log x, \quad x \to \infty.$$

The function $\pi(x)$ counts the number of primitive oriented closed geodesics of length $\leq \log x$. For the counting function $\varphi(x)$ which considers *all* oriented

closed geodesics we have, by (9.4.27),
$$\varphi(x) = \pi(x) + O(x^{1/2}).$$
Hence, the asymptotic formula in Theorem 9.6.1 also holds for φ.

Theorem 9.6.1 was first proved by Huber [2] with a marginally larger O-term. Later Heijal [2] and Randol [3] proved the theorem in its present form using Selberg's zeta function. Sarnak [1] gives a version which also holds for non-compact Riemann surfaces with finite area. Randol [7] gives a simpler proof which also works in higher dimensions. The proof which we shall give here is adapted from Sarnak [1] and Randol [7] and uses Selberg's trace formula with a test function g with compact support in the transform pair h, g.

9.6.2 Lemma. *The Chebyshev function ψ (Definition 9.4.13) satisfies*
$$\psi(x) = x + \sum_{1 > s_k \geq 3/4} \frac{1}{s_k} x^{s_k} + O(x^{3/4}).$$

Proof. We work with $t = \log x$. Let
$$g_t(u) = \begin{cases} 2\cosh\frac{u}{2} & \text{if } |u| \leq t \\ 0 & \text{if } |u| > t. \end{cases}$$

Using g_t, the Chebyshev function $\Psi(t) = \psi(x)$ can be written as
$$\Psi(t) = \sum_{\ell(\gamma) \leq t} \Lambda(\gamma) = \sum_{\gamma \in \mathscr{C}(M)} \Lambda(\gamma) \frac{g_t(\ell(\gamma))}{N_\gamma^{1/2} + N_\gamma^{-1/2}}.$$

A direct computation shows that the Fourier transform
$$\hat{g}_t(r) = (2\pi)^{-1/2} \int_{\mathbf{R}} e^{-iru} g_t(u) \, du$$
is given by
$$\hat{g}_t(r) = \left(\frac{2}{\pi}\right)^{1/2} \left(\frac{\sinh((\frac{1}{2} - ir)t)}{\frac{1}{2} - ir} + \frac{\sinh((\frac{1}{2} + ir)t)}{\frac{1}{2} + ir} \right).$$

To approximate g_t by smooth functions with compact support, we let $\chi : \mathbf{R} \to \mathbf{R}$ be a non-negative, even C^∞-function with support $[-1, 1]$ and satisfying
$$\int_{-1}^{1} \chi(u) \, du = (2\pi)^{1/2}.$$

We let

$$\chi_\varepsilon(u) = \varepsilon^{-1}\chi(\varepsilon^{-1}u), \quad \varepsilon > 0.$$

Note that

$$\int_R \chi_\varepsilon(u)\,du = (2\pi)^{1/2}, \quad \hat{\chi}_\varepsilon(r) = \hat{\chi}(\varepsilon r).$$

We define, using the convolution $*$,

$$g_{t,\varepsilon}(u) = (g_t * \chi_\varepsilon)(u) = (2\pi)^{-1/2}\int_R g_t(u-y)\chi_\varepsilon(y)\,dy,$$

$$h_{t,\varepsilon}(r) = \int_R e^{-ir}g_{t,\varepsilon}(u)\,du = (2\pi)^{1/2}\hat{g}_t(r)\hat{\chi}(\varepsilon r).$$

Our conventions for the Fourier transform and the convolution are such that $(f_1 * f_2)\hat{} = \hat{f}_1\hat{f}_2$ and such that $\hat{\hat{f}} = f$ if f is even. Since $g_{t,\varepsilon}$ has compact support, $h_{t,\varepsilon}$ and $g_{t,\varepsilon}$ form an admissible transform pair (cf. Example 9.5.2). With the abbreviation

$$\Psi_\varepsilon(t) = \sum_{\gamma \in \mathscr{C}(M)} \Lambda(\gamma)\frac{g_{t,\varepsilon}(\ell(\gamma))}{N_\gamma^{1/2} - N_\gamma^{-1/2}},$$

Selberg's trace formula becomes

$$\sum_{\lambda_k \leq 1/4} h_{t,\varepsilon}(r_k) + \sum_{\lambda_k > 1/4} h_{t,\varepsilon}(r_k) = \Psi_\varepsilon(t) + \frac{\text{area } M}{4\pi}\int_R rh_{t,\varepsilon}(r)\tanh(\pi r)\,dr.$$

Ψ_ε is close to Ψ, and Lemma 9.6.2 will be obtained by an estimate of the various terms in the formula.

For $\lambda_k < \frac{1}{4}$ we have $r_k = i\sqrt{\frac{1}{4} - \lambda_k} = i(s_k - \frac{1}{2})$ and

$$h_{t,\varepsilon}(r_k) = 2\hat{\chi}(\varepsilon r_k)\left(\frac{\sinh ts_k}{s_k} + \frac{\sinh(t(1-s_k))}{1-s_k}\right),$$

where for $k = 0$ the term $(\sinh(t(1-s_k)))/(1-s_k)$ must be replaced by t. Since $\hat{\chi}(\varepsilon r_k) = 1 + O(\varepsilon)$ for all $r_k \in i[0, \frac{1}{2}]$, then

$$h_{t,\varepsilon}(r_k) = (1 + O(\varepsilon))(\frac{1}{s_k}e^{ts_k} + O(e^{t/2})).$$

We write the second term as a Stieltjes integral

$$\sum_{\lambda_k > 1/4} h_{t,\varepsilon}(r_k) = \int_0^\infty h_{t,\varepsilon}(r)\,d\mu(r_k),$$

where $\mu(r)$ is the number of eigenvalues (counted with multiplicities) in the interval $[\frac{1}{4}, \frac{1}{4} + r^2]$, $r \geq 0$. Since $\hat{\chi}(\rho) = O((1 + |\rho|)^{-2})$ and since $\hat{g}_t$ satisfies

$\hat{g}_t(r) \leq c\,(1+r)^{-1}\,e^{t/2}$ for some constant c, we have

$$\left|\int_0^\infty h_{t,\varepsilon}(r)\,d\mu(r)\right| \leq \text{const } e^{t/2} \int_0^\infty (1+r)^{-1}(1+\varepsilon r)^{-2}d\mu(r).$$

We split the last integral into

$$\int_0^{1/\varepsilon} + \int_{1/\varepsilon}^\infty$$

and use integration by parts. Then, by Weyl's law, we observe that $\mu(r) = O(r^2)$. This yields an upper bound of order $O(\varepsilon^{-1}e^{t/2})$ for the right-hand side of the inequality. The same bound, up to a constant, is obtained for the integral containing $\tanh(\pi r)$ in the trace formula. To see this, it suffices to replace $d\mu(r)$ in the above arguments by $r\tanh(\pi r)\,dr$. Altogether we obtain

$$\Psi_\varepsilon(t) = \sum_{\lambda_k \leq 1/4} \frac{1}{s_k} e^{ts_k} + O(\varepsilon\, e^t + \varepsilon^{-1}e^{t/2}).$$

This is optimized with $\varepsilon = e^{-t/4}$. Observing that

$$g_{t-2\varepsilon,\varepsilon}(u-\varepsilon) \leq g_t(u) \leq g_{t+2\varepsilon,\varepsilon}(u+\varepsilon),$$

we conclude that

$$\sum_{\ell(\gamma) \leq t} \Lambda(\gamma)\tanh(\tfrac{1}{2}\ell(\gamma)) = \sum_{\lambda_k \leq 1/4} \frac{1}{s_k} e^{ts_k} + O(e^{3t/4}).$$

This implies that $\Psi(t) = O(e^t)$. From

$$\sum_{\ell(\gamma) \leq t} \Lambda(\gamma)\,e^{-\ell(\gamma)} = \int_0^t e^{-\tau}\,d\Psi(\tau) = O(t)$$

we finally get

$$\Psi(t) = \sum_{\lambda_k < 3/16} \frac{1}{s_k} e^{ts_k} + O(e^{3t/4}).$$

Lemma 9.6.2 is now proved. ◇

The remaining arguments for the proof of Theorem 9.6.1 are as in Section 9.4. That is, by (9.4.25), θ and Ψ have the same growth rate. Considering $\vartheta(x)$ instead of $\theta(t)$ (cf. Definition 9.4.13), we obtain

$$\vartheta(x) = \vartheta_1(x) + \vartheta_2(x),$$

where

$$\vartheta_1(x) = \sum_{\lambda_k < 3/16} \frac{1}{s_k} x^{s_k}, \quad \vartheta_2(x) = O(x^{3/4}).$$

This yields

$$\pi(x) = \int_1^x \frac{d\vartheta(\xi)}{\log \xi} = O(1) + \sum_{\lambda_k < 3/16} \text{li}(x^{s_k}) + O(\text{li}(x^{3/4})).$$

Theorem 9.6.1 is now proved. ◇

9.7 Lattice Points

The arguments of the preceding section can also be used to obtain error terms in Theorem 9.4.23. We shall follow the approach of Patterson [1]. As before we let $\Gamma : \mathbf{H} \to \mathbf{H}$ act such that $M = \Gamma \backslash \mathbf{H}$ is a compact Riemann surface. We lift the eigenfunctions $\varphi_0, \varphi_1, \ldots$ of the Laplacian of M to $\mathbf{H}$ in order to obtain Γ-automorphic eigenfunctions $\phi_0, \phi_1, \ldots$ on $\mathbf{H}$. Since

$$\int_M \varphi_0^2 \, dM = 1$$

and since φ_0 is constant, we have $\phi_0 = (\text{area } M)^{-1/2}$. As in Section 9.6 we set

$$s_k = \tfrac{1}{2} + \sqrt{\tfrac{1}{4} - \lambda_k} = \tfrac{1}{2} - ir_k, \quad \text{if } \lambda_k \leq \tfrac{1}{4},$$

where $\lambda_0, \lambda_1, \ldots$ are the eigenvalues of M corresponding to $\varphi_0, \varphi_1, \ldots$. For any pair of points $z, w \in \mathbf{H}$, we let $t \mapsto N(t; z, w)$ be the counting function

$$N(t; z, w) = \#\{T \in \Gamma \mid \text{dist}(z, Tw) \leq t\}$$

and we set $A = \text{area } M$.

9.7.1 Theorem. *Under the above hypotheses,*

$$N(t; z, w) = \frac{\pi}{A} e^t + \sum_{1 > s_k > 3/4} \pi^{1/2} \frac{\Gamma(s_k - \tfrac{1}{2})}{\Gamma(s_k + 1)} e^{ts_k} \phi_k(z) \phi_k(w) + O(e^{3t/4})$$

uniformly for $(z, w) \in \mathbf{H} \times \mathbf{H}$.

Here Γ denotes the gamma function. For some history and remarks we refer to the discussion at the end of this section.

For the proof of Theorem 9.7.1 we follow Patterson [1] using the eigenfunction expansion as in Theorem 9.3.11. The length trace formula of

Section 9.2 and Selberg's trace formula will not be needed here.

For convenience we make a parameter change, setting

$$\tau = 2\cosh t - 2$$

and

$$n(\tau) = n(\tau; z, w) := N(t; z, w).$$

The points $z, w \in \mathbf{H}$ will be kept fixed throughout the proof. The step from the counting function n to a continuous generating function $k : [0, \infty[\to \mathbf{R}$ is given by the following general principle. Let $\mathscr{P}$ be *any* countable set on which a weight function $\sigma : \mathscr{P} \to [0, \infty[$ is given (in our case, $\mathscr{P} = \Gamma$ and $\sigma(T) = 2 \cosh \mathrm{dist}(z, Tw) - 2$ for $T \in \Gamma$). Assume that the counting function

(1) $$n(\tau) = \#\{p \in \mathscr{P} \mid \sigma(p) \leq \tau\}$$

is finite for any $\tau > 0$. Fig. 9.7.1 illustrates the set $\{\sigma(p) \mid p \in \mathscr{P}, \sigma(p) \leq \tau\}$ drawn on the horizontal axis.

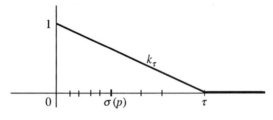

Figure 9.7.1

Now define the generating function

(2) $$k_\tau(s) = \begin{cases} 1 - \dfrac{s}{\tau} & \text{if } 0 \leq s \leq \tau \\ 0 & \text{if } s \geq \tau. \end{cases}$$

Integration by parts yields

$$\sum_{p \in \mathscr{P}} k_\tau(\sigma(p)) = \int_0^\tau k_\tau(s)\, dn(s) + n(0)$$

(3) $$= -\int_0^\tau k_\tau'(s)\, n(s)\, ds$$

$$= \frac{1}{\tau} \int_0^\tau n(s)\, ds.$$

For our purposes we now define

$$K(\rho) = k_\tau(2\cosh\rho - 2) = \begin{cases} \dfrac{2}{\tau}(\cosh u_\tau - \cosh\rho) & \text{if } 0 \le \rho \le u_\tau \\ 0 & \text{if } \rho \ge u_\tau, \end{cases}$$

where $u_\tau \ge 0$ is given by

$$\cosh u_\tau = \tfrac{1}{2}(\tau + 2).$$

Then $K(\text{dist}(z, Tw)) = k_\tau(\sigma(T))$ with $\sigma(T) = 2\cosh\text{dist}(z, Tw) - 2$.

For the induced kernel $\mathcal{K}_M$ on M, using Theorem 9.3.11 we obtain

(4) $$\mathcal{K}_M(x, y) = \sum_{k=0}^{\infty} h_\tau(r_k)\varphi_k(x)\varphi_k(y),$$

with uniform absolute convergence on $M \times M$. In this formula $h_\tau = h$ is the transform of K (with growth rate estimated below) as defined by Lemma 9.3.5, namely

$$h_\tau(r) = \int_{-\infty}^{+\infty} e^{iru} g(u)\, du,$$

where

$$g(u) = \sqrt{2} \int_{|u|}^{\infty} \frac{K(\rho)\sinh\rho}{\sqrt{\cosh\rho - \cosh u}}\, d\rho.$$

We let x and y be the images of z and w under the covering $\mathbf{H} \to \Gamma\backslash\mathbf{H}$. Then (4) becomes

(5) $$\mathcal{K}_\Gamma(z, w) = \sum_{k=0}^{\infty} h_\tau(r_k)\phi_k(z)\phi_k(w),$$

where by Definition 7.5.5,

$$\mathcal{K}_\Gamma(z, w) = \sum_{T \in \Gamma} K(\text{dist}(z, Tw)) = \sum_{T \in \Gamma} k_\tau(\sigma(T)), \quad z, w \in \mathbf{H}.$$

Applying the above formula (3) (to the set $\mathcal{P} = \Gamma$) we see that

(9.7.2) $$\frac{1}{\tau}\int_0^\tau n(s; z, w)\, ds = \sum_{k=0}^{\infty} h_\tau(r_k)\phi_k(z)\phi_k(w).$$

We now compute h_τ and estimate its growth rate, thereby justifying the above application of Theorem 9.3.11. For this we need some properties of the hypergeometric function which we list here for convenience (see M-O-S, chapter II or Whittaker-Watson [1], chapter XIV).

For $a, b, c \in \mathbf{C}$ and $z \in \mathbf{C} - \{\zeta \mid \zeta \in \mathbf{R}, \zeta \ge 1\}$, the hypergeometric function $F(a, b; c; z)$ is defined and holomorphic in z. In the case $|z| < 1$ and $c \ne 0, -1, -2$, the function F has the series expansion

$$F(a, b; c; z) = 1 + \frac{ab}{c} z + \frac{a(a+1)b(b+1)}{2! \, c(c+1)} z^2 + \cdots.$$

From this we see that F is symmetric in a and b. In the case $\mathcal{R}e\,c > \mathcal{R}e\,b > 0$, F has the integral representation

$$F(a, b; c; z) = \frac{\Gamma(c)}{\Gamma(b)\Gamma(c-b)} \int_0^1 t^{b-1}(1-t)^{c-b-1}(1-tz)^{-a}\, dt.$$

Numerous identities are known for F. We need the following two:

$$F(a, b; c; z) = (1-z)^{c-a-b} F(c-a, c-b; c; z),$$

and, if $a - b \notin \mathbf{Z}$:

$$F(a, b; c; z) = \frac{\Gamma(c)\Gamma(b-a)}{\Gamma(b)\Gamma(c-a)} (-z)^{-a} F(a, a-c+1; a-b+1; \tfrac{1}{z})$$

$$+ \frac{\Gamma(c)\Gamma(a-b)}{\Gamma(a)\Gamma(c-b)} (-z)^{-b} F(b, b-c+1; b-a+1; \tfrac{1}{z}).$$

(The formula in Whittaker-Watson [1], p. 289, contains a misprint.) With these formulae we compute h_τ. Substituting $s = \cosh \rho$, we first obtain

$$g(u) = \begin{cases} \dfrac{8\sqrt{2}}{3\tau}(\cosh u_\tau - \cosh u)^{3/2}, & |u| \leq u_\tau \\ 0, & \text{elsewhere.} \end{cases}$$

In the defining integral of h_τ (Lemma 9.3.5) we substitute

$$y = (e^u - e^{-u_\tau})/(e^{u_\tau} - e^{-u_\tau})$$

and obtain

$$h_\tau(r) = e^{(5/2 - ir)u_\tau} (\sinh u_\tau)^4 \frac{\pi}{2\tau} F(\tfrac{5}{2} - ir, \tfrac{5}{2}; 5; 1 - e^{2u_\tau}).$$

Using the above two identities, we transform this into

(9.7.3)
$$h_\tau(r) = \frac{\sqrt{8\pi}}{\tau} (\sinh u_\tau)^{3/2} \Big\{ \frac{\Gamma(ir)}{\Gamma(\tfrac{5}{2} + ir)} e^{iru_\tau} F(-\tfrac{3}{2}, \tfrac{5}{2}; 1 - ir; (1 - e^{2u_\tau})^{-1})$$

$$+ \frac{\Gamma(-ir)}{\Gamma(\tfrac{5}{2} - ir)} e^{-iru_\tau} F(-\tfrac{3}{2}, \tfrac{5}{2}; 1 + ir; (1 - e^{2u_\tau})^{-1}) \Big\}.$$

To estimate the growth rates we let $\delta > 0$ be so small that for all r_n of M, either $r_n = 0$ or $|r_n| > \delta$. For all eigenvalues $\lambda_n \neq \tfrac{1}{4}$ the corresponding r_n then lie on the set $[\delta, \infty[\, \cup \, i[\delta, \tfrac{1}{2}]$. For $r \in \mathbf{R}$ and $|r| \geq \delta$, Stirling's formula

(M-O-S, p.12) yields

$$\Gamma(ir)/\Gamma(ir+\tfrac{5}{2}) = O(|r|^{-5/2}).$$

By the series expansion of F, the following relation holds uniformly for $|Imr| \le 3/4$ and $\tau \ge 2$

$$F(-\tfrac{3}{2}, \tfrac{5}{2}; 1-ir; (1-e^{2u_\tau})^{-1}) = 1 + O(\tau^{-2}(1+|r|)^{-1}).$$

This yields the following growth rates (i) - (iii).

(i) If $\tau \ge 2$ and $r \in [\delta, \infty[$, then for some constant $c_1 > 0$,

$$|h_\tau(r)| \le c_1 \tau^{1/2}(1+|r|^2)^{-5/4}.$$

Here we see that $h = h_\tau$ indeed satisfies the hypothesis of Theorem 9.3.11. Moreover, since $\sum \lambda_k^{-1-\varepsilon} \phi_k(z)\phi_k(w)$ is uniformly absolutely convergent on $\mathbf{H} \times \mathbf{H}$ (Theorem 7.2.18), we also see that

$$\sum_{\lambda_k > 1/4} h_\tau(r_k)\phi_k(z)\phi_k(w) = O(\tau^{1/2})$$

uniformly on $\mathbf{H} \times \mathbf{H}$.

(ii) If $\tau \ge 2$, then for any fixed $r \in i[\delta, \tfrac{1}{2}]$,

$$h_\tau(r) = \pi^{1/2} \tau^{(1/2+|r|)} \frac{\Gamma(-ir)}{\Gamma(-ir+\tfrac{5}{2})} + O(\tau^{(1/2-|r|)}).$$

(iii) If $\tau \ge 2$, then

$$h_\tau(0) = \tfrac{8}{3}\tau^{1/2} \log \tau + O(\tau^{1/2}).$$

The last statement is obtained via (9.7.3) by taking $r \to 0$, $r > 0$ and by developing the functions $r \mapsto r\Gamma(ir)$ and $r \mapsto F(a, b, 1+ir; z)$ in a neighborhood of $r = 0$.

So far we have proved the following theorem in which A = area M.

9.7.4 Theorem. *The counting function $\tau \mapsto n(\tau; z, w)$ satisfies*

$$\frac{1}{\tau}\int_0^\tau n(s; z, w)\, ds = \frac{\pi}{2A}\tau + \sum_{1 > s_k > 1/2} \pi^{1/2} \frac{\Gamma(s_k - \tfrac{1}{2})}{\Gamma(s_k+2)} \tau^{s_k} \phi_k(z)\phi_k(w)$$

$$+ \sum_{s_k = 1/2} \tfrac{8}{3}\tau^{1/2} (\log \tau)\phi_k(z)\phi_k(w) + O(\tau^{1/2})$$

uniformly for $(z, w) \in \mathbf{H} \times \mathbf{H}$. ◇

The rest is a Tauberian differentiation argument which goes as follows.

9.7.5 Lemma. *Let a_k, $b \in \mathbf{R}$ and let $s_k \in [\frac{1}{2}, 1]$ for $k = 0, \ldots, q$. If*

$$\int_0^\tau n(s)\, ds = \sum_{k=0}^{q} a_k \tau^{1+s_k} + b\tau^{3/2} \log \tau + O(\tau^{3/2}),$$

where n is a non-decreasing function on $[0, \infty[$, then

$$n(\tau) = \sum_{k=0}^{q} a_k(1 + s_k)\tau^{s_k} + O(\tau^{3/4}).$$

Proof. By assumption, the following relation holds uniformly on the domain $\{(\tau, \alpha) \mid \tau > 0, -\tau \leq \alpha \leq \tau\}$:

$$\int_\tau^{\tau+\alpha} n(s)\, ds = \sum_{k=0}^{q+1} (f_k(\tau + \alpha) - f_k(\tau)) + O(\tau^{3/2}),$$

where $f_k(\tau) = a_k \tau^{1+s_k}$ for $k = 0, \ldots, q$ and $f_{q+1}(\tau) = b\tau^{3/2} \log \tau$.

By the mean value theorem, $f_k(\tau + \alpha) - f_k(\tau) = \alpha f_k'(\tau) + O(\alpha^2) f_k''(\tau)$. In our case, $f_k''(\tau) = O(1)$ as $\tau \to \infty$, for any k. Hence,

$$\int_\tau^{\tau+\alpha} n(s)\, ds = \alpha \sum_{k=0}^{q+1} f_k'(\tau) + O(\alpha^2) + O(\tau^{3/2}).$$

Since n is non-decreasing, then, for positive $\alpha \leq \tau$

$$\int_{\tau-\alpha}^{\tau} n(s)\, ds \leq \alpha n(\tau) \leq \int_\tau^{\tau+\alpha} n(s)\, ds.$$

It follows that

$$n(\tau) = \sum_{k=0}^{q+1} f_k'(\tau) + O(\alpha) + O(\alpha^{-1} \tau^{3/2}).$$

This is optimized for $\alpha = \tau^{3/4}$, and the lemma follows. ◇

Theorem 9.7.1 is an immediate consequence of Theorem 9.7.4 and Lemma 9.7.5. ◇

We conclude this section with a few remarks. The earliest article on hyperbolic lattice points seems to be the seemingly unnoticed note of Delsarte [1] in 1942. The details were not published but may today be found in the collected papers of Delsarte [2], pp. 829-845. The main result (in the above notation) is the formula

(9.7.6) $$n(\tau; z, w) = \pi\tau \sum_{k=0}^{\infty} F(\alpha_k, \beta_k; 2; -\frac{\tau}{4})\phi_k(z)\phi_k(w),$$

where

$$\alpha_k = \tfrac{1}{2} + (\tfrac{1}{4} - \lambda_k)^{1/2}, \ \beta_k = \tfrac{1}{2} - (\tfrac{1}{4} - \lambda_k)^{1/2}.$$

Delsarte also remarks that the small eigenvalues influence the asymptotic distribution of the lattice points. It should be mentioned that Delsarte states result (9.7.6) for discrete groups $\Gamma: \mathbf{H} \to \mathbf{H}$ with compact fundamental domain which may have fixed points.

Huber [2] gave an asymptotic formula without the error terms and Fricker [1] proved a theorem similar to Theorem 9.7.1 for Γ acting on $\mathbf{H}^3$, then applying the result to certain Diophantine equations [2]. Günther [2, 3] generalized the theorem to all rank one symmetric spaces of non-compact type.

The theorem in the above form is due to Patterson [1] and holds for all Riemann surfaces of finite area. An interesting variant of the theorem is given by Wolfe [1] and Randol [7] in which the pointwise deviation from the leading term is replaced by the variance.

Chapter 10

Wolpert's Theorem

In this chapter we prove Wolpert's theorem that a generic compact Riemann surface is determined up to isometry by its length or eigenvalue spectrum. A new element in the proof given here is the observation that a finite part of the length spectrum determines the entire spectrum. With this and with some explicit computations based on trigonometry the proof becomes considerably shorter than the original one.

10.1 Introduction

A compact Riemann surface of genus $g \geq 2$ is determined up to isometry by the lengths of a finite system of closed geodesics. More precisely, let F be a base surface and let $\mathcal{T}_g$ be the Teichmüller space of all marking equivalence classes of marked Riemann surfaces (S, φ), where S is a compact Riemann surface of genus g and $\varphi : F \to S$ is a marking homeomorphism (cf. Definitions 6.1.1 and 6.1.2). Consider a canonical curve system Ω on F consisting of $9g - 9$ simple closed curves as in Definition 6.1.6. For the present discussion it is only important to know that Ω is some well defined finite ordered set of closed curves with certain useful properties. For any $\beta \in \Omega$ and for any $S = (S, \varphi)$ we let $\beta(S)$ denote the simple closed geodesic in the free homotopy class of the curve $\varphi \circ \beta$ on S. In Corollary 6.2.8 we have seen that the marking equivalence class of (S, φ) is uniquely determined by the ordered set of all lengths $\ell\beta(S)$, $\beta \in \Omega$. (The lengths $\ell\beta(S)$ determine the Fenchel-Nielsen parameters.)

Hence, the position of an element in Teichmüller space is known if the lengths of sufficiently many closed geodesics are known, *provided we know which length belongs to which closed geodesic.*

In the length spectrum this information is not given. On the other hand, the length spectrum contains the lengths of *all* closed geodesics, and one may hope that this much larger amount of data may allow us to decode which length is attributed to which geodesic.

In [1] Gel'fand conjectured that this might be the case. As a first step towards the conjecture he proved that there exists no continuous deformation of a compact Riemann surface such that the spectrum remains fixed (cf. also Tanaka [1]). In [1] McKean showed that the number of pairwise non-isometric compact Riemann surfaces having the same spectrum is in fact finite. In [1] Wolpert succeeded to prove that the surfaces for which this number is different from 1 are located on a lower dimensional subvariety $\mathcal{V}_g$ of $\mathcal{T}_g$. This proved Gel'fand's conjecture in the generic case. At about the same time Marie France Vignéras showed that $\mathcal{V}_g$ is not empty. Hence Wolpert's result is optimal except that not much is known about $\mathcal{V}_g$ as yet.

Before we state Wolpert's theorem we review the definition of the length spectrum. In Definition 9.2.8 we considered the set $\mathcal{C}(M)$ of all *oriented* closed geodesics on the compact Riemann surface M. In the present chapter the orientation of a closed curve is irrelevant and we consider the non-oriented ones instead. We call parametrized closed geodesics

$$\gamma, \delta : \mathbf{S}^1 = \mathbf{R}^1 / [t \mapsto t + 2\pi] \to M$$

(parametrized with constant speed) *weakly equivalent* if and only if there exists a constant c such that either $\gamma(t) = \delta(t+c)$, $t \in \mathbf{S}^1$ or $\gamma(t) = \delta(-t+c)$, $t \in \mathbf{S}^1$. A *non-oriented closed geodesic* is the weak equivalence class of a closed parametrized geodesic. We could also introduce the non-oriented geodesics as the pairs of oppositely oriented oriented closed geodesics.

We recall that a closed geodesic is *primitive* if it is not the m-fold iterate of another closed geodesic for some $m \geq 2$ (Definition 1.6.4).

10.1.1 Definition. For any $S \in \mathcal{T}_g$ we let Lsp(S) denote the sequence of the lengths of all non-oriented primitive closed geodesics on S, arranged in ascending order.

By Huber's theorem (Theorem 9.2.9) and by Remark 9.2.13, we have Lsp(S) = Lsp(S') if and only if S and S' have the same spectrum of the Laplacian. In the present chapter we shall, however, not work with the eigenvalues.

In the following, "closed geodesic" always means "non-oriented closed geodesic". The "*length spectrum of S*" is, by definition, Lsp(S).

10.1.2 Definition. We define $\mathcal{V}_g$ to be the subset of all $S \in \mathcal{T}_g$ for which there exists $S' \in \mathcal{T}_g$ such that $\mathrm{Lsp}(S') = \mathrm{Lsp}(S)$ and such that S' is not isometric to S.

We do not expect that $\mathcal{V}_g$ is a closed subset of $\mathcal{T}_g$ although the proof for this is still missing (in Section 12.5 we have a family of non-isometric isospectral pairs which converges to an isometric pair but it looks difficult to prove that the limiting surface is not isospectral to *any* other surface).

Wolpert's theorem of [1, 3] is as follows.

10.1.3 Theorem. *$\mathcal{V}_g$ is a local real analytic proper subvariety of $\mathcal{T}_g$.*

The term "local" means that for any $x \in \mathcal{T}_g$ there exists a neighborhood U of x in $\mathcal{T}_g$ such that $\mathcal{V}_g \cap U$ is a finite union of sets $W - W^*$, where W is a real analytic subvariety of U and W^* is a real analytic subvariety of W.

In the proof of Theorem 10.1.3 we follow Wolpert's original approach. However, the following observation simplifies the global structure of the proof considerably.

10.1.4 Theorem. *Let $\mathcal{R}_g(\varepsilon) \subset \mathcal{T}_g$ denote the set of all compact Riemann surfaces of genus g and injectivity radius $\geq \varepsilon$ ($\varepsilon > 0$). Then there exists a real number $t = t(\varepsilon)$ such that for $S, S' \in \mathcal{R}_g(\varepsilon)$ we have $\mathrm{Lsp}(S) = \mathrm{Lsp}(S')$ if and only if $\mathrm{Lsp}(S) \cap [0, t(\varepsilon)] = \mathrm{Lsp}(S') \cap [0, t(\varepsilon)]$.*

Here $\mathrm{Lsp}(S) \cap [0, t(\varepsilon)]$ is the finite subsequence of $\mathrm{Lsp}(S)$ whose values lie in the interval $[0, t(\varepsilon)]$. The proof of the theorem will be given at the end of Section 10.3.

In Section 14.10 we shall prove the analogous theorem for the eigenvalues of the Laplacian.

10.2 Curve Systems

For the convenience of the reader we review some of the properties of Teichmüller space from earlier chapters.

Throughout this chapter we let G be the cubic graph given by Fig 10.2.1.

Every compact Riemann surface S of genus g may be obtained by gluing together $2g - 2$ Y-pieces with respect to the pattern of G, where each vertex

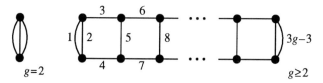

Figure 10.2.1

of G with its three half-edges stands for a Y-piece with its three boundary geodesics. To each edge c_k of G, $k = 1, \ldots, 3g-3$, corresponds a simple closed geodesic $\gamma_k(S)$ of S. The lengths of $\gamma_1(S), \ldots, \gamma_{3g-3}(S)$ may be prescribed arbitrarily. In addition to this, we have a twist parameter $\alpha_k(S)$ for each $\gamma_k(S)$ which also may be prescribed arbitrarily. The parameter $\alpha_k(S)$ describes how the two Y-pieces adjacent along $\gamma_k(S)$ are twisted against one another. The definition of $\alpha_k(S)$ is given in (3.3.2) and again in (3.6.1). The two vectors

$$L = (\ell\gamma_1(S), \ldots, \ell\gamma_{3g-3}(S)) \in \mathbf{R}_+^{3g-3}$$

and

$$A = (\alpha_1(S), \ldots, \alpha_{3g-3}(S)) \in \mathbf{R}^{3g-3}$$

determine S up to isometry. We shall write $\omega = (L, A)$, $S = S^\omega$ (as in Definition 6.2.5) and consider S^ω as a marked surface. In this way, two marked surfaces S^ω and $S^{\omega'}$ are considered different if $\omega = (L, A) \neq \omega' = (L', A')$, even if S^ω and $S^{\omega'}$ are isometric. In order to adapt this to Teichmüller theory, a particular homeomorphism

$$\varphi^\omega : F \to S^\omega$$

has been introduced in Definition 6.2.5 for each ω, where F is a fixed surface given by $F = S^{\omega_0}$ with $\omega_0 = (L_1, A_0)$, $L_1 = (1, \ldots, 1)$, $A_0 = (0, \ldots, 0)$. Below we shall recall some of the properties of the φ^ω which we need in the proofs of Theorems 10.1.3 and 10.1.4. In Theorems 6.2.7 and 6.3.2 we have seen that the set of all S^ω endowed with these marking homeomorphisms is a model of the Teichmüller space $\mathcal{T}_g$ in which the real analytic structure of $\mathcal{T}_g$ is that of $\mathbf{R}_+^{3g-3} \times \mathbf{R}^{3g-3}$. By abuse of notation we shall write

$$\mathcal{T}_g = \{ S^\omega \mid \omega = (L, A), L \in \mathbf{R}_+^{3g-3}, A \in \mathbf{R}^{3g-3} \}.$$

The homeomorphisms $\varphi^\omega : F \to S^\omega$ are defined such that $\varphi^\omega \circ \gamma_k(F) = \gamma_k(S^\omega)$, $k = 1, \ldots, 3g-3$. For each closed geodesic λ on F we denote by $\lambda(S^\omega)$ the unique closed geodesic in the free homotopy class of the closed curve $\varphi^\omega \circ \lambda$ on S^ω. For any finite or infinite ordered set Λ of closed geodesics $\lambda_1, \lambda_2, \ldots$ on F and for any $S = S^\omega \in \mathcal{T}_g$ we define the sequences

$$\Lambda(S) = (\lambda_1(S), \lambda_2(S), \ldots)$$
$$\ell\Lambda(S) = (\ell\lambda_1(S), \ell\lambda_2(S), \ldots).$$

By abuse of notation we shall write

$$\ell\Lambda(S) \subset \mathrm{Lsp}(S')$$

if after a suitable permutation of $\ell\Lambda(S)$, $\ell\Lambda(S)$ is a subsequence of $\mathrm{Lsp}(S')$. The following sequences will be used:

(10.2.1)
$$\Pi = (\beta_1, \beta_2, \ldots)$$
$$\Pi_k = (\beta_1, \ldots, \beta_k); k \in \mathbf{N}$$
$$\Omega = (\gamma_1, \ldots, \gamma_{3g-3}; \delta_1, \ldots, \delta_{3g-3}; \eta_1, \ldots, \eta_{3g-3})$$
$$\Sigma = \Omega \cup (\mu_1, \ldots, \mu_{2g-2}; \sigma_1, \ldots, \sigma_g).$$

They are defined as follows. $(\beta_1, \beta_2, \ldots)$ is the list of *all* non-oriented primitive closed geodesics on F, arranged so that $\ell\beta_1(F) \leq \ell\beta_2(F), \ldots$ (hence, $\ell\Pi(F) = \mathrm{Lsp}(F)$). The curve system $\Omega = \Omega_G$ is as in Definition 6.1.6. The $\mu_1, \ldots, \mu_{2g-2}$ are geodesics which have exactly one self-intersection ("figure eight geodesics"). They are chosen in such a way that the interior of each of the $2g-2$ Y-pieces of F contains exactly one such geodesic. The geodesics σ_1 and σ_g are the unique shortest (non-oriented) simple closed geodesics on F such that σ_1 intersects γ_1 and γ_2, and such that σ_g intersects γ_{3g-4} and γ_{3g-3}. For $j = 2, \ldots, g-1$ (if $g \geq 3$), σ_j is the unique shortest simple closed geodesic on F which intersects $\gamma_{3j-4}, \gamma_{3j-3}, \gamma_{3j-1}$ and γ_{3j-2} (Fig. 10.2.2).

The homeomorphisms $\varphi^\omega : F \to S^\omega$ have been defined in such a way that the following formulae hold (cf. Definition 6.1.6 in connection with Proposition 3.3.12).

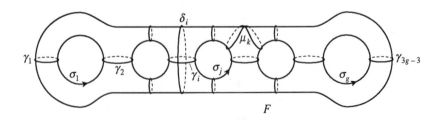

Figure 10.2.2

(10.2.2)
$$\cosh\{\tfrac{1}{2}\ell\delta_i(S)\} = u_i(L) + v_i(L)\cosh\{\tfrac{1}{2}\alpha_i(S)\,\ell\gamma_i(S)\}$$
$$\cosh\{\tfrac{1}{2}\ell\eta_i(S)\} = u_i(L) + v_i(L)\cosh\{\tfrac{1}{2}(1+\alpha_i(S))\ell\gamma_i(S)\},$$

where $S = S^\omega$ with $\omega = (L, A)$, and where u_i and v_i are analytic functions of L which do not depend on A, $i = 1, \ldots, 3g-3$. The computation of $\ell\mu_k(S)$ and $\ell\sigma_j(S)$ will follow in Section 10.5. We recall the following general fact (cf. Theorem 6.3.5).

10.2.3 Lemma. *The function $\omega \mapsto \ell\beta(S^\omega)$ is real analytic on $\mathcal{T}_g$ for each $\beta \in \Pi$.*

This is the key property in the proofs of Theorems 10.1.3 and 10.1.4. A further important property is that the φ^ω have uniformly controlled length distortion (cf. Theorem 6.4.2 or Lemmata 3.2.6 and 3.3.8).

10.2.4 Lemma. *Let $Q \subset \mathcal{T}_g$ be a compact subset. There exists a constant $q \geq 1$ which depends only on Q such that*

$$\frac{1}{q}\ell\beta(S) \leq \ell\beta(S') \leq q\,\ell\beta(S)$$

for any $S, S' \in Q$ and any $\beta \in \Pi$.

We also restate the Mahler compactness theorem (Theorem 6.6.5).

10.2.5 Lemma. *Let $\varepsilon > 0$. There exists a compact subset $Q(\varepsilon) \subset \mathcal{T}_g$ such that for each surface $S \in \mathcal{T}_g$ with injectivity radius $\geq \varepsilon$ there exists an isometric surface $S' \in Q(\varepsilon)$.*

In addition to the above we shall use the collar theorem of Chapter 4.

10.3 Finitely Many Lengths Determine the Length Spectrum

We now turn to the proof of Theorems 10.1.3 and 10.1.4. In order to make the global structure more transparent, some of the subsequent lemmata will not be stated in full generality.

Let $U \subset \mathcal{T}_g$ be an open neighborhood with compact closure which we fix during the proof of both Theorem 10.1.4 and 10.1.3. For Theorem 10.1.3 (in the next section), U is a neighborhood in which the analytic nature of $\mathcal{V}_g$ is tested; for Theorem 10.1.4 we let U be an open neighborhood in $\mathcal{T}_g$ which contains $Q(\varepsilon)$ (cf. Lemma 10.2.5).

By Lemma 10.2.4 there exists $\varepsilon_U > 0$ such that any surface $S \in U$ has injectivity radius $\geq \varepsilon_U$. We let $Q(\varepsilon_U)$ be a compact subset of $\mathcal{T}_g$ which contains all surfaces with injectivity radius $\geq \varepsilon_U$ (Lemma 10.2.5).

Note that whenever $\mathrm{Lsp}(S') = \mathrm{Lsp}(S)$ for $S' \in \mathcal{T}_g$ and for $S \in U$, then S and S' have the same length of the smallest closed geodesic and therefore the same injectivity radius. We may therefore assume without loss of generality that $S' \in Q(\varepsilon_U)$.

Let

$$D = \{ S^\omega \in \mathcal{T}_g \mid \tfrac{1}{10} \leq \ell_1, \ldots, \ell_{3g-3} \leq 10;\ -1 \leq \alpha_1, \ldots, \alpha_{3g-3} \leq 1 \}$$

and let C, C_1 be the compact closures of open *connected* sets $\mathring{C}, \mathring{C}_1 \subset \mathcal{T}_g$ such that

$$U \cup Q(\varepsilon_U) \cup D \subset \mathring{C} \subset C \subset \mathring{C}_1.$$

By Lemma 10.2.4 there exists $q \geq 1$ such that

(10.3.1) $$\frac{1}{q} \ell\beta(S) \leq \ell\beta(S') \leq q\,\ell\beta(S)$$

for any $S, S' \in C_1$ and any $\beta \in \Pi$. This q will remain fixed during the proof.

10.3.2 Lemma. *For any $k \in \mathbf{N}$ there exists $k^* \in \mathbf{N}$, $k^* \geq k$, which depends only on k and C_1, with the following property. If $S', S'' \in C_1$ and if $\rho : \Pi_k \to \Pi$ is an injection such that*

$$\ell\Pi_k(S') = \ell(\rho\Pi_k)(S'')$$

then $\rho\Pi_k \subset \Pi_{k^*}$.

In this lemma the equation $\ell\Pi_k(S') = \ell(\rho\Pi_k)(S'')$ is an equation between *sequences* but the statement $\rho\Pi_k \subset \Pi_{k^*}$ means that $\rho\Pi_k$ is a *subset* of Π_{k^*}. The statement simply means that if the length of $\Pi_k(S')$ occur in the length spectrum of S'' then the corresponding geodesics on S'' belong to Π_{k^*}.

Proof. Let $c_1 = \max\{\, \ell\beta(S) \mid S \in C_1, \beta \in \Pi_k \,\}$, and let $k^* \geq k$ be such that on the base surface F we have $\ell\beta_j(F) > qc_1$ for all $j > k^*$ (observe that F belongs to D). By (10.3.1) we then have $\ell\beta_j(S) > c_1$ for all $j > k^*$ and all $S \in C_1$. Since for $i \leq k$, $\ell(\rho\beta_i)(S'') = \ell\beta_i(S') \leq c_1$, it follows that $\rho\beta_i \in \Pi_{k^*}$, $i = 1, \ldots, k$. ◇

For the proof of Theorem 10.1.4 we define for each $k \in \mathbf{N}$

$$W_k^1 = \{(S, S') \in \mathring{C}_1 \times \mathring{C}_1 \mid \ell\Pi_k(S) \subset \mathrm{Lsp}(S') \text{ and } \ell\Pi_k(S') \subset \mathrm{Lsp}(S)\}.$$
$$W_k = W_k^1 \cap \mathring{C} \times \mathring{C}.$$

10.3.3 Lemma. W_k^1 *is a real analytic subvariety of* $\mathring{C}_1 \times \mathring{C}_1$ *for each k.*

Proof. Let Π_{k*} be as in Lemma 10.3.2. We set for each pair of injections $\rho, \rho': \Pi_k \to \Pi_{k*}$

$$W[\rho, \rho'] = \{(S, S') \in \mathring{C}_1 \times \mathring{C}_1 \mid \ell\Pi_k(S) = \ell(\rho\Pi_k)(S')$$
$$\text{and } \ell\Pi_k(S') = \ell(\rho'\Pi_k)(S)\}.$$

By Lemma 10.2.3, each $W[\rho, \rho']$ is a real analytic subvariety of $\mathring{C}_1 \times \mathring{C}_1$. By Lemma 10.3.2 we have

$$W_k^1 = \bigcup_{(\rho, \rho')} W[\rho, \rho'].$$

Since there are only finitely many pairs, W_k^1 is also real analytic. ◇

10.3.4 Lemma. *There exists* $K \in \mathbf{N}$ *such that* $W_{K+j} = W_K$ *for all* $j \geq 1$.

Proof. This follows from the fact that the ring of all real analytic functions on $\mathring{C}_1 \times \mathring{C}_1$ is Noetherian (Grauert-Remmert [1]) and that $C \times C$ in $\mathring{C}_1 \times \mathring{C}_1$ is compact. ◇

Proof of Theorem 10.1.4. Take U such that $Q(\varepsilon) \subset U$. Let K be as in Lemma 10.3.4 and define

$$t(\varepsilon) = \max\{\ell\beta(S) \mid S \in Q(\varepsilon), \beta \in \Pi_K\}.$$

If S, S' have injectivity radius $\geq \varepsilon$ then without loss of generality $S, S' \in Q(\varepsilon)$. If in addition $\mathrm{Lsp}(S) \cap [0, t(\varepsilon)] = \mathrm{Lsp}(S') \cap [0, t(\varepsilon)]$, then in particular $\ell\Pi_K(S) \subset \mathrm{Lsp}(S')$ and $\ell\Pi_K(S') \subset \mathrm{Lsp}(S)$, that is $(S, S') \in W_K$. By Lemma 10.3.4, $(S, S') \in W_{K+j}$ for any j, and therefore $\mathrm{Lsp}(S) = \mathrm{Lsp}(S')$. This proves the theorem. ◇

10.4 Generic Surfaces Are Determined by Their Spectrum

For the proof of Theorem 10.1.3 we fix K as in Lemma 10.3.4 chosen large enough so that

$$\Sigma \subset \Pi_K,$$

where Σ is the extension of the curve system Ω (cf. (10.2.1)) defined above. With the notation of Lemma 10.3.2 we let

(10.4.1) $$M = K^*, \quad N = M^*.$$

The effect of this is as follows. If $S, S' \in C$ are isospectral, then there exists a bijection $\rho : \Pi \to \Pi$ satisfying $\ell\Pi(S) = \ell(\rho\Pi)(S')$. By Lemma 10.3.2 we then have $\rho^{-1}(\Pi_K) \subset \Pi_M$ and $\rho(\Pi_M) \subset \Pi_N$. Hence,

10.4.2 Lemma. *If $S, S' \in C$ and $\mathrm{Lsp}(S) = \mathrm{Lsp}(S')$, then there exists an injection $\rho : \Pi_M \to \Pi_N$ satisfying*

(i) $$\rho(\Pi_M) \supset \Pi_K$$
(ii) $$\ell\Pi_M(S) = \ell(\rho\Pi_M)(S'). \qquad \diamond$$

The point is again that there are only finitely many injections $\rho : \Pi_M \to \Pi_N$.

10.4.3 Lemma. *Let $\rho : \Pi_M \to \Pi_N$ be an injection such that $\rho(\Pi_M) \supset \Pi_K$. The set*

$$V_\rho = \{ S \in \overset{\circ}{C} \mid \text{there exists } S^\rho \in \mathcal{T}_g \text{ such that } \ell\Pi_M(S) = \ell(\rho\Pi_M)(S^\rho) \}$$

is a real analytic subvariety of $\overset{\circ}{C}$. For each $S \in V_\rho$ the corresponding $S^\rho \in \mathcal{T}_g$ is uniquely determined by S. The subset

$$V_\rho^* = \{ S \in V_\rho \mid S^\rho \text{ is isometric to } S \}$$

is a real analytic subvariety of V_ρ.

Proof. That S^ρ, if it exists, is uniquely determined by S, follows from the fact that $\Omega \subset \Sigma \subset \Pi_K \subset \rho(\Pi_M)$ (cf. (10.2.1) and that

$$\ell\Omega(S^\rho) = \ell(\rho^{-1}\Omega)(S),$$

where $\ell\Omega(S^\rho)$ determines the parameters ω^ρ of $S^\rho = S^{\omega^\rho}$ (cf. (10.2.2)).

The first observation is that V_ρ is a closed subset in the relative topology of $\overset{\circ}{C}$. In fact, if $S \in \overset{\circ}{C}$ and if $S_n \to S$ for $S_n \in V_\rho$, then $\ell\Omega(S_n^\rho) = \ell(\rho^{-1}\Omega)(S_n) \to \ell(\rho^{-1}\Omega)(S)$. Since Ω is a canonical curve system, this means that S_n^ρ converges to a limiting surface $\bar{S}$. Now $\ell\Pi_M(S) = \lim \ell\Pi_M(S_n) = \lim \ell(\rho\Pi_M)(S_n^\rho)) = \ell(\rho\Pi_M)(\bar{S}))$, Hence, $S \in V_\rho$.

We proceed by proving that V_ρ is defined by real analytic mappings. We may assume that $V_\rho \neq \emptyset$. For $S \in V_\rho$ we let

$$\delta_i^* = \begin{cases} \delta_i & \text{if } \alpha_i(S^\rho) \neq 0 \\ \eta_i & \text{if } \alpha_i(S^\rho) = 0 \end{cases}, \quad i = 1, \ldots, 3g-3,$$

where $\alpha_1(S^\rho), \ldots, \alpha_{3g-3}(S^\rho)$ are the twist parameters of S^ρ. We further set

$$\Omega^* = (\gamma_1, \ldots, \gamma_{3g-3}; \delta_1^*, \ldots, \delta_{3g-3}^*).$$

The idea here is that the lengths of the geodesics δ_i^* do not have a local minimum at S^ρ and may therefore be used as moduli for the surfaces in an open neighborhood of S^ρ. It follows from (10.2.2) that in a sufficiently small neighborhood U_S of S in $\mathcal{T}_g$ there exists a uniquely determined *continuous* map

$$m_U : U_S \to \mathcal{T}_g$$

satisfying

$$\ell\Omega^*(m_U(S')) = \ell(\rho^{-1}\Omega^*)(S'), \quad S' \in U_S.$$

Note in particular that $m_U(S) = S^\rho$. By (10.2.2) and Lemma 10.2.3, m_U is a real analytic mapping. Moreover, for all $S' \in (V_\rho \cap U_S)$ the corresponding surface S'^ρ satisfies $S'^\rho = m_U S'$ and we may describe $V_\rho \cap U_S$ in the following way,

$$V_\rho \cap U_S = \{ S' \in U_S \mid \ell\Pi_M(S') = \ell(\rho\Pi_M)(m_U S') \}.$$

Hence, $V_\rho \cap U_S$ is a real analytic subvariety of U_S. Now $S' \in (V_\rho \cap U_S)$ is isometric to $S'^\rho = m_U S'$ if and only if S'^ρ is the image of S' under a Teichmüller map, that is, if and only if there exists an injection $r : \Omega \to \Pi$ which is induced by a homeomorphism of the base surface F onto itself and for which

$$\ell\Omega(S') = \ell(r\Omega)(m_U S').$$

By Lemma 10.3.2, $r\Omega \subset \Pi_M$. This shows that the set R^* of all possible such injections r is finite. For the set V_ρ^* as in the lemma, this implies that the intersection

$$V_\rho^* \cap U_S = \bigcup_{r \in R^*} \{ S' \in V_\rho \cap U_S \mid \ell\Omega(S') = \ell(r\Omega)(m_U S') \}$$

is a real analytic subvariety of $V_\rho \cap U_S$. Since V_ρ is closed in the relative topology of $\overset{\circ}{C}$, Lemma 10.4.3 is now proved. ◇

10.4.4 Lemma. *Let $S \in V_\rho$ and let S^ρ be as in Lemma 10.4.3. Then $\mathrm{Lsp}(S) = \mathrm{Lsp}(S^\rho)$.*

Proof. Since $\rho\Pi_M \supset \Pi_K$, we have $\ell\Pi_K(S^\rho) \subset \ell\rho\Pi_M(S^\rho) = \ell\Pi_M(S) \subset \mathrm{Lsp}(S)$. Conversely, since $\Pi_K \subset \Pi_M$, we also have $\ell\Pi_K(S) \subset \mathrm{Lsp}(S^\rho)$. Therefore, $(S, S^\rho) \in W_K$. By Lemma 10.3.4, $(S, S^\rho) \in W_{K+j}$ for all j, that is, S and S^ρ are isospectral. ◇

Since V_ρ is a real analytic subvariety of $\overset{\circ}{C}$ and since $\overset{\circ}{C}$ is connected (by definition) we either have $V_\rho = \overset{\circ}{C}$ or else $\dim V_\rho < \dim \mathcal{T}_g$.

Now comes the crucial point.

10.4.5 Lemma. *Let ρ, V_ρ and S^ρ be as in Lemma 10.4.3. If $\dim V_\rho = \dim \mathcal{T}_g$, then ρ is induced by a homeomorphism of the base surface F and S^ρ is isometric to S for any $S \in V_\rho$.*

Before proving Lemma 10.4.5 we shall finish the proof of Wolpert's theorem. Let R be the set of all injections $\rho : \Pi_M \to \Pi_N$ satisfying $\rho\Pi_M \supset \Pi_K$ and $\dim V_\rho < \dim \mathcal{T}_g$. Set

$$V = \bigcup_{\rho \in R} (V_\rho - V_\rho^*).$$

By Lemma 10.4.3 and by the finiteness of R, V is a local real analytic subvariety of $\overset{\circ}{C}$. By Lemma 10.4.4, $V \subset \mathcal{V}_g \cap \overset{\circ}{C}$ so that for our initial neighborhood U we have $V \cap U \subset \mathcal{V}_g \cap U$. Conversely, if $S \in \mathcal{V}_g \cap U$ then there exists S' not isometric to S satisfying $\mathrm{Lsp}(S') = \mathrm{Lsp}(S)$. It follows that S and S' have the same injectivity radius $\geq \varepsilon_U$ (because they have the same length of the smallest closed geodesic) so that we may assume without loss of generality that $S' \in Q(\varepsilon_U) \subset \overset{\circ}{C}$ (the set $Q(\varepsilon_U)$ is defined in Lemma 10.2.5). By Lemma 10.4.2 and Lemma 10.4.5, $S \in V \cap U$. Hence,

$$V \cap U = \mathcal{V}_g \cap U.$$

The proof of Theorem 10.1.3 is now reduced to the proof of Lemma 10.4.5 which will be given in the next section. ◇

10.5 Decoding the Moduli

We now restate and prove Lemma 10.4.5 in a more independent form. The proof contains a way of detecting the moduli of S among the lengths in $\mathrm{Lsp}(S)$. For this we use the curve system Σ as in (10.2.1). $D \subset \overset{\circ}{C}$ will again be the domain

$$D = \{ S^\omega \in \mathcal{T}_g \mid \tfrac{1}{10} \leq \ell_1, \ldots, \ell_{3g-3} \leq 10;\ -1 \leq \alpha_1, \ldots, \alpha_{3g-3} \leq 1 \}.$$

Recall that Teichmüller mappings (Definition 6.5.2) map surfaces to isometric ones. We now have the following characterization of Teichmüller mappings (Wolpert [3]).

10.5.1 Proposition. *Let $m : D \to \mathcal{T}_g$ and let $\rho : \Sigma \to \Pi$ be mappings such that*

$$\ell(\rho\Sigma)(m(S)) = \ell\Sigma(S)$$

for all $S \in D$. Then ρ is induced by a homeomorphism of the base surface F and m is the restriction to D of a Teichmüller mapping.

Proof. We abbreviate

$$\rho(\beta) =: \tilde{\beta} \text{ for } \beta \in \Sigma, \quad m(S) =: \tilde{S} \text{ for } S \in D.$$

The strategy is to detect the nature of the curves in $\rho\Sigma$ by examining the lengths which they assume on the different surfaces $\tilde{S}$. We proceed in several steps.

(i) *$\tilde{\gamma}_1, \ldots, \tilde{\gamma}_{3g-3}$ are simple and pairwise disjoint.*

Proof. Set $L = (\frac{1}{2}, \ldots, \frac{1}{2})$ and $A = (0, \ldots, 0) = A_0$. Then $\ell\tilde{\gamma}_i(\tilde{S}) = \frac{1}{2}$, $i = 1, \ldots, 3g-3$, and (i) follows immediately from the collar theorem. (If $\tilde{\gamma}_1(\tilde{S}), \ldots, \tilde{\gamma}_{3g-3}(\tilde{S})$ are simple and pairwise disjoint then the same holds for $\gamma_1, \ldots, \gamma_{3g-3}$ by Theorem 1.6.6(iii) and Theorem 1.6.7). ◇

In the next statement, F is the base surface for the marking homeomorphisms as in Section 10.2. Note that we have defined D in such a way that $F \in D$.

(ii) *If γ_i, γ_j and γ_k are the boundary geodesics of a Y-piece of F which carries μ_n, then $\tilde{\gamma}_i(\tilde{S})$, $\tilde{\gamma}_j(\tilde{S})$ and $\tilde{\gamma}_k(\tilde{S})$ are the boundary geodesics of a Y-piece in $\tilde{S}$ which carries $\tilde{\mu}_n(\tilde{S})$.*

With (i) and (ii) together we already see that the restriction of ρ to the system $(\gamma_1, \ldots, \gamma_{3g-3})$ is induced by a homeomorphism of F.

Proof of (ii). The formula for the length $\ell\mu_n(S)$ is as follows, possibly after a permutation of the indices i, j, k.

$$\cosh \tfrac{1}{2}\ell\mu_n(S) = 2 \cosh \tfrac{1}{2}\ell\gamma_i(S) \cosh \tfrac{1}{2}\ell\gamma_j(S) + \cosh \tfrac{1}{2}\ell\gamma_k(S).$$

(cf. (4.2.3)). For $L = (\frac{1}{4}, \ldots, \frac{1}{4})$ this yields the numerical value

$$\ell\mu_n(S) = \ell(\tilde{\mu}_n(\tilde{S})) = 3.553\ldots$$

On the other hand, the collars $\mathscr{C}(\tilde{\gamma}_i(\tilde{S}))$ have width $2w(\tilde{\gamma}_i(\tilde{S})) > 5$ (cf. Theorem 4.1.1). It follows that $\tilde{\mu}_n(\tilde{S})$ lies in the interior of one of the Y-pieces of $\tilde{S}$ which are defined by the partition $\tilde{\gamma}_1(\tilde{S}), \ldots, \tilde{\gamma}_{3g-3}(\tilde{S})$, say $\tilde{\mu}_n(\tilde{S}) \subset \tilde{Y}_n$.

For fixed ν let $\ell\gamma_\nu(S)$ be variable and keep the remaining $\ell\gamma_k(S)$ fixed. Then on $\tilde{S}$ only $\ell\tilde{\gamma}_\nu(\tilde{S})$ is variable and $\ell\tilde{\mu}_n(\tilde{S})$ is non-constant if and only if $\ell\mu_n(S)$ is non-constant. Since $\ell\tilde{\mu}_n(\tilde{S})$ depends only on the boundary lengths of $\tilde{Y}_n$, it follows that $\tilde{\gamma}_\nu(\tilde{S})$ is a boundary geodesic of $\tilde{Y}_n$ if and only if $\gamma_\nu(S)$ is a boundary geodesic of the Y-piece in S which carries $\mu_n(S)$. This proves (ii). ◇

At present we know that $\tilde{S}$ is composed of the same Y-pieces as S and this with respect to the same graph G. It remains to prove that the twist parameters are also the same.

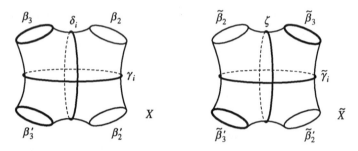

Figure 10.5.1

Let X be the surface of signature $(0, 4)$ formed by the two Y-pieces adjacent along γ_i. Denote by β_2, β_3, β_2', β_3' its boundary geodesics (Fig. 10.5.1) labelled in such a way that γ_i separates the pair β_2, β_3 from the pair β_2', β_3' and such that δ_i separates the pair β_2, β_2' from the pair β_3, β_3' (all β's are, of course, among $\gamma_1, \ldots, \gamma_{3g-3}$). The adjacent Y-pieces of $\tilde{S}$ along $\tilde{\gamma}_i(\tilde{S})$ (with respect to the partition $\tilde{\gamma}_1(\tilde{S}), \ldots, \tilde{\gamma}_{3g-3}(\tilde{S})$) form a surface $\tilde{X}$ (some of the boundary geodesics of $\tilde{X}$ may coincide on $\tilde{S}$) with boundary components $\tilde{\beta}_2$, $\tilde{\beta}_3$, $\tilde{\beta}_2'$ and $\tilde{\beta}_3'$, where $\tilde{\gamma}_i$ separates the pair $\tilde{\beta}_2$, $\tilde{\beta}_3$ from the pair $\tilde{\beta}_2'$, $\tilde{\beta}_3'$. All this is a consequence of (ii).

(iii) $\tilde{\delta}_i(\tilde{S})$ and $\tilde{\eta}_i(\tilde{S})$ are simple closed geodesics on $\tilde{X}$. Each of these geodesics intersects $\tilde{\gamma}_i(\tilde{S})$ in exactly two points and separates the pair $\tilde{\beta}_2(\tilde{S})$, $\tilde{\beta}_2'(\tilde{S})$ from the pair $\tilde{\beta}_3(\tilde{S})$, $\tilde{\beta}_3'(\tilde{S})$.

Proof. Start with $S = S^\omega$, where $\omega = (\ell_1, \ldots, \ell_{3g-3}, 0, \ldots, 0)$ with $\ell_i = 8$ and all other $\ell_k = 1/8$. Formulae (10.2.2) in the more explicit form of Proposition 3.3.11 yield the numerical value $\ell\delta_i(S) = 1.091$. As a function of the twist parameter α_i, this value is a global minimum. Hence, $\ell\tilde{\delta}_i(\tilde{S}) = 1.091$, and it follows from the collar theorem (Theorem 4.1.1) that $\tilde{\delta}_i(\tilde{S})$ is simple. Moreover, for $k \neq i$ the collars around $\tilde{\gamma}_k(\tilde{S})$, are so long that $\tilde{\delta}_i(\tilde{S}) \cap \tilde{\gamma}_k(\tilde{S}) = \emptyset$. Since $\ell\delta_i(S) = \ell\tilde{\delta}_i(\tilde{S})$ and since $\ell\delta_i(S)$ is a minimum, it follows that in the present case X and $\tilde{X}$ are isometric and that $\tilde{\delta}_i(\tilde{S})$ intersects $\tilde{\gamma}_i(\tilde{S})$ in exactly two points.

Now change $L = (\ell_1, \ldots, \ell_{3g-3})$ so that

$$\ell\gamma_i(S) = 3, \quad \ell\beta_3(S) = \ell\beta_3'(S) = 4, \quad \ell\beta_2(S) = \ell\beta_2'(S) = 1.$$

With the same formulae as before we compute

$$\ell\delta_i(S) = \ell\tilde{\delta}_i(\tilde{S}) = 7.004\ldots.$$

On the other hand, let ζ be a simple closed geodesic on $\tilde{X}$ which intersects $\tilde{\gamma}_i(\tilde{S})$ in exactly two points and separates the pair $\tilde{\beta}_3$, $\tilde{\beta}_3'$ from the pair $\tilde{\beta}_2$, $\tilde{\beta}_2'$ as in Fig. 10.5.1. The minimal possible length of ζ would be obtained when $\tilde{\alpha}_i(\tilde{S}) = \frac{1}{2} +$ integer, and the formulae mentioned above yield 7.239 for this minimum. Hence,

$$\ell\zeta(\tilde{S}) \geq 7.239\ldots$$

independently of the twist parameter $\tilde{\alpha}_i(\tilde{S})$. This proves that $\tilde{\delta}_i(\tilde{S})$ separates $\tilde{\beta}_2$ and $\tilde{\beta}_2'$ from $\tilde{\beta}_3$ and $\tilde{\beta}_3'$. Replacing $\alpha_i = 0$ in ω by $\alpha_i = -1$ we prove the same properties for $\tilde{\eta}_i$. This proves (iii). ◇

The next point is more difficult. Each η_i is obtained from δ_i by a positively oriented Dehn twist along γ_i (Definitions 3.3.9 and 3.3.10). Taking $A = (0, \ldots, 0)$ in $\omega = (L, A)$, we conclude from (iii) and from the fact that $\ell\delta_i(S) = \ell\tilde{\delta}_i(\tilde{S})$ and $\ell\eta_i(S) = \cdot\ell\tilde{\eta}_i(\tilde{S})$ that

(iv) $\quad \tilde{\eta}_i = \tilde{D}_i^{\pm 1}(\tilde{\delta}_i), \quad i = 1, \ldots, 3g - 3,$

where $\tilde{D}_i$ is a positively oriented Dehn twist along $\tilde{\gamma}_i$. In order to prove that the curve system $\tilde{\Omega} = \rho(\Omega)$ is obtained by a homeomorphism of F, it remains to prove that the exponents in (iv) are the same for all i. (Since the expected homeomorphism may be orientation reversing, this exponent may be -1.)

Suppose on the contrary that the exponents in (iv) change sign. Then they change their sign along one of the curves $\sigma_1, \ldots, \sigma_g$ (cf. Fig. 10.2.1). Let us therefore examine the nature of $\tilde{\sigma}_1, \ldots, \tilde{\sigma}_g$. Taking $L = (8, \ldots, 8)$, $A = (0, \ldots, 0)$ in $S = S^\omega$ we compute using the hexagon formulae of Chapter 2:

$$\ell\sigma_1(S) = \tfrac{1}{2}\ell\sigma_2(S) = \ldots = \tfrac{1}{2}\ell\sigma_{g-1}(S) = \ell\sigma_g(S) = 0.549\ldots.$$

Since the $\tilde{\sigma}_k(\tilde{S})$ also have length $0.549\ldots$, they are simple by virtue of Theorem 4.2.1. Now fix j and vary α_j in a small neighborhood. Then $\ell\sigma_k(S) = \ell\sigma_k(\tilde{S})$ is non-constant if and only if σ_k intersects γ_j. This also holds if and only if $\tilde{\sigma}_k$ intersects $\tilde{\gamma}_j$ since by (iii), $\tilde{\alpha}_j(\tilde{S})$ is the only twist parameter on $\tilde{S}$ which is variable. Hence, $\tilde{\sigma}_k$ intersects $\tilde{\gamma}_j$ if and only if σ_k intersects γ_j.

Looking at the lengths again, we see immediately that $\tilde{\sigma}_k$ intersects $\tilde{\gamma}_j$ at most once. The $\tilde{\sigma}_1, \ldots, \tilde{\sigma}_g$ have therefore the same configuration as the $\sigma_1, \ldots, \sigma_g$. Now take $L = (1, \ldots, 1)$ and $A = (\frac{1}{4}, \ldots, \frac{1}{4})$.

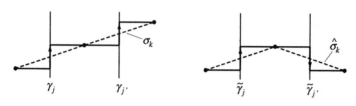

Figure 10.5.2

If we assume that the exponents in (iv) change sign, then we find σ_k intersecting some γ_j and $\gamma_{j'}$ as in Fig. 10.5.2 such that for the corresponding geodesics $\tilde{\gamma}_j(\tilde{S})$ and $\tilde{\gamma}_{j'}(\tilde{S})$ we have $\tilde{\alpha}_j(\tilde{S}) = \frac{1}{4}$ and $\tilde{\alpha}_{j'}(\tilde{S}) = -\frac{1}{4}$. As shown in Fig. 10.5.2, we then find a piecewise geodesic curve $\hat{\sigma}_k(\tilde{S})$ in the free homotopy class of $\tilde{\sigma}_k(\tilde{S})$ of length $\ell\hat{\sigma}_k(\tilde{S}) = \ell\sigma_k(S)$. It follows that $\ell\tilde{\sigma}_k(\tilde{S}) < \ell\sigma_k(S)$, a contradiction. It is now established that all $\tilde{D}_i$ in (iv) have the same exponent.

There exists an orientation reversing homeomorphism of F keeping the curves $\gamma_1, \ldots, \gamma_{3g-3}$ and $\delta_1, \ldots, \delta_{3g-3}$ fixed and mapping $\eta_1, \ldots, \eta_{3g-3}$ to $D_1^{-1}(\delta_1), \ldots, D_1^{-1}(\delta_{3g-3})$ (recall that in this chapter as in Chapter 6, the closed geodesics are non-oriented). The existence of this homeomorphism together with (i) - (iv) shows that the injection $\Sigma \to \tilde{\Sigma} = \rho(\Sigma)$ is induced by a homeomorphism of F. In particular, the system $\tilde{\Omega} := \rho(\Omega)$ is a canonical curve system.

There exists a unique Teichmüller mapping μ satisfying $\ell\Omega(S) = \ell\tilde{\Omega}(\mu(S))$. By Corollary 6.2.8, any $S' \in \mathcal{T}_g$ is uniquely determined by $\ell\tilde{\Omega}(S')$. Since $\ell\tilde{\Omega}(\mu(S)) = \ell\Omega(S) = \ell\tilde{\Omega}(m(S))$, we conclude that $m(S) = \mu(S)$ for all $S \in D$. Proposition 10.5.1 is now proved. ◇

Chapter 11

Sunada's Theorem

By Wolpert's theorem almost all compact Riemann surfaces are determined up to isometry by their length spectrum or, equivalently, by their spectrum of the Laplacian. On the other hand, there are surfaces which are only determined up to finitely many. The first such examples were given by Marie France Vignéras [1, 2] based on quaternion lattices. Later, in 1984 Sunada [3] found a general construction of isospectral manifolds, based on finite groups. In Chapter 12 we shall use Sunada's construction to give numerous examples of isospectral Riemann surfaces. In the present chapter we prove Sunada's theorem for arbitrary Riemannian manifolds and study its combinatorial aspects in detail. It will turn out that graphs are again a useful tool. Here they occur in the form of Cayley graphs.

After looking at some relevant examples of finite groups in Section 11.2 we prove Sunada's theorem in Section 11.3. Then we study Cayley graphs and finite coverings.

In the final two sections we present a different approach to some of Sunada's examples. This approach is combinatorial in nature, and the two sections may be read independently of the rest of the chapter. Both sections contain the example of an isospectral non-isometric pair of compact Riemann surfaces, where the isospectrality can be proved by *direct checking*. For further combinatorial aspects of Sunada's construction we refer to Bérard [3].

11.1 Some History

Sunada's theorem has its origins in number theory. If f is an algebraic number field and if $p \in \mathbf{N}$ is a prime number, then p has a unique prime ideal decomposition with respect to $\mathfrak{f}$. Let $d_1(p), \ldots, d_\nu(p)$ be the degrees of the prime ideal factors, arranged such that $d_1(p) \leq \ldots \leq d_\nu(p)$ and call the finite sequence

$$\ell[\mathfrak{f}](p) = (d_1(p), \ldots, d_\nu(p))$$

the *length of p with respect to* $\mathfrak{f}$. How much information about $\mathfrak{f}$ can we read out of these lengths? In 1880 Kronecker [1] conjectured that if $\mathfrak{f}_1, \mathfrak{f}_2$ are two number fields such that

$$\ell[\mathfrak{f}_1](p) = \ell[\mathfrak{f}_2](p), \text{ for all } p \geq p_0$$

for sufficiently large p_0, then $\mathfrak{f}_1$ and $\mathfrak{f}_2$ are isomorphic. A shorter form of Kronecker's conjecture would be: "isospectral number fields are isomorphic".

In 1925, after previous work by Hurwitz [2], Bauer, and others, Gassmann [1] showed that the existence of two isospectral non-isomorphic number fields is equivalent to the existence of a pair of *finite* groups with a particular property (cf. Definition 11.1.1). Gassmann then constructed such a pair (Example 11.2.1) giving a counterexample to Kronecker's conjecture. Many more such examples have been studied since. Number fields which are not uniquely determined by their length spectrum are a point of interest and are sometimes called *non-solitary*.

In 1984 Sunada discovered that the covering technique which leads from Gassmann equivalent groups to non-solitary number fields works also in Riemannian geometry and leads to isospectral manifolds. In contrast to the earlier sporadic examples of isospectral manifolds (Milnor [1], Vignéras [1, 2], Ejiri [1], Kuwabara [1], Ikeda-Yamamoto [1], Ikeda [1], Urakawa [1], Gordon-Wilson [1] and others), Sunada's examples are based on a general construction principle. In particular, they are not limited to any form of homogeneity. More recently, De Turck and Gordon [1] extended Sunada's method to infinite groups. They also showed that in this extended form, almost all of the presently known examples of pairs of isospectral manifolds fit into the same combinatorial scheme. The references on this subject, up to 1988, may be found in the review article [2] of Bérard.

We now state Sunada's theorem which leads from certain pairs of finite groups to pairs of isospectral manifolds. The groups act by isometries on a given compact or non-compact Riemannian manifold, the examples are the quotients. In the following definition we denote by # the *cardinality* of a set. For any group G we denote the conjugacy class of an element $g \in G$ by $[g]$,

$$[g] = \{\sigma g \sigma^{-1} \mid \sigma \in G\}.$$

11.1.1 Definition. Let G be a finite group. Two subgroups H_1, H_2 of G are called *almost conjugate* or *Gassmann equivalent* if for all $g \in G$

$$\#([g] \cap H_1) = \#([g] \cap H_2).$$

Observe that conjugate subgroups are almost conjugate. Non-conjugate examples will be given in Section 11.2. Some of them will not be isomorphic as groups. Note that almost conjugate subgroups have the same cardinality.

We now state Sunada's theorem in the following form.

11.1.2 Theorem. *Let M be a complete Riemannian manifold and let G be a finite group acting on M by isometries with at most finitely many fixed points. If H_1 and H_2 are almost conjugate subgroups of G acting freely on M, then the quotients $H_1 \backslash M$ and $H_2 \backslash M$ are isospectral.*

Remarks. In the statement of the theorem the isospectrality holds for both the length and the eigenvalue spectrum. For this we remark that if M is non-compact, then one has to be careful with the definition of the eigenvalue spectrum (cf. Theorem 11.3.5), and if M is not a manifold of negative curvature, then one has to be careful with the definition of the length spectrum (cf. Theorem 11.3.2).

Theorem 11.1.2 does not claim that the quotients $H_1 \backslash M$ and $H_2 \backslash M$ are non-isometric. For manifolds of variable curvature Sunada [3] and Buser [9] give techniques to make the metric of M "bumpy" enough so that the non-isometry of $H_1 \backslash M$ and $H_2 \backslash M$ can be inferred from the non-conjugacy of H_1 and H_2. In the case of compact Riemann surfaces a useful technique is to use small closed geodesics in connection with the collar theorem. This is, for instance, carried out at the end of Section 11.4. General criteria for the non-isometry of isospectral pairs of compact Riemann surfaces will be given in Section 12.7.

11.2 Examples of Almost Conjugate Groups

The examples in this section are from Gassmann [1], Gerst [1], Komatsu [1] and Perlis [1].

11.2.1 Example. (The symmetric group $\mathfrak{S}_6$). This example is due to Gassmann [1] and is the first known example in the literature. It is also the easiest example. Let $\mathfrak{S}_6$ be the group of all permutations of the set $\{1, 2, 3, 4, 5, 6\}$. Denote by $(a, b) \in \mathfrak{S}_6$ the cyclic permutation which interchanges a

with b and leaves the remaining elements fixed. For $(a, b), (c, d) \in \mathfrak{S}_6$ we let $(a, b)(c, d)$ denote the product $((c, d)$ followed by $(a, b))$. *The following subgroups H_1, H_2 of $\mathfrak{S}_6$ are almost conjugate but not conjugate.*

$$H_1 = \{id;\ (1, 2)(3, 4);\ (1, 3)(2, 4);\ (1, 4)(2, 3)\}$$
$$H_2 = \{id;\ (1, 2)(3, 4);\ (1, 2)(5, 6);\ (3, 4)(5, 6)\}.$$

Proof. Every element in $(H_1 \cup H_2) - \{id\}$ acts in the same way: it changes a string $(ab\ cd\ ef)$ into a string $(ba\ dc\ ef)$. Since $\mathfrak{S}_6$ permutes all possible strings, all elements in $(H_1 \cup H_2) - \{id\}$ belong to the same conjugacy class. Hence, H_1 and H_2 satisfy the condition of Definition 11.1.1. Now H_1 acts with two fixed points and H_2 acts without fixed points. This shows that H_1 and H_2 are not conjugate in $\mathfrak{S}_6$.

The index of H_1 and H_2 in $\mathfrak{S}_6$ is 180.

11.2.2 Example. (The semi direct product $\mathbf{Z}_8^* \ltimes \mathbf{Z}_8$). This example is due to Gerst [1]. We let $\mathbf{Z}_8 = \mathbf{Z}/8\mathbf{Z} = \{0, 1, 2, 3, 4, 5, 6, 7\}$ be the additive group and $\mathbf{Z}_8^* = \{1, 3, 5, 7\}$ be the multiplicative group of integers modulo 8. The semi direct product

$$\mathbf{Z}_8^* \ltimes \mathbf{Z}_8 = \{(a, b)\ |\ a = 1, 3, 5, 7;\ b = 0, 1, \ldots, 7\}$$

is a group of order 32 with the multiplication rule

$$(a, b)(a', b') := (aa', ab' + b)$$

(all integers mod 8). $\mathbf{Z}_8^* \ltimes \mathbf{Z}_8$ has the following isomorphic abelian subgroups of index 8.

$$H_1 = \{(1, 0);\ (3, 0);\ (5, 0);\ (7, 0)\}$$
$$H_2 = \{(1, 0);\ (3, 4);\ (5, 4);\ (7, 0)\}.$$

They are almost conjugate but not conjugate.

Proof. Writing

$$\sigma = (a, b) = (1, b)(a, 0), \qquad \sigma^{-1} = (a, 0)(1, -b)$$

we find $\sigma(m, 0)\sigma^{-1} = (1, b)(m, 0)(1, -b) = (m, b(1 - m))$. From this we see that H_1 and H_2 are not conjugate. To see that H_1, H_2 are almost conjugate we use the fact that $\mathbf{Z}_8^*$ is an abelian factor. It is then easily checked that the elements in H_1 resp. in H_2 are pairwise non-conjugate and that for each element in H_1 there exists a conjugate in H_2. ◊

11.2.3 Example. (Groups with prime exponent). This example is due to Komatsu [1]. Let $p \geq 3$ be a prime number and let H_1, H_2 be finite groups with the same cardinality, say with $\#H_1 = \#H_2 = n$. Assume that both groups have exponent p (i.e. $x^p = 1$ for all $x \in H_1$ and all $x \in H_2$). Interpret H_1 and H_2 as subgroups of the symmetric group $\mathfrak{S}_n$, where H_1 and similarly H_2 is identified with the set $\{1, \ldots, n\}$ in some fixed way, and where the permutation corresponding to $g \in H_i$, $i = 1, 2$, is obtained via its action on H_i by left multiplication.

Under these hypotheses, H_1 and H_2 are almost conjugate in $\mathfrak{S}_n$.

Proof. We have to show that

$$\#([h] \cap H_1) = \#([h] \cap H_2)$$

for all $h \in H_1 \cup H_2$. For $h = id$ there is nothing to prove. If $h \neq id$ then h is a permutation of $\{1, \ldots, n\}$ which has no fixed point. Since h has prime order p, there exists a partition of $\{1, \ldots, n\}$ into pairwise disjoint subsets of cardinality p such that h operates by cyclic permutation on each of these subsets. This proves that all $h \in (H_1 \cup H_2) - \{id\}$ belong to the same conjugacy class in $\mathfrak{S}_n$, and therefore $\#([h] \cap H_1) = \#([h] \cap H_2)$ for all $h \in \mathfrak{S}_n$. ◊

Obviously, H_1 and H_2 are conjugate in $\mathfrak{S}_n$ if and only if H_1 is isomorphic to H_2.

Examples of non-isomorphic groups which satisfy the above hypotheses are, for instance, given by the Heisenberg groups: let $k, p \in \mathbf{N}$, $k \geq 3$, $p \geq k$, p prime. Define

$$H_1 = \left\{ \begin{pmatrix} 1 & & * \\ & \ddots & \\ 0 & & 1 \end{pmatrix} \,\middle|\, * \in \mathbf{Z}_p \right\}$$

to be the group of all upper triangular $(k \times k)$ matrices with coefficients from $\mathbf{Z}_p = \mathbf{Z}/p\mathbf{Z}$ and all diagonal elements equal to 1. Each such matrix g may be written in the form

$$g = a + I,$$

where I is the identity matrix and a is an upper triangular matrix with zero diagonal. Since a and I commute, the binomial formula yields

$$g^p = (a + I)^p = \sum_{i=0}^{p} \binom{p}{i} a^i.$$

In $\mathbf{Z}_p$ we have $\binom{p}{i} = 0$ for $i = 1, \ldots, p-1$. As the matrix a is k-nilpotent and $p \geq k$, we have $a^p = 0$. Hence, $g^p = I$. It follows that H_1 has exponent p and order p to the power $\frac{1}{2}k(k-1)$. The same holds for

$$H_2 = \mathbf{Z}_p \times \ldots \times \mathbf{Z}_p, \quad (\tfrac{1}{2}k(k-1) \text{ factors}).$$

As H_2 is abelian and H_1 is not, these groups are not isomorphic.

For our final example we use another definition of almost conjugacy which is also useful for the proof of Sunada's theorem.

11.2.4 Lemma. *Let G be a finite group. Define for any subgroup $H \subset G$ and for any $\sigma \in G$*

$$\chi_H(\sigma) = \frac{1}{\#H} \#\{g \in G \,|\, g\sigma g^{-1} \in H\}.$$

Two subgroups H_1, H_2 of G are almost conjugate if and only if $\chi_{H_1} = \chi_{H_2}$.

Proof. Let $C_\sigma = \{\alpha \in G \,|\, \alpha\sigma\alpha^{-1} = \sigma\}$ be the centralizer of σ in G. For any $g, h \in G$ we have

$$g\sigma g^{-1} = h\sigma h^{-1} \Leftrightarrow \sigma = (g^{-1}h)\sigma(g^{-1}h)^{-1} \Leftrightarrow h \in gC_\sigma.$$

This shows that for fixed $\sigma' = g\sigma g^{-1}$

$$\#\{h \in G \,|\, h\sigma h^{-1} = \sigma'\} = \#(gC_\sigma) = \#C_\sigma.$$

Therefore,

$$\chi_H(\sigma) = \frac{\#C_\sigma}{\#H} \#([\sigma] \cap H).$$

Since $\chi_H(id) = \#G/\#H$ and since almost conjugate subgroups have the same cardinality, the lemma follows. ◇

For a number of interesting properties of the next example we refer to the book of Weinstein [1] and to the Atlas of Finite Groups (Conway et al. [1]). The particular case SL(3, 2) is described in detail in Perlis [1].

11.2.5 Example. (The groups SL(n, p)). Let p be a prime number and let $n \geq 2$ be an arbitrary integer. SL(n, p) is the group of all $(n \times n)$-matrices with coefficients in $\mathbf{Z}_p$ and with determinant 1. A good exercise (which will be needed below) is to prove that for $n \geq 2$, SL(n, p) acts transitively on the vector space $(\mathbf{Z}_p)^n$. In our notation, elements of $(\mathbf{Z}_p)^n$ are columns and SL(n, p) acts by multiplication from the left.

We consider the two subgroups

$$H_1 = \left\{ g \in \text{SL}(n, p) \,\Big|\, g = \begin{pmatrix} * & * & \cdots & * \\ 0 & & & \\ \vdots & & * & \\ 0 & & & \end{pmatrix} \right\}$$

and
$$H_2 = \left\{ g \in \mathrm{SL}(n,p) \mid g = \begin{pmatrix} * & 0 & \cdots & 0 \\ * & & & \\ \vdots & & * & \\ * & & & \end{pmatrix} \right\}.$$

H_1 and H_2 are almost conjugate. If $n \geq 3$ they are non-conjugate.

Proof. We shall prove that $\chi_{H_1}(\sigma) = \chi_{H_2}(\sigma)$ for all $\sigma \in \mathrm{SL}(n,p)$. Since we have $g \sigma g^{-1} \in H_2$ if and only if $(g^T)^{-1} \sigma^T g^T \in H_1$, where ^{T} denotes the transpose, it suffices to prove that

$$\chi_{H_1}(\sigma) = \frac{1}{\#H_1} \#\{ g \in \mathrm{SL}(n,p) \mid g \sigma^T g^{-1} \in H_1 \} = \chi_{H_1}(\sigma^T).$$

Now observe that for each $\lambda \in \mathbf{Z}_p$ the matrices $(\sigma - \lambda I)$ and $(\sigma - \lambda I)^T = (\sigma^T - \lambda I)$ have the same rank (I is the identity matrix). Hence, σ and σ^T have the same eigenvalues. Moreover, for each eigenvalue they have the same dimension of the corresponding eigenspace. Since the field $\mathbf{Z}_p$ is a finite set, each k-dimensional subspace of $(\mathbf{Z}_p)^n$ consists of p^k elements, and we conclude that σ and σ^T have the same number of eigenvectors. Denoting this number by m we get $m(\sigma) = m(\sigma^T)$. It suffices therefore to prove that $\chi_{H_1}(\sigma)$ depends only on $m(\sigma)$.

First we remark that $h \in H_1$ if and only if $e_1 := (1, 0, \ldots, 0)^T$ is an eigenvector of h. Let Γ_1 be the subgroup of all $h \in H_1$ for which $he_1 = e_1$. By our remark,

$$g \sigma g^{-1} \in H_1 \iff g^{-1} e_1 \text{ is an eigenvector of } \sigma.$$

Now let e be an eigenvector of σ. Since $\mathrm{SL}(n,p)$ acts transitively on $(\mathbf{Z}_p)^n$ (the above exercise), there exists $\gamma \in \mathrm{SL}(n,p)$ such that $e = \gamma^{-1} e_1$. If $g \in \mathrm{SL}(n,p)$, then $g^{-1} e_1 = \gamma^{-1} e_1$ if and only if $g \in \Gamma_1 \gamma$. Hence, the cardinality of all g satisfying $g^{-1} e_1 = e$ is $\#\Gamma_1$. This proves that

$$\chi_{H_1}(\sigma) = \frac{1}{\#H_1} m(\sigma) \#\Gamma_1.$$

We have now proved that H_1 and H_2 are almost conjugate.

To see that H_1 and H_2 are non-conjugate if $n \geq 3$, we simply observe that for $n \geq 3$ the elements of H_1 have a common eigenvector, and the elements of H_2 do not. (Use that $\mathrm{SL}(n-1, p)$ acts transitively on $(\mathbf{Z}_p)^{n-1}$.) ◊

A particularly interesting case which will be used in Chapter 12 is the group $\mathrm{SL}(3,2)$. This group has a number of remarkable properties (cf. Weinstein [1], Example 4.12). It is isomorphic to $\mathrm{SL}(2,7)$ and, up to isomorphism, the only simple group of order 168. H_1 and H_2 have index 7 in $\mathrm{SL}(3,2)$. Perlis

[1] has shown that this is the smallest possible index of almost conjugate non-conjugate subgroups in any group (see also Guralnick [1]).

We shall later need two lemmata about generators of SL(3, 2). They are as follows, where H_1 is as in Example 11.2.5.

11.2.6 Lemma. *Let A, B, C be the following elements of* SL(3, 2).

$$A = \begin{pmatrix} 0 & 1 & 1 \\ 0 & 1 & 0 \\ 1 & 0 & 0 \end{pmatrix}, \quad B = \begin{pmatrix} 1 & 0 & 0 \\ 0 & 0 & 1 \\ 0 & 1 & 1 \end{pmatrix}, \quad C = \begin{pmatrix} 0 & 1 & 1 \\ 1 & 1 & 1 \\ 0 & 1 & 0 \end{pmatrix}.$$

Then the following hold, where I is the identity matrix.
  (i) $A^4 = B^3 = C^7 = I$, $C = ABA^{-1}B^{-1}$.
  (ii) *The cosets $H_1, H_1 C, \ldots, H_1 C^6$ are pairwise disjoint and fill out* SL(3, 2).
  (iii) *A and C generate* SL(3, 2).
  (iv) *A and B generate* SL(3, 2).

Proof. (i) By direct computation.

(ii) C and all C^m with $C^m \neq I$ have the prime order 7 which is not a divisor of $\# H_1$. Since for $m = 1, \ldots, 6$, $C^m \notin H_1$, the cosets $H_1 C^0, \ldots, H_1 C^6$ are pairwise different, and since $7 \cdot 24 = \# $SL(3, 2), they fill out the group.

(iii) A quick though not elegant way is as follows. Set $M = CAC^{-1}$ and $N = C^{-1}AC^3$. The numerical values are

$$C^{-1} = \begin{pmatrix} 1 & 1 & 0 \\ 0 & 0 & 1 \\ 1 & 0 & 1 \end{pmatrix}, \quad M = \begin{pmatrix} 1 & 1 & 1 \\ 0 & 1 & 1 \\ 0 & 0 & 1 \end{pmatrix}, \quad N = \begin{pmatrix} 1 & 1 & 0 \\ 0 & 0 & 1 \\ 0 & 1 & 1 \end{pmatrix}.$$

Thus, $M, N \in H_1$. Products of the form $(M^p N^q)^r$ yield (quite rapidly) more than 12 elements in H_1. Hence, M and N generate H_1. Now (iii) follows from (ii).

(iv) This follows from (iii) and (i). ◇

11.2.7 Lemma. *Let*

$$\alpha = \begin{pmatrix} 1 & 0 & 1 \\ 1 & 1 & 0 \\ 0 & 1 & 0 \end{pmatrix}, \quad \beta = \begin{pmatrix} 0 & 0 & 1 \\ 1 & 0 & 0 \\ 1 & 1 & 0 \end{pmatrix}.$$

  (i) *α, β and $\alpha\beta$ are elements of order 7.*
  (ii) *α and β generate* SL(3, 2).

Proof. (i) by computation. (ii) $\beta^5 = C$ and $\beta^5 \alpha \beta^5 = A$, where C and A are the same generators as in Lemma 11.2.6. ◇

11.3 Proof of Sunada's Theorem

Let G be a finite group acting on the complete (not necessarily compact) Riemannian manifold M by isometries and with at most finitely many fixed points. G need not be the full isometry group of M. Assume that H_1 and H_2 are almost conjugate subgroups of G which act without fixed points. We then have the normal coverings

(11.3.1)
$$\begin{array}{ccc} & M & \\ \swarrow & & \searrow \\ M_1 = H_1\backslash M & & M_2 = H_2\backslash M \,. \end{array}$$

Sunada's theorem for the closed geodesics states that M_1 and M_2 have the same length spectrum. If M is an arbitrary closed Riemannian manifold, we no longer have a one-to-one correspondence between the closed geodesics and the free homotopy classes. We restate therefore Sunada's theorem in a particular form which takes care of this.

In the following we shall work with non-oriented geodesics. Recall that a parametrized closed geodesic $\gamma : \mathbf{S}^1 \to N$ on a Riemannian manifold N is called *primitive* if it is not the covering of a shorter closed geodesic. Two parametrized closed geodesics are considered *weakly equivalent* if they differ by a parameter change of type $t \mapsto \varepsilon t + \text{const}$, $t \in \mathbf{S}^1$, where ε is either 1 or -1. The weak equivalence classes are the *non-oriented closed geodesics*. As we shall only work with these, we shall drop the adjective "non-oriented". In order to have a short notation we shall call the weak equivalence class of a primitive parametrized closed geodesic a *prime geodesic*. Observe that with this notation, the prime geodesic represented by $\gamma : \mathbf{S}^1 \to N$ may be identified with the point set $\gamma(\mathbf{S}^1)$.

We denote by $\mathcal{P}(N)$ the set of all prime geodesics of N.

11.3.2 Theorem. *Let M_1 and M_2 be complete Riemannian manifolds as in (11.3.1). Then there exists a bijection $\phi : \mathcal{P}(M_1) \to \mathcal{P}(M_2)$ satisfying $\ell(\gamma) = \ell(\phi(\gamma))$ for all $\gamma \in \mathcal{P}(M_1)$.*

Proof. The proof is essentially that of Sunada [3]. However, since we allow G to have fixed points, we shall not work with the quotient $G\backslash M$.

For $i = 1, 2$ the natural projections $\pi_i : M \to M_i = H_i\backslash M$ are normal Riemannian coverings. We say that a geodesic γ in M is a *prime lift* of prime geodesic α in M_i if γ is a prime geodesic and $\pi_i(\gamma) = \alpha$. If γ is a prime lift of α, then the length of γ is an integer multiple of the length of α. We remark that γ and γ' in M are lifts of the same α if and only if $\gamma' = h(\gamma)$ for some

$h \in H_i$.

We say that prime geodesics α and β in the disjoint union $M_1 \cup M_2$ are *correlated* if there exist prime lifts γ and δ of α and β respectively such that $\delta = g(\gamma)$ for some $g \in G$. From the preceding remark it follows that correlation is an equivalence relation. Note that correlated geodesics need not have the same length. We shall denote by $[\alpha]$ the equivalence class of α and set

$$[\alpha]_i = \{\alpha' \in [\alpha] \mid \alpha' \subset M_i\}, \quad i = 1, 2.$$

For the proof of Theorem 11.3.2 we show that there exists, for each $[\alpha]$, a length-preserving one-to-one mapping from $[\alpha]_1$ onto $[\alpha]_2$.

For any prime geodesic η in M and for any subgroup $F \subset G$ we denote by F_η the subgroup of all $f \in G$ satisfying $f(\eta) = \eta$. Since G acts with at most finitely many fixed points, F_η is isomorphic to the group $\{(f|\eta) \mid f \in F\}$.

Let γ in M be a prime lift of α. Then G_γ is isomorphic to a finite subgroup of $\mathrm{Is}(\mathbf{S}^1)$. This implies that G_γ is either a cyclic group or a dihedral group. If G_γ is cyclic, we let σ denote a generator of G_γ. If G_γ is dihedral, then it contains a cyclic subgroup G'_γ of index 2 and we let σ be a generator of G'_γ. We then have the disjoint union $G_\gamma = G'_\gamma \cup G'_\gamma \rho$, where ρ is an element of order 2 with at least one fixed point on γ. We let $\mathrm{ord}(\sigma)$ denote the order of σ. The number

$$\delta := \ell(\gamma)/\mathrm{ord}(\sigma)$$

will be called the *displacement length* of σ.

We let H be one of the subgroups H_1, H_2 with the natural projection $\pi_H : M \to M_H := H\backslash M$ and write

$$[\alpha]_H = \{\alpha' \in [\alpha] \mid \alpha' \subset M_H\}.$$

Next, we let $\tau_1, \ldots, \tau_r \in G$ be such that any $g(\gamma)$, $g \in G$, coincides with some $h\tau_i(\gamma)$ for a *unique* $i \in \{1, \ldots, r\}$. The images $\alpha_i := \pi_H(\tau_i(\gamma))$, $i = 1, \ldots, r$, are pairwise distinct and fill out $[\alpha]_H$:

(1) $$[\alpha]_H = \{\alpha_1, \ldots, \alpha_r\}, \quad \#[\alpha]_H = r.$$

Finally, we abbreviate $\tau_i(\gamma)$ by γ_i. Since H operates without fixed points, each H_{γ_i} is a cyclic group. The displacement length of any generator of H_{γ_i} is an integer multiple $\Lambda_i \delta$ (because $\tau_i \sigma \tau_i^{-1}$ is a generator of a maximal cyclic subgroup of G_{γ_i}). It follows that

(2) $$\ell(\alpha_i) = \Lambda_i \delta.$$

For any positive integer m, $\tau_i \sigma^m \tau_i^{-1}$ is an element of G_{γ_i} and belongs therefore to H if and only if it belongs to H_{γ_i}. From this we get the following.

(3) $$\tau_i \sigma^m \tau_i^{-1} \in H \Leftrightarrow \Lambda_i \,|\, m,$$

where $\Lambda_i \,|\, m$ means that m is an integer multiple of Λ_i. We next claim that any $g \in G$ can be written in a *unique* way in the form

(4) $$g = h\tau_i \sigma^j \kappa$$

with $h \in H$, $i \in \{1, \ldots, r\}$, $j \in \{0, \ldots, \Lambda_i - 1\}$ and $\kappa = id$ if G_γ is cyclic, and $\kappa \in \{id, \rho\}$ if G_γ is dihedral with ρ as above. For the proof we consider the case that G_γ is dihedral and leave the cyclic case to the reader. On γ we let c be an arc of length $\delta/2$ such that one endpoint of c is a fixed point of ρ. Then c is a fundamental domain for the action of G on the set

$$\mathscr{C} = \bigcup_{g \in G} g(\gamma).$$

For given $i \in \{1, \ldots, r\}$, the images $\sigma^j(c)$ and $\sigma^j \rho(c)$, $j = 0, \ldots, \Lambda_i - 1$ fill out an arc c_i of length $\Lambda_i \delta$ without overlapping. The image $\tau_i(c_i)$ on γ_i is a fundamental domain for the action of H on the set

$$\mathscr{C}_i = \bigcup_{h \in H} h(\gamma_i).$$

Finally, $\mathscr{C}_1, \ldots, \mathscr{C}_r$ fill out $\mathscr{C}$ without overlapping (except for finitely many intersection points). Hence, the elements listed in (4) are pairwise different and the images of c fill out $\mathscr{C}$. Since c is a fundamental domain for the action of G on $\mathscr{C}$, this proves that the elements in (4) fill out G.

In view of (4), we have, for any positive integer m,

$$\chi_H(\sigma^m) = \frac{1}{\#H} \#\{h\tau_i \sigma^j \kappa \in G \mid h\tau_i \sigma^j \kappa \sigma^m (h\tau_i \sigma^j \kappa)^{-1} \in H\}$$

$$= \#\{\tau_i \sigma^j \kappa \mid \tau_i \sigma^j \kappa \sigma^m \kappa^{-1} \sigma^{-j} \tau_i^{-1} \in H, 1 \le i \le r, 0 \le j \le \Lambda_i - 1, \kappa \in \{id, \rho\}\},$$

where again we look at dihedral G_γ. Since $\rho \sigma^m \rho^{-1} = \sigma^{-m}$ and since $\tau_i \sigma^m \tau_i^{-1} \in H$ if and only if $\tau_i \sigma^{-m} \tau_i^{-1} \in H$, we obtain together with (3)

(5) $$\chi_H(\sigma^m) = 2^\# \sum_{\substack{i=1 \\ \tau_i \sigma^m \tau_i^{-1} \in H}}^{r} \Lambda_i = 2^\# \sum_{\substack{i=1 \\ \Lambda_i | m}}^{r} \Lambda_i.$$

Here $2^\# = 2$ if G_γ is dihedral and $2^\# = 1$ if G_γ is cyclic. We apply this to H_1 and H_2 writing $[\alpha]_1 = \{\alpha'_1, \ldots, \alpha'_{r'}\}$ and $[\alpha]_2 = \{\alpha''_1, \ldots, \alpha''_{r''}\}$, where $r' = \#[\alpha]_1$ and $r'' = \#[\alpha]_2$. In view of (2) there are integers $\Lambda'_1, \ldots, \Lambda'_{r'}$ and $\Lambda''_1, \ldots, \Lambda''_{r''}$ such that

$$\ell(\alpha'_i) = \Lambda'_i \delta \quad \text{and} \quad \ell(\alpha''_j) = \Lambda''_j \delta$$

for $i = 1, \ldots, r'$ and $j = 1, \ldots, r''$. By (5) and Lemma 11.2.4, we have, for any positive integer m,

$$\sum_{\substack{i=1 \\ \Lambda'_i \mid m}}^{r'} \Lambda'_i = \sum_{\substack{i=1 \\ \Lambda''_i \mid m}}^{r''} \Lambda''_i.$$

From this we see that any length appears as often in $[\alpha]_1$ as in $[\alpha]_2$. This proves Theorem 11.3.2 ◇

We next prove Sunada's theorem for the eigenfunctions of the Laplacian. The proof consists of an appropriate splitting of G, H_1 and H_2 into cosets. The proof is the same as in Sunada [3] except that we work with finite dimensional real vector spaces in order to avoid the introduction of trace class operators.

Consider a finite group G acting on a finite dimensional vector space V by linear transformations. The action of $g \in G$ on $v \in V$ will be denoted by gv, and we assume the rule $(g'g)v = g'(gv)$ for any $g', g \in G$. (Below, V will be an eigenspace of the Laplacian and G acts via isometries of the underlying manifold.) For any subgroup $H \subset G$ we denote by V^H the pointwise H-invariant subspace of H:

$$V^H = \{v \in V \mid hv = v \text{ for any } h \in H\}.$$

For $h \in H$, we set $[h]' := \{\alpha h \alpha^{-1} \mid \alpha \in H\}$, the conjugacy class with respect to H.

Now let $A : V \to V$ be a linear transformation which commutes with the action of G. (A will later be the identity operator restricted to an eigenspace.) Since A commutes with G, the restriction $A \mid V^H$ maps V^H into itself. In the next lemma, tr denotes the trace of a linear transformation of a finite dimensional vector space. The trace $\text{tr}(A \mid V^H)$ is with respect to the vector space V^H, the traces $\text{tr}(hA)$ and $\text{tr}(gA)$ are with respect to V.

11.3.3 Lemma. (Elementary trace formula).

$$\text{tr}(A \mid V^H) = \sum_{[h]' \subset H} \frac{\#[h]'}{\#H} \text{tr}(hA)$$

$$= \frac{1}{\#H} \sum_{[g] \subset [G]} \#([g] \cap H) \text{tr}(gA).$$

Proof. The linear mapping P given by

$$P(v) = \frac{1}{\#H} \sum_{h \in H} hv, \quad v \in V,$$

maps V onto V^H and acts as the identity on V^H. Taking any vector space basis of V^H and extending it to a basis of V, we see that $\operatorname{tr}(PA) = \operatorname{tr}(A \mid V^H)$. Hence,

$$\operatorname{tr}(A \mid V^H) = \frac{1}{\#H} \sum_{h \in H} \operatorname{tr}(hA).$$

This yields the first equation. Next, for any $h \in H$, $[h]'$ is a subset of H and for $g \in G$ we either have $[h]' \subset [g]$ or $[h]' \cap [g] = \emptyset$. Since the conjugacy classes $[g]$ are pairwise disjoint, we have

$$\sum_{[h]' \subset H} = \sum_{[g] \subset G} \sum_{[h]' \subset [g] \cap H}.$$

Moreover,

$$\sum_{[h]' \subset [g] \cap H} \#[h]' = \#([g] \cap H).$$

The lemma follows. ◇

11.3.4 Corollary. *If H_1 and H_2 are Gassmann equivalent subgroups of G, then*

$$\operatorname{tr}(A \mid V^{H_1}) = \operatorname{tr}(A \mid V^{H_2}).$$ ◇

Let again N be a complete Riemannian manifold and let Δ denote the Laplacian on N. The manifold need not be compact. For any real number λ we have the eigenspace $E_\lambda(N)$ consisting of all smooth L^2-functions u on N satisfying the differential equation $\Delta u = \lambda u$. We do not consider the question for which λ the dimension of $E_\lambda(N)$ is positive and whether there are any positive dimensional eigenspaces at all; we also may have $\dim E_\lambda(N) = \infty$. Sunada's theorem for the eigenfunctions is as follows.

11.3.5 Theorem. *Let M_1 and M_2 be complete Riemannian manifolds as in (11.3.1). Then M_1 and M_2 are isospectral in the following sense. For any $\lambda \in \mathbf{R} - \{0\}$ we have $\dim E_\lambda(M_1) = \dim E_\lambda(M_2)$.*

Proof. Let E be any finite dimensional subspace of $E_\lambda(M_1)$. Let $u_1, \ldots, u_k$ be a basis of E and let $v_1, \ldots, v_k$ be lifts of $u_1, \ldots, u_k$ on M. The images gv_κ, $g \in G$, $\kappa = 1, \ldots, k$, span a finite dimensional G-invariant subspace $V \subset L^2(M)$, and each $v \in V$ is an eigenfunction of the Laplacian with eigenvalue λ. Now let A be the identity operator on V. Then $\operatorname{tr}(A \mid V^{H_i}) = \dim V^{H_i}$, $i = 1, 2$, and by Corollary 11.3.4, $\dim V^{H_1} = \dim V^{H_2}$, where $\dim V^{H_1} \geq k$. Since the natural projection of V^{H_2} into $E_\lambda(M_2)$ is injective this proves that $\dim E_\lambda(M_2) \geq k$. Hence, $\dim E_\lambda(M_2) \geq \dim E_\lambda(M_1)$. Similarly we have $\dim E_\lambda(M_1) \geq \dim E_\lambda(M_2)$. ◇

11.3.6 Remark. A glance at the preceding proof shows that Theorem 11.3.5 remains valid if, in contrast to the above, $E_\lambda(N)$ is defined as the space of all C^2-functions u on N (not necessarily in L^2) satisfying $\Delta u = \lambda u$. For a more general version of Theorem 11.3.5 we refer to Bérard [3].

11.4 Cayley Graphs

We now describe ways to obtain isospectral examples out of almost conjugate groups. The surfaces in this section need not have constant curvature, the constructions which follow can also be carried out in higher dimensions. For ease of exposition, however, we shall not always consider the most general cases.

Let G be a finite group with subgroups H_1 and H_2. In order to obtain a surface on which G acts by isometries we may proceed in an algebraic or in a combinatorial way. The algebraic approach is more common in differential geometry, and we first show how it leads to the combinatorial construction in a natural way.

We begin with a compact orientable surface M_0 whose fundamental group $\pi_1(M_0)$ admits a *surjective* homomorphism

$$\omega : \pi_1(M_0) \to G.$$

If ω is given, then $\pi_1(M_0)$ has the normal subgroup $\omega^{-1}(\{id\})$, and there is a corresponding compact Riemannian covering surface M on which G acts as the group of covering transformations so that $M_0 = G \backslash M$. Restricting this action to H_1 and H_2 we obtain the quotients

$$M_1 = H_1 \backslash M, \quad M_2 = H_2 \backslash M.$$

If H_1 and H_2 are almost conjugate, then, by Sunada's theorem, M_1 and M_2 are isospectral. At the end of this section we shall describe ways of obtaining non-isometric M_1 and M_2.

A simple way to obtain a surjective homomorphism $\omega : \pi_1(M_0) \to G$ is as follows. Choose any set of generators $A_1, \ldots, A_n$ of G, all different from the identity. Let M_0 be any compact orientable surface of genus $h \geq n$. The fundamental group of M_0 has canonical generators $a_1, b_1, \ldots, a_h, b_h$ with the single defining relation

$$\prod_{v=1}^{h} a_v b_v a_v^{-1} b_v^{-1} = id$$

(cf. Section 6.7). Set $\omega(a_v) = A_v$, $v = 1, \ldots, n$, and $\omega(a_{n+1}) = \ldots = \omega(a_h) =$

$\omega(b_1) = \ldots = \omega(b_h) = id$. Then ω extends to a homomorphism of $\pi_1(M_0)$ onto G.

We now show how graphs come in. (The description is only meant to serve as a motivation for Definition 11.4.1.) For simplicity we assume that M_0 is a Riemann surface. We also assume that $A_i \neq A_j^{\pm 1}$ for $i \neq j$.

Let $\alpha_1, \ldots, \alpha_n$ be simple closed geodesics in the free homotopy classes of the curves $b_1, \ldots, b_n$ (not $a_1, \ldots, a_n$). Cut M_0 open along $\alpha_1, \ldots, \alpha_n$. Since $a_1, \ldots, b_h$ are canonical generators, the resulting bordered surface $\mathcal{B}$ is connected. We lift $\mathcal{B}$ into the covering surface M in order to obtain a fundamental domain $\tilde{\mathcal{B}}$ for the action of G such that the projection $M \to M_0 = G \backslash M$ maps $\tilde{\mathcal{B}}$ isometrically onto $\mathcal{B}$. Hence, M is obtained by pasting together $\#G$ copies of $\mathcal{B}$ which are permuted by G. If $g, g' \in G$ satisfy $g' = gA_i$, then in M, $g'(\tilde{\mathcal{B}})$ is pasted to $g(\tilde{\mathcal{B}})$ along the geodesic $g(\tilde{\alpha}_i)$, where $\tilde{\alpha}_i$ is one of the two boundary geodesics of $\tilde{\mathcal{B}}$ which are lifts of α_i. Interpreting the different domains $g(\tilde{\mathcal{B}})$, $g \in G$, as vertices of a graph whose edges are the lifts of $\alpha_1, \ldots, \alpha_h$ in M, we arrive at the following definition.

11.4.1 Definition. Let G be a finite group with generators $A_1, \ldots, A_n$ which are all different from the identity (but not necessarily pairwise distinct). A graph $\mathcal{G} = \mathcal{G}[A_1, \ldots, A_n]$ is called the *Cayley graph* of G with respect to the generators $A_1, \ldots, A_n$ if there exists a one-to-one mapping j from G onto the vertex set of $\mathcal{G}$ such that for any $g, g' \in G$ the following holds. If u is the number of generators A in the sequence $A_1, \ldots, A_n$ satisfying $gA = g'$, and if v is the number of generators A satisfying $g'A = g$, then the vertices $j(g)$ and $j(g')$ are connected by exactly $u + v$ edges.

As an introduction to Cayley graphs we refer to the last chapter of Bollobás [1]. In the following, however, it will be sufficient to know the definition.

If we want to draw pictures, we represent each element g of G by a dot named g with $2n$ half-edges $a_1, \bar{a}_1, \ldots, a_n, \bar{a}_n$. Then, for each ordered pair $(g, g') \in G \times G$ and for $i = 1, \ldots, n$ we connect half-edge a_i of dot g with half-edge $\bar{a}_i$ of dot g' if and only if $gA_i = g'$.

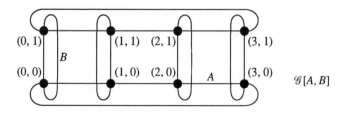

Figure 11.4.1

11.4.2 Example. Let G be the product $G = \mathbf{Z}_4 \times \mathbf{Z}_2$ with generators $A = (1, 0)$ and $B = (0, 1)$. The Cayley graph $\mathcal{G}[A, B]$ is shown in Fig. 11.4.1. The horizontal edges correspond to multiplication from the right by A, the vertical edges correspond to multiplication from the right by B.

Graphs may be enhanced by adding attributes. For the present purposes the attributes will be *color* and *orientation*. If e is an edge connecting the distinct vertices P and Q we may orient e by saying that *e goes from P to Q* or *from Q to P*. Formally, an oriented edge may be defined as the ordered triplet (P, e, Q). In the figures, the orientation is indicated by an arrow pointing from P to Q. We do not orient edges which connect a vertex with itself. If $\mathcal{G}[A_1, \ldots, A_n]$ is the Cayley graph of G with respect to the generators $A_1, \ldots, A_n$, the edge which connects $j(g)$ with $j(gA_i)$ will be oriented from $j(g)$ to $j(gA_i)$, $g \in G$, $i = 1, \ldots, n$. A graph is called *oriented* if some or all of its edges are oriented.

Color is obtained by grouping edges into equivalence classes. Edges of the same class will be said *to have the same color* or *to be of the same type*. In a Cayley graph the edges which correspond to right multiplication by generator A_i will be said to be *of type A_i*. In Fig. 11.4.1 the horizontal edges are of type A, the vertical edges are of type B.

When are two graphs $\mathcal{A}$ and $\mathcal{B}$ "the same"? We shall say that $\phi : \mathcal{A} \to \mathcal{B}$ is an *isomorphism* if ϕ is a one-to-one mapping from the vertex set of $\mathcal{A}$ onto the vertex set of $\mathcal{B}$ such that whenever vertices P and Q of $\mathcal{A}$ are connected by exactly k edges, then $\phi(P)$ and $\phi(Q)$ are connected by exactly k edges. If $\mathcal{A}$ and $\mathcal{B}$ have color and orientation, an isomorphism $\phi : \mathcal{A} \to \mathcal{B}$ need not preserve this additional structure. But if it does, we shall say that ϕ is a *strong isomorphism*. Hence, for a strong isomorphism the number of edges of a given type from P to Q is always equal to the number of edges of the same type from $\phi(P)$ to $\phi(Q)$. Analogous properties are requested when only part of the edges are colored and/or oriented. Graphs $\mathcal{A}$ and $\mathcal{B}$ are *isomorphic* (respectively, *strongly isomorphic*) if there exists an isomorphism (respectively, strong isomorphism) from $\mathcal{A}$ to $\mathcal{B}$.

Let again $\mathcal{G} = \mathcal{G}[A_1, \ldots, A_n]$ be the Cayley graph of a finite group G. Each $h \in G$ acts on $\mathcal{G}$ by left multiplication: $g' = gA_i$, if and only if $hg' = hgA_i$. It follows that $G : \mathcal{G} \to \mathcal{G}$ acts by strong isomorphisms. In fact, G is the full group of strong isomorphisms from $\mathcal{G}$ onto $\mathcal{G}$. The example of Fig. 11.4.1 shows that there may be additional non-strong isomorphisms.

For each subgroup $H \subset G$ we define the *quotient graph* $H \backslash \mathcal{G}$ as follows. Each coset Hg, $g \in G$ is a vertex with emanating half-edges $a_1, \bar{a}_1, \ldots, a_n, \bar{a}_n$. For any ordered pair (Hg, Hg') half-edge a_i of vertex Hg and half-edge $\bar{a}_i$ of vertex Hg' together form an edge if and only if $g' = hgA_i$ for some $h \in H$.

If they form an edge, then the edge is said to be *of type* A_i and the orientation is *from Hg to* HgA_i, provided Hg and HgA_i are distinct.

11.4.3 Example. We consider again the graph of Fig. 11.4.1. Let

$$H_1 = \{(0, 0); (1, 0); (2, 0); (3, 0)\} \text{ and } H_2 = \{(0, 0); (0, 1)\}.$$

The quotients $H_1\backslash \mathcal{G}$ and $H_2\backslash \mathcal{G}$ are shown in Fig. 11.4.2. Color and orientation are not indicated in the figure.

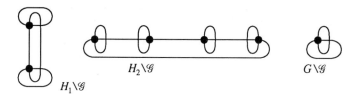

Figure 11.4.2

11.4.4 Theorem. *Let* H_1 *and* H_2 *be almost conjugate subgroups of G and let* $\mathcal{G}$ *be a Cayley graph of G. Then* $H_1\backslash \mathcal{G}$ *and* $H_2\backslash \mathcal{G}$ *are strongly isomorphic if and only if* H_1 *and* H_2 *are conjugate.*

Proof. Clearly, conjugate subgroups have strongly isomorphic quotients. Now assume that $H_1\backslash \mathcal{G}$ and $H_2\backslash \mathcal{G}$ are strongly isomorphic. Then there exists a bijective mapping ϕ from the cosets of H_1 to the cosets of H_2 such that whenever $\phi(H_1 g) = H_2 g'$ and whenever A is one of the generators, then we also have $\phi(H_1 g A) = H_2 g' A$. By induction on the generators we find $\phi(H_1 g f) = H_2 g' f$ for all $f \in G$. Now suppose $\phi(H_1) = H_2 g_0$. For $h_1 \in H_1$ we then have

$$H_2 g_0 = \phi(H_1) = \phi(H_1 h_1) = H_2 g_0 h_1,$$

that is, $g_0 h_1 g_0^{-1} \in H_2$ for all $h_1 \in H_1$. By the symmetry of the statement with respect to H_1 and H_2 this proves the theorem. ◊

For later applications we work out three examples.

11.4.5 Example. Let $G = \mathbb{Z}_8^* \ltimes \mathbb{Z}_8$ as in Example 11.2.2 with the almost conjugate subgroups

$$H_1 = \{(1, 0); (3, 0); (5, 0); (7, 0)\}$$
$$H_2 = \{(1, 0); (3, 4); (5, 4); (7, 0)\}.$$

The following are generators of G.

$$A = (5, 0), \quad B = (3, 0), \quad C = (1, 1).$$

For $k = 0, \ldots, 7$ the elements $C^k = (1, k)$ are pairwise non-equivalent mod H_i, $i = 1, 2$. We may therefore represent the cosets in the form $H_i C^0, \ldots, H_i C^7$. For each k there are unique exponents $a_i(k)$ and $b_i(k)$ such that

$$C^k A \in H_i C^{a_i(k)}, \quad C^k B \in H_i C^{b_i(k)}, \quad i = 1, 2.$$

The exponents are given by the following table.

k	0	1	2	3	4	5	6	7
$a_1(k)$	0	5	2	7	4	1	6	3
$b_1(k)$	0	3	6	1	4	7	2	5
$a_2(k)$	4	1	6	3	0	5	2	7
$b_2(k)$	4	7	2	5	0	3	6	1

The corresponding graphs are shown in the next figure. The horizontal edges are of type C, the dashed edges are of type B and the remaining edges are of type A. Dot k in $H_i \backslash \mathcal{G}$ represents $H_i C^k$, $i = 1, 2$; $k = 0, \ldots, 7$. The graphs are not isomorphic.

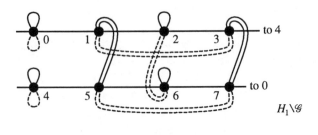

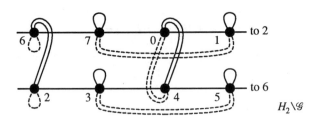

Figure 11.4.3

11.4.6 Example. We let G, H_1 and H_2 be as in Example 11.4.5, but now with generators

$$A = (5, 0), \quad B = (7, 5), \quad C = (1, 1).$$

The corresponding table is the following.

k	0	1	2	3	4	5	6	7
$a_1(k)$	0	5	2	7	4	1	6	3
$b_1(k)$	3	2	1	0	7	6	5	4
$a_2(k)$	4	1	6	3	0	5	2	7
$b_2(k)$	3	2	1	0	7	6	5	4

The graphs are shown in Fig. 11.4.4. They are isomorphic but not strongly isomorphic.

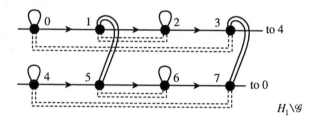

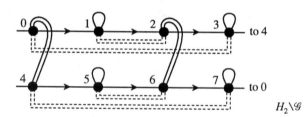

Figure 11.4.4

11.4.7 Example. We consider the group SL(3, 2) with the same subgroups H_1 and H_2 as in Example 11.2.5. By Lemma 11.2.6, the following are generators

$$A = \begin{pmatrix} 0 & 1 & 1 \\ 0 & 1 & 0 \\ 1 & 0 & 0 \end{pmatrix}, \quad B = \begin{pmatrix} 1 & 0 & 0 \\ 0 & 0 & 1 \\ 0 & 1 & 1 \end{pmatrix}.$$

The commutator $C = ABA^{-1}B^{-1}$ has order 7. As before we write cosets in the form H_iC^k, $i = 1, 2$; $k = 0, \ldots, 6$, although C is not in the list of generators. Note that $C^k \notin H_i$ unless $k \equiv 0 \pmod 7$. The cosets are thus pairwise distinct. For each k there are unique exponents $\alpha_i(k)$ and $\beta_i(k)$ such that

$C^k A \in H_i C^{\alpha_i(k)}$ and $C^k B \in H_i C^{\beta_i(k)}$, $i = 1, 2$; $k = 0, \ldots, 6$. These exponents are given by the following table.

k	0	1	2	3	4	5	6
$\alpha_1(k)$	3	1	6	0	5	2	4
$\beta_1(k)$	0	2	5	6	3	1	4
$\alpha_2(k)$	1	6	5	0	4	2	3
$\beta_2(k)$	0	4	5	1	3	6	2

The graphs $H_1 \backslash \mathcal{G}$ and $H_2 \backslash \mathcal{G}$ are given in Fig. 11.4.5. Dot k stands for $H_i C^k$. Dashed lines represent edges of type B, the remaining ones are of type A. The graphs are not isomorphic.

The example will be used in Section 12.4; a slightly different form of it will be described in Section 12.5.

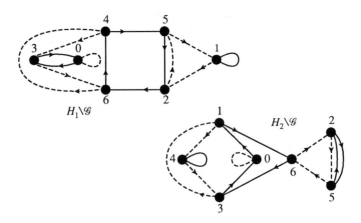

Figure 11.4.5

Theorem 11.4.4 provides a simple construction of non-isometric isospectral pairs. For $i = 1, 2$, we replace each vertex of $H_i \backslash \mathcal{G}$ with emanating half-edges $a_1, \bar{a}_1, \ldots, a_n, \bar{a}_n$ by a copy of a fixed building block $\mathcal{B}$ with boundary geodesics $\alpha_1, \bar{\alpha}_1, \ldots, \alpha_n, \bar{\alpha}_n$ and paste the blocks together according to $H_i \backslash \mathcal{G}$.

In order to paste the boundaries consistently, we start with a closed surface M_0 of genus $h \geq n$, and cut it open into a connected surface $\mathcal{B}$ along n simple closed geodesics, where each such geodesic is marked by the initial point of the given parametrization. If we paste $\#G$ blocks according to $\mathcal{G}$ in such a way that that the initial points of the parametrization are again pasted together, then we obtain a surface M with an isometric action of G and such that $M_0 = G \backslash M$. If we paste $\#G / \#H_1$ copies according to $H_1 \backslash \mathcal{G}$ and $H_2 \backslash \mathcal{G}$ in the same

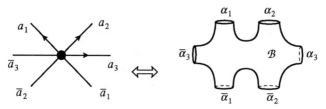

Figure 11.4.6

manner, we obtain the surfaces $M_1 = H_1\backslash M$ and $M_2 = H_2\backslash M$. By Sunadas's theorem, M_1 and M_2 are isospectral.

How can we obtain non-isometric M_1 and M_2? The answer is given by Theorem 11.4.4: it suffices to provide the building block $\mathcal{B}$ with such a metric that the colored oriented graphs $H_1\backslash \mathcal{G}$ and $H_2\backslash \mathcal{G}$ can be reconstructed from the *intrinsic* geometry of M_1 and M_2. If we admit variable curvature, this can be achieved by making the metric sufficiently bumpy (see Sunada [3] and Buser [9]). In the case of constant curvature we use the collar theorem. The idea can be seen in the following example (for a more systematic construction see Chapter 12).

Paste together $2(n-1)$ Y-pieces as shown in Fig. 11.4.7 to obtain a Riemann surface $\mathcal{B}$ of signature $(0, 2n)$ with boundary geodesics $\bar{\alpha}_1, \alpha_2, \bar{\alpha}_2, \ldots, \bar{\alpha}_n$, α_n, α_1 and the intrinsic geodesics $\eta_1, \ldots, \eta_{2n-3}$ such that $\ell(\alpha_j) = \ell(\bar{\alpha}_j)$, $j = 1, \ldots, n$ and such that

$$\ell(\alpha_1) < \ldots < \ell(\alpha_n) < \ell(\eta_1) = \ldots = \ell(\eta_{2n-3}) = 1.$$

In $\mathcal{B}$ and in M_1 and M_2 we let all twist parameters be zero. By the collar theorem, the only geodesics on M_1 and M_2 which have length less or equal 1 are those corresponding to $\alpha_1, \ldots, \alpha_n, \eta_2, \ldots, \eta_n$. Hence, we can reconstruct the building blocks by cutting M_1 and M_2 open along the closed geodesics of length less or equal 1. Color and orientation can be obtained by looking at the lengths of the boundary geodesics and at the way in which the boundary geodesics are arranged on $\mathcal{B}$.

Figure 11.4.7

11.5 Transplantation of Eigenfunctions

Some of Sunada's examples are particularly simple so that the isospectrality can be seen directly. In this and in the next section we deal with such an example. Although the example is modeled over $\mathbf{Z}_8^* \ltimes \mathbf{Z}_8$ with generators and quotient graphs as in Example 11.4.6, we shall not refer to this nor to any of the preceding sections. Instead, we give a description ab ovo, so that the example will be understandable without background in almost conjugate groups.

In order to prove the isospectrality we use a technique which we call *transplantation* (Buser [8]). Bérard [3] has investigated transplantation more generally, giving deeper insight into the combinatorial nature of Sunada's examples.

Let $\mathcal{B}$ be a compact Riemann surface of signature $(0, 5)$ with boundary

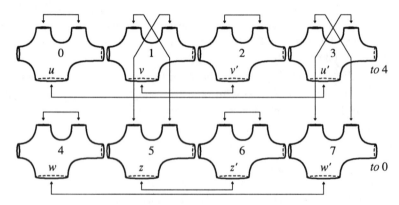

Figure 11.5.1

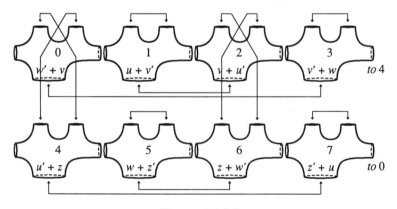

Figure 11.5.2

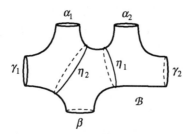

Figure 11.5.3

geodesics α_1, α_2, β, γ_1 and γ_2 as shown in Fig. 11.5.3. $\mathcal{B}$ consists of three Y-pieces which have been pasted together along the geodesics η_1 and η_2 with twist parameter $\frac{1}{4}$. We assume that

$$\ell(\alpha_1) = \ell(\alpha_2) < \ell(\beta) < \ell(\gamma_1) = \ell(\gamma_2) < \ell(\eta_1) < \ell(\eta_2) \leq 1.$$

By the collar theorem any other simple closed geodesic on $\mathcal{B}$ is longer than 1. It follows that the only isometry $\mathcal{B} \to \mathcal{B}$ is the identity.

Now we paste together 8 copies (building blocks) $\mathcal{B}_0, \ldots, \mathcal{B}_7$ of $\mathcal{B}$ with twist parameter 0 according to Fig. 11.5.1 and 11.5.2 to obtain the compact Riemann surfaces M_1 and M_2 respectively. Building block $\mathcal{B}_k$ on M_i will be denoted by $\mathcal{B}_k(M_i)$. Observe the pasting along the copies of γ_1 and γ_2: $\mathcal{B}_3(M_i)$ is pasted to $\mathcal{B}_4(M_i)$ and $\mathcal{B}_7(M_i)$ to $\mathcal{B}_0(M_i)$.

The closed geodesics on M_1 and M_2 which correspond to α_1, α_2, β, γ_1 and γ_2 will be called *the geodesics of type α, β and γ* respectively.

M_1 and M_2 are non-isometric.

Proof. By the collar theorem, any simple closed geodesic on M_1 and M_2 different from the copies of α_1, α_2, β, γ_1, γ_2, η_1, η_2 is longer than 1. Hence, any isometry $\phi : M_1 \to M_2$ sends building blocks to building blocks. More precisely, there exists $m \in \{0, \ldots, 7\}$ such that $\phi(\mathcal{B}_k(M_1)) = \mathcal{B}_{k+m}(M_2)$, $k = 0, \ldots, 7$ (indices mod 8). The way in which the boundary geodesics α_1 and α_2 are pasted together implies that $m \in \{1, 3, 5, 7\}$. Since $\mathcal{B}_1(M_1)$ and $\mathcal{B}_2(M_1)$ are pasted together along copies of β, it follows that $\phi(\mathcal{B}_1(M_1))$ and $\phi(\mathcal{B}_2(M_1))$ are pasted together along copies of β. This is only possible if $m \in \{0, 4\}$. Hence, ϕ does not exist. ◇

M_1 and M_2 have the same spectrum of the Laplacian.

Proof. For $\lambda \in \mathbf{R}$ we denote by $E_i(\lambda)$ the eigenspace of the Laplacian on M_i corresponding to λ (dim $E_i(\lambda)$ is non-zero if and only if λ is an eigenvalue). We shall prove that dim $E_1(\lambda) =$ dim $E_2(\lambda)$ for any λ. For this we shall "transplant" eigenfunctions from M_1 onto M_2 and vice-versa.

Let $J : M_i \to M_i$ be the isometry defined by the property that $J(\mathcal{B}_k(M_i)) = \mathcal{B}_{k+4}(M_i)$. In Fig. 11.5.1 and Fig. 11.5.2, J interchanges the upper and the lower row. Let

$$E_i^+(\lambda) = \{f \in E_i(\lambda) \mid f \circ J = f\},$$
$$\qquad\qquad i = 1, 2,$$
$$E_i^-(\lambda) = \{f \in E_i(\lambda) \mid f \circ J = -f\}.$$

$E_i(\lambda)$ is the orthogonal sum $E_i(\lambda) = E_i^+(\lambda) \oplus E_i^-(\lambda)$, and it suffices to prove that

$$\dim E_1^+(\lambda) = \dim E_2^+(\lambda), \quad \dim E_1^-(\lambda) = \dim E_2^-(\lambda).$$

For the transplantation we let

$$M_i' = \text{int } \mathcal{B}_0(M_i) \cup \ldots \cup \text{int } \mathcal{B}_7(M_i),$$

for $i = 1, 2$, where int denotes the interior. For $m = 0, \ldots, 7$ we let $\varphi_m : M_1' \to M_2'$ be the uniquely determined isometry which sends any int $\mathcal{B}_k(M_1)$ onto int $\mathcal{B}_{k+m}(M_2)$, (indices mod 8). (Recall that $\mathcal{B}$ has trivial isometry group.) For any smooth function f on M_1 we define

$$f^{T, m} = f \circ \varphi_m^{-1}.$$

$f^{T, m}$ is the "transplanted function on M_2' shifted forward by m blocks". In a similar way we transplant functions from M_2 to M_1 setting

$$g_{T, m} = g \circ \varphi_m$$

for any smooth function g on M_2. Note that $(f^{T, m})_{T, m} = f$ on M_1' and $(g_{T, m})^{T, m} = g$ on M_2'. We are interested in those transplantations or linear combinations thereof which can be extended smoothly onto M_1 and M_2. The following is clear from the pasting scheme for any smooth function f.

(1) $f^{T, m}$ can be extended smoothly along any geodesic of type γ.

(2) If m is odd, then $f^{T, m}$ can be extended smoothly along any geodesic of type α.

(3) If $f \circ J = f$, then $f^{T, 0}$ can be extended smoothly onto M_2.

The analogous statements hold for the transplantations from M_2 to M_1. The next statement is more involved.

(4) $f^{T, 1} + f^{T, 7}$ can be extended smoothly onto M_2.

Proof. It remains to prove that $f^{T,1} + f^{T,7}$ can be extended smoothly along the geodesics of type β. We denote by u, v, w and z the restrictions of f to the collars around the geodesics of type β in M_1 as shown in Fig. 11.5.1. Fig. 11.5.2 already shows that the sums of the transplanted functions "fit" along the boundaries of type β. Let us, nevertheless, give the details, for example for the collar around β in blocks $\mathcal{B}_1(M_2)$ and $\mathcal{B}_2(M_2)$.

We denote by $\mathcal{C}_k(M_i)$ the half-collar around β of $\mathcal{B}_k(M_i)$ and consider the natural isometry $\psi_{k\ell} : \mathcal{C}_k(M_1) \to \mathcal{C}_\ell(M_2)$ which preserves the parametrization of β. Then

$$f^{T,1} + f^{T,7} = f \circ \psi_{01}^{-1} + f \circ \psi_{21}^{-1} \quad \text{on} \quad \mathcal{C}_1(M_2) - \beta$$

and

$$f^{T,1} + f^{T,7} = f \circ \psi_{12}^{-1} + f \circ \psi_{32}^{-1} \quad \text{on} \quad \mathcal{C}_2(M_2) - \beta.$$

Now ψ_{01} and ψ_{32} are the restrictions of an isometry $\mu : \mathcal{C}_0(M_1) \cup \mathcal{C}_3(M_1) \to \mathcal{C}_1(M_2) \cup \mathcal{C}_2(M_2)$, and the mappings ψ_{21} and ψ_{12} are the restriction of an isometry $\eta : \mathcal{C}_1(M_1) \cup \mathcal{C}_2(M_1) \to \mathcal{C}_1(M_2) \cup \mathcal{C}_2(M_2)$. On $\mathcal{C}_1(M_2) \cup \mathcal{C}_2(M_2)$ we get

$$f^{T,1} + f^{T,7} = f \circ \mu^{-1} + f \circ \eta^{-1}.$$

Here the right-hand side is smooth in a neighborhood of β. ◊

The preceding argument also shows that if f is an eigenfunction of the Laplacian with eigenvalue λ, then the same holds for $f^{T,1} + f^{T,7}$. Note also that if $f \circ J = -f$, then the same holds for $f^{T,1} + f^{T,7}$.

By (3) we have a linear injection from $E_1^+(\lambda)$ to $E_2^+(\lambda)$ given by $f \mapsto f^{T,0}$. By (4) and by the above remarks we have a linear mapping $L : E_1^-(\lambda) \to E_2^-(\lambda)$ given by $L(f) = f^{T,1} + f^{T,7}$. If $L(f) = 0$, then $f \circ \varphi_0^{-1} \circ \varphi_2 = -f$ and it follows that $f \circ J = f$. Since we also have $f \circ J = -f$, we conclude that $f = 0$. Hence, L is injective. This shows that $\dim E_1(\lambda) \leq \dim E_2(\lambda)$, and similarly we prove $\dim E_2(\lambda) \leq \dim E_1(\lambda)$. ◊

We remark that a similar transplantation works on the examples modelled over the graph of Example 11.4.5. Here one needs two boundary geodesics β_1, β_2 for $\mathcal{B}$, and $L : E_1^-(\lambda) \to E_2^-(\lambda)$ is given by $L(f) = f^{T,7} - f^{T,5}$.

11.6 Transplantation of Closed Geodesics

In the example of the preceding section, we may also prove the length isospectrality in a direct manner, again by some sort of transplantation. Here the transplantation is as follows.

First we observe that there is a length-preserving one-to-one correspondence between the boundary geodesics of the building blocks in M_1 and M_2. We may therefore restrict ourselves to those closed geodesics which intersect the boundaries of the building blocks only finitely many times.

Let δ on M_1 be such a geodesic. Parametrize it on the interval $[0, 1]$ such that $p = \delta(0) = \delta(1)$ is an interior point of one of the building blocks, say $\mathcal{B}_k(M_1)$. If δ is entirely contained in $\mathcal{B}_k(M_1)$ we set $\delta = \delta_1$. Otherwise we decompose δ into a sequence of subarcs: $\delta = \delta_1 \delta_2 \ldots \delta_N$ such that δ_1 runs from p to the boundary of $\mathcal{B}_k(M_1)$, each δ_i, $i = 2, \ldots, N-1$ is contained in one of the building blocks with the endpoints on the boundary, and δ_N runs from the boundary of $\mathcal{B}_k(M_1)$ back to p.

Using criterion (11.6.1) below we select the initial block $\mathcal{B}_{k*}(M_2)$ and define

$$\delta_1^* = \varphi_{kk*} \circ \delta_1,$$

where $\varphi_{kk*} : \text{int } \mathcal{B}_k(M_1) \to \text{int } \mathcal{B}_{k*}(M_2)$ is the natural identification. To complete the definition we extend δ_1^* continuously onto the boundary of the block $\mathcal{B}_{k*}(M_2)$.

Assume that δ_2 is contained in $\mathcal{B}_\ell(M_1)$. We let $\mathcal{B}_{\ell*}(M_2)$ be the block such that δ_1^* and $\delta_2^* := \varphi_{\ell\ell*} \circ \delta_2$ meet smoothly at the endpoint of δ_1^*. This is illustrated in Fig. 11.6.1.

Then, if δ_3 lies on $\mathcal{B}_m(M_1)$ we let $\mathcal{B}_{m*}(M_2)$ be the block such that δ_2^* and $\delta_3^* := \varphi_{mm*} \circ \delta_3$ meet smoothly at the endpoint of δ_2^*, and so on. The geodesic

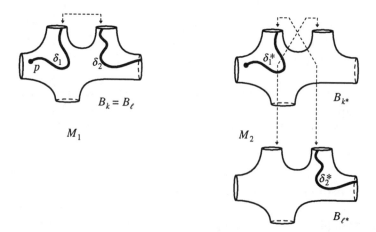

Figure 11.6.1

$$\delta^* := \delta_1^* \ldots \delta_N^*$$

has the same length as δ, and we have to prove that δ^* is again a smooth closed geodesic. Note that the transplantation can also be applied to non-closed geodesics. The criterion for the choice of the initial block on M_2 is the following.

(11.6.1) $\quad k^* = \begin{cases} k & \text{if } \#\alpha \text{ is even,} \\ k+1 & \text{if } \#\alpha \text{ is odd and } \#\beta \text{ is even,} \\ k+2 & \text{if } \#\alpha \text{ and } \#\beta \text{ are odd.} \end{cases}$

In this definition, $\#\alpha$ and $\#\beta$ denote the number of intersection points of δ with the geodesics of type α and type β respectively (cf. the preceding section). If we draw δ in Fig. 11.5.1, then δ "jumps" from one building block to another whenever it crosses the boundary of a building block. Jumps due to a crossing of a boundary geodesic of type α (or β) will be called α-jumps (respectively, β-jumps). Thus, $\#\alpha$ is the number of α-jumps and $\#\beta$ is the number of β-jumps. Fig. 11.6.1 shows an α-jump.

In the next lemma, δ is either a smooth closed geodesic or a geodesic arc of finite length.

11.6.2 Lemma. *δ^* is a smooth closed geodesic if and only if δ is a smooth closed geodesic.*

Proof. We think of two observers running simultaneously along δ and δ^* with the same speed. For $t \in [0, 1]$ we let $\mathcal{B}_{\ell(t)}(M_1)$ be the block occupied by the first observer at time t and $\mathcal{B}_{m(t)}(M_2)$ the block occupied by the second observer. The difference $m(t) - \ell(t)$ (mod 8) will be called the *shift* between the two observers. In order to prove the lemma, it suffices to show that the final shift (at time $t = 1$) coincides with the initial shift. Note that the shift does not change when δ crosses a geodesic of type γ.

Case 1. $\#\alpha$ is even. The initial shift is 0. We claim that at any time the shift is either 0 or 4. In fact, if the shift is 0 or 4, then any β-jump leaves the shift invariant and any α-jump changes the shift by 4 (check with Fig. 11.5.1 and 11.5.2). Since $\#\alpha$ is even, the final shift is again 0.

Case 2. $\#\alpha$ is odd and $\#\beta$ is even. The initial shift is 1. We claim that at any time the shift is either 1 or -1 (mod 8). In fact, if the shift is 1 or -1, then any α-jump leaves the shift invariant and any β-jump changes it from 1 to -1 or from -1 to 1. Since the number of β-jumps is even, the final shift is 1.

Case 3. $\#\alpha$ and $\#\beta$ are odd. The initial shift is 2. We claim that at any time the shift is either 2 or -2 (mod 8). In fact, if the shift is 2 or -2, then any α-jump and any β-jump changes it by 4. Since $\#\alpha + \#\beta$ is an even number, the final shift is 2. ◇

Obviously the transplantation can be reversed and Lemma 11.6.2 implies that the reversed transplantation maps closed geodesics to closed geodesics. Hence, we have a one-to-one correspondence between the closed geodesics of a given length on M_1 and M_2.

We remark that the transplantation with initiation (11.6.1) works also for surfaces modelled over the graph of Example 11.4.5 except that the proof of Lemma 11.6.2 needs a minor modification in case 2.

Chapter 12

Examples of Isospectral Riemann Surfaces

In Section 11.4 we glued together building blocks according to the pattern of the graphs $H_1\backslash \mathcal{G}, H_2\backslash \mathcal{G}$, where $\mathcal{G}$ is the Cayley graph of a finite group G and H_1, H_2 are almost conjugate subgroups of G. This technique will be applied to geodesic polygons of the hyperbolic plane to construct a large number of isospectral Riemann surfaces. The main result (Theorem 12.5.1) is the inequality dim $\mathcal{V}_g > 0$ for all $g \geq 4$, where $\mathcal{V}_g$ is the subvariety in Teichmüller space consisting of those Riemann surfaces which are not determined by their spectrum.

In Section 12.1 we describe the pasting of polygons. Sections 12.2 - 12.5 give the examples. In Section 12.6 we show that for infinitely many g there are examples of $g^{1/3}$ pairwise isospectral non-isometric Riemann surfaces of genus g. (An upper bound for this number will be given in Chapter 13.) In the final section we give some general sufficient conditions under which Sunada's construction yields non-isometric pairs of Riemann surfaces.

12.1 Cayley Graphs and Hyperbolic Polygons

Let (G, H_1, H_2) be a triplet as in Chapter 11, where G is a finite group and H_1 and H_2 are almost conjugate subgroups of G. We choose a set of generators $A_1, \ldots, A_n$ of G, and consider the corresponding Cayley graph

$$\mathcal{G} = \mathcal{G}[A_1, \ldots, A_n].$$

Then G operates on $\mathcal{G}$ by isomorphisms and we let

(12.1.1) $\qquad \mathcal{G}_1 = H_1\backslash \mathcal{G}, \quad \mathcal{G}_2 = H_2\backslash \mathcal{G}$

be the quotients with respect to the subgroups H_1 and H_2 of G. Each vertex of $\mathcal{G}_1$ and $\mathcal{G}_2$ has half-edges $a_1, \bar{a}_1, \ldots, a_n, \bar{a}_n$ corresponding to the generators $A_1, \ldots, A_n$, and the graph $\mathcal{G}_i$ is defined in the following way: if $(H_i g, H_i g')$ is an ordered pair of vertices of $\mathcal{G}_i$, $i = 1, 2$, such that $H_i g A_k = H_i g'$, then the half-edge a_k of vertex $H_i g$ and the half-edge $\bar{a}_k$ of vertex $H_i g'$ form together an edge of graph $\mathcal{G}_i$. We say that this edge is *of type A_k*.

Figure 12.1.1

The graph $\mathcal{G}$ is defined in the same way, but with H_i replaced by $\{id\}$. Now consider a geodesic polygon (domain) $\mathcal{D}$ in the hyperbolic plane, with N sides which are grouped into n families. Each family consists of an even number of sides, and will correspond to one of the generators $A_1, \ldots, A_n$ of G. The sides in the k-th family are denoted by

$$\alpha_{k1}, \bar{\alpha}_{k1}, \ldots, \alpha_{km}, \bar{\alpha}_{km},$$

where $m = m(k)$, $k = 1, \ldots, n$.

We parametrize each side on the interval $[0, 1]$ with constant speed and with positive boundary orientation. For each k and each μ, sides $\alpha_{k\mu}$ and $\bar{\alpha}_{k\mu}$ are assumed to have the same length.

In order to obtain the triplet M, M_1, M_2 of compact Riemann surfaces as in Sunada's theorem (cf. (11.3.1)), we let, for each $g \in G$, $\mathcal{D}_g$ be a copy of $\mathcal{D}$ and glue the copies together according to the Cayley graph

$$\mathcal{G} = \mathcal{G}[A_1, \ldots, A_n].$$

The pasting is as follows: if $g, g' \in G$ and if $g A_k = g'$, then for $\mu = 1, \ldots, m(k)$, side $\alpha_{k\mu}$ of block $\mathcal{D}_g$, denoted by $\alpha_{k\mu}[\mathcal{D}_g]$, is glued to side $\bar{\alpha}_{k\mu}[\mathcal{D}_{g'}]$ of block $\mathcal{D}_{g'}$ via the identification

(12.1.2)
$$\alpha_{k\mu}[\mathcal{D}_g](t) = \bar{\alpha}_{k\mu}[\mathcal{D}_{g'}](1 - t), \quad t \in [0, 1],$$

$$\mu = 1, \ldots, m(k), \ k = 1, \ldots, n.$$

The resulting surface M has a smooth hyperbolic structure if and only if the sum of the angles which come together at a common vertex is always equal to 2π.

As in Chapter 11, G acts on M by isometries. This action is the same as the action of G by strong isomorphisms on $\mathcal{G}$. If we restrict the action to the subgroups H_1 and H_2, we obtain the surfaces $M_1 = H_1\backslash M$ and $M_2 = H_2\backslash M$. The following is clear.

12.1.3 Lemma. *$H_1\backslash M$ and $H_2\backslash M$ are isometric to the surfaces obtained by gluing together $\#G/\#H_1$ copies of $\mathcal{D}$ with respect to the graphs $H_1\backslash \mathcal{G}$ and $H_2\backslash \mathcal{G}$.* ◊

It follows from Sunada's theorem that M_1 and M_2 are isospectral.

In the next section we describe conditions for the geometry of $\mathcal{D}$ such that M, M_1 and M_2 are smooth. For the proof of the non-isometry of M_1 and M_2, we shall also work with the surface $M_0 = G\backslash M$ obtained by pasting together the sides of $\mathcal{D}$ with respect to $G\backslash \mathcal{G}$. This surface, in contrast, need not be smooth. We use the following terminology. (We already briefly considered cone-like singularities in the proof of Hurwitz' theorem 6.5.9.)

12.1.4 Definition. Let S be a closed surface obtained by pasting together the sides of hyperbolic polygons $\mathcal{D}_1, \ldots, \mathcal{D}_\ell$ with respect to some pasting scheme. If the vertices $p_1, \ldots, p_k$ of $\mathcal{D}_1, \ldots, \mathcal{D}_\ell$ together yield the point p of S, and if the sum ϑ of the interior angles at $p_1, \ldots, p_k$ is different from 2π, then p is called a *singular point* or *cone-like singularity* of S with the interior angle ϑ, and S is called a *hyperbolic surface with cone-like singularities*. If $\vartheta = 2\pi/k$, where k is an integer, $k \geq 2$, then we say that p has *order* k.

In the examples of Sections 12.3 - 12.5 the non-isometry of M_1 and M_2 will be proved by direct checking. Some general sufficient criteria will follow in Section 12.7.

12.2 Smoothness

We describe an algorithm, running on the vertices of $\mathcal{D}$, which decides the smoothness of M, M_1 and M_2.

Let $p_1, \ldots, p_N$ be the vertices of $\mathcal{D}$. We shall group these vertices into equivalence classes or *cycles* $[p], [p'], \ldots$, and define the conjugacy class $[\mathfrak{g}_p]$ of a corresponding element $\mathfrak{g}_p \in G$. This conjugacy class will only depend on the equivalence class $[p]$ of p.

At each vertex p of $\mathcal{D}$ we have an interior angle ϑ. By abuse of notation, ϑ will also denote the circle sector of radius ε at p, where ε is chosen so small that the various circle sectors of $\mathcal{D}$ do not overlap each other. Since $\mathcal{D}$ lies in the hyperbolic plane, each circle sector may be considered oriented, and has

Figure 12.2.1

therefore a well-defined left-hand side (ℓ) and a right-hand side (r) as shown in Fig. 12.2.1.

We recall that for $k = 1, \ldots, n$, the family of sides $\alpha_{k1}, \bar{\alpha}_{k1}, \ldots, \alpha_{km}, \bar{\alpha}_{km}$, $m = m(k)$, corresponds to the generator A_k of G.

12.2.1 Algorithm. (To determine the cycle $[p]$ of vertex p). Rename p by p_1. Let ϑ_1 be the circle sector of $\mathcal{D}$ at p_1.

Now let $\lambda \geq 1$, and assume that $\vartheta_1, \ldots, \vartheta_\lambda$ with vertices $p_1, \ldots, p_\lambda$ have been determined. If the left-hand side of ϑ_λ is side $\alpha_{k\mu}$ of $\mathcal{D}$, then define

$$g_\lambda := A_k,$$

and let $\vartheta_{\lambda+1}$ with vertex $p_{\lambda+1}$ be the circle sector of $\mathcal{D}$ whose right-hand side is $\bar{\alpha}_{k\mu}$. If the left-hand side of ϑ_λ is side $\bar{\alpha}_{k\mu}$ of $\mathcal{D}$, then define

$$g_\lambda := A_k^{-1},$$

and let $\vartheta_{\lambda+1}$ with vertex $p_{\lambda+1}$ be the circle sector whose right-hand side is $\alpha_{k\mu}$.

If $p_{\lambda+1} \neq p_1$, then continue, otherwise set $\ell := \lambda$ and stop. ◇

Using this algorithm we define

(12.2.2) $\qquad [p] = \{p_1, \ldots, p_\ell\} = $ cycle of p,

(12.2.3) $\qquad \mathfrak{g}_p := g_1 \cdots g_\ell \in G,$

(12.2.4) $\qquad \mathrm{ord}[p] = $ order of $\mathfrak{g}_p$.

Two cycles which differ by cyclic permutation will be considered equal. The number ℓ occurring in (12.2.2) is called the *length of the cycle*. Observe that if we start at a different vertex of the cycle, then the above algorithm leads to the same cycle. Observe also that the conjugacy class $[\mathfrak{g}_p]$ does not depend on the representative p of the cycle because $[g_1 \cdots g_\ell] = [g_2 \cdots g_\ell g_1] = \cdots$. It follows that $\mathrm{ord}[p]$ depends only on the cycle.

12.2.5 Example.

The following example will later be used to construct isospectral pairs of Riemann surfaces of genus 5. Here we shall use the example to illustrate the above algorithm. G is the group $\mathbf{Z}_8^* \ltimes \mathbf{Z}_8$ as in Example 11.2.2 with the generators

$$A_1 = (1, 1), A_2 = (5, 0), A_3 = (5, 4), A_4 = (7, 5).$$

The domain is as in Fig. 12.2.2.

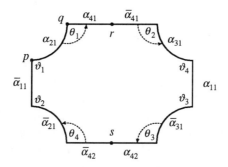

Figure 12.2.2

We have the cycles $[p]$, $[q]$, $[r]$, $[s]$. The reader may check that

$$\begin{aligned}
\mathfrak{g}_p &= A_2 A_1^{-1} A_3^{-1} A_1 &&= (1, 0); &&\text{ord}[p] = 1, \\
\mathfrak{g}_q &= A_4 A_3 A_4 A_2^{-1} &&= (1, 0); &&\text{ord}[q] = 1, \\
\mathfrak{g}_r &= A_4^{-1} &&= (7, 5); &&\text{ord}[r] = 2, \\
\mathfrak{g}_s &= A_4^{-1} &&= (7, 5); &&\text{ord}[s] = 2.
\end{aligned}$$

We return to the general case. Observe that $G \backslash M$ is obtained by gluing together the corresponding sides of *one* building block. The circle sectors belonging to cycle $[p]$ yield, therefore, a circular neighborhood of a point p^* in $G \backslash M$. (Here $G \backslash M$ need not be a smooth surface, and p^* may be a cone-like singularity.)

Now consider a lift $\tilde{p}$ of p^* in M (with respect to the natural projection $M \to G \backslash M$).

M is tessellated by copies $\mathcal{D}_g$ of $\mathcal{D}$, $g \in G$. Furthermore, on M side $\alpha_{k\mu}$ of $\mathcal{D}_g$ coincides with side $\bar{\alpha}_{k\mu}$ of $\mathcal{D}_{g'}$ if and only if $gA_k = g'$ (see (12.1.2)). From this we obtain the following picture, where p is as in (12.2.2).

In $G \backslash M$ the circle sectors $\vartheta_1, \ldots, \vartheta_\ell$ are arranged around vertex p^* in exactly this order and with ϑ_ℓ again adjacent to ϑ_1.

At $\tilde{p}$ in M, lifts $\tilde{\vartheta}_1, \ldots, \tilde{\vartheta}_\ell$ of $\vartheta_1, \ldots, \vartheta_\ell$ are arranged in the same order, except that $\tilde{\vartheta}_\ell$ is the neighbor of a lift $\tilde{\vartheta}_1'$ of ϑ_1 which may be different from $\tilde{\vartheta}_1$. The following holds.

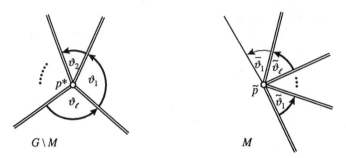

Figure 12.2.3

(12.2.6) *Each ϑ_λ has exactly ord[p] lifts at $\tilde{p}$.*

Proof. Let $\tilde{\vartheta}_1$ belong to block $\mathcal{D}_g$, where $g \in G$. Assume that $\alpha_{k\mu}$ (respectively, $\bar{\alpha}_{k\mu}$) is the left-hand side of ϑ_1. Algorithm 12.2.1 then defines $g_1 := A_k$ (respectively, $g_1 := A_k^{-1}$), and ϑ_2 becomes the circle sector of $\mathcal{D}$ whose right-hand side is $\bar{\alpha}_{k\mu}$ (respectively, $\alpha_{k\mu}$). At the same time the tessellation of M is such that side $\alpha_{k\mu}$ (respectively, $\bar{\alpha}_{k\mu}$) of $\mathcal{D}_g$ coincides with side $\bar{\alpha}_{k\mu}$ of $\mathcal{D}_{g'}$, where $g' = gA_k = gg_1$ (respectively, with side $\alpha_{k\mu}$ of $\mathcal{D}_{g'}$, where $g' = gA_k^{-1} = gg_1$). Hence, $\tilde{\vartheta}_2$ is the copy of ϑ_2 on block $\mathcal{D}_{g'}$, where $g' = gg_1$. In a similar way, $\tilde{\vartheta}_3$ is the copy of ϑ_3 on block $\mathcal{D}_{g''}$, where $g'' = gg_1g_2$, and so on. Finally, $\tilde{\vartheta}_\ell$ is the copy of ϑ_ℓ on $\mathcal{D}_{g^*}$, where $g^* = gg_1 \ldots g_{\ell-1}$, and the next circle sector (which is again a lift of ϑ_1) is the copy of ϑ_1 on $\mathcal{D}_h$, where $h = gg_1 \ldots g_\ell = g\mathfrak{g}_p$. We denote this copy by $\bar{\vartheta}_1$.

The action of G on M is defined in such a way that for $\gamma' = g\mathfrak{g}_p g^{-1}$ we have

$$\gamma'(\mathcal{D}_g) = \mathcal{D}_{g\mathfrak{g}_p} = \mathcal{D}_h,$$

and, moreover, $\bar{\vartheta}_1 = \gamma'(\tilde{\vartheta}_1)$. In particular, γ' fixes $\tilde{p}$. Since the union set $\tilde{\vartheta}_1 \cup \ldots \cup \tilde{\vartheta}_\ell$ is a fundamental domain for the cyclic subgroup $G_{\tilde{p}}$ of G which fixes point $\tilde{p}$ (recall that G has only finitely many fixed points so that $G_{\tilde{p}}$ is indeed a cyclic group), we see that γ' is a generator of $G_{\tilde{p}}$, and therefore $G_{\tilde{p}}$ has order ord(γ') = ord$(\mathfrak{g}_p)$ = ord$[p]$, as claimed. $\diamond$

As an immediate consequence we have that *M is smooth at $\tilde{p}$ if and only if the following condition is satisfied.*

(12.2.7) $\quad\quad\quad (\vartheta_1 + \ldots + \vartheta_\ell)\,\text{ord}[p] = 2\pi.$

For the smoothness of $M_1 = H_1 \backslash M$ (and similarly for M_2), we look at the image $\hat{p}$ of $\tilde{p}$ under the natural projection $M \to H_1 \backslash M$. Here we have the necessary and sufficient condition that $\tilde{p}$ be not a fixed point of H_1 (since we already assume that M is smooth at $\tilde{p}$). Since γ' generates $G_{\tilde{p}}$ (see above),

Ch.12, §2] Smoothness 317

this is equivalent to saying that no power $(\gamma')^\nu$ belongs to H_1 for $\nu = 1, \ldots,$ ord$[p] - 1$. For any image $\sigma(\tilde{p})$, $\sigma \in G$, the isotropy subgroup of G which fixes $\sigma(p)$ is generated by $\sigma\gamma'\sigma^{-1}$. Hence, for M_1 to be smooth at all lifts of vertex $p^* \in G \backslash M$ (recall that $\tilde{p}$ is a lift of p^*) we have the following necessary and sufficient condition.

(12.2.8) $\qquad [(\mathfrak{g}_p)^\nu] \cap H_1 = \varnothing$ for $\nu = 1, \ldots,$ ord$[p] - 1$.

Since H_1 and H_2 are almost conjugate, this condition holds for H_1 if and only if it holds for H_2. Together with (12.2.7) and (12.2.8) we obtain the following smoothness condition.

12.2.9 Proposition. *In the construction of Section 12.1 the triplet M, M_1, M_2 consists of smooth surfaces if and only if, for each cycle $[p] = \{p_1, \ldots, p_\ell\}$ determined by Algorithm 12.2.1, the angles $\vartheta_1, \ldots, \vartheta_\ell$ and the associated conjugacy class $[\mathfrak{g}_p]$ satisfy the following conditions.*

(1) $\qquad (\vartheta_1 + \ldots + \vartheta_\ell) = 2\pi/\text{ord}[p],$

(2) $\qquad [(\mathfrak{g}_p)^\nu] \cap H_1 = \varnothing$ *for* $\nu = 1, \ldots,$ ord$[p] - 1$. $\qquad \diamond$

We compute the genus via the Euler characteristic. $H_1 \backslash M$ (and similarly $H_2 \backslash M$) is tessellated with $[G : H_1]$ copies of $\mathcal{D}$, where $[G : H_1]$ is the index of H_1 in G. The N sides of $\mathcal{D}$ give rise to $[G : H_1]N/2$ edges of $H_1 \backslash M$. Since every cycle $[p]$ has angle sum $2\pi/\text{ord}[p]$, we need ord$[p]$ cycles for one vertex of M_1. Hence there are $[G : H_1] \sum 1/\text{ord}[p]$ vertices, where the summation is over all cycles of $\mathcal{D}$. This yields the following.

12.2.10 Proposition. *If the geodesic polygon $\mathcal{D}$ has N sides then in the above construction the surfaces $M_1 = H_1 \backslash M$ and $M_2 = H_2 \backslash M$ have genus*

$$g = 1 + \frac{1}{2}[G : H_1]\left(\frac{N}{2} - 1 - \sum_{[p]} \frac{1}{\text{ord}[p]}\right). \qquad \diamond$$

12.2.11 Example. We check the smoothness and compute the genus of Example 12.2.5. For condition (2) of Proposition 12.2.9 we have to prove that no conjugate of the element $(7, 5)$ belongs to H_1. To this end we note that, for $\sigma = (a, b) \in G$, the inverse is

$$\sigma^{-1} = (a, -ab).$$

Hence, $(a, b)(m, 0)(a, b)^{-1} = (a, b)(m, 0)(a, -ab) = (m, -mb + b)$. The set of all conjugates of all elements in H_1 is therefore as follows.

$$H' = \{(m, -mb + b) \mid m \in \mathbb{Z}_8^*, b \in \mathbb{Z}_8\}.$$

Here $-mb + b$ is even and therefore $(7, 5) \notin H'$, as claimed.

In order to satisfy smoothness condition (1), we let the angles at r and s be π and the remaining angles $\pi/2$. We obtain $g = 5$.

12.3 Examples over $\mathbf{Z}_8^* \ltimes \mathbf{Z}_8$

In this section we construct non-isometric isospectral pairs, for all odd $g \geq 5$ and all even $g \geq 12$. The group is $\mathbf{Z}_8^* \ltimes \mathbf{Z}_8$, and the generators are

(12.3.1) $A_1 = (1, 1)$, $A_2 = (5, 0)$, $A_3 = (5, 4)$, $A_4 = (7, 5)$, $A_5 = (1, 4)$.

The generator A_5 will be absent in the examples of genus 5 and 12. The example of genus 5 has already been described in Examples 12.2.5 and 12.2.11. For the even and odd cases we use two slightly different types of domains. Each domain depends on an integer parameter $n \geq 0$, and we also distinguish between even and odd n. The combinatorial pattern of the examples is shown in Fig. 12.3.2.

We begin with the domain $\mathcal{D}$ of Fig. 12.3.1, which consists of 4 copies of a right-angled geodesic pentagon $rpqq'p'$ and $2n$ copies of a right-angled geodesic pentagon $uvww'v'$, where $v = 2q$. By Lemma 2.3.5, such pentagons exist for any values of u and v satisfying $\sinh u \sinh v > 1$. For simplicity, we fix $u = v = 1$ so that $q = 1/2$. The parameter r is free, and $\mathcal{D}$ now depends on r. The formulae of Theorem 2.3.4 imply that p and p' converge to infinity as $r \to 0$, and q' converges to q'_0, where q'_0 is determined by the equation $\sinh q'_0 \sinh \frac{1}{2} = 1$. Since the pentagons with sides u and v remain invariant as $r \to 0$, it follows that if $r > 0$ is sufficiently small, then $\mathcal{D}$ has the following property.

(12.3.2) *Any curve on $\mathcal{D}$ which connects non-adjacent sides has length $\geq 2r$. Equality holds if and only if the curve is one of the sides of $\mathcal{D}$ of length $2r$.* ◇

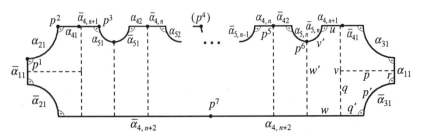

Figure 12.3.1

We subdivide some of the sides of $\mathcal{D}$ by introducing an additional vertex at the midpoint, as shown in the figure. The number of vertices of $\mathcal{D}$ becomes

(12.3.3) $$N = 4n + 10.$$

The sides $\alpha_{k\mu}$ and $\bar{\alpha}_{k\mu}$, $\mu = 1, \ldots, m = m(k)$, correspond to the generators A_k, $k = 1, \ldots, 5$. They are arranged as shown in Fig. 12.3.1.

Observe that $\alpha_{k\mu}$ and $\bar{\alpha}_{k\mu}$ indeed have the same length. The smoothness of the resulting surfaces is checked in the same way as in Example 12.2.11. For the convenience of the reader, we list the sequence of generators and their product $g_1 \ldots g_\ell = g_p$ for some of the vertices (cf. (12.2.3)). The j-th row of the table contains the sequence of generators belonging to the cycle which begins at vertex p^j in Fig. 12.3.1 (p^3 to p^6 are absent if $n = 0$; p^3 is absent if $n = 1$; p^4 is absent if n is even.). The symbol $\#_e(n)$ denotes the number of cycles which have the same sequence as the given one if n is even, $\#_o(n)$ denotes this number if n is odd.

generators		order	$\#_e(n)$	$\#_o(n)$
$A_2 A_1^{-1} A_3^{-1} A_1$	$= (1, 0)$	1	1	1
$A_4 A_3 A_4 A_2^{-1}$	$= (1, 0)$	1	1	1
$A_5 A_4 A_5 A_4$	$= (1, 0)$	1	$n/2$	$(n-1)/2$
$A_5 A_4$	$= (7, 1)$	2	0	1
$A_4^{-1} A_4^{-1}$	$= (1, 0)$	1	$n/2$	$(n+1)/2$
A_5^{-1}	$= (1, 4)$	2	n	n
A_4^{-1}	$= (7, 5)$	2	2	1

Conditions (1) and (2) of Proposition 12.2.9 are easily checked (cf. Example 12.2.11). By Proposition 12.2.10 the genus is

$$g = 5 + 2n; \quad n = 0, 1, 2, \ldots.$$

By Sunada's theorem, the surfaces M_1 and M_2 obtained by gluing copies of $\mathcal{D}$, with respect to the graphs $\mathcal{G}_1 = H_1 \backslash \mathcal{G}$ and $\mathcal{G}_2 = H_2 \backslash \mathcal{G}$, are isospectral. To prove that M_1 and M_2 are not isometric, we cut the surfaces open along all closed geodesics of length $2r$. It will turn out that M_1 remains connected and M_2 falls into two connected components.

In view of (12.3.2), all closed geodesics on M_1 and M_2 of length $2r$ are among those consisting of copies of side α_{11} of $\mathcal{D}$. The gluing pattern for the adjacent sides α_{21}, $\bar{\alpha}_{21}$, α_{31} and $\bar{\alpha}_{31}$ is shown in Fig. 12.3.2. The figure shows the graphs $\mathcal{G}_i = H_i \backslash \mathcal{G}$, $i = 1, 2$, with respect to the generators (12.3.1). The dashed edges correspond to A_2 (for the pasting of the copies of α_{21} and

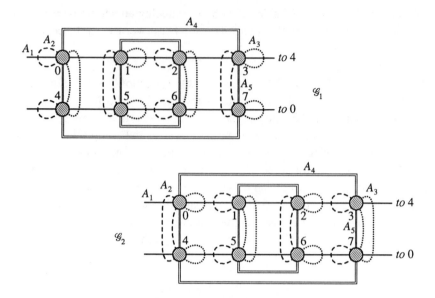

Figure 12.3.2

$\bar{\alpha}_{21}$), the dotted edges correspond to A_3 (for the pasting of the copies of α_{31} and $\bar{\alpha}_{31}$). The graph has valency 10.

We observe that, in M_1, the copies of α_{11} on blocks 1, 3, 5 and 7 define a set Γ_1 of four closed geodesics of length $2r$. In M_2, these copies define a set Γ_1' of two closed geodesics of length $4r$. In M_2, the copies of α_{11} on blocks 0, 2, 4 and 6 define a set Γ_2 of four closed geodesics of length $2r$, and, in M_1, these copies define a set Γ_2' of two closed geodesics of length $4r$.

In view of (12.3.2) we have, for $i = 1, 2$,

(12.3.4)   *Γ_i is the set of all closed geodesics of length $2r$ on M_i.*

From Fig. 12.3.2 we see that Γ_2 separates M_2, but Γ_1 does not separate M_1. That is, M_2 has a separating set consisting of four closed geodesics of length $2r$, but M_1 has no such set. Hence, M_1 and M_2 are not isometric.   ◊

The examples with even genus are obtained in a similar way, with the same group and the same generators as before. The polygon is shown in Fig. 12.3.3 (with $n = 2$, for simplicity). The middle section consists of $2n$ copies of a rectangular geodesic pentagon, and is the same as before. The outer sections consist of 4 copies of a trirectangle with acute angle α, 4 copies of a trirectangle with acute angle β, and 4 copies of a trirectangle with acute angle γ. The outer sections have again a horizontal axis of symmetry. The angles α, β and γ satisfy the condition

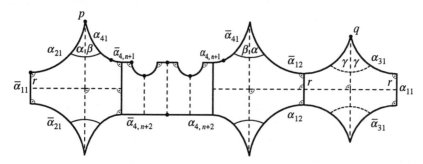

Figure 12.3.3

(12.3.5) $\qquad \alpha + \beta = \pi/16, \qquad \gamma = \pi/16.$

The remaining angles are right angles. It is not difficult to see that such polygons exist for arbitrarily small r. In contrast to the preceding examples, there is a second pair of sides of $\mathcal{D}$ (α_{12} and $\bar{\alpha}_{12}$) with a common perpendicular of length $2r$. It turns out that these perpendiculars do not yield closed geodesics of length $2r$ on M_1 and M_2.

The sequences of generators of the cycles are the same as before, with two exceptions: the sequence $A_4 A_3 A_4 A_2^{-1}$ is replaced by $A_4 A_1^{-1} A_4 A_2^{-1}$ (belonging to the cycle of vertex p), and we have the additional sequence $A_3 A_1$ (belonging to the cycle of vertex q). The corresponding cycles have order 8, and are shown in the figure.

Smoothness and non-isometry are checked in the same way as before (and as in Example 12.2.11). The polygon has $N = 4n + 12$ vertices, and the genus of M_1 and M_2 is $g = 12 + 2n$, $n = 0, 1, 2, \ldots$.

12.4 Examples over SL(3, 2)

In the next two sections we describe the examples of Brooks and Tse [1], which are modeled over the group SL(3, 2). We shall use Cayley graphs based on different choices of generators. The examples will have genus $g = 8 + 7k$, $g = 4 + 3k$, for $k = 0, 1, \ldots$, and $g = 6$. The two almost conjugate subgroups of $G = $ SL(3, 2) are again the following,

$$H_1 = \left\{ \text{all} \begin{pmatrix} 1 & * & * \\ 0 & * & * \\ 0 & * & * \end{pmatrix} \in G \right\}, \quad H_2 = \left\{ \text{all} \begin{pmatrix} 1 & 0 & 0 \\ * & * & * \\ * & * & * \end{pmatrix} \in G \right\}.$$

The first series of examples is obtained via the generators A and B as in

Example 11.4.7. The corresponding graphs $\mathcal{G}_1$ and $\mathcal{G}_2$ are shown in Fig. 11.4.5. They are not isomorphic.

We replace each vertex of $\mathcal{G}_1$ and $\mathcal{G}_2$ by a compact Riemann surface $\mathcal{B}$ of signature $(k, 4)$, whose boundary geodesics have length δ for small positive δ. As described at the end of Section 11.4, we take $\mathcal{B}$ in such a way that, on the resulting Riemann surfaces M_1 and M_2, the only closed geodesics of length δ are those arising from the boundary of $\mathcal{B}$. Graphs $\mathcal{G}_1$ and $\mathcal{G}_2$ can be reconstructed via the instruction "cut M_i open along all closed geodesics of length δ". Since $\mathcal{G}_1$ and $\mathcal{G}_2$ are not isomorphic, M_1 and M_2 are not isometric. These examples have genus $g = 8 + 7k$, $k = 0, 1, \ldots$.

For the next series we use the generators A, B and C^κ, where as in Lemma 11.2.6,

$$A = \begin{pmatrix} 0 & 1 & 1 \\ 0 & 1 & 0 \\ 1 & 0 & 0 \end{pmatrix}, \quad B = \begin{pmatrix} 1 & 0 & 0 \\ 0 & 0 & 1 \\ 0 & 1 & 1 \end{pmatrix}, \quad C = \begin{pmatrix} 0 & 1 & 1 \\ 1 & 1 & 1 \\ 0 & 1 & 0 \end{pmatrix},$$

and κ is a suitable exponent. Here A and B are as in Example 11.4.7. To obtain a picture of the corresponding graphs $\mathcal{G}_1$ and $\mathcal{G}_2$, we use Fig. 11.4.5 and draw, for $j = 0, \ldots, 6$, an additional edge from vertex j to vertex $j + \kappa$ (mod 7).

Our first goal is to construct an example of genus 4. The remaining examples will be obtained with minor modifications. For genus 4 we set $\kappa = 1$.

In order to define the building block $\mathcal{D}$ which is again a geodesic polygon in the hyperbolic plane, we first consider a trirectangle with acute angle $\pi/14$ and opposite sides of lengths $\delta/2$ and $\ell/2$. This exists for any $\delta > 0$, and ℓ is determined by the equation

$$\sinh \frac{\delta}{2} \sinh \frac{\ell}{2} = \cos \pi/14.$$

We glue 4 copies together as in Fig. 12.4.1 to obtain a hyperbolic rectangle $\mathcal{D}'$ whose center p is a cone-like singularity. If we cut $\mathcal{D}'$ open along a geodesic arc γ from p to one of the vertices as shown in Fig. 12.4.1, we obtain a geodesic hexagon $\mathcal{D}$ with sides α, γ, $\bar{\gamma}$, β, $\bar{\alpha}$ and $\bar{\beta}$. The angles between α and γ, and between $\bar{\gamma}$ and β, add up to $\pi/2$. The values of the remaining angles are given in the figure. Sides α and $\bar{\alpha}$ have length δ, sides β and $\bar{\beta}$ have length ℓ.

We let α, β and γ correspond to the edges of $\mathcal{G}_1$ and $\mathcal{G}_2$ of type A, B and C respectively, and glue copies $\mathcal{D}_0, \ldots, \mathcal{D}_6$ of $\mathcal{D}$ together according to $\mathcal{G}_1$ and $\mathcal{G}_2$ as described in Section 12.1.

To prove that the resulting surfaces M_1 and M_2 are smooth, we note that the vertex p of $\mathcal{D}$ has cycle $[p] = \{p\}$. The associated conjugacy class as

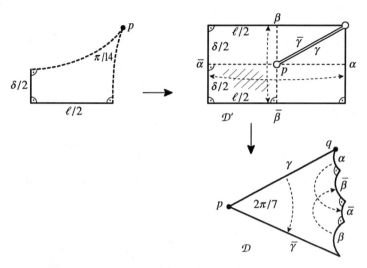

Figure 12.4.1

given by (12.2.3) is $[\mathfrak{g}_p] = [C]$, and so the order is $\operatorname{ord}[p] = 7$ (cf. Lemma 11.2.6). Since for any $\sigma \in G$, and any $v = 1, \ldots, 6$, $\sigma C^v \sigma^{-1}$ has order 7, and since $\#H_1 = 24$, we have $[C^v] \cap H_1 = \emptyset$ for $v = 1, \ldots, 6$. Hence, p satisfies conditions (1) and (2) of Proposition 12.2.9. The remaining vertices form a cycle of length $\ell = 5$ with the associated conjugacy class $[\mathfrak{g}_q]$ given by the equation $[\mathfrak{g}_q] = [ABA^{-1}B^{-1}C^{-1}] = [I]$, so that they also satisfy the smoothness conditions. By Proposition 12.2.10, the genus is $g = 4$.

To prove that M_1 and M_2 are non-isometric, we seek a geometric quantity which distinguishes M_1 from M_2. (Since $G \backslash M$ has one of the exceptional signatures ([1; 7]) listed in (12.7.1), we cannot apply the non-isometry criteria of Section 12.7 here.)

From the gluing pattern and from the right angles we see that, on M_1 and on M_2, side $\bar{\alpha}$ of block $\mathcal{D}_0$ yields a simple closed geodesic η of length δ. The sides $\bar{\alpha}$ on $\mathcal{D}_j$ for $j \neq 0$, together yield two simple closed geodesics η' and η'' of length 3δ.

η, η' and η'' are the only primitive closed geodesics of length $\leq 3\delta$ on M_1 and on M_2.

Proof. We consider the surface M_0 of genus 1 with a cone-like singularity obtained by gluing together the sides of $\mathcal{D}$ with respect to $SL(3, 2) \backslash \mathcal{G}$, where $\mathcal{G}$ is the Cayley graph of $SL(3, 2)$ with respect to the generators A, B and C. The pasting scheme is given by the arrows in Fig. 12.4.1. It follows that M_1

and M_2 are covering surfaces of M_0.

The coverings $M_1 \to M_0$ and $M_2 \to M_0$ are length-preserving and branched over the singular points of M_0. However, they are not Galois (i.e. not normal) coverings. Therefore, a primitive closed geodesic on M_0 may have primitive closed lifts in M_i with different lengths, $i = 1, 2$.

Any closed geodesic on M_i is mapped onto a closed geodesic of the same length on M_0. Now M_0 is also obtained by gluing together the sides of $\mathcal{D}'$ as shown in Fig. 12.4.1. From this we see that if δ is sufficiently small, then there exists exactly one primitive closed geodesic η_0 of length δ on M_0, and any other primitive closed geodesic on M_0 has length $> 3\delta$. Since η, η' and η'' are the only primitive closed geodesics on M_i which are mapped onto η_0, this proves the claim. ◇

Next, we observe that there exists exactly one simple geodesic arc of length ℓ on M_0 perpendicular to η_0 at both endpoints. It follows that on M_1 and on M_2 there are exactly two simple common perpendiculars of length ℓ between η and the union $\eta' \cup \eta''$. On M_1 the two perpendiculars form a closed geodesic. On M_2 this is not the case. Hence, M_1 and M_2 are not isometric. ◇

To get similar examples with larger genus, we let $k \in \mathbf{N} - 7\mathbf{N}$ and glue to-

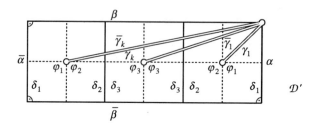

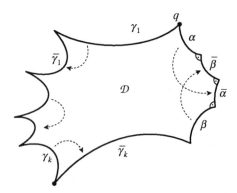

Figure 12.4.2

gether $4k$ trirectangles as shown in Fig. 12.4.2 (for $k = 3$). The pasting yields a hyperbolic rectangle $\mathcal{D}'$ with k cone-like singularities. If $k = 1$, we take all four trirectangles with acute angle $\varphi_3 = \pi/14$ and opposite sides of lengths $\delta/2$ and $\ell/2$ just as before. If $k \geq 2$ we take the two leftmost and the two rightmost trirectangles with acute angles φ_1, the four adjacent trirectangles with acute angle φ_2 and the remaining $4(k-2)$ trirectangles with acute angle φ_3, where

$$\varphi_1 + \varphi_2 = 2\varphi_3 = \pi/7.$$

The opposite sides are denoted by δ_i and ℓ_i, where δ_i is small and ℓ_i is large. From the trigonometry of trirectangles we see that we may choose these trirectangles in such a way that

$$\delta_1 \leq \frac{1}{10}\delta_2, \quad \delta_2 = \delta_3$$

with arbitrarily small δ_3. The rectangle $\mathcal{D}'$ has a horizontal axis of symmetry.

We cut $\mathcal{D}'$ open along k geodesic arcs from the singular points to a given vertex as shown in Fig. 12.4.2. This yields a geodesic polygon $\mathcal{D}$ in the hyperbolic plane with $N = 2k + 4$ sides $\alpha, \gamma_1, \bar{\gamma}_1, \ldots, \gamma_k, \bar{\gamma}_k, \beta, \bar{\alpha}, \bar{\beta}$.

We let $\kappa \in \{1, \ldots, 6\}$ be the number satisfying $\kappa k \equiv 1 \pmod{7}$ and use the generators A, B, C^{κ} with A, B and C as before. Then we glue together copies $\mathcal{D}_0, \ldots, \mathcal{D}_6$ according to $H_1 \backslash \mathcal{G}$ and $H_1 \backslash \mathcal{G}$, where $\mathcal{G} = \mathcal{G}[A, B, C^{\kappa}]$. The proofs for the smoothness and non-isometry, of the resulting surfaces M_1 and M_2, are the same as before (k vertices of $\mathcal{D}$ have order 7 with a cycle of length $\ell = 1$, the remaining vertices form a cycle of length $k + 4$ with the associated conjugacy class $[\mathfrak{g}_q] = [ABA^{-1}B^{-1}C^{-\kappa k}] = [I]$). By Proposition 12.2.10, the genus is $g = 1 + 3k$, $k \in \mathbf{N} - 7\mathbf{N}$.

12.5 Genus 6

This is the most difficult example and we include it only for completeness. By Lemma 11.2.7, the matrices

$$A = \begin{pmatrix} 1 & 0 & 1 \\ 1 & 1 & 0 \\ 0 & 1 & 0 \end{pmatrix}, \quad B = \begin{pmatrix} 0 & 0 & 1 \\ 1 & 0 & 0 \\ 1 & 1 & 0 \end{pmatrix}, \quad C = \begin{pmatrix} 1 & 1 & 1 \\ 1 & 0 & 1 \\ 1 & 0 & 0 \end{pmatrix}$$

are generators of $G = \mathrm{SL}(3, 2)$, and have order 7. We check that $C = AB$. For $k = 0, \ldots, 6$ we have uniquely determined exponents $a_i(k)$ and $b_i(k)$ satisfying

$$C^k A \in H_i C^{a_i(k)}, \quad C^k B \in H_i C^{b_i(k)}; \quad i = 1, 2.$$

They are given by the following table.

k	0	1	2	3	4	5	6
$a_1(k)$	4	5	1	0	6	3	2
$b_1(k)$	4	3	0	6	1	2	5
$a_2(k)$	4	6	5	2	1	0	3
$b_2(k)$	6	5	4	0	1	3	2

The surfaces $M_i = H_i \backslash M$, $i = 1, 2$, will have the property that the quotient $G \backslash M$ is a surface X of genus 0 with 4 cone-like singularities of order 7, that is, with interior angle $2\pi/7$ at each singular point.

Since the general non-isometry criteria of Section 12.7 do not apply to the present example ($G \backslash M$ has one of the exceptional signatures ([0; 7,7,7,7]) listed in (12.7.1)), we have to describe the geometry of X in considerable detail. The geometry will be similar to that of a 4-holed sphere in Section 3.3.

To construct X we start with a trirectangle with acute angle $\pi/7$ and opposite sides $\varepsilon/4$ and $\lambda/4$. Here $\varepsilon > 0$ is arbitrarily small and λ is defined by the equation

$$\sinh \frac{\lambda}{4} \sinh \frac{\varepsilon}{4} = \cos \pi/7.$$

Four copies together yield a surface Y as in Fig. 12.5.1, which is a topological disk with two cone-like singularities of order 7 and a boundary geodesic δ of length ε.

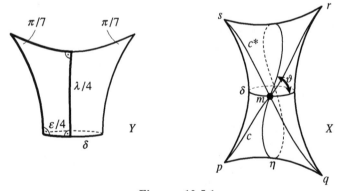

Figure 12.5.1

Two copies of Y glued together along δ yield X. As for pairs of pants, we have a twist parameter along δ as an additional degree of freedom. In order to reduce the order of the isometry group of X as much as possible, we let this twist parameter be in the interval $]0, \frac{1}{4}[$. (If the twist parameter is zero, then M_1 and M_2 are isometric.) The isometry group of X is now Klein's 4-group

Ch.12, §5] Genus 6 327

acting by orientation-preserving isometries.

X has properties similar to those of the X-pieces in Chapter 3. For instance, if ε is small, then any primitive closed geodesic on X different from δ (and δ^{-1}) is much longer than δ. We take ε so small that the primitive closed geodesics different from δ have length greater than 3ε.

We also have the family of simple closed geodesics in X intersecting δ in two opposite points. We let η be the smallest such geodesic. Its length

$$\ell = \ell(\eta)$$

lies between λ and $\lambda + 2\varepsilon$ for some λ. We denote by $\vartheta \in {]}0, \pi/2[$ the smaller of the two angles between η and δ. This angle is the same at both intersection points. We obtain the following characterization of η.

η is the unique simple closed geodesic on X intersecting δ in two opposite points and forming the angle ϑ at both intersection points.

We denote the singular points by p, q, r, s, as in Fig. 12.5.1 (the pairs p, q and r, s are separated from each other by δ, the pairs p, s and q, r are separated from each other by η). The shortest connections pq, qr, rs and sp decompose X into two isometric geodesic quadrangles $\mathcal{F}$ and $\mathcal{F}'$ which cover X without overlapping. $\mathcal{F}'$ is the image of $\mathcal{F}$ under an orientation-preserving isometry. Since the twist parameter lies in the interval ${]}0, \frac{1}{4}[$, there is no orientation reversing isometry sending $\mathcal{F}$ to $\mathcal{F}'$. Finally we let

$$c = pr \quad \text{and} \quad c^* = qs$$

be the two diagonals of $\mathcal{F}$. By symmetry they intersect each other in their midpoints, and this intersection point is at the same time the intersection point m of δ and η on $\mathcal{F}$. (cf. Fig. 12.5.1). Since the twist parameter is different from 0 and small, the diagonals c and c^* have different length. This implies that the only non-trivial isometry of $\mathcal{F}$ (and $\mathcal{F}'$) onto itself is the half-turn around its center.

In order to obtain a building block for M_1 and M_2, we cut X open along the arcs c, rq and rs to obtain a geodesic hexagon $\mathcal{D}$ as in Fig. 12.5.2 with sides $\gamma, \bar{\gamma}, \beta, \bar{\beta}, \bar{\alpha}, \alpha$. The diagonals in $\mathcal{D}$ from p to q, and from p to s, correspond respectively to the arcs pq and ps of X; sides $qr = \bar{\beta}$ and $rs = \bar{\alpha}$ of $\mathcal{D}$ correspond to the arcs qr and rs of X. Recall, from the isometry between $\mathcal{F}$ and $\mathcal{F}'$, that pq and rs have the same length and that ps and qr have the same length.

Now we proceed as in Section 12.4. We let α, β and γ correspond to the above generators A, B and C of SL(3, 2), and glue together the copies $\mathcal{D}_0$, $\mathcal{D}_1, \ldots, \mathcal{D}_6$ of $\mathcal{D}$ according to the gluing pattern given by the above table. By

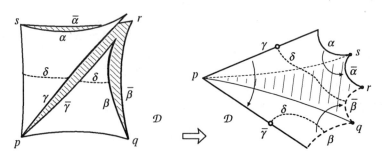

Figure 12.5.2

Sunada's theorem again, the resulting surfaces M_1 and M_2 are isospectral.

We check the smoothness condition of Section 12.2. The conjugacy classes of SL(3, 2) associated to the vertices p, q, r and s of $\mathcal{D}$, given by Algorithm 12.2.1 and by (12.2.3), are $[\mathfrak{g}_p] = [C]$, $[\mathfrak{g}_q] = [B]$, $[\mathfrak{g}_r] = [B^{-1}C^{-1}A] = [B^{-2}]$ and $[\mathfrak{g}_s] = [A^{-1}]$. These are classes of elements of order 7 and do not intersect H_1, because 7 does not divide the order of H_1. We check that the angle sum of each cycle is $2\pi/7$. By Proposition 12.2.9, M_1 and M_2 are smooth. By Proposition 12.2.10 the genus is 6.

The proof that M_1, M_2 are not isometric is involved. The difficulty may be explained by the fact that if X has twist parameter 0, then the two surfaces are isometric. Hence, there is no striking difference between M_1 and M_2.

We follow our earlier idea to reconstruct the graphs $\mathcal{G}_1$ and $\mathcal{G}_2$. Unfortunately, we are only able to reconstruct the tessellation of M_1 and M_2 with the copies of $\mathcal{F}$ instead of $\mathcal{D}$. We use therefore a modification of the earlier methods.

Let $\mathcal{P}_1$ and $\mathcal{P}_2$ be the hyperbolic geodesic polygons shown in Fig. 12.5.3 and 12.5.4, obtained by gluing together the copies $\mathcal{D}_0, \ldots, \mathcal{D}_6$ of $\mathcal{D}$ along the copies of sides γ and $\bar{\gamma}$ according to $\mathcal{G}_1$ and $\mathcal{G}_2$. Fig. 12.5.3 shows the tessellation of $\mathcal{P}_1$ with the copies of $\mathcal{D}$. Fig. 12.5.4 shows the tessellation of $\mathcal{P}_2$ with the copies of $\mathcal{F}$. The gluing of the sides of $\mathcal{P}_1$ and $\mathcal{P}_2$ is given by the labelling. The code is as follows. The dashed sides are copies of β and $\bar{\beta}$. They are labelled bold face, and correspond to the edges (of the graph) of type B. For $k = 0, \ldots, 6$, side **k** (a copy of β) is pasted to side **k'** (a copy of $\bar{\beta}$). The remaining sides are copies of α and $\bar{\alpha}$ and correspond to the edges of type A. For $k = 0, \ldots, 6$, side k (a copy of α) is pasted to side k' (a copy of $\bar{\alpha}$). These identifications correspond with the table given at the beginning of this section.

Recall that M_1 and M_2 are covering surfaces of X. Recall, also, that the coverings $M_1 \to X$ and $M_2 \to X$ are not Galois coverings.

Ch.12, §5]   Genus 6   329

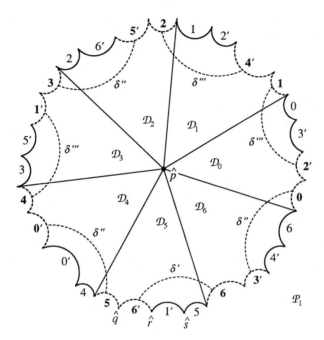

Figure 12.5.3

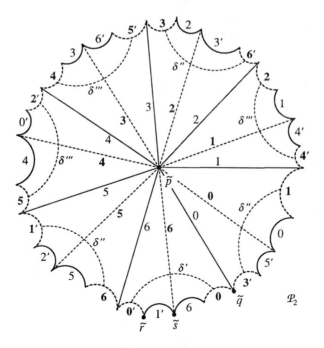

Figure 12.5.4

From the gluing pattern it follows that geodesic δ in X has a simple closed lift δ', of length $\varepsilon = \ell(\delta)$, and two simple closed lifts δ'', δ''' of length 3ε in M_1 and in M_2 (in Fig. 12.5.2, δ is marked by the dotted line on X; on $\mathcal{D}$ this line appears separated into two connected components; Fig. 12.5.3 and Fig. 12.5.4 show the copies of these components in $\mathcal{P}_1$ and $\mathcal{P}_2$). Since any simple closed geodesic in X different from δ has length greater than 3ε, the lifts δ', δ'' and δ''' are the only simple closed geodesics of length $\ell \leq 3\varepsilon$ on M_1 and M_2. This implies that δ', δ'' and δ''' can be reconstructed out of the intrinsic geometry of M_1 and M_2.

By drawing similar lines in Fig. 12.5.2, 3 and 4, we see that the geodesic η in X has a simple closed lift η' of length $7\ell(\eta)$. Since η intersects δ in two opposite points, η' cuts the union $\delta' \cup \delta'' \cup \delta'''$ into segments of length $\varepsilon/2$, where at each intersection the angle is ϑ. It follows from the above mentioned properties of η that on M_1 and M_2, η' is the *unique* simple closed geodesic intersecting $\delta' \cup \delta'' \cup \delta'''$ in this way. Hence, η' can also be reconstructed out of the intrinsic geometry of M_1 and M_2.

At each intersection point $\tilde{m}$ of $\eta' \cap (\delta' \cup \delta'' \cup \delta''')$ we draw geodesic arcs $\tilde{c}$ and $\tilde{c}^*$, of length $\ell(c)$ and $\ell(c^*)$ (cf. Fig. 12.5.1), centered at $\tilde{m}$, such that the various angles between $(\delta' \cup \delta'' \cup \delta''')$, $\tilde{c}$ and $\tilde{c}^*$ at $\tilde{m}$ are the same as the corresponding angles between δ, c and c^* at m in X, and such that $\tilde{c}$ is in the interior of the angle ϑ. This condition determines the arcs $\tilde{c}$ and $\tilde{c}^*$ *uniquely*, and our drawing instruction uses only the intrinsic geometry of M_1 and M_2. We now see immediately the following.

The tessellations of M_1 and M_2 with the copies of $\mathcal{F}$ can be reconstructed out of the intrinsic geometry of M_1 and M_2.

Consequently, any isometry $\phi : M_1 \to M_2$ sends the tessellation of M_1 to the tessellation of M_2. Also, ϕ sends the unique geodesic δ' of length ε on M_1 onto the unique geodesic δ' of length ε on M_2. The geodesic η' on M_1 intersects δ' in opposite points and, at both points, under the angle ϑ. The same holds for δ' and η' on M_2. Since ϑ is different from $\pi/2$, this implies that ϕ *preserves the orientation*. (The orientations are defined in such a way that the coverings $M_1 \to X$ and $M_2 \to X$ are orientation preserving.) Taking all these properties into account we see that there remain four possibilities:

If $\hat{p} \in M_1$ denotes the lift of p (with respect to the covering $M_1 \to X$) which is the center of $\mathcal{P}_1$, and if $\tilde{p}$, $\tilde{q}$, $\tilde{r}$ and $\tilde{s}$ in M_2 are the lifts of p, q, r and s as shown in Fig. 12.5.4, then we have one of the following cases:

$$\phi(\hat{p}) = \tilde{p}, \qquad \phi(\hat{p}) = \tilde{q}, \qquad \phi(\hat{p}) = \tilde{r}, \qquad \phi(\hat{p}) = \tilde{s}.$$

In order to see that neither case is compatible with the given tessellations, we draw, for $x = \tilde{p}, \tilde{q}, \tilde{r}, \tilde{s}$, a figure of the polygon $\mathcal{P}_2(x)$ formed by the 14 copies of $\mathcal{F}$ around x in M_2 (Fig. 12.5.4 shows $\mathcal{P}_2(\tilde{p})$). With the notation of Fig. 12.5.4, the sides of $\mathcal{P}_2(x)$, listed counterclockwise, are given by the list below, where the primes have been omitted. The lists begin with the presumed images of $6'1'$. The sequences of sides in this list can also be obtained via Algorithm 12.2.1, applied to $\tilde{p}, \tilde{q}, \tilde{r}$ and $\tilde{s}$.

$\mathcal{P}_2(\tilde{p})$: **0 1 6 0 3 5 0 1 4 4 1 2 6 3 2 3 5 6 3 4 2 0 4 5 1 2 5 6**

$\mathcal{P}_2(\tilde{q})$: **6 6 1 6 5 5 3 2 1 1 0 4 3 3 4 1 0 0 2 5 4 4 6 3 2 2 5 0**

$\mathcal{P}_2(\tilde{r})$: **6 6 0 6 3 3 5 4 1 1 4 3 2 2 6 5 0 0 3 2 5 5 1 0 4 4 2 1**

$\mathcal{P}_2(\tilde{s})$: **0 0 6 0 2 2 1 4 5 5 4 2 1 1 0 3 6 6 5 1 3 3 2 6 4 4 3 5**

On the other hand, $\mathcal{P}_1 = \mathcal{P}_1(\hat{p})$ has the following sequence.

$\mathcal{P}_1(\hat{p})$: **6** 1 5 6 3 4 6 0 2 3 0 1 4 2 1 2 5 6 2 3 1 5 3 4 0 0 4 5

This sequence is not compatible with any of the preceding four, and hence, none of the above four cases for ϕ is possible. This proves that M_1 and M_2 are not isometric. ◊

Remark. If we reverse the order of the sequence of $\mathcal{P}_1(\hat{p})$ and start the cycle with **6 5 1 6**, we get the sequence

6 5 1 6 5 4 0 0 4 3 5 1 3 2 6 5 2 1 2 4 1 0 3 2 0 6 4 3.

This is now compatible with the sequence of $\mathcal{P}_2(\tilde{p})$. If we admit that X has twist parameter zero, then $\mathcal{F}$ has an orientation reversing symmetry, and, indeed, there exists an orientation reversing isometry $\phi : M_1 \to M_2$.

The examples of Sections 12.3 - 12.5 cover any genus $g \geq 4$. In each of these examples there is a parameter which may vary freely in some open interval. Hence, we have the following theorem in which $\mathcal{V}_g$ is the subvariety of Teichmüller space, as in Wolpert's theorem, consisting of those Riemann surfaces which are not determined by their spectrum.

12.5.1 Theorem. dim $\mathcal{V}_g > 0$ *for any* $g \geq 4$. ◊

An analysis on the number of free parameters shows that the examples modeled over $H_1 \backslash G$ and $H_2 \backslash G$ fill out a subvariety of $\mathcal{T}_g$ of dimension roughly dim $\mathcal{T}_g / [G : H_1]$, where $[G : H_1]$ is the index of H_1 in G. Perlis [1] has shown that the index, of Gassmann equivalent non-conjugate subgroups in any group, is always larger or equal 7.

By Wolpert's theorem, $\dim \mathcal{V}_g \leq \dim \mathcal{T}_g - 1$. No better upper bound for $\dim \mathcal{V}_g$ is known as yet.

Haas [1] and Buser-Semmler [1] have shown that no pairs of non-isometric isospectral one-holed tori occur ("genus 1.5"). We conjecture that this can be extended to genus 2.

In genus 3, the examples over SL(3, 2) turn out to be isometric. However, if we admit variable curvature, then we may use bumpy metrics so that the examples become non-isometric (Brooks-Tse [1]). We do not know whether $\mathcal{V}_3$ is empty or not.

12.6 Large Families

In Chapter 13 we shall prove that the cardinality of a set of pairwise isospectral non-isometric Riemann surfaces of genus g has an upper bound which depends only on g. In the present section we give simple examples to show that such a bound cannot be independent of g. The cardinality of the set of examples obtained in this section is roughly $g^{1/3}$. More involved examples with cardinality of order $g^{1/2}$ have been given by Tse [1, 2].

We let G' be a given group with Gassmann equivalent non-conjugate subgroups H_1, H_2, such as, for instance, $G' = \mathrm{SL}(3, 2)$ with the subgroups as in the preceding sections. For fixed $m = 1, 2, \ldots$, we consider the direct product

$$G = G \times \ldots \times G \ (m \text{ factors}).$$

For any function

$$f : \{1, \ldots, m\} \to \{1, 2\}$$

we define

$$H_f = H_{f(1)} \times \ldots \times H_{f(m)} \subset G.$$

Since conjugation in G cannot exchange factors, the various H_f are pairwise non-conjugate. Observe that elements $g = (g_1, \ldots, g_m)$ and $g' = (g'_1, \ldots, g'_m)$ of G are conjugate if and only if each factor g'_k is a conjugate of g_k, $k = 1, \ldots, m$. We have therefore

$$\#([g] \cap H_f) = \prod_{k=1}^{m} \#([g_k] \cap H_{f(k)}).$$

For fixed g, the right-hand side is independent of f. This proves that the various H_f are pairwise Gassmann equivalent.

Next, we let $A_1, \ldots, A_n$ be generators of G', and let $\mathcal{G}$ be the Cayley graph of G with respect to the nm generators $A_{\nu\mu} = (1, \ldots, A_\nu, \ldots, 1)$ (A_ν is at the μ-th position). For any f, we set $\mathcal{G}_f = H_f \backslash \mathcal{G}$ and glue together $[G : H_f] = \#G/\#H_f$ copies of a compact Riemann surface $\mathcal{B}$ of signature $(0, 2nm)$ as described at the end of Section 11.4. The resulting surfaces M_f are pairwise isospectral. The criteria of the next section will show that for a generic choice of $\mathcal{B}$ these surfaces are pairwise non-isometric.

To obtain a large family, we take $G' = \mathrm{SL}(3, 2)$ with the generators as in Example 11.4.7. In this case, the surfaces M_f consist of 7^m blocks of signature $(0, 4m)$. This yields 2^m surfaces of genus $g = 1 + 7^m(2m - 1)$, and we obtain the following result.

12.6.1 Theorem. *For infinitely many g there are examples of $g^{1/3}$ pairwise non-isometric isospectral compact Riemann surfaces of genus g.* ◊

12.7 Criteria For Non-Isometry

We prove two theorems stating that non-isometry in Sunada's construction is a generic property. The first result (Theorem 12.7.5) is due to Sunada [3], its proof uses a result of Margulis about arithmetic groups. The second result (Theorem 12.7.6) has a more elementary proof, but holds in less generality. It states that once a particular pair of isospectral examples in Sunada's construction is non-isometric, then we automatically have a several parameter family of pairs of such examples.

Let again (G, H_1, H_2) be a triplet as in Definition 11.1.1, where G is a finite group and H_1 and H_2 are non-conjugate almost conjugate subgroups of G. We assume that G acts on a given compact Riemann surface M by orientation-preserving isometries, possibly with fixed points. The quotient $M_0 = G \backslash M$ is a compact Riemann surface, of a certain genus $h \geq 0$, with a number of cone-like singularities, say with k singular points of orders $\nu_1, \ldots, \nu_k$. Of course we may have $k = 0$. The sequence

$$\sigma = [h; \nu_1, \ldots, \nu_k]$$

is called the *signature* of M_0. If $k = 0$ we shall write $\sigma = [h; -]$. More general signatures of surfaces with cusps and boundaries will not be considered. M_0 is the quotient $M_0 = \Gamma_0 \backslash \mathbf{H}$, where $\mathbf{H}$ is the hyperbolic plane and Γ_0 is a discrete subgroup of $\mathrm{PSL}(2, \mathbf{R}) = \mathrm{Is}^+(\mathbf{H})$. The sequence $\sigma = [h; \nu_1, \ldots, \nu_k]$ is also called the *signature of Γ_0*.

For the first result we use certain results about Riemann surfaces with cone-like singularities, which we adopt from the literature. For the second result we restrict ourselves again to smooth Riemann surfaces.

Let $\Gamma_0 \subset \text{PSL}(2, \mathbf{R})$ be a discrete subgroup of signature $\sigma = [h; v_1, \ldots, v_k]$. Then Γ_0 has a presentation with $2h + k$ generators $B_1, \ldots, B_{2h}, E_1, \ldots, E_k$ and $k + 1$ relations

$$\left(\prod_{v=1}^{h} B_{2v-1} B_{2v} B_{2v-1}^{-1} B_{2v}^{-1}\right) E_1 \ldots E_k = 1, \quad E_1^{v_1} = \ldots = E_k^{v_k} = 1,$$

where $B_1, \ldots, B_{2h}$ are hyperbolic elements, and $E_1, \ldots, E_k$ are elliptic elements. We denote by $\mathfrak{G}_\sigma$ the abstract group defined by this presentation. Every discrete subgroup Γ of $\text{Is}^+(\mathbf{H})$, with signature σ, is an *action* of $\mathfrak{G}_\sigma$ on $\mathbf{H}$, that is, there exists an isomorphism of groups

$$\rho : \mathfrak{G}_\sigma \to \Gamma.$$

Two actions Γ' and Γ'', together with choices of generators $B_1', \ldots, E_k'$ and $B_1'', \ldots, E_k''$ (as above), are called *marking equivalent* if they are conjugate in $\text{Is}(\mathbf{H})$ under an isometry which sends B_1' to $B_1'', \ldots, E_k'$ to E_k''. This is in accordance with Definition 6.8.3.

We let $\mathcal{F}_\sigma$ denote the set of all marking equivalence classes.

It is well known (see e.g. Keen [2]) that $\mathcal{F}_\sigma$ has a natural topology in which it is a cell, except for the signatures $\sigma = [0; v_1, v_2, v_3]$. In these latter cases $\mathcal{F}_\sigma$ is a point.

An equivalent definition of $\mathcal{F}_\sigma$ would be to call actions Γ' and Γ'', together with the choices of generators, *marking equivalent* if they are conjugate in $\text{Is}^+(\mathbf{H}) = \text{PSL}(2, \mathbf{R})$ instead of $\text{Is}(\mathbf{H})$, and to define $\mathcal{F}_\sigma$ as the identity component of the set of all marking equivalence classes (with respect to the natural topology).

For $\sigma = [h; -]$, $h \geq 2$, $\mathcal{F}_\sigma$ coincides with $\mathcal{F}_h$ as in Definition 6.8.3, and the natural topology is the same as that in Section 6.8. By Theorem 6.8.13, $\mathcal{F}_h$ is just a different interpretation of the Teichmüller space $\mathcal{T}_h$. We may therefore understand $\mathcal{F}_\sigma$ as the *Teichmüller space* of all marking equivalence classes of Riemann surfaces of signature σ.

There are five types of signatures where any action of $\mathfrak{G}_\sigma$ on $\mathbf{H}$ yields a surface with a non-trivial isometry group. These types are

(12.7.1) $\quad [0; v_1, v_2, v_3], \quad [0; v, v, \mu, \mu], \quad [1; v], \quad [1; v, v], \quad [2; -],$

where v_1, v_2, v_3, v and μ are any integers greater than 1. For any other signature we have the following generalization of Theorem 6.5.7.

12.7.2 Theorem. *If signature σ is not as in (12.7.1), then there exists an open dense subset $\mathcal{F}'_\sigma \subset \mathcal{F}_\sigma$ such that the quotient $\Gamma_0 \backslash \mathbf{H}$ has the trivial isometry group for all $\Gamma_0 \in \mathcal{F}'_\sigma$.*

For a proof we refer to Singerman [1]. ◇

For the sake of clarity we redescribe Sunada's construction in the following form. Let (G, H_1, H_2) be the triplet as as in Definition 11.1.1. Let $\mathfrak{G}_\sigma$ be an abstract group together with a *surjective* homomorphism

$$w : \mathfrak{G}_\sigma \to G.$$

For any action $\Gamma_0 = \rho(\mathfrak{G}_\sigma) \subset \mathrm{PSL}(2, \mathbf{R})$ of $\mathfrak{G}_\sigma$ we have the corresponding surjective homomorphism

$$\omega : \Gamma_0 \to G, \qquad \omega := w \circ \rho^{-1}.$$

We define the following discrete subgroups of $\mathrm{PSL}(2, \mathbf{R})$.

(12.7.3)
$$\Gamma_0 = \omega^{-1}(G)$$
$$\Gamma_1 = \omega^{-1}(H_1), \quad \Gamma_2 = \omega^{-1}(H_2)$$
$$\Gamma = \omega^{-1}(\{id\}).$$

They are canonically isomorphic to the fundamental groups of the surfaces

(12.7.4)
$$M = \Gamma \backslash \mathbf{H}$$
$$M_1 = \Gamma_1 \backslash \mathbf{H} \qquad M_2 = \Gamma_2 \backslash \mathbf{H}$$
$$M_0 = \Gamma_0 \backslash \mathbf{H}.$$

All our examples of isospectral surfaces obtained in Chapters 11 and 12, by pasting together hyperbolic building blocks, have this form.

Since we are only interested in smooth isospectral surfaces, we have to add a restriction to our present construction. Recall that M_1 and M_2 are smooth if and only if Γ_1 and Γ_2 act freely on $\mathbf{H}$, that is, if and only if Γ_1 and Γ_2 contain no elements of finite order. Accordingly, we shall only admit those surjective homomorphisms $w : \mathfrak{G}_\sigma \to G$ for which $w^{-1}(H_1)$ and $w^{-1}(H_2)$ contain no elements of finite order other than the identity. If this condition is satisfied we shall say that $(w, \mathfrak{G}_\sigma; G, H_1, H_2)$ is an *admissible sequence* for Sunada's construction.

Once this sequence is given, we obtain an isospectral pair M_1, M_2 for any choice of the action $\Gamma_0 = \rho(\mathfrak{G}_\sigma)$. The groups Γ_0 fill out the Teichmüller space $\mathcal{F}_\sigma$, and we are interested in the cases where the corresponding surfaces M_1

and M_2 are non-isometric. The following theorem is due to Sunada [3]. The word "generic" will be explained in the proof.

12.7.5 Theorem. *Let* $(w, \mathfrak{G}_\sigma; G, H_1, H_2)$ *be an admissible sequence for Sunada's construction, and assume that the signature σ is not one of those listed in* (12.7.1). *Then for a generic choice of* $\Gamma_0 = \rho(\mathfrak{G}_\sigma) \in \mathcal{F}_\sigma$, *the surfaces M_1 and M_2 are non-isometric.*

Proof. This proof uses properties of arithmetic groups, for which we refer the reader to the book of Zimmer [1].

We know that $\mathcal{F}_\sigma$ is homeomorphic to $\mathbf{R}^n$ for some n, and that only countably many $\Gamma_0 \in \mathcal{F}_\sigma$ are arithmetic. By a theorem of Margulis, the group

$$\text{Comm}(\Gamma) = \{\alpha \in \text{Is}(\mathbf{H}) \mid \alpha \Gamma \alpha^{-1} \text{ is commensurable with } \Gamma\}$$

is discrete if Γ is not arithmetic. Here Γ and Γ' are said to be *commensurable* if $\Gamma \cap \Gamma'$ has finite index in Γ and in Γ'. If $\Gamma_0 \in \mathcal{F}'_\sigma$, where $\mathcal{F}'_\sigma$ is as in Theorem 12.7.2, then Γ_0 cannot be contained in any larger discrete subgroup of $\text{PSL}(2, \mathbf{R})$, because $\Gamma_0 \backslash \mathbf{H}$ has trivial isometry group. It follows that if $\mathcal{F}^*_\sigma$ denotes the set of all non-arithmetic groups in $\mathcal{F}'_\sigma$, then

$$\text{Comm}(\Gamma_0) = \Gamma_0 \text{ for any } \Gamma_0 \in \mathcal{F}^*_\sigma.$$

Now let $\Gamma, \Gamma_1, \Gamma_2$ and Γ_0 be the groups as in (12.7.3), with corresponding surfaces M, M_1, M_2 and M_0 as in (12.7.4). Assume that $\Gamma_0 \in \mathcal{F}^*_\sigma$. For $i = 0, 1, 2$, we have normal coverings $M \to M_i$ with deck transformation groups

$$H_1 = \Gamma_1/\Gamma, \quad H_2 = \Gamma_2/\Gamma, \quad G = \Gamma_0/\Gamma.$$

If M_1 is isometric to M_2, then there exists an isometry $\alpha \in \text{Is}(\mathbf{H})$ such that $\Gamma_2 = \alpha \Gamma_1 \alpha^{-1}$. It follows that Γ_2 is a finite index subgroup of Γ_0 *and also of* $\alpha \Gamma_0 \alpha^{-1}$. Hence, Γ_0 and $\alpha \Gamma_0 \alpha^{-1}$ are commensurable and $\alpha \in \text{Comm}(\Gamma_0)$. Since $\text{Comm}(\Gamma_0) = \Gamma_0$, it follows that $\alpha \in \Gamma_0$. Therefore, Γ_1 and Γ_2 are conjugate in Γ_0, and H_1 is a conjugate in G of H_2, a contradiction. Hence, M_1 and M_2 are non-isometric for all $\Gamma_0 \in \mathcal{F}^*_\sigma$. ◊

In the second theorem (Theorem 12.7.6 below) we restrict ourselves to the Teichmüller space $\mathcal{T}_h$ of the (smooth) Riemann surfaces of genus h, $h \geq 2$. $\mathfrak{G}_h$ denotes the abstract group, with generators $b_1, \ldots, b_{2h}$, defined by the single relation

$$\prod_{v=1}^{h} b_{2v-1} b_{2v} b_{2v-1}^{-1} b_{2v}^{-1} = 1.$$

We let F be the base surface for the marking homeomorphisms of $\mathcal{T}_h$; and we choose canonical generators $\beta_1, \ldots, \beta_{2h}$ of the fundamental group of F (at some given point of F), as described in Section 6.7. We denote by $x = (M_0(x), \varphi_x)$ a variable element of $\mathcal{T}_h$ ($\varphi_x : F \to M_0(x)$ is the marking homeomorphism). Via Theorem 6.8.13 we have, for each $x \in \mathcal{T}_h$, a Fuchsian group $\Gamma_0(x) = \psi \circ \pi(x) = \tau^{-1}(x) \in \mathcal{F}_h$, where the notation is as in Theorem 6.8.13. More precisely, we have a mapping

$$(x, a) \mapsto \rho_x(a) \in \text{PSL}(2, \mathbf{R}), \; x \in \mathcal{T}_h, \; a \in \mathfrak{G}_h$$

with the following properties. (i) For fixed x, the mapping

$$\rho_x : \mathfrak{G}_h \to \Gamma_0(x) := \rho_x(\mathfrak{G}_h) \subset \text{PSL}(2, \mathbf{R})$$

is a group isomorphism. (ii) For fixed $a \in \mathfrak{G}_h$, the mapping $x \mapsto \rho_x(a)$ is a real analytic mapping from $\mathcal{T}_h$ into $\text{PSL}(2, \mathbf{R})$. (iii) For any x, the elements $B_1 := \rho_x(b_1), \ldots, B_{2h} := \rho_x(b_{2h})$ are normal canonical generators of $\Gamma_0(x)$ (cf. Section 6.8). (*Normal* means that the axis of B_1 is oriented from ∞ to 0 and the axis of B_2 is oriented from σ_2 to τ_2, where $\sigma_2, \tau_2 \in \mathbf{R}$ with $\sigma_2 < 0 < \tau_2$ and $\sigma_2 \tau_2 = -1$.)

Consider again the surjective homomorphism $w : \mathfrak{G}_h \to G$, where G is the finite group in the triplet (G, H_1, H_2), and consider the inverse images

$$\mathfrak{G}_{h1} = w^{-1}(H_1), \quad \mathfrak{G}_{h2} = w^{-1}(H_2).$$

Then $\Gamma_1(x) := \rho_x(\mathfrak{G}_{h1})$ and $\Gamma_2(x) := \rho_x(\mathfrak{G}_{h2})$ are subgroups of $\Gamma_0(x)$ whose signature, say $(g; -)$, is independent of x. Setting $\omega_x = w \circ \rho_x^{-1}$ we have, for any $x \in \mathcal{T}_h$, the surjective group homomorphism

$$\omega_x : \Gamma_0(x) \to G.$$

The groups

$$\Gamma_1(x) = \omega_x^{-1}(H_1), \quad \Gamma_2(x) = \omega_x^{-1}(H_2)$$

$$\Gamma(x) := \omega_x^{-1}(\{id\})$$

are as in (12.7.3), and the surfaces

$$M(x) = \Gamma(x) \backslash \mathbf{H}$$

$$M_1(x) = \Gamma_1(x) \backslash \mathbf{H} \quad M_2(x) = \Gamma_2(x) \backslash \mathbf{H}$$

are as in (12.7.4). By Sunada's theorem, $M_1(x)$ and $M_2(x)$ are isospectral.

We now prove a variant of Theorem 12.7.5 (g is the common genus of the surfaces $M_i(x)$, $x \in \mathcal{T}_h$, $i = 1, 2$).

12.7.6 Theorem. *Assume that in the above construction the surfaces $M_1(x_0)$ and $M_2(x_0)$ are non-isometric for at least one $x_0 \in \mathcal{T}_h$. Then there exists an open subset $U \subset \mathcal{T}_h$ such that $M_1(x)$ and $M_2(x)$ are non-isometric for any $x \in U$, and the set of all surfaces $M_1(x)$ forms a $(6h-6)$-dimensional local real analytic subvariety of $\mathcal{T}_g$ consisting of pairwise non-isometric surfaces.* ◇

Proof. We first prove the following.

(12.7.7) *The mapping $x \mapsto M_i(x) = \Gamma_i(x)\backslash \mathbf{H}$ is a real analytic embedding of $\mathcal{T}_h$ into $\mathcal{T}_g$, $i = 1, 2$.*

In view of Remark 6.8.10 and Theorem 6.8.13, it suffices to show that the mapping $x \mapsto \Gamma_i(x)$ is a real analytic embedding of $\mathcal{T}_h$ into $\mathcal{F}_g$. We show this for $i = 1$.

Let $B_1, \ldots, B_{2h}$ be the above normal canonical generators of $\Gamma_0(x)$. We may understand $B_1, \ldots, B_{2h}$ as elements of $SL(2, \mathbf{R})$ with positive traces whose components are analytic functions of x.

We fix x temporarily. The canonical isomorphism between $\Gamma_0(x)$ and the fundamental group of $M_0(x) = \Gamma_0(x)\backslash\mathbf{H}$ sends $B_1, \ldots, B_{2h}$ to curves $\beta_1, \ldots, \beta_{2h}$ which form a canonical dissection of $M_0(x)$. From the topology of these curves we see that there are exponents ℓ_ν, m_ν and n_ν, $\nu = 1, \ldots, h$, such that the curves $\beta_{2\nu-1}^{\ell_\nu}$ and $\beta_{2\nu}^{m_\nu}\beta_{2\nu-1}^{-n_\nu}$ have lifts, in the covering surface $M_1(x) = \Gamma_1(x)\backslash\mathbf{H}$, which are simple closed loops with a common base point. These lifts can be extended to a canonical dissection $\alpha_1, \ldots, \alpha_{2g}$ of $M_1(x)$. We label the curves of this dissection in such a way that α_1 is a lift of $\beta_1^{\ell_1}$ and α_2 is a lift of $\beta_2^{m_1}\beta_1^{-n_1}$. Each pair $\beta_{2\nu-1}^{\ell_\nu}, \beta_{2\nu}^{m_\nu}\beta_{2\nu-1}^{-n_\nu}$ has a pair of lifts $\alpha_{2\kappa-1}, \alpha_{2\kappa}$ for some $\kappa = \kappa(\nu)$. This yields the following.

For $\nu = 1, \ldots, h$, the elements $B_{2\nu-1}^{\ell_\nu}$ and $B_{2\nu}^{m_\nu}B_{2\nu-1}^{-n_\nu}$ of $\Gamma_0(x)$ belong to $\Gamma_1(x)$, and can be extended to a system of canonical generators $A_1, \ldots, A_{2g}$ of $\Gamma_1(x)$ with the following properties.

(i) $A_1 = B_1^{\ell_1}$ and $A_2 = B_2^{m_1}B_1^{-n_1}$.

(ii) *For each ν, there exists an integer $\kappa = \kappa(\nu)$ such that $A_{2\kappa-1} = B_{2\nu-1}^{\ell_\nu}$ and $A_{2\kappa} = B_{2\nu}^{m_\nu}B_{2\nu-1}^{-n_\nu}$.*

(iii) *Each A_k is a product $A_k = B_{i_1}^{\varepsilon_1}\ldots B_{i_s}^{\varepsilon_s}$, with $s = s(k)$, $i_\sigma \in \{1, \ldots, 2h\}$ and $\varepsilon_\sigma \in \{-1, 1\}$ for $\sigma = 1, \ldots, s(k)$.*

(iv) *The axis of A_1 is the imaginary axis oriented from ∞ to 0, the axis of A_2 intersects the imaginary axis at some point $z = i\eta(x)$ and is oriented from $\mathcal{R}e\, z < 0$ to $\mathcal{R}e\, z > 0$.*

Since $\rho_x : \mathfrak{G}_h \to \Gamma_0(x)$ is an isomorphism of groups, the elements $a_1, \ldots, a_{2g} \in \mathfrak{G}_h$, defined by $a_k = b_{i_1}^{\varepsilon_1} \ldots b_{i_s}^{\varepsilon_s}$, are generators of $\mathfrak{G}_{h1}$, and we have

(v) $\qquad A_k = \rho_x(a_k), k = 1, \ldots, 2g.$

Now let $x \in \mathcal{T}_g$ again be variable, and fix the exponents and integer valued functions occurring in (i) - (iv). Then $A_1, \ldots, A_{2g}$ are functions of x defined by (v), for which (i) - (iv) remain valid; and for any x they are canonical generators of a Fuchsian group $\Gamma^*(x) \in \mathcal{T}_g$. However, since $A_1, \ldots, A_{2g}$ are defined by (v), this group satisfies $\Gamma^*(x) = \rho_x(\mathfrak{G}_{h1}) = \Gamma_1(x)$. Therefore, (v) defines generators of $\Gamma_1(x)$.

The A_k are not normal canonical generators since the axes of A_1 and A_2 intersect each other at $z = i\eta(x)$, instead of $z = i$. We conjugate them therefore with

$$L = \begin{pmatrix} \eta^{1/2}(x) & 0 \\ 0 & \eta^{-1/2}(x) \end{pmatrix},$$

defining $A'_k = L^{-1} A_k L$ for $k = 1, \ldots, 2g$. Note that $\eta(x)$ is a positive valued real analytic function of x, $x \in \mathcal{T}_h$. Now $A'_1, \ldots, A'_{2g}$ are normal canonical generators of $\Gamma_1(x)$ and, seen as elements of $SL(2, \mathbf{R})$, have positive traces.

For $k = 1, \ldots, 2g$ we let $\sigma'_k(x), \tau'_k(x)$ and $\lambda'_k(x)$ be the invariants of A'_k, as in Section 6.8, such that $(\sigma'_3(x), \ldots, \sigma'_{2g}(x), \tau'_3(x), \ldots, \tau'_{2g}(x), \lambda'_3(x), \ldots, \lambda'_{2g}(x))$ is the sequence of Bers coordinates of $\Gamma_1(x)$ (cf. Definition 6.8.5). By (iii) and formulae (6.8.14), these invariants are analytic functions of x. Hence, the mapping $x \mapsto \Gamma_i(x) \in \mathcal{F}_g$ is real analytic. By (ii) and (6.8.14) there exist real analytic functions which, for $v = 1, \ldots, h$, compute the invariants $\sigma_{2v-1}, \tau_{2v-1}, \lambda_{2v-1}$ of B_{2v-1} in terms of $\sigma'_{2\kappa-1}(x), \tau'_{2\kappa-1}(x), \lambda'_{2\kappa-1}(x)$, and similarly the invariants $\sigma_{2v}, \tau_{2v}, \lambda_{2v}$ of B_{2v} in terms of $\sigma'_{2\kappa}(x), \tau'_{2\kappa}(x), \lambda'_{2\kappa}(x)$. Hence, the mapping $x \mapsto \Gamma_i(x)$ is an embedding of $\mathcal{T}_h$ into $\mathcal{F}_g$. This proves (12.7.7).

By Theorem 6.5.4, two surfaces $S, S' \in \mathcal{T}_g$ are isometric if and only if there exists a mapping μ in the Teichmüller modular group $\mathcal{M}_g$ satisfying $S' = \mu(S)$. Since $\mathcal{M}_g$ acts properly discontinuously on $\mathcal{T}_g$, it follows that if $M_1(x)$ and $M_2(x)$ are non-isometric, then there exist open neighborhoods U_i of $M_i(x)$ in $\mathcal{T}_g$, $i = 1, 2$, such that any pair $(S, S') \in U_1 \times U_2$ is non-isometric. Now let $\mathcal{T}$ be the image of $\mathcal{T}_h$ under the embedding $x \mapsto M_1(x)$ as in (12.7.7), and let $\Omega = \mathcal{T} \cap U_1$. By Lemma 6.5.8 there exists a proper real analytic subvariety $\mathcal{I}$ of Ω such that, if $M_1(x) \in \Omega - \mathcal{I}$, then there exists an open neighborhood U of x in $\mathcal{T}_h$ such that $M_1(x')$ and $M_1(x'')$ are non-isometric for any $x', x'' \in U$. This proves the theorem. ◇

Chapter 13

The Size of Isospectral Families

McKean [1] has shown that only finitely many compact Riemann surfaces have a given spectrum. In this chapter we give an explicit bound which depends only on the genus. The proof covers Sections 13.2 - 13.4. In the first section we state the result and give a proof of McKean's theorem in order to outline the approach. In the first section we also review the necessary material from earlier chapters.

13.1 Finiteness

13.1.1 Theorem. *Let S^* be a compact Riemann surface of genus $g \geq 2$. At most $\exp(720\, g^2)$ pairwise non-isometric compact Riemann surfaces are isospectral to S^*.*

13.1.2 Remarks. (i) In Section 12.6 we gave examples with roughly $g^{1/3}$ isospectral non-isometric pairs. Hence, there exists no g-independent upper bound. It would be interesting to know whether there are examples with exponentially many pairs.

(ii) By Weyl's asymptotic law (Theorem 9.2.14), compact Riemann surfaces of different genus cannot be isospectral. Hence, we may restrict ourselves to work with a fixed genus.

(iii) It is interesting to observe that the bound in Theorem 13.1.1 is a universal upper bound for the number of occurrences of the spectrum of a compact Riemann surface. In fact, the dimension of a compact Riemannian manifold M is determined by the spectrum of the Laplacian, and in dimension 2 the spectrum of M determines whether or not for given $\kappa \in \mathbf{R}$, M has constant curvature κ (see for instance Berger-Gauduchon-Mazet [1]). Hence, if S^* is a compact Riemann surface of genus $g \geq 2$ and if M is any compact Riemannian manifold isospectral to S^*, then M is also a compact Riemann surface of genus g. The spectrum of S^* occurs therefore at most $\exp(720\, g^2)$ times among the spectra of *all* compact Riemannian manifolds.

(iv) Berry [1], Pesce [1] and Wolpert [2] have an analog of Theorem 13.1.1 for flat tori. Berry's result is in dimension 3 and states that at most 15 different tori can be isospectral to a given one. Wolpert proves the finiteness of isospectral sets of flat tori in all dimensions, and Pesce improves this result giving explicit bounds depending on the volume and the injectivity radius of the torus. ◊

We now collect the necessary material from earlier chapters and give a short proof of McKean's theorem that only finitely many compact Riemann surfaces of genus g are isospectral to S^*.

The proof is based on two main tools. One is Bers' theorem (Theorems 5.1.2 and 5.2.3) which states that any compact Riemann surface of genus g can be decomposed into pairs of pants along pairwise disjoint simple closed geodesics of length $\leq L_g$, where L_g is a constant depending only on g. In Chapter 5 we obtained the explicit bound $L_g \leq 26(g-1)$.

The other tool is the explicit description of the compact Riemann surfaces in the form $S = F(G, L, A)$ in Section 3.6 together with the fact that these surfaces are determined, up to isometry, by the lengths of certain closed geodesics. The idea is that if S is isospectral to S^*, then the lengths of the determining geodesics occur in the length spectrum of S^* and this is possible in only finitely many ways.

The surface $S = F(G, L, A)$ is obtained by pasting together $2g - 2$ pairs of pants with respect to the pattern of a cubic graph G with $2g - 2$ vertices. The boundary geodesics of the pants yield closed geodesics $\gamma_1(S), \ldots, \gamma_{3g-3}(S)$ on S, the *parameter geodesics*, where

$$L = (\ell_1, \ldots, \ell_{3g-3}) = (\ell(\gamma_1(S)), \ldots, \ell(\gamma_{3g-3}(S))).$$

The parameter

$$A = (\alpha_1, \ldots, \alpha_{3g-3}) = (\alpha_1(S), \ldots, \alpha_{3g-3}(S))$$

is the sequence of *twist parameters* of S. In order to give a precise meaning to sentences such as "how does the length of geodesic β on S vary as S varies?", we introduced in Section 6.2 for each $F(G, L, A)$ a particular homeomorphism $\varphi = \varphi(G, L, A) : F \to F(G, L, A)$, where F is a fixed base surface. We let $S(G, L, A)$ denote the surface $F(G, L, A)$ together with this homeomorphism. (In Section 6.2 we used the notation S^ω.) For any homotopically non-trivial closed curve β on F and for any $S = S(G, L, A)$ marked with $\varphi = \varphi(G, L, A)$ we denote by $\beta(S)$ the closed geodesic in the homotopy class of $\varphi \circ \beta$. We shall say that $\beta(S)$ is "β on S". When S remains fixed, we sometimes write $\beta(S) = \beta$ in order to simplify the notation.

The marking homeomorphisms $\varphi = \varphi(G, L, A)$ have been designed in such a way that for any β the lengths $\ell(\beta(S(G, L, A)))$ are real analytic functions of (L, A) (Theorem 6.3.5). Moreover, if

$$\Omega_G = \{\gamma_1, \ldots, \gamma_{3g-3}, \delta_1, \ldots, \delta_{3g-3}, \eta_1, \ldots, \eta_{3g-3}\}$$

is the curve system on F as introduced in Definition 6.1.6, then the curves $\gamma_1(S), \ldots, \gamma_{3g-3}(S)$ coincide with the parameter geodesics of $S = S(G, L, A)$, and for fixed L, each twist parameter $\alpha_i(S)$ is computable in terms of $\ell(\delta_i(S))$ and $\ell(\eta_i(S))$, and $|\alpha_i(S)|$ is computable in terms of $\ell(\delta_i(S))$ alone (cf. Proposition 3.3.11).

Finally, by Bers' theorem, any compact Riemann surface of genus g is isometric to $S(G, L, A)$ for some graph G and some L and A satisfying $0 < \ell_i \leq L_g$, $-\frac{1}{2} \leq \alpha_i \leq \frac{1}{2}$, $i = 1, \ldots, 3g-3$.

As in Chapter 10, we denote by Lsp(S) the sequence of the lengths of all non-oriented primitive closed geodesics on S, arranged in ascending order. The finiteness theorem without bounds is as follows (McKean [1]).

13.1.3 Theorem. *Any set of pairwise isospectral non-isometric compact Riemann surfaces is finite.*

Proof. Fix S^* and let λ be the length of the smallest non-trivial closed geodesic on S^*. If Lsp(S) = Lsp(S^*), then there exists a cubic graph G such that S is isometric to a surface in $\mathcal{F}_\lambda(G)$, where

$$\mathcal{F}_\lambda(G) = \{S(G, L, A) \mid \lambda \leq \ell_i \leq L_g, -\tfrac{1}{2} \leq \alpha_i \leq \tfrac{1}{2}; i = 1, \ldots, 3g-3\}.$$

Since the number of cubic graphs G with $2g - 2$ vertices is finite, it suffices to prove that the set

$$\mathcal{F}_\lambda^*(G) = \{S \in \mathcal{F}_\lambda(G) \mid \text{Lsp}(S) = \text{Lsp}(S^*)\}$$

is finite. Let $\mu = \max\{\ell(\beta(S)) \mid \beta \in \Omega_G, S \in \mathcal{F}_\lambda(G)\}$ and let χ be the cardinality of Lsp(S^*) $\cap$ $[0, \mu]$. Then for $i = 1, \ldots, 3g-3$, at most χ values of $\ell(\gamma_i(S))$ and at most χ values of $\ell(\delta_i(S))$ are possible for $S \in \mathcal{F}_\lambda^*(G)$. Since $|\alpha_i(S)|$ is determined by $\ell(\delta_i(S))$, this shows that $\mathcal{F}_\lambda^*(G)$ is finite ◇

In the preceding proof it would not be difficult to find explicit bounds for μ and χ in terms of λ. The problem is to make it independent of λ.

We now begin with the proof of Theorem 13.1.1. As working tools we use the collar theorem (in the form of Theorems 4.1.1, 4.1.6 and 4.2.1) and the trigonometric formulae for the right-angled triangles, trirectangles and pentagons.

By Theorem 4.1.6, S^* has at most $3g - 3$ simple closed geodesics of length ≤ 1. We let m and n be such that $3g - 3 - m$ is the number of simple closed geodesics of length ≤ 1, and $3g - 3 - n$ is the number of simple closed geodesics of length $\leq \exp(-4g)$ on S^*. (The cumbersome borderline $\exp(-4g)$ has its roots in the quantity ε_k which will be introduced in (13.2.4); our arguments in the case of large ε_k will not be the same as for small ε_k.)

By the collar theorem (Theorem 4.1.1), any simple closed geodesic γ of length $\ell(\gamma)$ on S^* has an open neighborhood

$$\mathscr{C}(\gamma) = \{p \in S^* \mid \mathrm{dist}(p, \gamma) \leq w(\gamma)\},$$

the *collar* around γ, of *width*

$$w(\gamma) = \mathrm{arcsinh}\{1/\sinh(\tfrac{1}{2}\ell(\gamma))\}$$

which is homeomorphic to an annulus. Any closed geodesic β which intersects γ transversally intersects both boundary components of $\mathscr{C}(\gamma)$ and has length $\ell(\beta) \geq 2w(\gamma)$. The following is clear.

(13.1.4) *If γ is a simple closed geodesics of length $\ell(\gamma) \leq e^{-4g}$, then its collar has width $w(\gamma) > 4g$.*

For the proof of Theorem 13.1.1 we proceed in four steps. Recall that any compact Riemann surface S isospectral to S^* is isometric to one of the surfaces $S(G, L, A)$. In the first step we estimate the number of graphs and the number of possible L-parameters. In the second step (Section 13.2) we estimate the number of possible twist parameters α_i for $i \leq n$. Since here the lengths of the geodesics δ_i have a uniform upper bound, this step is straightforward.

In the third step (Section 13.3) we introduce particularly well adapted geodesics ϑ_i to measure the twist parameters α_i for $i = n + 1, \ldots, 3g - 3$. The correct choice of the ϑ_i is the heart of the proof. In the final step (Section 13.4) we estimate the number of those geodesics on S^* whose lengths are the length of some $\vartheta_i(S)$ where S is isospectral to S^*.

(13.1.5) *If $\mathrm{Lsp}(S) = \mathrm{Lsp}(S^*)$, then S is isometric to a surface $S(G, L, A)$, where*

$$-\tfrac{1}{2} \leq \alpha_1, \ldots, \alpha_{3g-3} \leq \tfrac{1}{2}$$

and
$$26(g-1) \geq \ell_1, \ldots, \ell_m > 1 \geq \ell_{m+1}, \ldots, \ell_n > e^{-4g} \geq \ell_{n+1}, \ldots, \ell_{3g-3}.$$

Proof. Since S and S^* are isospectral, S has exactly $3g - 3 - m$ primitive closed geodesics of length ≤ 1. By Theorem 4.2.1, these geodesics are simple and by Theorem 4.1.6, they are pairwise disjoint. By Theorem 5.2.3, they belong to a partition with lengths $\leq 26(g-1)$ and the statement follows. ◊

(13.1.6) *At most g^{3g} different cubic graphs G occur in (13.1.5).*

Proof. This is part of Theorem 3.5.3. ◊

From now on we fix one of the graphs G occurring in (13.1.5) and define
$$\mathcal{F}(G) = \{ S(G, L, A) \mid 0 < \ell_i < 26(g-1), -\tfrac{1}{2} \leq \alpha_i \leq \tfrac{1}{2}; i = 1, \ldots, 3g-3 \},$$
$$\mathcal{F}^*(G) = \{ S \in \mathcal{F}(G) \mid \mathrm{Lsp}(S) = \mathrm{Lsp}(S^*) \}.$$

By Lemma 6.6.4 there are at most $(g-1)\exp(26(g-1)+6)$ different simple closed geodesics of length $\leq 26(g-1)$ on S^*. If $S = S(G, L, A) \in \mathcal{F}^*(G)$, then each ℓ_i of L has the length of one of these geodesics on S^*. Taking our bound to the power $3g - 3$, we obtain the following.

(13.1.7) *At most $\exp(79g^2)$ different values of L are possible for the surfaces $S = S(G, L, A) \in \mathcal{F}^*(G)$.*

13.2 Parameter Geodesics of Length $> \exp(-4g)$

From now on we fix one such value L and define
$$\mathcal{F}(G, L) = \{ S(G, L, A) \mid -\tfrac{1}{2} \leq \alpha_i \leq \tfrac{1}{2}; i = 1, \ldots, 3g-3 \},$$
$$\mathcal{F}^*(G, L) = \{ S \in \mathcal{F}(G, L) \mid \mathrm{Lsp}(S) = \mathrm{Lsp}(S^*) \}.$$

In $\mathcal{F}(G, L)$ the twist parameters $\alpha_1, \ldots, \alpha_{3g-3}$ are variables and the length parameters $\ell_1, \ldots, \ell_{3g-3}$ are constants. We let the numbering of the parameter geodesics $\gamma_1, \ldots, \gamma_{3g-3}$ be such that the inequalities (13.1.5) hold. For the geodesics $\delta_1, \ldots, \delta_{3g-3}$ (cf. Definition 6.1.6) which measure the twist parameters up to the minus sign, we claim the following.

(13.2.1) $\ell(\delta_j(S)) \leq 78g - 6$ *for any* $S \in \mathcal{F}(G, L)$, $j = 1, \ldots, n$.

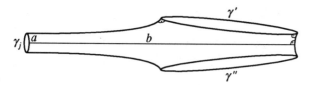

Figure 13.2.1

Proof. Let Y with boundary geodesics γ_j, γ' and γ'', be one of the Y-pieces of the given partition of S. The perpendiculars between the boundary geodesics decompose Y into two isometric right-angled geodesic hexagons (Proposition 3.1.5). Dropping the common perpendicular b from γ_j to the opposite side in both hexagons, we decompose Y into four right-angled pentagons as shown in Fig. 13.2.1.

In this decomposition, γ' and γ'' fall into two arcs of equal length, and γ_j falls into four arcs of two different lengths. We let a be the length which satisfies $a \geq \frac{1}{4}\ell(\gamma_j)$. We label γ' and γ'' in such a way that, as in Fig. 13.2.1, γ' contains a side of the pentagon with sides a, b. Formula (i) of Theorem 2.3.4 yields

$$\sinh a \sinh b = \cosh \tfrac{1}{2}\ell(\gamma').$$

Since $\ell(\gamma') \leq 26(g-1)$ we get

$$b \leq b_j := \operatorname{arcsinh}\{(\cosh 13(g-1))/\sinh \tfrac{1}{4}\ell(\gamma_j)\}.$$

This inequality holds for both Y-pieces adjacent at γ_j, and since the twist parameters are between $-\frac{1}{2}$ and $\frac{1}{2}$, it follows that δ_j is homotopic to a curve of length $\leq 2\ell(\gamma_j) + 4b_j$. Now (13.2.1) follows from the inequality $\exp(-4g) \leq \ell(\gamma_j) \leq 26(g-1)$ via an elementary estimate (distinguish the cases $\ell(\gamma_j) \leq 1$, $1 \leq \ell(\gamma_j) \leq g$ and $\ell(\gamma_j) \geq g$). ◊

If $S \in \mathscr{F}^*(G, L)$, then $\ell(\delta_j(S))$ appears in $\mathrm{Lsp}(S^*)$. By Lemma 6.6.4 there are at most $(g-1)\exp(78g)$ spectral lines in $\mathrm{Lsp}(S^*) \cap [0, 78g - 6]$. Since $\ell(\delta_j(S))$ determines the twist parameter $\alpha_j(S)$ (up to the minus sign), we get the following.

(13.2.2) *At most $\exp(79g)$ values of $\alpha_j(S)$ occur in $\mathscr{F}^*(G, L)$, $j = 1, \ldots, n$.*

We estimate the cardinality of $\mathscr{F}^*(G, L)$ and assume that $n < 3g - 3$ and that $\mathscr{F}^*(G, L)$ is not empty (otherwise we are done). Let $\bar{S}$ be a fixed surface in $\mathscr{F}^*(G, L)$ and let $\bar{\alpha}_1, \ldots, \bar{\alpha}_{3g-3}$ be its twist parameters. According to our convention, $\bar{\alpha}_1, \ldots, \bar{\alpha}_n$ correspond to the parameter geodesics of $\bar{S}$ of length >

$\exp(-4g)$. We now keep the sequence $\bar{\alpha}_1, \ldots, \bar{\alpha}_n$ fixed and define

$$\mathcal{F}(G, L, \bar{\alpha}_1, \ldots, \bar{\alpha}_n) = \{ S \in \mathcal{F}(G, L) \mid \alpha_1(S) = \bar{\alpha}_1, \ldots, \alpha_n(S) = \bar{\alpha}_n \},$$
$$\mathcal{F}^*(G, L, \bar{\alpha}_1, \ldots, \bar{\alpha}_n) = \{ S \in \mathcal{F}(G, L, \bar{\alpha}_1, \ldots, \bar{\alpha}_n) \mid \mathrm{Lsp}(S) = \mathrm{Lsp}(S^*) \}.$$

There exists no a priori upper bound for $\ell(\delta_i(S))$ for $i > n$; and as α_i runs through the interval $[-\frac{1}{2}, \frac{1}{2}]$, $\ell(\delta_j)$ sweeps out an interval in which there may be arbitrarily many spectral lines of $\mathrm{Lsp}(S^*)$. We shall therefore abandon the δ_j and measure the twist parameters with geodesics which are better adapted. We shall also work with a carefully defined hierarchy among the parameter geodesics $\gamma_{n+1}, \ldots, \gamma_{3g-3}$.

Assume by induction that $i \geq n$ and that a sequence of twist parameters $\bar{\alpha}_1, \ldots, \bar{\alpha}_i$ belonging to some surface in $\mathcal{F}^*(G, L)$ has been fixed, where $\bar{\alpha}_1, \ldots, \bar{\alpha}_i$ contains $\bar{\alpha}_1, \ldots, \bar{\alpha}_n$ as a subsequence, and assume that the sets

(13.2.3) $\quad \mathcal{F}_i = \mathcal{F}(G, L, \bar{\alpha}_1, \ldots, \bar{\alpha}_i), \quad \mathcal{F}_i^* = \mathcal{F}^*(G, L, \bar{\alpha}_1, \ldots, \bar{\alpha}_i)$

have been defined, where

$$\mathcal{F}_i^* = \{ S \in \mathcal{F}_i \mid \mathrm{Lsp}(S) = \mathrm{Lsp}(S^*) \}.$$

In $\mathcal{F}_i$ we have to relabel the parameter geodesics $\gamma_{i+1}, \ldots, \gamma_{3g-3}$ before defining $\mathcal{F}(G, L, \bar{\alpha}_1, \ldots, \bar{\alpha}_{i+1})$. To this end we set

$$t_k := \min\{ \ell(\vartheta(S)) \mid S \in \mathcal{F}_i \text{ and } \vartheta(S) \text{ is a closed geodesic on } S \text{ which intersects } \gamma_k \text{ transversally} \},$$

(13.2.4)

$$\varepsilon_k := (8g)^4 \ell^2(\gamma_k), \quad k = i+1, \ldots, 3g-3.$$

Since $\ell(\gamma_k) \leq \exp(-4g)$ we have

(13.2.5) $\quad\quad \varepsilon_k < 1/100, \quad k = i+1, \ldots, 3g-3$.

In $\mathcal{F}_i$ we now renumber the $\gamma_{i+1}, \ldots, \gamma_{3g-3}$ such that the following *condition* is satisfied:

(13.2.6) $\quad\quad t_{i+1} + \varepsilon_{i+1} \leq t_k + \varepsilon_k, \quad k = i+2, \ldots, 3g-3$.

We would like to remark that, although ε_k/t_k may be arbitrarily small, and although to some extent the coefficient $(8g)^4$ in the definition of ε_k is rough, the reader should not replace the condition "$t_{i+1} + \varepsilon_{i+1} \leq t_k + \varepsilon_k$" by the condition "$t_{i+1} \leq t_k$", for otherwise the proof of (13.4.3) below will not work (cf. Remark 13.4.4).

Condition (13.2.6) will allow us to prove the following.

(13.2.7) *At most* $\exp(213g)$ *values of* $\alpha_{i+1}(S)$ *occur in* $\mathcal{F}_i^*$.

In conjunction with (13.1.6), (13.1.7) and (13.2.2) we then have at most $\exp(720g^2)$ different sets $\mathcal{F}^*(G, L, \bar{\alpha}_1, \ldots, \bar{\alpha}_{3g-3})$. Since by (13.1.5), any S isospectral to S^* occurs in one of these sets, and since each set consists of exactly one element, this will prove Theorem 13.1.1.

13.3 Measuring the Twist Parameters

For the proof of (13.2.7) we may assume without loss of generality that $S^* \in \mathcal{F}_i^*$ (otherwise $\mathcal{F}_i^* = \emptyset$). As remarked earlier, if $k \in \{i+1, \ldots, 3g-3\}$, then the lengths of the geodesic $\delta_k(S)$, $S \in \mathcal{F}_i$, sweep out an interval which may contain arbitrarily many spectral lines of $\mathrm{Lsp}(S^*)$. We introduce therefore a set of new geodesics which also measure the twist parameters and whose lengths sweep out intervals in which the number of spectral lines has an upper bound depending only on g.

(13.3.1) *For each* $k \in \{i+1, \ldots, 3g-3\}$ *there exists a Riemann surface* $S_k \in \mathcal{F}_i$ *and a simple closed geodesic* $\vartheta_k(S_k)$ *intersecting* $\gamma_k(S_k)$ *transversally such that* $\ell(\vartheta_k(S_k)) = t_k$.

Proof. Let $S_0 \in \mathcal{F}_i$ be defined by $\alpha_{i+1}(S_0) = \ldots = \alpha_{3g-3}(S_0) = 0$, and let $\vartheta(S_0)$ be a closed geodesic on S_0 which intersects $\gamma_k(S_0)$ transversally. As $\alpha_{i+1}(S), \ldots, \alpha_{3g-3}(S)$ run through the interval $[-\frac{1}{2}, \frac{1}{2}]$, S runs through $\mathcal{F}_i$ and $\ell(\vartheta(S))$ is a continuous function which ranges in an interval of width less than $\ell(\gamma_{i+1}) + \ldots + \ell(\gamma_{3g-3})$. Observe that only finitely many closed geodesics ϑ' exist such that $\ell(\vartheta'(S)) \leq \ell(\vartheta(S_0))$ for some $S \in \mathcal{F}_i$. Hence, there exists $S_k \in \mathcal{F}_i$ and a closed geodesic $\vartheta_k(S_k)$ which intersects $\gamma_k(S_k)$ transversally such that $\ell(\vartheta_k(S_k)) = t_k$. It remains to show that ϑ_k is simple.

To this end let us walk along ϑ_k starting at an intersection point with γ_k. In order to return to γ_k we have to leave the collar $\mathcal{C}(\gamma_k)$. There are two cases.

Case 1. Departure and first return to γ_k take place on opposite sides of γ_k (Fig. 13.3.1).

We claim that in this case the tour *ends* at the first return. Suppose not. We then would have to leave $\mathcal{C}(\gamma_k)$ a second time. Since the width $w(\gamma_k)$ of the collar is larger than $\ell(\gamma_k)$, it would be possible to construct a homotopically non-trivial closed curve ϑ_k^* shorter than ϑ_k and also leaving and reentering $\mathcal{C}(\gamma_k)$ on opposite sides of γ_k. The closed geodesic in the free homotopy

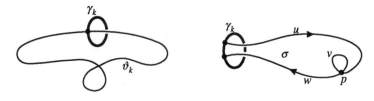

Figure 13.3.1   Figure 13.3.2

class of ϑ_k^* intersects γ_k transversally and is shorter than ϑ_k, a contradiction. This proves our claim.

If ϑ_k is not simple, then it contains a closed loop which we may cut off and obtain a shorter homotopically non-trivial closed curve which leaves and reenters $\mathscr{C}(\gamma_k)$ on opposite sides of γ_k. Again, this is impossible, and we obtain the simplicity of ϑ_k. This completes the proof in case 1.

Case 2. Departure and first return to γ_k take place on the same side of γ_k (Fig. 13.3.2).

We let σ be the corresponding arc on ϑ_k. Since $\ell(\gamma_k) < w(\gamma_k)$, a similar argument as in the first case shows that now ϑ_k intersects γ_k in exactly two points and that $\vartheta_k = \sigma\sigma'$, where σ' is a second arc, also with departure and first return to γ_k on the same side of γ_k. The two arcs intersect each other only at their endpoints on γ_k, otherwise ϑ_k would contain a closed loop which is freely homotopic to a closed geodesic of the type considered in the first case. Thus, it remains to prove that σ (and similarly σ') is simple and has two different endpoints.

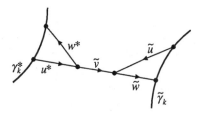

Figure 13.3.3

Suppose this is not the case. Then there exists a point p which is either a transversal self-intersection point of σ or else the common initial and endpoint of σ. The arc is a product $\sigma = uvw$, where u is an arc from the initial point of σ to p, v is a geodesic loop at p, and w is the arc from p to the endpoint of σ. (If p is the common initial and endpoint, then u and w are point curves.) Since v is not contractible, either uw or $uv^{-1}w$ is a curve which is not homotopic with fixed endpoints to an arc on γ_k. In fact, if uw is homotopic to an arc on γ_k, then any lift of $uv^{-1}w$ in the universal covering connects different lifts of γ_k as shown in Fig. 13.3.3. Hence, either $uw\sigma'$ or $uv^{-1}w\sigma'$ is a closed curve which is freely homotopic to a closed geodesic shorter than ϑ_k and interesting γ_k transversally, a contradiction. (13.3.1) is now proved. ◇

(13.3.2) $t_k \leq \ell(\vartheta_k(S)) \leq 4(w(\gamma_k) + \log 7g)$ *for any* $S \in \mathcal{F}_i$ *and for any* $k = i + 1, \ldots, 3g - 3$.

Proof. Let S_k be as in (13.3.1). Cut S_k open along γ_k to obtain a bordered surface S'_k (not necessarily connected) with boundary geodesics γ_k^1 and γ_k^2.

For small $t > 0$ the distance set

$$Z^\nu(t) = \{ p \in S'_k \mid \text{dist}(p, \gamma_k^\nu) < t \}, \quad \nu = 1, 2,$$

is isometric to the annulus $\gamma_k^\nu \times [0, t[$ endowed with the hyperbolic metric $ds^2 = d\rho^2 + \cosh^2\rho \, d\sigma^2$ and has area

$$\text{area } Z^\nu(t) = \ell(\gamma_k) \sinh t.$$

By the collar theorem this holds for any $t \leq w(\gamma_k)$. Since $\ell(\gamma_k) \leq \exp(-4g)$, we have, up to negligible errors, area $Z^\nu(w(\gamma_k)) = 2$. Now let $t \geq w(\gamma_k)$ and let t grow until it reaches a limiting value t_ν beyond which $Z^\nu(t)$ is no longer isometric to $\gamma_k^\nu \times [0, t[$. Since $Z^\nu(t_\nu)$ has area less than $4\pi(g-1)$, we obtain up to negligible errors

$$t_\nu \leq w(\gamma_k) + \log 2\pi(g - 1).$$

If $Z^1(t_1) \cap Z^2(t_2) \neq \emptyset$, we find a curve of length $\leq t_1 + t_2$ on S'_k which connects γ_k^1 and γ_k^2. On S_k again, this curve can be extended to a closed curve by adding an arc on γ_k. In the free homotopy class of this new curve we have a closed geodesic ϑ of length $\ell(\vartheta) \leq t_1 + t_2 + \frac{1}{2}\ell(\gamma_k)$ which intersects γ_k in exactly one point. If $Z^1(t_1) \cap Z^2(t_2) = \emptyset$, then we find, for each ν, two geodesic arcs of length $2t_\nu$ emanating perpendicularly from γ_k^ν and meeting each other smoothly. The two arcs together form an arc σ^ν with both endpoints on γ_k^ν not homotopic (with fixed endpoints) to an arc on γ_k^ν. This remains true if σ^1 and σ^2 are interpreted as arcs on S_k. Hence, adding suitable arcs on γ_k, we obtain a closed curve on S_k which is freely homotopic to a geodesic ϑ of length $\ell \leq 2(t_1 + t_2) + \ell(\gamma_k)$ which intersects γ_k twice. Since ϑ_k is not longer than ϑ we obtain the following result.

$$\ell(\vartheta_k(S_k)) \leq 4(w(\gamma_k) + \log 2\pi(g - 1)).$$

In order to estimate $\ell(\vartheta_k(S))$ for *all* $S \in \mathcal{F}_i$, we observe that $\vartheta_k = \vartheta_k(S)$ intersects the union $\gamma_{i+1} \cup \ldots \cup \gamma_{3g-3}$ at most $3g - 3 - i + 1 \leq 3g - 3$ times.

In fact, if ϑ_k intersects γ_k twice, then it intersects none of the γ_j for $j = i + 1, \ldots, 3g - 3$, for otherwise its length ℓ would satisfy

$$\ell \geq 4w(\gamma_k) + 2w(\gamma_j) > 4(w(\gamma_k) + \log 2\pi(g - 1))$$

(cf. (13.1.4)). If ϑ_k intersects γ_k in only one point, then it intersects any of the γ_j at most once, $j = i + 1, \ldots, 3g - 3$, since otherwise we would be able to construct a shorter geodesic than ϑ_k which also intersects γ_k in exactly one point (recall that $\ell(\gamma_j) < w(\gamma_j)$). Hence, $\vartheta_k = \vartheta_k(S)$ intersects the union $\gamma_{i+1}(S) \cup \ldots \cup \gamma_{3g-3}(S)$ at most $3g - 3$ times. If now the twist parameters $\alpha_{i+1}(S), \ldots, \alpha_{3g-3}(S)$ run through the interval $[-\frac{1}{2}, \frac{1}{2}]$, then $\ell\vartheta_k(S)$ varies by at most $(3g - 3) \exp(-4g)$. This proves (13.3.2) ◇

(13.3.3) *If $S \in \mathcal{F}_i$ and if $\zeta(S) \notin \{\gamma_{i+1}(S), \ldots, \gamma_{3g-3}(S)\}$ is a primitive closed geodesic of length $\ell(\zeta(S)) \leq 1 + t_{i+1}$, then $\zeta(S)$ intersects the union $\gamma_{i+1}(S) \cup \ldots \cup \gamma_{3g-3}(S)$ at most twice.*

Proof. If $\zeta(S)$ has more than two intersections points, then it runs through at least 3 collars, and in view of (13.1.4) and (13.3.2), its length satisfies the inequality

$$\ell(\zeta(S)) \geq 6 \min\{w(\gamma_k) \mid k = i + 1, \ldots, 3g - 3\}$$
$$\geq 2 + \min\{t_k \mid k = i + 1, \ldots, 3g - 3\}.$$

By (13.2.5) and (13.2.6), $\min\{t_k \mid k = i + 1, \ldots, 3g - 3\} > t_{i+1} - 1$, and $\ell(\zeta(S)) > 1 + t_{i+1}$. ◇

(13.3.4) *Let $S \in \mathcal{F}_i$. If there exists a closed geodesic $\zeta(S)$ such that $\ell(\zeta(S)) \leq \frac{1}{4} + t_{i+1}$ which intersects γ_j and γ_k for some $j \in \{n + 1, \ldots, 3g - 3\}$ and some $k \in \{i + 1, \ldots, 3g - 3\}$, then $\ell(\gamma_k) \leq (8g)^2 \ell(\gamma_j)$.*

Proof. Since ζ runs through the collars $\mathscr{C}(\gamma_j)$ and $\mathscr{C}(\gamma_k)$, we have in view of (13.2.5) and (13.2.6)

$$2(w(\gamma_j) + w(\gamma_k)) \leq \ell(\zeta(S)) \leq \tfrac{1}{4} + t_{i+1} \leq \tfrac{1}{3} + t_k.$$

Since $t_k \leq (w(\gamma_k) + \log 7g)$, (see (13.3.2)), we obtain

$$w(\gamma_j) - w(\gamma_k) \leq \tfrac{1}{6} + 2 \log 7g$$

and (13.3.4) follows. ◇

(13.3.5) $\quad t_{i+1} \leq \ell(\vartheta_{i+1}(S)) \leq t_{i+1} + \frac{1}{2}\varepsilon_{i+1}$ for all $S \in \mathcal{F}_i$.

Proof. We restrict ourselves to the case where ϑ_{i+1} intersects γ_k for some $k \in \{i+2, \ldots, 3g-3\}$. The case where ϑ_{i+1} intersects only γ_{i+1} is similar and easier. Observe by (13.3.3) that in the present case ϑ_{i+1} intersects no further γ_ℓ for $\ell \geq i+2$, $\ell \neq k$. The length of $\vartheta_{i+1}(S)$ is therefore uniquely determined by the twist parameters $\alpha_{i+1} = \alpha_{i+1}(S)$ and $\alpha_k = \alpha_k(S)$. Recall ((13.2.4) and (13.3.1)) that $\ell(\vartheta_{i+1}(S_{i+1})) = t_{i+1}$ is the minimal possible length of a closed geodesic on S_{i+1} which intersects $\gamma_k(S_{i+1})$ transversally. Hence, $\vartheta_{i+1}(S_{i+1})$ intersects $\gamma_{i+1}(S_{i+1})$ and $\gamma_k(S_{i+1})$ perpendicularly. We estimate how much $\ell(\vartheta_{i+1}(S))$ increases if, starting from this position, we alter $\alpha_{i+1}(S)$ and $\alpha_k(S)$ by at most ± 1.

To this end we first consider the geodesic arc μ on ϑ_{i+1} contained in the collar $\mathscr{C}(\gamma_k)$ connecting boundary points p' and p''. Since $\vartheta_{i+1}(S_{i+1})$ intersects $\gamma_k(S_{i+1})$ perpendicularly, we have

$$\ell(\mu) = 2w(\gamma_k).$$

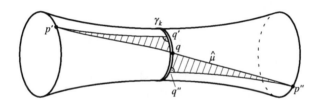

Figure 13.3.4

Replace $\alpha_k(S_{i+1})$ by $\alpha_k(S_{i+1}) + \varphi$, $|\varphi| \leq 1$, to obtain a new surface $\hat{S}$. Let $\hat{\vartheta}_{i+1}(\hat{S})$ be the broken geodesic in the free homotopy class of $\vartheta_{i+1}(\hat{S})$ which outside $\mathscr{C}(\gamma_k)$ coincides with $\vartheta_{i+1}(S_{i+1})$ and inside $\mathscr{C}(\gamma_k)$ consists of a geodesic arc $\hat{\mu}$ which connects p' and p'' as shown in Fig. 13.3.4. Let q' and q'' be the feet of the perpendiculars from p' and p'' to γ_k. By symmetry, $\hat{\mu}$ intersects γ_k in the midpoint q between q' and q'' and forms two isometric right-angled triangles $p'q'q$ and $p''q''q$, where $p'q' = p''q'' = w(\gamma_k)$, $qq' = qq'' = |\varphi|\,\ell(\gamma_k) \leq \ell(\gamma_k)$. Formula (i) of Theorem 2.2.2 yields

$$\cosh \tfrac{1}{2}\ell(\hat{\mu}) \leq \cosh w(\gamma_k) \cosh \tfrac{1}{2}\ell(\gamma_k).$$

Since γ_k is small, we compute with negligible errors using (13.3.4),

$$\tfrac{1}{2}(\ell(\hat{\mu}) - \ell(\mu)) = \tfrac{1}{2}\ell(\hat{\mu}) - w(\gamma_k) \leq \tfrac{1}{8}\ell^2(\gamma_k) \leq \tfrac{1}{8}(8g)^4\,\ell^2(\gamma_{i+1}).$$

In a second step we vary $\alpha_{i+1}(\hat{S})$ to obtain S and argue similarly, however without involving the factor $(8g)^4$. Then we replace the broken geodesic

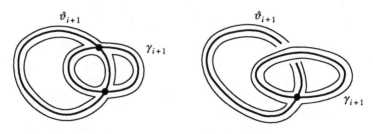

Figure 13.3.5   **Figure 13.3.6**

obtained so far by the closed geodesic $\vartheta_{i+1}(S)$. In view of the definition of ε_{i+1} (in (13.2.4)) we now have (13.3.5).   ◇

(13.3.6) *If ϑ_{i+1} intersects none of the geodesics $\gamma_{i+2}, \ldots, \gamma_{3g-3}$, then a given value of $\ell(\vartheta_{i+1}(S))$ is obtained for at most two values of $\alpha_{i+1}(S)$, as $S \in \mathscr{F}_i$.*

Proof. If ϑ_{i+1} intersects γ_{i+1} twice, then a small ε-neighborhood E of $\vartheta_{i+1} \cup \gamma_{i+1}$ has the topological signature $(0, 4)$ (Fig. 13.3.5). If ϑ_{i+1} intersects γ_{i+1} only once, then E has signature $(1, 1)$ (Fig. 13.3.6). By (13.3.3), other cases do not occur.

Since the angle sum of a hyperbolic geodesic quadrangle is less than 2π, none of the boundary components of E is contractible. Replace each boundary component β of E by the closed geodesic β^* in its free homotopy class. By the Baer-Zieschang theorem on isotopies (Theorem A.3), the geodesics β^* form the boundary of a geodesically bordered surface E^* of signature $(0, 4)$ or $(1, 1)$ which is isometrically immersed in S and contains ϑ_{i+1} and γ_{i+1} in its interior. (The interior of E^* is embedded but some of the boundary geodesics of E^* may coincide on S). Moreover, since for $j = i+2, \ldots, 3g-3$ we have $\beta \cap \gamma_j = \emptyset$ for each boundary component β of E, it follows from the minimal intersection property of geodesics (Theorem 1.6.7) that $\beta^* \cap \gamma_j = \emptyset$ for each boundary geodesic of E^*, and by Theorem A.3 we have $E^* \cap \gamma_j = \emptyset$ $j = i+2, \ldots, 3g-3$. Hence, the geometry of E^* is not affected by $\alpha_{i+2}, \ldots, \alpha_{3g-3}$, and it follows that for $S \in \mathscr{F}_i$ the twist parameter along $\gamma_{i+1}(S)$ with respect to E^* differs from $\alpha_{i+1}(S)$ only by an additive constant. This proves (13.3.6).   ◇

If ϑ_{i+1} intersects some γ_k, $k \geq i+2$, then the length of ϑ_{i+1} no longer determines α_{i+1} up to finitely many values. We shall therefore introduce an additional geodesic ϑ'_{i+1}. By (13.3.3), ϑ_{i+1} intersects no further γ_j, $j \geq 2$, and $\vartheta_{i+1} \cap \gamma_{j+1}$ and $\vartheta_{i+1} \cap \gamma_k$ consist of only one point (each). A small open

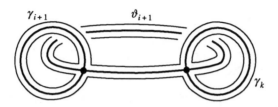

Figure 13.3.7

neighborhood F of $\vartheta_{i+1} \cup \gamma_{i+1} \cup \gamma_k$ has signature $(1, 2)$ (Fig. 13.3.7).

As before, there exists a geodesically bordered surface F^* of signature $(1, 2)$ carrying ϑ_{i+1}, γ_{i+1} and γ_k in its interior such that the twist parameters of F^* along γ_{i+1} and γ_k differ from α_{i+1} and α_k only by additive constants (as S varies through $\mathscr{F}_i$).

To define ϑ'_{i+1} we use the intersection point of ϑ_{i+1} and γ_{i+1} as the initial point of a parametrization so that ϑ_{i+1} and γ_{i+1} become closed loops. This allows us to form the product $\vartheta_{i+1}\gamma_{i+1}$, and we let ϑ'_{i+1} be the closed geodesic in the free homotopy class of $\vartheta_{i+1}\gamma_{i+1}$. The length of $\vartheta'_{i+1}(S)$ is the same as the length of $\vartheta_{i+1}(S')$, where S' is the surface obtained from S by replacing $\alpha_{i+1}(S)$ with $\alpha_{i+1}(S) + 1$ (or $\alpha_{i+1}(S) - 1$), depending on how $\gamma_{i+1}(S)$ is oriented). As in the proof of (13.3.5), this affects $\ell(\vartheta_{i+1})$ by at most $\ell^2(\gamma_{i+1})$ so that by (13.3.5) we have

(13.3.7) $\quad t_{i+1} \leq \ell(\vartheta'_{i+1}(S)) \leq t_{i+1} + \varepsilon_{i+1}$ for all $S \in \mathscr{F}_i$.

We also have the following analog of (13.3.6).

(13.3.8) *If ϑ_{i+1} intersects γ_{i+1} and γ_k, where $k \geq i + 2$, then any value of the pair $(\ell\vartheta_{i+1}(S), \ell\vartheta'_{i+1}(S))$ is obtained for at most two values of $\alpha_{i+1}(S)$, as S runs through $\mathscr{F}_i$.*

Proof. Consider the simple closed geodesic η on F^* which intersects γ_k twice but does not intersect γ_{i+1} and ϑ_{i+1}. The length of η depends only on α_k. By (13.3.6), a given value of $\ell(\eta)$ is only possible for two values of α_k.

Cut F^* open along η and let Q be the resulting connected component which carries ϑ_{i+1} and γ_{i+1} in its interior. By the minimal intersection property of geodesics, ϑ'_{i+1} lies in the interior of Q. It remains to prove that the length of η is determined by the lengths of ϑ_{i+1} and ϑ'_{i+1}.

Cut Q open along γ_{i+1} to obtain a geodesically bordered surface Y of signature $(0, 3)$ with boundary components η, γ'_{i+1}, γ''_{i+1}. Drop the common perpendicular a on Y from γ'_{i+1} to γ''_{i+1}. (Fig. 13.3.8 shows γ'_{i+1}, a and γ''_{i+1} in

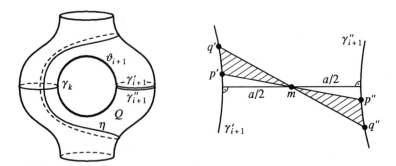

Figure 13.3.8

the universal covering.) Then $\ell(\gamma'_{i+1})$, $\ell(\gamma''_{i+1})$ and a determine the length of η. It remains to prove that a is determined by ϑ_{i+1} and ϑ'_{i+1}. By symmetry, ϑ_{i+1} intersects ϑ'_{i+1} in the midpoint m of a and we obtain two isometric geodesic triangles $mp'q'$ and $mp''q''$, where

$$mq' = \tfrac{1}{2}\ell(\vartheta'_{i+1}), \quad mp' = \tfrac{1}{2}\ell(\vartheta_{i+1}), \quad p'q' = \tfrac{1}{2}\ell(\gamma_{i+1}).$$

Since these data determine the triangles and their interior angles uniquely, the data also determine a. This proves (13.3.8). ◊

13.3.9 Definition. We let $\mathcal{L}$ be the set of all primitive closed geodesics on S^* which have their length in the interval $[t_{i+1}, t_{i+1} + \varepsilon_{i+1}]$ and which intersect at least one of the geodesics $\gamma_{i+1}, \ldots, \gamma_{3g-3}$.

(13.3.10) Let $S \in \mathcal{F}_i^*$. Let $\vartheta(S)$ be a closed geodesic on S intersecting $\gamma_{i+1}(S)$ and assume that $\ell(\vartheta(S)) \in [t_{i+1}, t_{i+1} + \varepsilon_{i+1}]$. Then there exists $\zeta \in \mathcal{L}$ satisfying $\ell(\vartheta(S)) = \ell(\zeta(S^*))$.

Proof. Let $\mathrm{Lsp}^{(i)}(S)$ be the subsequence of $\mathrm{Lsp}(S)$ which corresponds to all those primitive closed geodesics on S which intersect at least one of the geodesics $\gamma_{i+1}, \ldots, \gamma_{3g-3}$. As S runs through $\mathcal{F}_i$, only $\mathrm{Lsp}^{(i)}(S)$ varies. Since $S^* \in \mathcal{F}_i^*$, we conclude that for $S \in \mathcal{F}_i^*$ the equality $\mathrm{Lsp}(S) = \mathrm{Lsp}(S^*)$ holds if and only if $\mathrm{Lsp}^{(i)}(S) = \mathrm{Lsp}^{(i)}(S^*)$. Hence, if $\mathrm{Lsp}(S) = \mathrm{Lsp}(S^*)$, then $\ell(\vartheta(S)) \in \mathrm{Lsp}^{(i)}(S^*)$ and (13.3.10) follows. ◊

(13.3.11) $\mathcal{L}$ has cardinality $\#\mathcal{L} \leq \exp(106g)$.

The proof of (13.3.11) will be our fourth and final step. Let us first show

how (13.3.11) implies Theorem 13.1.1.

We begin with the case that ϑ_{i+1} intersects one of the geodesics $\gamma_{i+2}, \ldots,$ γ_{3g-3}. For all $S \in \mathscr{F}_i^*$ we have $\ell(\vartheta_{i+1}(S))$, $\ell(\vartheta'_{i+1}(S)) \in [t_{i+1}, t_{i+1} + \varepsilon_{i+1}]$, ((13.3.5) and (13.3.7)). By (13.3.10), there exist ζ and $\zeta' \in \mathscr{L}$ such that $\ell(\vartheta_{i+1}(S)) = \ell(\zeta(S^*))$ and $\ell(\vartheta'_{i+1}(S)) = \ell(\zeta'(S^*))$. By (13.3.8), the two values determine $\alpha_{i+1}(S)$ up to two possible values. By (13.3.11), this yields the estimate $\# \mathscr{F}_i^* \leq 2 \exp^2(106g)$. This yields (13.2.7) and it proves the theorem in the present case.

Now assume that ϑ_{i+1} intersect only γ_{i+1}. Then we have the better estimate $\# \mathscr{F}_i^* \leq 2 \exp(106g)$, ((13.3.7) and (13.3.8)). By (13.3.3), this covers all the cases, and the proof of Theorem 13.1.1 is now reduced to the proof of (13.3.11).

13.4 Parameter Geodesics of Length $\leq \exp(-4g)$

Each geodesics $\zeta \in \mathscr{L}$ (cf. Definition 13.3.9) intersects the collar of some parameter geodesic γ_k whose length has no a priori lower bound. If we alter ζ by adding a Dehn twist along γ_k, the length of ζ changes arbitrarily little. We now show that the interval $[t_{i+1}, t_{i+1} + \varepsilon_{i+1}]$ is small enough so that the cardinality of $\mathscr{L}$ has an a priori upper bound.

Recall that $\gamma_{m+1}, \ldots, \gamma_{3g-3}$ are all the simple closed geodesics of length ≤ 1 on S^*. Their collars $\mathscr{C}(\gamma_j)$ have width

$$w(\gamma_j) \geq \operatorname{arcsinh}\{1/\sinh \tfrac{1}{2}\} = 1.406\ldots, \qquad j = m+1, \ldots, 3g-3.$$

For $j = m+1, \ldots, 3g-3$ we introduce the *reduced collars*

(13.4.1) $\qquad \hat{\mathscr{C}}(\gamma_j) = \{p \in S^* \mid \operatorname{dist}(p, \gamma_j) \leq w(\gamma_j) - \tfrac{1}{2}\}$

and define

$$\hat{S} = \text{closure of } (S^* - (\hat{\mathscr{C}}(\gamma_{m+1}) \cup \ldots \cup \hat{\mathscr{C}}(\gamma_{3g-3}))).$$

$\hat{S}$ need not be connected. By the collar theorem, the reduced collars $\hat{\mathscr{C}}(\gamma_j)$ are pairwise disjoint. The boundary components of $\hat{\mathscr{C}}(\gamma_j)$ will be denoted by ∂'_j and ∂''_j. They are pairwise disjoint simple closed boundary curves of $\hat{S}$. On $\mathscr{C}(\gamma_j) \subset S^*$ we draw a geodesic arc μ_j of length $2w(\gamma_j)$ which connects the two boundary components of $\mathscr{C}(\gamma_j)$ with each other. Then μ_j intersects ∂'_j, γ_j and ∂''_j perpendicularly.

We let p'_j and p''_j be the intersection points of μ_j with ∂'_j and ∂''_j. (cf. Fig. 13.4.1).

13.4.2 Lemma. (i) *The injectivity radius of S^* at p is $\geq \frac{1}{2}$ for any $p \in \hat{S}$.*

(ii) *On S^* the points $p'_{m+1}, p''_{m+1}, \ldots, p'_{3g-3}, p''_{3g-3}$ have pairwise distances ≥ 1.*

(iii) *If $x \in \partial'_j$ and $y \in \partial''_j$, then $\mathrm{dist}(x, p'_j) \leq \frac{3}{4}$ and $\mathrm{dist}(y, p''_j) \leq \frac{3}{4}$ for $j = m+1, \ldots, 3g-3$.*

Proof. Statement (i) is an immediate consequence of Theorem 4.1.6. Statement (ii) follows from the definition of the reduced collars. Statement (iii) follows from formula (v) of Theorem 2.3.1: the maximal possible distance δ for $x \in \partial'_j$ is assumed when x and p'_j are the vertices of a trirectangle with sides $(w(\gamma_j) - \frac{1}{2})$, $\frac{1}{4}\ell(\gamma_j)$, $\bullet$, $\frac{1}{2}\delta$. The formula is

$$\sinh \tfrac{1}{2}\delta = \sinh(\tfrac{1}{4}\ell(\gamma_j)) \cosh(w(\gamma_j) - \tfrac{1}{2}).$$

Observe that δ is a monotone increasing function of $\ell(\gamma_j)$. For $\ell(\gamma_j) = 1$ we obtain $\delta = 0.712 \ldots$. ◇

(13.4.3) *Let $\zeta \in \mathcal{L}$ and let $\eta \subset \zeta$ be a subarc contained in $\mathcal{C}(\gamma_j)$ which connects the two boundary components of $\mathcal{C}(\gamma_j)$ with each other. If $\eta \neq \mu_j$, then $\#(\eta \cap \mu_j) \leq 3(8g)^4 - 100$, $j = m+1, \ldots, 3g-3$.*

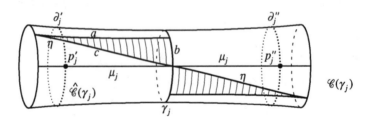

Figure 13.4.1

Proof. Assume $\eta \neq \mu_j$ and set $N = \#(\eta \cap \mu_j)$. Drop the perpendiculars a of length $w(\gamma_j)$ from either endpoint of η to γ_j to obtain two isometric right-angled triangles abc which are isometrically immersed in $\mathcal{C}(\gamma_j)$ (in Fig. 13.4.1 these triangles are embedded, but in general, b may be longer than γ_j and the triangle appears wrapped around $\mathcal{C}(\gamma_j)$). If we lift η and γ_j into the universal covering we get an ordinary right-angled triangle abc in $\mathbf{H}$, and the immersion of this triangle into $\mathcal{C}(\gamma_j)$ is given by the universal covering map. We have $c = \frac{1}{2}\ell(\eta)$, and b satisfies

$$\tfrac{1}{2}(N-1)\ell(\gamma_j) \leq b \leq \tfrac{1}{2}(N+1)\ell(\gamma_j).$$

By formula (i) of Theorem 2.2.2, we have $\cosh c = \cosh a \cosh b$. For $x =$

$c - a$ we obtain $e^x \geq \cosh c/\cosh a = \cosh b \geq 1 + \frac{1}{2}b^2$. This yields the following inequality,

$$e^x \geq 1 + \tfrac{1}{8}(N-1)^2 \ell^2(\gamma_j),$$

where $x = \frac{1}{2}\ell(\eta) - w(\gamma_j)$. Now let η' be the shortest geodesic arc in $\mathscr{C}(\gamma_j)$ which connects the endpoints of η, and consider the corresponding triangle $ab'c'$. We have $b' \leq \frac{1}{4}\ell(\gamma_j)$ and since $\ell(\gamma_j) < 1 < w(\gamma_j)$, we obtain

$$y \leq \tfrac{1}{16}\ell^2(\gamma_j),$$

where $y = \frac{1}{2}\ell(\eta') - w(\gamma_j)$. Now replace η by η' and let ζ' be the closed geodesic in the free homotopy class of the resulting curve. Then

$$\ell(\zeta') \leq \ell(\zeta) - 2x + 2y.$$

Recall that $\zeta \in \mathscr{L}$ intersects γ_k for some $k \in \{i+1, \ldots, 3g-3\}$. (If $j \geq i+1$, we take $k = j$.) Hence, ζ' intersects γ_j and γ_k. Since

$$t_{i+1} + \varepsilon_{i+1} \leq t_k + \varepsilon_k$$

(cf. (13.2.6)), and since $\zeta \in \mathscr{L}$ has length $\ell(\zeta) \in [t_{i+1}, t_{i+1} + \varepsilon_{i+1}]$, it follows from the definition of t_k (in (13.2.4)) that

$$t_k \leq \ell(\zeta') \leq t_k + \varepsilon_k - 2x + 2y.$$

Together with (13.2.4) we get

$$x \leq \tfrac{1}{2}\varepsilon_k + y \leq \tfrac{1}{2}(8g)^4 \ell^2(\gamma_k) + \tfrac{1}{16}\ell^2(\gamma_j).$$

This shows, first of all, that x is small, and the above lower bound for e^x can be simplified as follows.

$$x \geq \tfrac{1}{9}(N-1)^2 \ell^2(\gamma_j).$$

Observe that

$$\ell(\gamma_k) \leq (8g)^2 \ell(\gamma_j).$$

If $j \in \{m+1, \ldots, n\}$, this is trivial because in this case $\ell(\gamma_k) \leq \ell(\gamma_j)$. If $j \geq n+1$, the inequality follows from (13.3.4).

The last three inequalities yield (13.4.3). ◊

13.4.4 Remark. It is easy to give examples where $t_{i+1} = t_j$, for some $j > i+1$, and where ϑ_{i+1} intersects γ_{i+1} twice but ϑ_j intersects γ_j only once. In this case we have approximately $\ell(\gamma_j) = \ell^2(\gamma_{i+1})$, and a single Dehn twist along γ_{i+1} "costs" about the same as $\text{int}(1/\ell(\gamma_{i+1}))$ Dehn twists along γ_j. This shows that we cannot prove (13.4.3) (in any form) if the condition

"$t_{i+1} + \varepsilon_{i+1} \leq t_k + \varepsilon_k$" in (13.2.6) is replaced by the condition "$t_{i+1} \leq t_k$".

13.4.5 Definition. We let $\mathscr{L}_1 \subset \mathscr{L}$ be the subset of those closed geodesics on S^* which intersect some γ_k twice, $k \in \{i+1, \ldots, 3g-3\}$. We let $\mathscr{L}_2 \subset \mathscr{L}$ be the set of all $\zeta \in \mathscr{L} - \mathscr{L}_1$ which intersect some γ_k, $k \in \{i+1, \ldots, 3g-3\}$ and some γ_j, $j \in \{m+1, \ldots, 3g-3\}$ satisfying $\ell(\gamma_j) \leq (8g)^2 \ell(\gamma_k)$. Finally we set $\mathscr{L}_3 = \mathscr{L} - (\mathscr{L}_1 \cup \mathscr{L}_2)$.

(13.4.6) $$\# \mathscr{L}_1 \leq (14g)^{17}.$$

Proof. Let $\zeta \in \mathscr{L}_1$ intersect γ_k twice for some $k \geq i+1$. By (13.3.3), ζ intersects γ_k in exactly two points and is a product of four arcs: $\zeta = \eta_1 \sigma_1 \eta_2 \sigma_2$, where η_1 and η_2 are contained in $\mathscr{C}(\gamma_k)$ connecting the boundary components of $\mathscr{C}(\gamma_k)$ with each other. Since $\zeta \in \mathscr{L}$, it follows from (13.2.6) that $\ell(\zeta) \leq t_k + \varepsilon_k$, so that by (13.3.2)

$$\ell(\sigma_1), \ell(\sigma_2) \leq 4 \log 7g + \varepsilon_k.$$

By Lemma 13.4.2(iii), we may replace ζ by a broken geodesic $\hat{\zeta}$ homotopic to ζ which is a product $\hat{\zeta} = \hat{\eta}_1 \hat{\sigma}_1 \hat{\eta}_2 \hat{\sigma}_2$, where the curves $\hat{\eta}_1$ and $\hat{\eta}_2$ are contained in $\hat{\mathscr{C}}(\gamma_k)$ (the reduced collar) connecting p_k' with p_k'' (Fig. 13.4.1 with k instead of j), and where

$$\ell(\hat{\sigma}_\nu) \leq 4 \log 7g + \tfrac{3}{2} + \varepsilon_k,$$
$$\nu = 1, 2,$$
$$\#(\hat{\eta}_\nu \cap \mu_k) \leq 3(8g)^4, \text{ if } \hat{\eta}_\nu \neq \mu_k.$$

The second claim follows from (13.4.3). We count the possibilities for $\hat{\zeta}$. Since by Lemma 13.4.2 the lifts of p_k' and p_k'' in the universal covering of S^* have pairwise distances ≥ 1, there cannot be more arcs $\hat{\sigma}_1$ and $\hat{\sigma}_2$ than pairwise disjoint disks of radius $\tfrac{1}{2}$ in a disk of radius $4 \log 7g + 2 + \varepsilon_k$. Hence, there are at most $(17g)^4$ possibilities for $\hat{\sigma}_1$, and the same holds for $\hat{\sigma}_2$. Since a given cardinality of $\hat{\eta}_\nu \cap \mu_k$ determines $\hat{\eta}_\nu$ up to two, the above estimates and the fact that there are at most $3g-3$ different γ_k show that there are at most $(14g)^{17}$ possibilities for ζ. ◊

(13.4.7) $$\# \mathscr{L}_2 \leq (12g)^{26}.$$

Proof. Fix $k \in \{i+1, \ldots, 3g-3\}$ and $j \in \{m+1, \ldots, 3g-3\}$ such that

$$\ell(\gamma_j) \leq (8g)^2 \ell(\gamma_k).$$

Up to negligible errors we get $w(\gamma_j) \geq w(\gamma_k) - 2 \log 8g$. As in the proof of (13.4.6), $\zeta \in \mathscr{L}_2$ splits into an arc in $\mathscr{C}(\gamma_k)$, an arc in $\mathscr{C}(\gamma_j)$ and into two additional arcs, where by (13.3.2) and (13.2.5) the lengths of the latter add

(13.4.8) $\#\mathcal{L}_3 \leq \exp(105g)$.

Proof. Let $\zeta \in \mathcal{L}_3$. Then ζ runs exactly once across one of the reduced collars $\hat{\mathscr{C}}(\gamma_k)$, $k \in \{i+1, \ldots, 3g-3\}$ (cf. (13.4.1)), and we have the inclusion $\ell(\zeta) \in [t_{i+1}, t_{i+1} + \varepsilon_{i+1}]$. We subdivide ζ into $\hat{\eta}$-arcs and σ-arcs: each $\hat{\eta}$-arc is contained in some $\hat{\mathscr{C}}(\gamma_j)$ with $j \in \{m+1, \ldots, 3g-3\}$, and connects the two boundary components of $\hat{\mathscr{C}}(\gamma_j)$ with each other; the σ-arcs are the remaining arcs on ζ.

(1) *ζ has at most $3g-3$ $\hat{\eta}$-arcs.*

In fact, if ζ runs twice across some $\hat{\mathscr{C}}(\gamma_j)$, then $\mathscr{C}(\gamma_j)$ has width $w(\gamma_j) \geq 1$, and γ_j has length $\ell(\gamma_j) < 1$. This would allow us to construct a new geodesic ζ' of length $\ell(\zeta') < \ell(\zeta) - 1 < t_k$ which also intersects γ_k exactly once, a contradiction.

(2) *If ζ intersects γ_j for some $j \in \{m+1, \ldots, 3g-3\}$, $j \neq k$, then $j < i+1$, and ζ intersects μ_j at most 3 times.*

(cf. Fig. 13.4.1 for the definition of μ_j). In fact, since $\ell(\zeta) < t_{i+1} + \frac{1}{4}$, (13.3.4) implies that $j \leq n+1 < i+1$. Since $\zeta \in \mathcal{L}_3$, we get $\ell^2(\gamma_j) \geq \varepsilon_k$. We can now repeat the proof of (13.4.3) to get (2).

(3) *Each σ-arc of ζ has length ≥ 1.*

This holds because $\hat{\mathscr{C}}(\gamma_j)$ has width $w(\gamma_j) - \frac{1}{2}$, $j = m+1, \ldots, 3g-3$ (cf. (13.4.1)).

(4) *At each point of a σ-arc, the injectivity radius is $\geq \frac{1}{2}$.*

If not, then Lemma 13.4.2 would imply that a subarc σ' of some σ-arc enters one of the collars $\mathscr{C}(\gamma_j)$ with $j \geq m+1$, then intersects $\hat{\mathscr{C}}(\gamma_j)$ and leaves $\mathscr{C}(\gamma_j)$ without intersecting γ_j, as shown in Fig. 13.4.2.

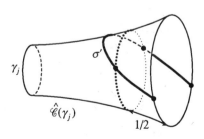

Figure 13.4.2

However, a simple computation shows that σ' has positive distance from $\hat{\mathscr{C}}(\gamma_j)$ unless it has self-intersections. But σ' cannot have self-intersections because this allows us to construct a new geodesic ζ' of length $< t_k$ which still intersects γ_k transversally, a contradiction.

(5) *The total length of all σ-arcs is $\leq 17(g-1)$.*

For the proof we subdivide each σ-arc into arcs of length $\lambda_\sigma \geq 1$ with λ_σ as close to 1 as possible. Let $s_1, \ldots, s_\ell$ be the set of all endpoints of all arcs in this subdivision of of all σ-arcs. The total length of the σ-arcs is less than ℓ. We also have

$$\mathrm{dist}(s_\nu, s_\mu) \geq 1 - \varepsilon_k, \qquad \nu \neq \mu,$$

because a shorter distance would again allow us to construct a closed geodesic ζ' of length $< t_k$ which intersects γ_k transversally. The disks of radius $r = \frac{1}{2}(1 - \varepsilon_k)$ around $s_1, \ldots, s_\ell$ are pairwise disjoint and have area equal to $2\pi(\cosh r - 1)$. Since the sum of these areas cannot exceed $2\pi(g-1)$, this proves (5).

In the next step we let $\mathcal{P} \subset \hat{S}$ be a maximal set of points of pairwise distances ≥ 1 which for $j = m+1, \ldots, 3g-3$ contains the points p'_j and p''_j as in Lemma 13.4.2 (Fig. 13.4.1). Then we construct a homotopic curve $\hat{\zeta}$ which connects points of $\mathcal{P}$ in the following way (Fig. 13.4.3).

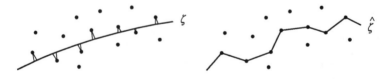

Figure 13.4.3

Divide each σ-arc σ of ζ into $N_\sigma \leq 1 + \ell(\sigma)$ arcs of length ≤ 1 using division points $q_1^\sigma, \ldots, q_{N_\sigma}^\sigma$. Then q_1^σ and $q_{N_\sigma}^\sigma$ lie on the boundary of $\hat{S}$. Connect q_1^σ and similarly $q_{N_\sigma}^\sigma$ with that point of $\{p'_{m+1}, \ldots, p''_{3g-3}\}$ which lies on the same boundary component. Connect each remaining q_ν^σ with the point of $\mathcal{P}$ which lies next to it (Fig. 13.4.3). We let $p_1^\sigma, \ldots, p_{N_\sigma}^\sigma$ be the set of the joined points. By Lemma 13.4.2 and by the maximality of $\mathcal{P}$, the joining arcs $q_\nu^\sigma p_\nu^\sigma$ have length ≤ 1. Renumber all points successively along ζ in the form $q_1, \ldots, q_N$ and $p_1, \ldots, p_N$. The new curve $\hat{\zeta}$ is defined as the sequence of the geodesic arcs $p_\nu p_{\nu+1}$, $\nu = 1, \ldots, N$, $p_{N+1} := p_1$, where $p_\nu p_{\nu+1}$ denotes the geodesic arc in the homotopy class (with fixed endpoints) of the curve $p_\nu q_\nu q_{\nu+1} p_{\nu+1}$ in which $q_\nu q_{\nu+1}$ is the arc on ζ from q_ν to $q_{\nu+1}$. The arcs of $\hat{\zeta}$

are of the following 3 types.

(a) An arc from p'_k to p''_k in $\hat{\mathscr{C}}(\gamma_k)$ having less than $3(8g)^4$ transversal intersections with μ_k (cf. (13.4.3)).

(b) Arcs from p'_j to p''_j in $\hat{\mathscr{C}}(\gamma_j)$, $j \geq m+1$, $j \neq k$, having at most 5 transversal intersections with μ_j (cf. (2) above).

(c) Arcs of length ≤ 3 which connect points of $\mathcal{P}$.

Since $\zeta \in \mathscr{L}_3$, there is exactly one arc of type (a). By (1), ζ has at most $3g-4$ arcs of type (b), and by (5) and (3) at most $17g$ arcs of type (c).

How many $\hat{\zeta}$ are possible? If we draw a geodesic $\hat{\zeta}$ beginning with the arc of type (a), we have at most $6(3g-3)(8g)^4$ different ways to begin. For the remaining at most $20g$ arcs of type (b) and (c) we have at each step at most 132 choices, since at most 122 points of $\mathcal{P}$ lie at distance ≤ 3 of a given point in $\mathcal{P}$. These are at most $\exp(105g)$ possibilities for $\hat{\zeta}$. This proves (13.4.8). ◊

Together with (13.4.7) and (13.4.6) we get the estimate (13.3.11) and Theorem 13.1.1 is proved. ◊

Chapter 14

Perturbations of the Laplacian in Teichmüller Space

An important consequence of the real analytic structure of Teichmüller space is the analyticity of the Laplacian. In Buser-Courtois [1], for instance, this was used to show that the spectrum of a compact Riemann surface is determined by a finite part. That article also contains an outline proof of the analytic structure of the Laplacian. In this chapter we give a detailed account. The main theorems are contained in Sections 14.7 and 14.9. The finiteness result follows in Section 14.10. Sections 14.1 - 14.5 provide the necessary material from the perturbation theory of linear operators. These sections are based on Kato's book [1] but are written in a self-contained style. Section 14.6 then links the general theory with Teichmüller space, where for convenience we recall the necessary facts from Chapter 6.

All analyticity arguments are reduced to a few elementary properties of holomorphic functions of several complex variables such as the Cauchy integral formula and Weierstrass' convergence theorem. We use chapter 1 of Grauert-Fritzsche [1] and chapter 1 of Narasimhan [1] as a general reference for holomorphic functions of several complex variables.

The review of the basic notation for linear operators and sesquilinear forms is postponed to the beginning of Section 14.3.

14.1 The Hilbert Spaces H_0 and H_1

We let M be a compact Riemannian manifold which will be kept fixed throughout this section. $C^\infty(M)$ denotes the space of all *complex valued* smooth functions on M. For $f, g \in C^\infty(M)$ we define an inner product

$(f, g)_0$, a positive semi definite symmetric sesquilinear form $q[f, g]$ and the corresponding 1-product $(f, g)_1$ as follows.

$$(f, g)_0 := \int_M f\bar{g}\, dM, \qquad q[f, g] := \int_M \langle \operatorname{grad} f, \operatorname{grad} \bar{g} \rangle\, dM$$
(14.1.1)
$$(f, g)_1 := (f, g)_0 + q[f, g].$$

Here $\langle\, ,\, \rangle$ is the Riemannian metric tensor on M, grad is the associated gradient and dM is the volume element. The bar denotes complex conjugation. The only theorem from analysis on manifolds which will be needed in this chapter is Green's formula for the Laplace operator Δ. In the above notation, Green's formula is as follows

(14.1.2) $(\Delta f, g)_0 = q[f, g] = (f, \Delta g)_0, \quad f, g \in C^\infty(M).$

$C^\infty(M)$ is a pre-Hilbert space with respect to $(\, ,\,)_0$ and also with respect to $(\, ,\,)_1$. The corresponding norms are denoted by $\|\ \|_0$ and $\|\ \|_1$:

$$\|f\|_0 = \sqrt{(f, f)_0}, \qquad \|f\|_1 = \sqrt{(f, f)_1}.$$

We complete these pre-Hilbert spaces via Cauchy sequences.

14.1.3 Definition. For $v = 0, 1$ we let $H_v = H_v(M)$ be the completion of the pre-Hilbert space $(C^\infty(M), (\, ,\,)_v)$.

For metric quantities (convergence, etc.) we use the prefix v to indicate that the quantity is understood with respect to H_v. Thus, "v-convergence" means convergence with respect to the norm $\|\ \|_v$, etc.

Every $u \in H_v$ is the v-equivalence class $[f_k]_v$ of a v-Cauchy sequence $\{f_k\} \subset C^\infty(M)$, $k = 1, 2, \ldots$, where, by definition, two sequences $\{f_k\}$ and $\{g_k\}$ are v-equivalent if and only if $\|f_k - g_k\|_v \to 0$ as $k \to \infty$. For $u = [f_k]_v \in H_v$ and $w = [g_k]_v \in H_v$ the inner product is

$$(u, w)_v = \lim_{k \to \infty} (f_k, g_k)_v.$$

For $u, w \in H_1$ the extension of the sesquilinear form q is

$$q[u, w] = \lim_{k \to \infty} q[f_k, g_k].$$

The corresponding quadratic form will also be denoted by q:

$$q[u] := q[u, u], \qquad u \in H_1.$$

We first show that there is a natural injection of H_1 into H_0. Since $\|\ \|_0 \leq \|\ \|_1$, every 1-Cauchy sequence is a 0-Cauchy sequence, and 1-equivalence implies

0-equivalence. For any 1-equivalence class $[f_k]_1 \in H_1$, the following is therefore an image in H_0, independent of the choice of the representative $\{f_k\}$

$$i([f_k]_1) = [f_k]_0.$$

14.1.4 Proposition. $i : H_1 \to H_0$ *is a linear injection.*

Proof. Linearity is clear. To prove injectivity, we must show that if $\{f_k\} \subset C^\infty(M)$ is a 1-Cauchy sequence such that $\|f_k\|_0 \to 0$, then $\|f_k\|_1 \to 0$ as $k \to \infty$.

Let $\omega = \lim_{k \to \infty} \|f_k\|_1$. Fix $\varepsilon > 0$. There exists $N \in \mathbf{N}$ such that for all $n \geq N$, $\|f_n - f_N\|_1 < \varepsilon$. From the fact that $\|f_k\|_0 \to 0$ and the inequality $|(f_n, \Delta f_N)_0| \leq \|f_n\|_0 \|\Delta f_N\|_0$, it follows that $(f_n, \Delta f_N)_0 \to 0$ as $n \to \infty$. By Green's formula (14.1.2), $\mathfrak{q}[f_n, f_N] \to 0$ as $n \to \infty$. Now

$$\varepsilon^2 > \|f_n - f_N\|_1^2 = \|f_n\|_1^2 + \|f_N\|_1^2 - 2\mathcal{R}e\{(f_n, f_N)_0 + \mathfrak{q}[f_n, f_N]\}.$$

As $n \to \infty$, the right-hand side converges to $\omega^2 + \|f_N\|_1^2$. Hence, $\omega^2 < \varepsilon^2$ for all $\varepsilon > 0$. This proves the proposition. ◇

The pre-Hilbert space $(H_1, (\,,\,)_0)$ will from now on be considered as a dense subspace of the Hilbert space H_0 with respect to the injection i.

In Section 14.6 we shall pull back smooth functions via piecewise smooth homeomorphisms. For this we also need the following function space.

14.1.5 Definition. $C^L(M)$ is the space of all Lipschitz-continuous functions $f : M \to \mathbf{C}$ which are C^1-differentiable outside some compact subset $\Gamma_f \subset M$ of measure zero.

For $f \in C^L(M)$ with corresponding set $\Gamma = \Gamma_f$, grad f is a continuous gradient vectorfield with bounded norm on $M - \Gamma$. Thus, for $f, g \in C^L(M)$ the forms (14.1.1) are well defined.

14.1.6 Lemma. $C^\infty(M)$ *is 1-dense in* $C^L(M)$.

Proof. For $f \in C^L(M)$ with corresponding set Γ, a sequence $f_1, f_2, \ldots$ in $C^\infty(M)$ must be constructed such that $\|f_k - f\|_1 \to 0$. Using a partition of unity on M, we may assume that the support of f lies in the neighborhood U of a coordinate system (U, φ) with $\varphi : U \to \varphi(U) \subset \mathbf{R}^n$, where $n = \dim M$. We set $\gamma := \varphi(\Gamma)$, $F := f \circ \varphi^{-1}$ on $\varphi(U)$, $F := 0$ elsewhere. Then there exists a constant L such that

(1) $$|F(x) - F(y)| \leq L|x - y|, \quad x, y \in \mathbf{R}^n,$$

and F is C^1-differentiable on $\mathbf{R}^n - \gamma$. It suffices now to find a sequence $F_1, F_2, \ldots$ of smooth functions on $\mathbf{R}^n$ with the following properties, where ∂_i denotes the first partial derivative with respect to the i-th coordinate:

(2) $$F_k \to F \text{ uniformly on } \mathbf{R}^n,$$

(3) $$|\partial_i(F_k - F)| \leq \text{const on } \mathbf{R}^n - \gamma, \quad i = 1, \ldots, n,$$

(4) $\partial_i F_k \to \partial_i F$ uniformly on compact subsets of $\mathbf{R}^n - \gamma$, $\quad i = 1, \ldots, n$.

We use smoothing operators. Let $\vartheta : [0, \infty[\to \mathbf{R}$ be a smooth function satisfying $\vartheta(t) = 1$ for $0 \leq t \leq \frac{1}{2}$, $\vartheta(t) \geq 0$ for $\frac{1}{2} \leq t \leq 1$, $\vartheta(t) = 0$ for $t \geq 1$. For $k = 1, 2, \ldots$ we set $\vartheta_k(x) = c_k \vartheta(k|x|)$, $x \in \mathbf{R}^n$, where c_k is a constant such that

(5) $$\int_{\mathbf{R}^n} \vartheta_k(y) \, dy = 1,$$

and $|x|$ is the distance in $\mathbf{R}^n$ from x to the origin. Now we define

$$F_k(x) = \int_{\mathbf{R}^n} \vartheta_k(x - y) F(y) \, dy, \quad x \in \mathbf{R}^n.$$

Since ϑ_k is smooth with compact support and since F is bounded and continuous, F_k is smooth. We prove (2) - (4). Property (2) follows immediately from the uniform continuity of F and from (5). To prove (3) we first note from (1) that $|\partial_i F| \leq L$ on $\mathbf{R}^n - \gamma$. Then we write $\partial_i F_k$ as follows, where T_h denotes the parallel translation $x \mapsto T_h(x) = x + h e_i$, $x \in \mathbf{R}^n$, and e_i is from the standard basis $\{e_1, \ldots, e_n\}$ of $\mathbf{R}^n$:

(6)
$$\partial_i F_k(x) = \lim_{h \to 0} \frac{1}{h} \int_{\mathbf{R}^n} (\vartheta_k(T_h x - y) - \vartheta_k(x - y)) F(y) \, dy$$
$$= \lim_{h \to 0} \frac{1}{h} \int_{\mathbf{R}^n} \vartheta_k(x - y) (F(T_h y) - F(y)) \, dy.$$

Then using (1) and (5), we obtain $|\partial_i F_k(x)| \leq L$. Since by (1), $|\partial_i F(x)| \leq L$ for all $x \in \mathbf{R}^n - \gamma$, this proves (3). For the proof of (4) we let B be a compact subset of $\mathbf{R}^n - \gamma$. Let $\varepsilon > 0$. By the uniform continuity of $\partial_i F$ on compact subsets of $\mathbf{R}^n - \gamma$, there exists δ with $0 < \delta < \text{dist}(B, \gamma)$ such that $|\partial_i F(y) - \partial_i F(x)| \leq \varepsilon$ for all $x, y \in \mathbf{R}^n$ with $x \in B$ and $|x - y| < \delta$. By (6) and (5) then

$$\partial_i F_k(x) - \partial_i F(x) = \int_{\mathbf{R}^n} \vartheta_k(x - y)(\partial_i F(y) - \partial_i F(x)) dy.$$

Now let $k \geq 1/\delta$. Then $\vartheta_k(x - y) = 0$ if $|x - y| \geq \delta$, and, by using (5) again, we see that $|\partial_i F_k(x) - \partial_i F(x)| \leq \varepsilon$, for any $x \in B$. The lemma is proved. ◇

If $f \in C^L(M)$ and if $\{f_k\}$ is an infinite sequence in $C^\infty(M)$ with $\|f_k - f\|_1 \to 0$, then $\{f_k\}$ is a 1-Cauchy sequence. Its 1-equivalence class $[f_k]_1 \in H_1(M)$ is uniquely defined by f. Setting $j(f) = [f_k]_1$ gives an isometric embedding

(14.1.7) $$j : C^L(M) \to H_1(M)$$

of the pre-Hilbert space $(C^L(M), (\,,\,)_1)$ into the Hilbert space $H_1(M)$. We shall from now on consider $C^L(M)$ as a dense subspace of $H_1(M)$ with respect to this embedding. This gives the inclusions

(14.1.8) $$C^\infty(M) \subset C^L(M) \subset H_1(M) \subset H_0(M).$$ ◇

14.2 The Friedrichs Extension of the Laplacian

In Chapter 7 we used a family of integral operators to prove that the Laplacian has a complete orthonormal sequence of eigenfunctions in H_0. More precisely, we showed that there exists a 0-orthonormal sequence of real valued C^∞-eigenfunctions $\varphi_0, \varphi_1, \ldots$ which is complete in $C^\infty(M)$ with respect to the 0-norm. The corresponding eigenvalues are non-negative and satisfy $0 = \lambda_0 < \lambda_1 \leq \lambda_2 \leq \ldots$, with $\lambda_k \to \infty$ as $k \to \infty$. (The sign conventions for the Laplacian are such that $\Delta \varphi_k = \lambda_k \varphi_k$.)

We use the sequence $\varphi_0, \varphi_1, \ldots$ to give a different characterization of the spaces $H_0 = H_0(M)$ and $H_1 = H_1(M)$ and then extend the Laplacian from $C^\infty(M)$ to a larger linear subspace H_2 of H_0.

Since $C^\infty(M)$ is dense in H_0, the sequence $\{\varphi_k\}$ is complete in H_0 too, and so every $u \in H_0$ has a Fourier series expansion

$$u = \sum_{k=0}^{\infty} a_k \varphi_k, \text{ with } a_k = (u, \varphi_k)_0 \text{ and } \sum_{k=0}^{\infty} |a_k|^2 = \|u\|_0^2 < \infty.$$

14.2.1 Proposition. $H_1 = \{ u = \sum_{k=0}^{\infty} a_k \varphi_k \mid \sum_{k=0}^{\infty} \lambda_k |a_k|^2 < \infty \}$.

Proof. We abbreviate the right-hand side by H_R. For $u \in H_R$ the partial sums

$$s_n = \sum_{k=0}^{n} a_k \varphi_k$$

belong to $C^\infty(M)$. It follows therefore by Green's formula and by the orthonormality of the eigenfunctions φ_k for $n > m$ that

$$q[s_n - s_m] = (\Delta(s_n - s_m), s_n - s_m)_0 = \sum_{k=m+1}^{n} \lambda_k |a_k|^2.$$

Since $u \in H_R$, it follows that the partial sums s_n form a 1-Cauchy sequence, and therefore $H_R \subset H_1$. It remains to show that H_R is 1-closed in H_1 and that its 1-orthogonal complement is zero.

To show that H_R is 1-closed let $v \in H_1$ be in the 1-closure of H_R and consider a sequence $v_1, v_2, \ldots$ in H_R which 1-converges to v. Each v_r has a Fourier series expansion $v_r = \Sigma a_{r,k} \varphi_k$ in H_0. We modify the sequence $v_1, v_2, \ldots$ in H_R in such a way that it still converges to v, but such that for each v_r only finitely many $a_{r,k}$ are different from 0. Then $v_r \in C^\infty(M)$ and, by Green's formula

$$\|v_r\|_1^2 = \sum_{k=0}^{\infty} (1 + \lambda_k) |a_{r,k}|^2.$$

Since $\|v_r - v\|_1 \to 0$ as $r \to \infty$, it follows that $\|v_r - v\|_0 \to 0$ and therefore $a_{r,k} = (v_r, \varphi_k)_0 \to (v, \varphi_k)_0 =: a_k$, for each k. Since $\|v_r\|_1 \to \|v\|_1$, then for any n

$$\sum_{k=0}^{n} (1 + \lambda_k) |a_k|^2 \leq \|v\|_1^2.$$

Since $\lambda_k \to \infty$ as $k \to \infty$, this shows that $v \in H_R$, and thus H_R is closed.

Now let $v \in C^\infty(M)$ be 1-orthogonal to H_R. Then $(v, \varphi_k)_1 = 0$ for all k, where, by Green's formula,

$$(v, \varphi_k)_1 = (v, \varphi_k)_0 + q[v, \varphi_k] = (v, \varphi_k)_0 + (v, \Delta\varphi_k)_0 = (1 + \lambda_k)(v, \varphi_k)_0.$$

By the completeness of the sequence $\{\varphi_k\}$ in H_0 this implies that $v = 0$. Since $C^\infty(M)$ is 1-dense in H_1, we have thus shown that the 1-orthogonal complement of H_R in H_1 is zero. ◇

We shall extend the Laplacian to the following linear subspace of H_1.

14.2.2 Definition. $H_2 = H_2(M) = \{u = \sum_{k=0}^{\infty} a_k \varphi_k \mid \sum_{k=0}^{\infty} \lambda_k^2 |a_k|^2 < \infty\}$.

If $v = \sum_{k=0}^{\infty} b_k \varphi_k \in C^\infty(M)$, then $\Delta v \in C^\infty(M)$. Also, for any k

$$(\Delta v, \varphi_k)_0 = (v, \Delta\varphi_k)_0 = \lambda_k (v, \varphi_k)_0 = \lambda_k b_k.$$

Since $C^\infty(M) \subset H_0$, then

$$\Delta v = \sum_{k=0}^{\infty} \lambda_k b_k \varphi_k, \quad \sum_{k=0}^{\infty} \lambda_k^2 |b_k|^2 < \infty.$$

This shows that

(14.2.3) $$C^{\infty}(M) \subset H_2 \subset H_1.$$

Moreover, the following operator Q is an extension of Δ.

14.2.4 Definition. We let Q be the linear operator with domain $D(Q) = H_2 = H_2(M)$ defined by

$$Q(u) = \sum_{k=0}^{\infty} \lambda_k a_k \varphi_k \quad \text{for } u = \sum_{k=0}^{\infty} a_k \varphi_k \in H_2.$$

Q is called the *Friedrichs extension* of Δ.

14.2.5 Proposition. (i) *Q is a maximal self-adjoint extension of Δ in H_0.*
  (ii) *$(Qu, v)_0 = q[u, v]$ for all $u \in H_2$ and all $v \in H_1$.*
  (iii) *$D(Q)$ is the set of all $u \in H_1$ for which there exists $u' \in H_0$ satisfying $(u', v)_0 = q[u, v]$ for all $v \in H_1$.*

Proof. Let $u, v \in H_0$ be given by

$$u = \sum_{k=0}^{\infty} a_k \varphi_k, \quad v = \sum_{k=0}^{\infty} b_k \varphi_k,$$

and denote by u_n, v_n the partial sums

$$u_n = \sum_{k=0}^{n} a_k \varphi_k, \quad v_n = \sum_{k=0}^{n} b_k \varphi_k.$$

We first prove (ii). If $u, v \in H_1$, then $\{u_n\}_{n=1}^{\infty}$ and $\{v_n\}_{n=1}^{\infty}$ are 1-Cauchy sequences, and therefore

$$q[u, v] = \lim_{n \to \infty} q[u_n, v_n] = \lim_{n \to \infty} (\Delta u_n, v_n)_0 = \sum_{k=0}^{\infty} \lambda_k a_k \bar{b}_k.$$

If $u \in H_2$ and $v \in H_0$, then

$$Qu = \sum_{k=0}^{\infty} \lambda_k a_k \varphi_k \quad \text{and} \quad (Qu, v)_0 = \sum_{k=0}^{\infty} \lambda_k a_k \bar{b}_k = q[u, v].$$

We next prove (i). If $u, v \in H_2$, then $(Qu, v)_0 = q[u, v] = (u, Qv)_0$. Hence, Q is a symmetric operator. If $v \in H_0$, and if there exists v^* of the form

$$v^* = \sum_{k=0}^{\infty} b_k^* \varphi_k \in H_0$$

satisfying $(Qw, v)_0 = (w, v^*)_0$ for all $w \in H_2$, then in particular $\lambda_k \bar{b}_k = (Q\varphi_k, v)_0 = (\varphi_k, v^*)_0 = \bar{b}_k^*$ for $k = 0, 1, \ldots$. This implies that

$$\sum_{k=0}^{\infty} \lambda_k^2 |b_k|^2 = \sum_{k=0}^{\infty} |b_k^*|^2 < \infty,$$

that is, that $v \in H_2$. Hence, the domain of the adjoint of Q is H_2, and therefore Q is self-adjoint. A similar argument shows that if $u \in H_0$, and if there exists $u' \in H_0$ satisfying $(u', \varphi_k)_0 = (u, \Delta\varphi_k)_0$ for $k = 0, 1, \ldots$, then $u \in H_2$. This proves (i).

The last argument also proves that the right-hand side in (iii) is contained in $H_2 = D(Q)$. The converse inclusion follows from (ii). ◇

Since $\lambda_0 = 0$ is an eigenvalue of Q, Q is not invertible. But $Q + 1$ is invertible (see the beginning of Section 14.3 for notation), and $(Q + 1)^{-1}$ is a bounded operator. Later we shall need the square root of $(Q + 1)^{-1}$. It is defined as follows.

(14.2.6) $\quad K(u) = \sum_{k=0}^{\infty} (1 + \lambda_k)^{-1/2} a_k \varphi_k \quad$ for $u = \sum_{k=0}^{\infty} a_k \varphi_k \in H_0$.

This is a bounded operator with the property that the three mappings

$$K : H_0 \to H_1, \quad K : H_1 \to H_2, \quad K^2 : H_0 \to H_2$$

are vector space isomorphisms. Moreover,

(14.2.7) $\quad (Q+1) \circ K^2 = id$ on $H_0, \quad K^2 \circ (Q+1) = id$ on H_2.

The following properties are easily checked.

(14.2.8) $\quad (u, v)_0 = (Ku, Kv)_1 = (Ku, Kv)_0 + q[Ku, Kv], \quad u, v \in H_0,$

(14.2.9) $\quad q[K\varphi_k] = \dfrac{\lambda_k}{\lambda_k + 1}, \quad k = 0, 1, \ldots,$

(14.2.10) $\quad q[Ku] \leq \|u\|_0^2, \quad u \in H_0,$

(14.2.11) $\quad (Ku, v)_0 = (u, Kv)_0, \quad u, v \in H_0.$

14.3 A Representation Theorem

The Friedrichs extension Q of Δ represents the quadratic form q in the sense that $(Qu, v)_0 = \mathfrak{q}[u, v]$ for $u \in H_2$, $v \in H_1$ (cf. Proposition 14.2.5). In this section we shall prove that if $\mathfrak{t}$ is a perturbation of q sufficiently close to q (in a sense to be made precise), then there exists a suitable perturbation T of Q which represents $\mathfrak{t}$. (For the general form of the representation theorem we refer to Kato [1] theorem VI-2.1).

We first recall some definitions. For this we let X with norm $\| \ \|$ and convergence $\to$ be an arbitrary Banach space.

A linear operator *in* X is a linear mapping

$$T : D(T) \to X,$$

where $D(T)$ is a linear subspace (in general not a closed one) of X called the *domain* of T. The set

$$R(T) = \{Tu \mid u \in D(T)\}$$

is called the *range* of T.

If $D(T) = X$ and $\|T\| := \sup\{\|Tu\| \mid \|u\| = 1\} < \infty$, then T is a *bounded operator* and $\|T\|$ is its *norm*.

An operator $T : D(T) \to X$ is *closed* if the following holds. Whenever $\{x_n\}_{n=1}^\infty \subset D(T)$ is a sequence such that $\{Tx_n\}_{n=1}^\infty$ is a Cauchy sequence and such that $x_n \to x$ for some $x \in X$, then $x \in D(T)$ and $Tx_n \to Tx$.

We denote by $\mathscr{B}(X)$ the set of all bounded operators and with $\mathscr{C}(X)$ the set of all closed operators in X.

If T and S are operators in X, then the domain of their sum is defined as $D(T + S) = D(T) \cap D(S)$. We shall use the notation "$T + S$" only if one of the two operators is bounded. If $\zeta \in \mathbf{C}$ is a complex number and if T is an operator, then $T + \zeta$ denotes the operator $T + \zeta \mathrm{id}$.

An operator A is said *to have a bounded inverse* if A is one-to-one with range $R(A) = X$ and if its inverse A^{-1} belongs to $\mathscr{B}(X)$.

A *sesquilinear form* in X is a mapping $\mathfrak{t} : D(\mathfrak{t}) \times D(\mathfrak{t}) \to \mathbf{C}$ which is linear in the first and conjugate linear in the second argument. $D(\mathfrak{t})$ is a linear subspace of X (not in general closed) called the *domain* of $\mathfrak{t}$.

For $\mathfrak{t}$ a sesquilinear form we usually abbreviate $\mathfrak{t}[u, u]$ by $\mathfrak{t}[u]$.

14.3.1 Exercise. (i) *If $T \in \mathscr{C}(X)$ and if $B \in \mathscr{B}(X)$, then $T + B \in \mathscr{C}(X)$ with $D(T + B) = D(T)$.*

(ii) *If $A \in \mathscr{B}(X)$ and A is one-to-one, then $A^{-1} : R(A) \to X$ is a closed operator.*

(iii) *If $T \in \mathscr{C}(X)$ and if $B, B^{-1} \in \mathscr{B}(X)$, then $BT \in \mathscr{C}(X)$.*

(iv) *For any compact manifold M, the Friedrichs extension Q of Δ (cf. Definition 14.2.4) is a closed operator in $H_0(M)$.* (Hint: use (14.2.7).)

14.3.2 Lemma. *If T is a linear operator in X which has a bounded inverse, and if A is a bounded operator with $\|A\| < \|T^{-1}\|^{-1}$, then $T + A$ has a bounded inverse.*

Proof. Observe that $u \in D(T)$ if and only if $(1 + T^{-1}A)u \in D(T)$. Hence, we are allowed to write $T + A = T(1 + T^{-1}A)$. Set $C = -T^{-1}A$. Then $C \in \mathscr{B}(X)$ and $\|C\| < 1$. Hence, the series $1 + C + C^2 + \ldots$ converges in $\mathscr{B}(X)$ and defines a bounded operator C' with the property $(1 - C)C' = C'(1 - C) = id$.

If $u \in D(T)$, then $C'T^{-1}(T + A)u = u$, that is, $T + A$ is one-to-one. If $u \in X$, then $(1 - C)C'T^{-1}u = T^{-1}u \in D(T)$. Hence, $C'T^{-1}u \in D(T)$ and now $(T + A)C'T^{-1}u = T(1 - C)C'T^{-1}u = u$. Hence, by the above observation, $T + A$ has range X, and $C'T^{-1}$ is a bounded inverse of $T + A$. ◇

We now consider sesquilinear forms $\mathfrak{t}$ which are small perturbations of $\mathfrak{q}$ (cf. the hypothesis of Theorem 14.3.5 below). The goal is to represent these sesquilinear forms by closed operators T. Since we will require that $T - \zeta$ have a bounded inverse for all $\zeta \in \mathbf{C}$ which are not eigenvalues of Q, we first consider the following situation. Let $\zeta \in \mathbf{C}$ be any complex number different from the eigenvalues $\lambda_0, \lambda_1, \ldots$. We then define the bounded operator $Z = 1 - (1 + \zeta)K^2$. It is characterized by the property

$$(14.3.3) \qquad Z\varphi_k = \frac{\lambda_k - \zeta}{\lambda_k + 1}\varphi_k, \qquad k = 0, 1, \ldots,$$

where $\{\varphi_0, \varphi_1, \ldots\}$ is the complete orthonormal sequence of eigenfunctions of Q with respect to $\lambda_0, \lambda_1, \ldots$. Since $\lambda_k - \zeta \neq 0$ for all k, Z has a bounded inverse defined by

$$(14.3.4) \qquad Z^{-1}\varphi_k = \frac{\lambda_k + 1}{\lambda_k - \zeta}\varphi_k, \qquad k = 0, 1, \ldots .$$

Observe that $\|Z\| \geq 1$ and $\|Z^{-1}\| \geq 1$.

14.3.5 Theorem. *Let $\mathfrak{t}$ be a sesquilinear form in H_0 with domain $D(\mathfrak{t}) = H_1$ and assume that for all $u, v \in H_1$*

$$|\mathfrak{t}[u, v] - \mathfrak{q}[u, v]| \leq c\mathfrak{q}^{1/2}[u]\mathfrak{q}^{1/2}[v],$$

where c is a positive constant, $c < 1$. Then there exist operators $B \in \mathscr{B}(H_0)$ and $T \in \mathscr{C}(H_0)$ with the following properties.

(i) $\mathfrak{t}[Ku, Kv] = (Bu, v)_0$ *for all $u, v \in H_0$, and B is uniquely determined* *by this property.*

(ii) $B + K^2$ has a bounded inverse.

(iii) $T = K^{-1}BK^{-1}$; the domain of T is $\mathsf{D}(T) = K(B + K^2)^{-1}(H_1)$, and this is a 1-dense subset of H_1.

(iv) $\mathfrak{t}[u, v] = (Tu, v)_0$ for all $u \in \mathsf{D}(T)$ and all $v \in H_1$.

(v) If $\zeta \in \mathbf{C}$, and if $B_\zeta := B - \zeta K^2$ has a bounded inverse, then $T - \zeta$ has bounded inverse $(T - \zeta)^{-1} = KB_\zeta^{-1}K \in \mathscr{B}(H_0)$.

(vi) If $\zeta \in \mathbf{C} - \{\lambda_0, \lambda_1, \ldots\}$ and if in the hypothesis of the theorem $c < \inf_k |\lambda_k - \zeta|/(\lambda_k + 1)$, then $B - \zeta K^2$ has a bounded inverse.

We remark that if in the above theorem $\mathfrak{t} = \mathfrak{q}$, then $B = 1 - K^2$ and $T = K^{-2} - 1 = Q$ (cf. (14.2.7,8,11) and Proposition 14.2.5).

Proof. (i) The function β given by

$$\beta[u, v] := \mathfrak{t}[Ku, Kv] = (\mathfrak{t} - \mathfrak{q})[Ku, Kv] + \mathfrak{q}[Ku, Kv]$$
$$= (\mathfrak{t} - \mathfrak{q})[Ku, Kv] + ((1 - K^2)u, v)_0$$

(cf. (14.2.8,11)) is a sesquilinear form with domain H_0. By (14.2.10) it satisfies the inequality

$$|\beta[u, v]| \le (c + 1) \|u\|_0 \|v\|_0, \quad u, v \in H_0.$$

The bounded operator B and its uniqueness are therefore given by the Riesz representation theorem.

(ii) and (vi) Observe that (ii) is a special case of (vi) for $\zeta = -1$. To prove (vi) we consider the difference $A = B - (1 - K^2)$, (which vanishes if $\mathfrak{t} = \mathfrak{q}$). A is a bounded operator with norm

$$\|A\| = \sup_1 |(Au, v)_0| = \sup_1 |(\mathfrak{t} - \mathfrak{q})[Ku, Kv]| \le c$$

(cf. (14.2.10)), where the supremum is taken over all $u, v \in H_0$ with $\|u\|_0 = \|v\|_0 = 1$. By hypothesis, $c < \|Z^{-1}\|^{-1}$, where Z^{-1} is as in (14.3.3). Recalling that $Z = 1 - (1 + \zeta)K^2$, we obtain $B - \zeta K^2 = Z + A$ with $\|A\| < \|Z^{-1}\|^{-1}$. By Lemma 14.3.2, $B - \zeta K^2$ has a bounded inverse.

(iii) Since $(B + K^2)^{-1}$ is a bounded operator with range H_0, and since H_1 is 0-dense in H_0, $(B + K^2)^{-1}(H_1)$ is 0-dense in H_0. By (14.2.8), K is a Hilbert space isometry from $(H_0, (\,,\,)_0)$ onto $(H_1, (\,,\,)_1)$. Therefore, $\mathsf{D} := K(B + K^2)^{-1}(H_1)$ is 1-dense in H_1.

The operator $T' := K^{-1}(B + K^2)K^{-1}$ is defined on D. For $u \in \mathsf{D}$ we have $(B + K^2)K^{-1}u \in H_1$ and $K^2(K^{-1}u) \in H_1$. Hence,

$$BK^{-1}u \in H_1 \text{ for all } u \in \mathsf{D}.$$

It follows that $K^{-1}BK^{-1}u$ is well-defined for all $u \in D$. Setting $D(T) := D$ and $Tu = K^{-1}BK^{-1}u$ for $u \in D(T)$, we obtain an operator T with the property $D(T) = D(T')$, and

$$T + 1 = T'.$$

By Exercise 14.3.1(ii), $T' \in \mathscr{C}(H_0)$ and by Exercise 14.3.1(i), $T \in \mathscr{C}(H_0)$.

(iv) If $u \in D(T)$, and if $v \in H_1$, then there exists $u', v' \in H_0$ satisfying $u = K(B + K^2)^{-1}Ku'$, $v = Kv'$. Note that $u' = T'u$. From (i) and (14.2.11) it follows that

$$\begin{aligned} \mathfrak{t}[u, v] &= (B(B + K^2)^{-1}Ku', v')_0 \\ &= (Ku' - K^2(B + K^2)^{-1}Ku', v')_0 \\ &= (u' - u, Kv')_0 \\ &= (Tu, v)_0. \end{aligned}$$

(v) Since $B = B_\zeta + \zeta K^2$, then $(B + K^2)B_\zeta^{-1} = 1 + (1 + \zeta)K^2 B_\zeta^{-1}$ and therefore $(B + K^2)B_\zeta^{-1}(H_1) \subset H_1$. Similarly, writing $B_\zeta = B + K^2 - (1 + \zeta)K^2$, we see that $B_\zeta(B + K^2)^{-1}(H_1) \subset H_1$. Both statements together yield $B_\zeta^{-1}(H_1) = (B + K^2)^{-1}(H_1)$ and therefore $KB_\zeta^{-1}(H_1) = K(B + K^2)^{-1}(H_1) = D(T)$. It follows that $KB_\zeta^{-1}K : H_0 \to D(T)$ is one-to-one and onto.

Hence, if $w \in D(T)$, then $B_\zeta K^{-1}w \in H_1 = D(K^{-1})$, and we may write $(T - \zeta)w = K^{-1}BK^{-1}w - \zeta w = K^{-1}(B_\zeta + \zeta K^2)K^{-1}w - \zeta w = K^{-1}B_\zeta K^{-1}w$. It follows that $(T - \zeta)KB_\zeta^{-1}Ku = u$ for all $u \in H_0$. This concludes the proof that $KB_\zeta^{-1}K$ is the bounded inverse of $T - \zeta$. ◇

14.4 Resolvents and Projectors

In this section we again let X be a Banach space. Norm and convergence in X are denoted by $\| \; \|$ and $\to$. We recall that by definition, the assertion "$A^{-1} \in \mathscr{B}(X)$" implies that A has range $R(A) = X$.

14.4.1 Definition. Let $T \in \mathscr{C}(X)$ be a closed operator in X. The set

$$\rho(T) = \{\zeta \in \mathbb{C} \mid (T - \zeta)^{-1} \in \mathscr{B}(X)\}$$

is called the *resolvent set* of T, and $\mathbb{C} - \rho(T)$ is called the *spectrum* of T. The family of bounded operators

$$R(\zeta) = R(\zeta, T) := (T - \zeta)^{-1}, \quad \zeta \in \rho(T),$$

is called the *resolvent of T*.

14.4.2 Lemma. (Resolvent equation). *Let $R(\zeta)$ be the resolvent of T and let $\zeta_1, \zeta_2 \in \rho(T)$. Then*

$$R(\zeta_1) - R(\zeta_2) = (\zeta_1 - \zeta_2) R(\zeta_1) R(\zeta_2).$$

In particular, $R(\zeta_1)$ commutes with $R(\zeta_2)$.

Proof. $R(\zeta_2)$ sends X to the domain $D(T - \zeta_2) = D(T - \zeta_1)$, and so the product $R(\zeta_1)((T - \zeta_1)R(\zeta_2))$ is well-defined on X and is equal to $R(\zeta_2)$. It follows that

$$\begin{aligned} R(\zeta_1) - R(\zeta_2) &= R(\zeta_1)(T - \zeta_2)R(\zeta_2) - R(\zeta_1)(T - \zeta_1)R(\zeta_2) \\ &= (\zeta_1 - \zeta_2)R(\zeta_1)R(\zeta_2). \end{aligned}$$ ◇

We also note that T commutes with the resolvent in the following sense.

14.4.3 Lemma. *Let $T \in \mathscr{C}(X)$ and let $\zeta \in \rho(T)$. For all $u \in D(T)$*

$$(T - \zeta)^{-1}Tu = T(T - \zeta)^{-1}u.$$

Proof.
$$\begin{aligned} (T - \zeta)^{-1}Tu &= (T - \zeta)^{-1}(Tu - \zeta u) + \zeta(T - \zeta)^{-1}u \\ &= (T - \zeta)(T - \zeta)^{-1}u + \zeta(T - \zeta)^{-1}u \\ &= T(T - \zeta)^{-1}u. \end{aligned}$$ ◇

14.4.4 Proposition. (Von Neumann series). *Let $T \in \mathscr{C}(X)$ and let $\zeta_0 \in \rho(T)$. If $|\zeta - \zeta_0| \|R(\zeta_0)\| < 1$, then $\zeta \in \rho(T)$ and*

$$R(\zeta) = \sum_{n=0}^{\infty} (\zeta - \zeta_0)^n R^{n+1}(\zeta_0).$$

In particular, $\rho(T)$ is an open subset of $\mathbb{C}$.

Proof. We denote by B_m the partial sum

$$B_m = \sum_{n=0}^{m} (\zeta - \zeta_0)^n R^{n+1}(\zeta_0),$$

$m = 1, 2, \ldots$. As $m \to \infty$, B_m converges in $\mathscr{B}(X)$ to some bounded operator B. From the resolvent equation $R(\zeta_0) = R(\zeta) - (\zeta - \zeta_0)R(\zeta)R(\zeta_0)$ we see that

$$(T - \zeta)B_m = \sum_{n=0}^{m} (\zeta - \zeta_0)^n (T - \zeta)R(\zeta_0)R^n(\zeta_0)$$

and further

$$(T - \zeta)B_m = \sum_{n=0}^{m} (\zeta - \zeta_0)^n (T - \zeta)R(\zeta)(1 - (\zeta - \zeta_0)R(\zeta_0))R^n(\zeta_0)$$
$$= \sum_{n=0}^{m} (\zeta - \zeta_0)^n (1 - (\zeta - \zeta_0)R(\zeta_0))R^n(\zeta_0)$$
$$= 1 - (\zeta - \zeta_0)^{m+1} R^{m+1}(\zeta_0).$$

Now let $u \in X$. Then $B_m u \in D(T - \zeta)$ for any m. As $m \to \infty$, $B_m u \to Bu$, and $(T - \zeta)B_m u = u - (\zeta - \zeta_0)^{m+1} R^{m+1}(\zeta_0) u \to u$. In particular, the sequence $\{(T - \zeta)B_m u\}$ is a Cauchy sequence. Since $T - \zeta$ is a closed operator, it follows that $Bu \in D(T - \zeta)$ and that $(T - \zeta)Bu = u$. Hence, $R(T - \zeta) = X$. By Lemma 14.4.3 applied to ζ_0, $(T - \zeta)B_m u = B_m(T - \zeta)u$ for $u \in D(T)$. It follows that $B(T - \zeta)u = u$ for all $u \in D(T)$, and therefore $(T - \zeta)$ is one-to-one, and B is its bounded inverse. ◇

14.4.5 Definition. A bounded operator $P \in \mathcal{B}(X)$ is a *projector* if $P^2 = P$.

14.4.6 Lemma. *Let $P \in \mathcal{B}(X)$ be a projector. Then $Pu = u$ for all $u \in P(X)$. If $Q \in \mathcal{B}(X)$ is a second projector such that $\|P - Q\| < 1$, then the restriction $Q \mid P(X) : P(X) \to Q(X)$ is an isomorphism of vector spaces. In particular, $\dim P(X) = \dim Q(X)$.*

Proof. Let $E = P(X)$. If $u \in E$, then $u = Pv$ for some $v \in X$ and therefore $Pu = P^2 v = Pv = u$. If $u \in E$ and if $u \neq 0$, then $\|u - Qu\| = \|Pu - Qu\| \leq \|P - Q\| \|u\| < \|u\|$. Hence, the restriction of Q to E has kernel zero, and therefore $\dim Q(E) = \dim E$. This shows that $\dim Q(X) \geq \dim P(X)$. Similarly we show that $\dim P(X) \geq \dim Q(X)$. ◇

In the following we consider curves in the complex plane. A curve is called *piecewise regular* if it consists of a finite number of smooth arcs, such that each arc has nowhere vanishing speed.

Let $T \in \mathcal{C}(X)$. To every piecewise regular closed curve Γ in the resolvent set $\rho(T)$ we associate a bounded operator P_Γ as follows

$$P_\Gamma := -\frac{1}{2\pi i} \int_\Gamma R(\zeta) d\zeta.$$

To see that the integral exists and defines a bounded operator, it suffices to note from the von Neumann series that the operator-valued function $\zeta \mapsto R(\zeta)$ is continuous and therefore uniformly continuous on Γ. Note that for all $u \in X$,

$$P_\Gamma u = -\frac{1}{2\pi i} \left[\int_\Gamma R(\zeta) d\zeta\right] u = -\frac{1}{2\pi i} \int_\Gamma R(\zeta) u \, d\zeta.$$

14.4.7 Lemma. *If Γ is homotopic in $\rho(T)$ to Γ', then $P_\Gamma = P_{\Gamma'}$.*

Proof. It suffices to show that if a piecewise regular curve γ is homotopic in $\rho(T)$ to a point, then $P_\gamma = 0$. Now for any $u \in X$ and for any bounded linear functional φ on X, it follows that

$$\varphi(P_\gamma u) = -\frac{1}{2\pi i} \int_\gamma \varphi(R(\zeta)u)d\zeta.$$

The von Neumann series shows that the complex-valued function $\zeta \mapsto \varphi(R(\zeta)u)$ is holomorphic on $\rho(T)$. By Cauchy's integral theorem, $\varphi(P_\gamma u) = 0$. Since this holds for any φ and any u, then $P_\gamma = 0$. ◇

From now on we let Γ in $\rho(T)$ be a piecewise regular *Jordan curve*. The complement $\mathbf{C} - \Gamma$ then consists of two open connected components. We denote by int Γ the bounded component and by ext Γ the unbounded component. In the following, Γ will always be parametrized with *positive* orientation so that

$$\int_\Gamma \frac{d\zeta}{\zeta - w} = \begin{cases} 2\pi i & \text{if } w \in \text{int } \Gamma \\ 0 & \text{if } w \in \text{ext } \Gamma \end{cases}.$$

To simplify the language, a closed Jordan curve will be called *admissible* if it is piecewise regular and positively oriented.

14.4.8 Theorem. *Let $T \in \mathscr{C}(X)$, let Γ be an admissible Jordan curve in $\rho(T)$ and let $P := P_\Gamma$ as above. Then*
  (i) *P is a projector,*
  (ii) $PR(\zeta) = R(\zeta)P$ *for all* $\zeta \in \rho(T)$,
  (iii) $P(X) \subset D(T)$,
  (iv) $PTu = TPu$ *for all* $u \in D(T)$,
  (v) $TPu = PTPu = -\dfrac{1}{2\pi i} \displaystyle\int_\Gamma \zeta R(\zeta) u \, d\zeta$ *for all* $u \in X$.

Proof. Let $\Gamma' \subset \text{int } \Gamma$ be a second admissible Jordan curve which is homotopic in $\rho(T)$ to Γ. Since $\rho(T)$ is open and since Γ is piecewise regular, such a curve is not difficult to find. By Lemma 14.4.7, $P = P_\Gamma = P_{\Gamma'}$. Hence,

$$P^2 = (2\pi i)^{-2} \int_\Gamma \left[\int_{\Gamma'} R(\zeta)R(z) \, dz \right] d\zeta.$$

From the resolvent equation and since $\Gamma' \subset \text{int } \Gamma$, we obtain

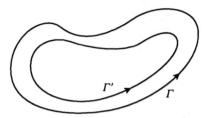

Figure 14.4.1

$$P^2 = (2\pi i)^{-2} \int_\Gamma \left[\int_{\Gamma'} \left[\frac{R(\zeta)}{\zeta - z} - \frac{R(z)}{\zeta - z} \right] dz \right] d\zeta$$

$$= (2\pi i)^{-2} \int_\Gamma R(\zeta) \left[\int_{\Gamma'} \frac{dz}{\zeta - z} \right] d\zeta - (2\pi i)^{-2} \int_{\Gamma'} R(z) \left[\int_\Gamma \frac{d\zeta}{\zeta - z} \right] dz$$

$$= -\frac{1}{2\pi i} \int_{\Gamma'} R(z)\, dz = P.$$

This proves (i). Statement (ii) holds because $R(\zeta)$ commutes with $R(z)$ for $\zeta, z \in \rho(T)$. To prove the remaining statements we approximate the defining integral by finite sums. Let Γ be parametrized in the form $t \mapsto \Gamma(t)$, $t \in [0, 1]$ with $\Gamma(0) = \Gamma(1)$. For each $n = 1, 2, \ldots$, we set $\zeta_k = \Gamma(k/n)$, $k = 0, \ldots, n$ (so that $\zeta_0 = \zeta_n$) and define

$$S_n = -\frac{1}{2\pi i} \sum_{k=0}^{n-1} R(\zeta_k)(\zeta_{k+1} - \zeta_k).$$

Then $S_n \to P$ as $n \to \infty$. Now let $u \in X$. Then $R(\zeta_k)u \in D(T - \zeta_k) = D(T)$ and therefore $S_n u \in D(T)$. Further,

$$TS_n u = -\frac{1}{2\pi i} \sum_{k=0}^{n-1} (\zeta_{k+1} - \zeta_k)(T - \zeta_k)R(\zeta_k)u - \frac{1}{2\pi i} \sum_{k=0}^{n-1} (\zeta_{k+1} - \zeta_k)\zeta_k R(\zeta_k)u.$$

The first sum vanishes, whereas for $n \to \infty$, the second sum converges to

$$w := -\frac{1}{2\pi i} \int_\Gamma \zeta R(\zeta)u\, d\zeta.$$

Since T is a closed operator and since $S_n u \to Pu$, it follows that $Pu \in D(T)$ and that $TPu = w$. This is (iii) and part of (v).

Now fix $\zeta \in \rho(T)$. By Lemma 14.4.2, $R(\zeta)$ commutes with S_n and therefore with P. For $u \in D(T)$ we conclude that $Pu = P(T - \zeta)^{-1}(T - \zeta)u = (T - \zeta)^{-1}P(T - \zeta)u \in D(T - \zeta)$. Applying $(T - \zeta)$ to both sides we see that

$(T-\zeta)Pu = P(T-\zeta)u$. This proves (iv). Finally, if $u \in X$, then $Pu \in D(T)$ by (iii). Since $P^2 = P$ and observing (iv) we obtain $TPu = TPPu = PTPu$. The theorem is proved. ◇

14.4.9 Exercise. *Prove that for all* $n \in \mathbb{N}$

$$T^n P = PT^n P = -\frac{1}{2\pi i} \int_\Gamma \zeta^n R(\zeta) d\zeta.$$

(This exercise will not be needed below.)

We relate P_Γ to the spectrum of T. Recall that $u \in D(T) - \{0\}$ is an eigenvector of T with eigenvalue $\lambda \in \mathbb{C}$ if $Tu = \lambda u$. Clearly, each eigenvalue belongs to the spectrum $\mathbb{C} - \rho(T)$, but the converse does not hold in general. For every eigenvalue λ of T we denote the corresponding eigenspace by $E(\lambda, T)$:

$$E(\lambda, T) = \{u \in D(T) \mid Tu = \lambda u\}.$$

We say that λ is an *isolated eigenvalue* of T if it is an eigenvalue and if there exists an open neighborhood U of $\lambda \in \mathbb{C}$ such that $U - \{\lambda\} \subset \rho(T)$.

14.4.10 Proposition. *Let T, Γ and $P = P_\Gamma$ be as above and assume that $u \in D(T) - \{0\}$ is an eigenfunction of T with eigenvalue λ. Then*

$$P_\Gamma u = \begin{cases} u & \text{if } \lambda \in \text{int } \Gamma \\ 0 & \text{if } \lambda \in \text{ext } \Gamma. \end{cases}$$

In particular, $E(\lambda, T) \subset P_\Gamma(X)$ if $\lambda \in \text{int } \Gamma$.

Proof. For every $z \in \rho(T)$, $(T-\zeta)u = (\lambda - \zeta)u$, and therefore $(T-\zeta)^{-1}u = (\lambda - \zeta)^{-1}u$. Hence,

$$P_\Gamma u = -\frac{1}{2\pi i}\left[\int_\Gamma \frac{1}{\lambda - \zeta} d\zeta\right]u.$$

The proposition follows. ◇

If λ is an isolated eigenvalue and if Γ is such that $\lambda \in \text{int } \Gamma$ but $\eta \in \text{ext } \Gamma$ for all other η in the spectrum of T, then we are tempted to conclude that $E(\lambda, T) = P_\Gamma(X)$. Here is a counterexample.

14.4.11 Example. Let $X = \mathbb{C}^2$ and let T be given by the following matrix (with respect to the standard basis of $\mathbb{C}^2$)

$$T = \begin{pmatrix} 1 & b \\ 0 & 1 \end{pmatrix}.$$

Then $\lambda = 1$ is an eigenvalue, and the resolvent set is $\rho(T) = \mathbf{C} - \{1\}$. The eigenspace $E(1, T)$ has dimension 2 if $b = 0$ and dimension 1 if $b \neq 0$. For $\zeta \in \rho(T)$ we compute

$$R(\zeta) = \frac{1}{(1-\zeta)^2} \begin{pmatrix} 1-\zeta & -b \\ 0 & 1-\zeta \end{pmatrix}.$$

For any admissible Jordan curve Γ with $1 \in \text{int } \Gamma$

$$P_\Gamma = \begin{pmatrix} 1 & 0 \\ 0 & 1 \end{pmatrix}.$$

Hence, $P_\Gamma(\mathsf{X}) \neq E(1, T)$ except for the case $b = 0$. The example also shows that an equality $E(\lambda, T) = P_\Gamma(\mathsf{X})$ may hold for a particular T and fail for T' arbitrarily close to T, even in very simple cases.

We now consider a particular case where equality *does* hold. In the following, a set or a sequence $A \subset \mathsf{X}$ is said to be *complete* if the subspace A' of all linear combinations of elements of A is dense in X. For linear subspaces $E_1, \ldots, E_n$ of X we denote by $E_1 + \ldots + E_n$ the smallest linear subspace of X which contains $E_1, \ldots, E_n$.

14.4.12 Theorem. *Let $T \in \mathscr{C}(\mathsf{X})$ and assume that T has a complete sequence $u_1, u_2, \ldots$ of eigenvectors with eigenvalues $\mu_1, \mu_2, \ldots$.*

(i) If η is an isolated eigenvalue of T, then $\eta = \mu_k$ for some k.

(ii) Let $\eta_1, \ldots, \eta_n$ be isolated eigenvalues and assume that the vector space $E := E(\eta_1, T) + \ldots + E(\eta_n, T)$ is finite dimensional. If Γ is an admissible Jordan curve in $\rho(T)$ such that $\eta_\nu \in \text{int } \Gamma$ for $\nu = 1, \ldots, n$ and such that $\mu_k \in \text{ext } \Gamma$ for all $\mu_k \notin \{\eta_1, \ldots, \eta_n\}$, then $P_\Gamma(\mathsf{X}) = E$.

Proof. We begin with (ii). By Proposition 14.4.10, $E \subset P_\Gamma(\mathsf{X})$. Now let $v \in P_\Gamma(\mathsf{X})$. There exists an infinite sequence $v_1, v_2, \ldots$ converging to v, where each v_i is a linear combination $v_i = \Sigma\, b_{ij}\, u_j$. By Proposition 14.4.10 again, $P_\Gamma v_i = \Sigma b_{ij}^* u_j$, where $b_{ij}^* = b_{ij}$ if $\mu_j \in \{\eta_1, \ldots, \eta_n\}$ and $b_{ij}^* = 0$ otherwise. Hence, $P_\Gamma v_i \in E$ for all i. Since P_Γ is a bounded operator, $P_\Gamma v_i$ converges to $P_\Gamma v = v$. Since E is finite dimensional, E is closed, and therefore $v \in E$.

To prove (i) we let Γ be an admissible Jordan curve in $\rho(T)$ with $\eta \in \text{int } \Gamma$ and with $\mu_k \in \text{ext } \Gamma$ for all $\mu_k \neq \eta$. Here $E := E(\eta, T)$ is closed because T is a closed operator. Hence, by the preceding argument, $E(\eta, T) = P_\Gamma(\mathsf{X})$. If η were different from $\mu_1, \mu_2, \ldots$, then all of the above b_{ij} would vanish, yielding $v = P_\Gamma v = 0$ for any $v \in E(\eta, T)$, a contradiction. ◇

14.5 Holomorphic Families

In this section we consider parametrized families $z \mapsto F(z) \in Y$ in a Banach space Y, where $z = (z_1, \ldots, z_m) \in U$ for some open subset $U \subset \mathbf{C}^m$. For any Banach space X we let X* denote its dual space, that is, the Banach space of all conjugate-linear bounded functionals on X. For $v \in X^*$ the action on $u \in X$ will be denoted by (u, v).

14.5.1 Definition. A function $w \mapsto F(w) \in Y$, $w \in U$ is *complex differentiable* at $z \in U$ if there exists an open neighborhood U' of z in U and *continuous* functions $\zeta \mapsto F_1(\zeta), \ldots, \zeta \mapsto F_m(\zeta) \in Y$, $\zeta \in U'$, such that

$$F(\zeta) = F(z) + \sum_{k=1}^{m} (\zeta_k - z_k) F_k(\zeta), \qquad \zeta = (\zeta_1, \ldots, \zeta_m) \in U'.$$

F is *holomorphic* in U if F is complex differentiable at z for all $z \in U$.

14.5.2 Exercise. (i) *Every holomorphic family is continuous.*

(ii) *If X is a Banach space with dual X*, and if $u(z)$ in X and $v(z)$ in X* are holomorphic families, then $(u(z), v(z))$ is a holomorphic function.*

(iii) *Let $Y = \mathscr{B}(X)$ be the Banach algebra of bounded operators over a Banach space X. If $T(z), S(z) \in \mathscr{B}(X)$ are holomorphic families, then $T(z)S(z)$ is a holomorphic family.*

(iv) *If $T(z) \in \mathscr{B}(X)$ and $u(z) \in X$ are holomorphic families, then $T(z)u(z)$ is a holomorphic family.*

We now give a different characterization of holomorphic families.

14.5.3 Theorem. *If $z \mapsto u(z) \in X$, $z \in U$, is a continuous family with the property that $(u(z), v)$ is a holomorphic function for all v in a dense subset of X*, then $u(z)$ is a holomorphic family.*

Proof. Fix $z = (z_1, \ldots, z_m) \in U \subset \mathbf{C}^m = \mathbf{C} \times \ldots \times \mathbf{C}$. For $k = 1, \ldots, m$, we let Γ_k be a positively oriented circle with center z_k in $\mathbf{C}$. We take the radii of these circles so small that the closure in $\mathbf{C}^m$ of int $\Gamma_1 \times \ldots \times$ int Γ_m is contained in U. Then the Cauchy integral formula (see for instance Grauert-Fritzsche [1], p. 11) yields for all $\zeta = (\zeta_1, \ldots, \zeta_m) \in$ int $\Gamma_1 \times \ldots \times$ int Γ_m,

$$(u(\zeta), v) = (2\pi i)^{-m} \int_{\Gamma_1} \ldots \int_{\Gamma_m} \frac{(u(\zeta'), v)}{(\zeta'_1 - \zeta_1) \ldots (\zeta'_m - \zeta_m)} d\zeta'_1 \ldots d\zeta'_m.$$

It follows that

$$(u(\zeta) - u(z), v) = (2\pi i)^{-m} \int_{\Gamma_1} \cdots \int_{\Gamma_m} (u(\zeta'), v) \frac{N}{D} d\zeta_1' \ldots d\zeta_m',$$

where $D := (\zeta_1' - \zeta_1) \ldots (\zeta_m' - \zeta_m)(\zeta_1' - z_1) \ldots (\zeta_m' - z_m)$,

$$N := \sum_{k=1}^{m} (\zeta_k - z_k) P_k,$$

and each P_k is a polynomial in the variables $\zeta_i, \zeta_i', z_i, i = 1, \ldots, m$. Hence,

$$(u(\zeta) - u(z), v) = \sum_{k=1}^{m} (\zeta_k - z_k)(2\pi i)^{-m} \int_{\Gamma_1} \cdots \int_{\Gamma_m} (u(\zeta'), v) \frac{P_k}{D} d\zeta_1' \ldots d\zeta_m'.$$

Now, for $k = 1, \ldots, m$, the family $u_k(\zeta), \zeta \in U', U' := \text{int } \Gamma_1 \times \ldots \times \text{int } \Gamma_m$, defined by

$$u_k(\zeta) := (2\pi i)^{-m} \int_{\Gamma_1} \cdots \int_{\Gamma_m} u(\zeta') \frac{P_k}{D} d\zeta_1' \ldots d\zeta_m'$$

is continuous and

$$(u(\zeta) - u(z), v) = (\sum_{k=1}^{m} (\zeta_k - z_k) u_k(\zeta), v).$$

(Note that the integral exists because $u(z)$ is a continuous family.) Since this holds for all v in a dense subset of X^*, then

$$u(\zeta) = u(z) + \sum_{k=1}^{m} (\zeta_k - z_k) u_k(\zeta).$$

This proves that u is complex differentiable at z for all $z \in U$. ◊

14.5.4 Theorem. *Let $z \mapsto T(z) \in \mathscr{B}(X), z \in U \subset \mathbf{C}^m$ be a continuous family of bounded operators. If for all u in a dense subset of X and all v in a dense subset of X^* the function $(T(z)u, v)$ is holomorphic, then the family $T(z)$ is holomorphic.*

Proof. The proof is similar to the preceding one and we use the same notation. First

$$(T(\zeta)u, v) = (2\pi i)^{-m} \int_{\Gamma_1} \cdots \int_{\Gamma_m} \frac{(T(\zeta')u, v)}{(\zeta_1' - \zeta_1) \ldots (\zeta_m' - \zeta_m)} d\zeta_1' \ldots d\zeta_m'.$$

Since the family $T(z)$ is continuous, the following integral exists

$$T_k(\zeta) = (2\pi i)^{-m} \int_{\Gamma_1} \cdots \int_{\Gamma_m} T(\zeta') \frac{P_k}{D} d\zeta_1' \ldots d\zeta_m'.$$

It defines a continuous family of bounded operators with the property

$$((T(\zeta) - T(z))u, v) = (\sum_{k=1}^{m} (\zeta_k - z_k) T_k(\zeta)u, v)$$

for all u in a dense subset of X and all v in a dense subset of X^*. Hence

$$T(\zeta) = T(z) + \sum_{k=1}^{m} (\zeta_k - z_k) T_k(\zeta). \qquad \diamond$$

The preceding theorem allows us to translate theorems from complex function theory to holomorphic families of operator-valued functions. The following theorem is frequently used.

14.5.5 Theorem. *If for $n = 1, 2, \ldots$ the families $z \mapsto T_n(z) \in \mathcal{B}(X)$, $z \in U$, are holomorphic and if $\|T_n(z) - T(z)\| \to 0$ uniformly on compact subsets of U, then the family $T(z)$ is holomorphic.*

Proof. We first observe from the locally uniform convergence that the family $T(z)$ is continuous. Now let $u \in X$, $v \in X^*$. Then $|((T_n(z) - T(z))u, v)| \le \|u\| \|v\| \|T_n(z) - T(z)\|$. Hence, the holomorphic functions $z \mapsto (T_n(z)u, v)$ converge uniformly on compact subsets of U to the function $f(z) = (T(z)u, v)$. By Weierstrass' theorem (cf. e.g. Narasimhan [1] p. 7), f is holomorphic, and by the preceding theorem, $T(z)$ is holomorphic. $\diamond$

14.5.6 Theorem. *Let $z \mapsto T(z) \in \mathcal{B}(X)$, $z \in U$, be a holomorphic family, let $z' \in U$ and assume that $T^{-1}(z') \in \mathcal{B}(X)$. Then there exists an open neighborhood U' of z' in U such that $T^{-1}(z) \in \mathcal{B}(X)$ for all $z \in U'$, and such that the family $z \mapsto T^{-1}(z)$ is holomorphic on U'.*

Proof. Let $S(z) = T(z') - T(z)$. By the continuity of the function $z \mapsto \|S(z)\|$ there exists an open neighborhood U' of z' in U such that $\|S(z)\| \le \frac{1}{2}\|T^{-1}(z')\|^{-1}$ for all $z \in U'$. Now, $T(z) = (1 - A(z))T(z')$ with $A(z) := S(z)T^{-1}(z')$ and $\|A(z)\| \le \frac{1}{2}$. By Theorem 14.5.5, the family $z \mapsto (1 - A(z))^{-1} = 1 + A(z) + A^2(z) + \ldots$ is holomorphic on U'. Thus, the family $z \mapsto T^{-1}(z')(1 - A(z))^{-1}$ is the bounded holomorphic inverse of $T(z)$ for $z \in U'$. $\diamond$

14.6 A Model of Teichmüller Space

Our goal is to study the Laplacians of the various surfaces $S \in \mathcal{T}_g$, where $\mathcal{T}_g$ is the Teichmüller space of the compact Riemann surfaces of genus g, ($g \ge 2$).

For this we shall pull back the structures from S onto some fixed base surface $S_0 \in \mathcal{T}_g$. The Laplacians then appear as certain closed operators in the Hilbert space $H_0(S_0)$.

To make this section independent we recall some facts from Chapter 6 and in particular from Section 6.7. The reader who is not familiar with Chapter 6 and with the notion of Teichmüller space may take the following lines as an ad hoc definition of Teichmüller space.

Let $\mathbf{D}$ be the disk model of the hyperbolic plane and let $\mathcal{P} \subset \mathbf{D}$ be a *convex geodesic polygon domain* with the consecutive sides $b_1, b_2, \bar{b}_1, \bar{b}_2, \ldots, b_{2g-1}, b_{2g}, \bar{b}_{2g-1}, \bar{b}_{2g}$, and angles $\zeta_1, \zeta_2, \bar{\zeta}_1, \bar{\zeta}_2, \ldots, \zeta_{2g-1}, \zeta_{2g}, \bar{\zeta}_{2g-1}, \bar{\zeta}_{2g}$ as in Fig. 6.7.1 and Fig. 14.6.1. The sides are geodesic arcs $t \mapsto b_n(t)$, $t \mapsto \bar{b}_n(t)$, $t \in [0, 1]$, $n = 1, \ldots, 2g$, parametrized with constant speed and with positive boundary orientation. $\mathcal{P}$ is called a *normal canonical polygon* if the following conditions are satisfied (ℓ denotes the length):

$$\ell(b_n) = \ell(\bar{b}_n), \quad n = 1, \ldots, 2g,$$
$$\zeta_1 + \bar{\zeta}_1 + \ldots + \zeta_{2g} + \bar{\zeta}_{2g} = 2\pi,$$
$$\zeta_1 + \zeta_2 = \bar{\zeta}_1 + \bar{\zeta}_2 = \pi.$$

Two normal canonical polygons $\mathcal{P}$ with sides $b_1, \ldots, \bar{b}_{2g}$ and $\mathcal{P}'$ with sides $b'_1, \ldots, \bar{b}'_{2g}$ are considered equal if and only if there exists an orientation-preserving isometry from $\mathcal{P}$ to $\mathcal{P}'$ which maps b_n to b'_n and $\bar{b}_n$ to $\bar{b}'_n$, $n = 1, \ldots, 2g$. We let $\mathcal{P}_g$ denote the set of all normal canonical polygons.

For each $\mathcal{P} \in \mathcal{P}_g$ we have a sequence $Z(\mathcal{P}) \in \mathbf{R}^{6g-6}$ defined by

$$z = Z(\mathcal{P}) = (\ell(b_3), \ldots, \ell(b_{2g}), \zeta_3, \ldots, \zeta_{2g}, \bar{\zeta}_3, \ldots, \bar{\zeta}_{2g}).$$

From Theorem 6.8.13 we have

14.6.1 Proposition. *For $\mathcal{P} \in \mathcal{P}_g$ the lengths of b_1 and b_2 and the angles $\zeta_1, \zeta_2, \bar{\zeta}_1, \bar{\zeta}_2$, are real analytic functions of $z = Z(\mathcal{P})$. The set*

$$\mathcal{Z}_g = \{ Z(\mathcal{P}) \mid \mathcal{P} \in \mathcal{P}_g \}$$

is a connected open subset of $\mathbf{R}^{6g-6}$. The mapping $\mathcal{P} \mapsto Z(\mathcal{P})$ from $\mathcal{P}_g$ to $\mathcal{Z}_g$ is one-to-one and onto. ◇

In the following we shall use $z \in \mathcal{Z}_g$ as the independent variable and denote by $\mathcal{P}[z]$ the unique polygon $\mathcal{P}$ in $\mathcal{P}_g$ with $Z(\mathcal{P}) = z$. When the need arises, dependence on z will be emphasized by denoting the the sides and angles of $\mathcal{P}[z]$ by $b_n[z], \bar{b}_n[z], \zeta_n[z], \bar{\zeta}_n[z]$, $n = 1, \ldots, 2g$. All analyticity arguments in the sequel will be understood with respect to the canonical analytic structure of the open subset $\mathcal{Z}_g \subset \mathbf{R}^{6g-6}$.

We relate $\mathcal{P}_g$ to the Teichmüller space $\mathcal{T}_g$. Let $\mathcal{P} \in \mathcal{P}_g$ with sides $b_n, \bar{b}_n$, $n = 1, \ldots, 2g$ be as above. We paste together b_n and $\bar{b}_n$ via the identifying equations

(14.6.2) $\quad b_n(t) = \bar{b}_n(1-t), \quad t \in [0, 1], \; n = 1, \ldots, 2g.$

The resulting quotient space $S = S(\mathcal{P})$ has a unique Riemann surface structure such that the canonical projection $\mathcal{P} \to S$ is an isometry in the interior of $\mathcal{P}$. By Theorem 6.8.13, every compact Riemann surface of genus g arises in the form $S = S(\mathcal{P})$ for some $\mathcal{P} \in \mathcal{P}_g$. For $\mathcal{P} = \mathcal{P}[z]$, $z \in \mathcal{Z}_g$, we set

$$S[z] := S(\mathcal{P}[z]).$$

Theorem 6.8.13 then states, more precisely, the following.

14.6.3 Proposition. *The mapping $z \to S[z]$ is a real analytic diffeomorphism from the open set $\mathcal{Z}_g \subset \mathbf{R}^{6g-6}$ onto the Teichmüller space $\mathcal{T}_g$.* ◊

We may therefore take the statement "$\mathcal{T}_g = \{ S[z] \mid z \in \mathcal{Z}_g \}$" as an ad hoc definition of Teichmüller space in this chapter.

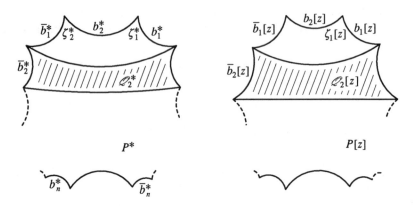

Figure 14.6.1

We next introduce particular homeomorphisms between the various polygons in $\mathcal{P}_g$. To this end we fix, once for all, a polygon $\mathcal{P}^* \in \mathcal{P}_g$ whose sides and angles are denoted by $b_n^*, \bar{b}_n^*, \zeta_n^*, \bar{\zeta}_n^*$. The corresponding surface

$$S^* := S[\mathcal{P}^*]$$

will serve as reference or base surface. We first decompose $\mathcal{P}^*$ into $2g - 1$ geodesic quadrilaterals $\mathcal{Q}_1^*, \ldots, \mathcal{Q}_{2g-1}^*$ as shown in Fig. 14.6.1 and then decompose all other $\mathcal{P}[z] \in \mathcal{P}_g$ *in the same manner*. The quadrilaterals on $\mathcal{P}[z]$ are denoted by $\mathcal{Q}_\kappa[z]$, $\kappa = 1, \ldots, 2g - 1$.

To define a homeomorphism from $\mathcal{P}^*$ to $\mathcal{P}[z]$ we use quadrilateral coordinates. They are defined as follows.

Let b, a, b', a' be the sides and let $w_1, \ldots, w_4$ be the vertices of a convex geodesic quadrilateral $\mathcal{Q} \subset \mathbf{D}$ as shown in Fig. 14.6.2 with all interior angles less than π. Each side is parametrized on the interval $[0, 1]$ with constant speed. The orientations are such that a and a' go from b to b' and such that b and b' go from a to a'. The parametrizations are extended to the interval $[-\varepsilon, 1 + \varepsilon]$ for small positive ε. We set

$$\mathcal{R} = \{(s, t) \in \mathbf{R}^2 \mid 0 \le s, t \le 1\}, \quad \mathcal{R}_\varepsilon = \{(s, t) \in \mathbf{R}^2 \mid -\varepsilon \le s, t \le 1 + \varepsilon\}.$$

For $(s, t) \in \mathcal{R}$, the geodesic connecting $a(s)$ with $a'(s)$ intersects the geodesic connecting $b(t)$ with $b'(t)$ in a uniquely determined point $x(s, t)$. Hence, we obtain a mapping

$$(s, t) \mapsto x(s, t) = x(s, t; w_1, \ldots, w_4) \in \mathcal{Q}, \quad (s, t) \in \mathcal{R}.$$

Since all interior angles of $\mathcal{Q}$ are smaller than π, this mapping can be extended to a diffeomorphism

(14.6.4) $\quad x : \mathcal{R}_\varepsilon \to \mathcal{Q}_\varepsilon = \mathcal{Q}_\varepsilon(w_1, \ldots, w_4) := x(\mathcal{R}_\varepsilon) \subset \mathbf{D}$

for some $\varepsilon > 0$. We call s and t the *quadrilateral coordinates* of the point $x(s, t) \in \mathcal{Q}_\varepsilon$.

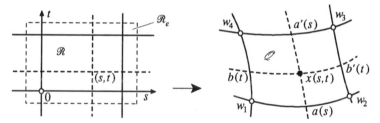

Figure 14.6.2

The trigonometric formulae of Chapter 2 imply that there exists $\varepsilon > 0$ and small open neighborhoods $U(w_1), \ldots, U(w_4)$ in $\mathbf{D}$ such that $\mathcal{Q}_\varepsilon(\omega_1, \ldots, \omega_4)$ is a convex quadrilateral for any $(\omega_1, \ldots, \omega_4) \in \Omega := U(w_1) \times \ldots \times U(w_4)$, and such that the mapping

$$(s, t, \omega_1, \ldots, \omega_4) \mapsto x(s, t; \omega_1, \ldots, \omega_4) \in \mathbf{D}$$

is real analytic on $\mathcal{R}_\varepsilon \times \Omega$. For given $(\omega_1, \ldots, \omega_4) \in \Omega$ we let

$$g_{ij}(s, t) = g_{ij}(s, t; \omega_1, \ldots, \omega_4), \quad (s, t) \in \mathcal{R}_\varepsilon$$

be the components of the hyperbolic metric tensor of $\mathcal{Q}_\varepsilon(\omega_1, \ldots, \omega_4)$ with respect to the quadrilateral coordinates, $1 \le i, j \le 2$. If ε is sufficiently small, then $\det(g_{ij}(s\,t;\omega_1,\ldots,\omega_4)) \geq \text{const} > 0$ on $\mathcal{R}_\varepsilon \times \Omega$, and the components

$$g^{jk}(s,t) = g^{jk}(s,t;\omega_1,\ldots,\omega_4)$$

of the inverse tensor are well defined on $\mathcal{R}_\varepsilon \times \Omega$. The g_{ij} and, by Cramer's rule, the g^{jk} are real analytic functions on $\mathcal{R}_\varepsilon \times \Omega$. For simplicity, we call both, g_{ij} and g^{jk} *the components* of the metric tensor.

Now again consider the quadrilaterals $\mathcal{Q}_\kappa^*$ and $\mathcal{Q}_\kappa[z]$ in $\mathcal{P}^*$ and in $\mathcal{P}[z]$, $z \in \mathcal{Z}_g$. For $\kappa = 1,\ldots,2g-1$, we label the sides of $\mathcal{Q}_\kappa^*$ and $\mathcal{Q}_\kappa[z]$ in a corresponding manner and then introduce quadrilateral coordinate mappings (cf. (14.6.4))

(14.6.5) $\qquad x_\kappa^* : \mathcal{R} \to \mathcal{Q}_\kappa^*, \quad x_\kappa[z] : \mathcal{R} \to \mathcal{Q}_\kappa[z],$

$\kappa = 1,\ldots,2g-1$ (cf. (14.6.4)). For each pair $\mathcal{Q}_\kappa^*, \mathcal{Q}_\kappa[z]$ we define a homeomorphism

$$\psi_\kappa[z] = x_\kappa[z] \circ (x_\kappa^*)^{-1} : \mathcal{Q}_\kappa^* \to \mathcal{Q}_\kappa[z].$$

From the definition of the quadrilateral coordinates follows the property that if u is any side of $\mathcal{Q}_\kappa^*$, parametrized with constant speed, then $\psi_\kappa[z] \circ u$ is also parametrized with constant speed. Hence, for $\kappa = 1,\ldots,2g-1$, the mappings $\psi_\kappa[z]$ coincide along any side belonging to two adjacent quadrilaterals, and there is a homeomorphism $\psi[z] : \mathcal{P}^* \to \mathcal{P}[z]$ satisfying

$$\psi[z] \circ x_\kappa^*(s\,t) = x_\kappa[z](s,t), \quad (s,t) \in \mathcal{R}, \quad \kappa = 1,\ldots,2g-1.$$

Moreover, $\psi[z]$ satisfies $\psi[z] \circ b_n^*(t) = b_n[z](t)$ and $\psi[z] \circ \bar{b}_n^*(t) = \bar{b}_n[z](t)$, $t \in [0,1]$, $n = 1,\ldots,2g$. Hence, $\psi[z]$ can also be interpreted as a homeomorphism

(14.6.6) $\qquad \psi[z] : S^* \to S[z],$

where $S[z] = S(\mathcal{P}[z])$ is the Riemann surface obtained by pasting together the sides of $\mathcal{P}[z]$ as above.

Now let $z_0 \in \mathcal{Z}_g$ be an arbitrary point which will be kept fixed. We abbreviate $S[z_0] = S_0$, $\mathcal{Q}_\kappa[z_0] = \mathcal{Q}_\kappa$ and $x_\kappa[z_0] = x_\kappa$. For $z \in \mathcal{Z}_g$ we consider the homeomorphism

(14.6.7) $\qquad \phi = \phi[z] := \psi[z] \circ (\psi[z_0])^{-1} : S_0 \to S[z].$

In quadrilateral coordinates, ϕ is given by

$$\phi[z] \circ x_\kappa(s,t) = x_\kappa[z](s,t), \quad (s,t) \in \mathcal{R}, \quad \kappa = 1,\ldots,2g-1.$$

From the properties of the mappings x (cf. (14.6.4) and (14.6.5)) and from Proposition 14.6.1 we have

14.6.8 Proposition. (i) $\phi : S_0 \to S[z]$ and $\phi^{-1} : S[z] \to S_0$ are Lipschitz-continuous everywhere and smooth outside a compact set of measure zero.

(ii) Denote by $g_\kappa[z]$ the pull-back on $\mathcal{Q}_\kappa$ via ϕ of the metric tensor of $\mathcal{Q}_\kappa[z]$, $\kappa = 1, \ldots, 2g - 1$, and let $g_{\kappa, ij}(s, t) = g_{\kappa, ij}(s, t; z)$ and $g_\kappa^{jk}(s, t) = g_\kappa^{jk}(s, t; z)$ be the components of $g_\kappa[z]$ with respect to the quadrilateral coordinates, $1 \leq i, j, k \leq 2$. Then these components are real analytic functions on $\mathcal{R} \times \mathcal{T}_g$ which can be extended analytically onto an open neighborhood of $\mathcal{R} \times \mathcal{T}_g$ in $\mathbf{R}^2 \times \mathcal{T}_g$. ◇

In general, the tensors $g_\kappa[z]$ do not match along the sides belonging to adjacent quadrilaterals. This causes no problem, for the tensors will only be used in connection with integration over S_0.

For each $S[z] \in \mathcal{T}_g$ we have the $(2g - 1)$-tuple of tensors $g[z] = (g_1[z], \ldots, g_{2g-1}[z])$ on S_0.

The family $\{ g[z] \mid z \in \mathcal{T}_g \}$ is our model of Teichmüller space.

Observe that we may construct this model on any given $S_0 \in \mathcal{T}_g$.

Let us now pull back the function spaces from $S = S[z]$ to S_0 via the above homeomorphism $\phi = \phi[z] : S_0 \to S[z]$. For $f \in C^\infty(S)$ we let

$$\phi_* f = f \circ \phi.$$

By Proposition 14.6.8 we have $\phi_* f \in C^L(S_0)$, and by (14.1.8), $C^L(S_0) \subset H_1(S_0)$. Since ϕ is Lipschitz-continuous, there exist positive constants c_0, c_1 such that $\|\phi_* f\|_\nu \leq c_\nu \|f\|_\nu$ for all $f \in C^\infty(S)$, $\nu = 0, 1$. Hence, ϕ_* maps every ν-Cauchy sequence in $C^\infty(S)$ onto a ν-Cauchy sequence in $H_1(S_0)$. We can therefore extend ϕ_* to a linear mapping $\phi_* : H_\nu(S) \to H_\nu(S_0)$, $\nu = 0, 1$, with the property that $\|\phi_* u\|_\nu \leq c_\nu \|u\|_\nu$ for all $u \in H_\nu(S_0)$. Since the same arguments hold for ϕ^{-1}, we obtain

14.6.9 Proposition. *For $\nu = 0, 1$ the mapping*

$$\phi_* : H_\nu(S) \to H_\nu(S_0)$$

is an isomorphism (but not an isometry) of Hilbert spaces with the property that for suitable constants c_0, c_1

$$\frac{1}{c_\nu} \|u\|_\nu \leq \|\phi_* u\|_\nu \leq c_\nu \|u\|_\nu, \quad u \in H_\nu(S).$$ ◇

Let us next pull back the sesquilinear forms and operators from S to S_0. We use the following notation. Subscript S indicates a quantity on $S = S[z]$, and subscript z indicates its pull-back on S_0. Thus $(\,,\,)_S$ is the inner product on $H_0(S)$, q_S the quadratic form q on $M = S$ as in (14.1.1) and Q_S the Friedrichs extension of the Laplacian on S as in Definition 14.2.4. The corresponding pull-backs on S_0 are denoted by ℓ_z, $\mathfrak{q}_z$ and Q_z:

$$(14.6.10) \quad \begin{aligned} \ell_z[u, v] &= (\phi_*^{-1}u, \phi_*^{-1}v)_S, & u, v &\in H_0(S_0), \\ \mathfrak{q}_z[u, v] &= \mathfrak{q}_S[\phi_*^{-1}u, \phi_*^{-1}v], & u, v &\in H_1(S_0), \\ Q_z u &= \phi_* \circ Q_S \circ \phi_*^{-1} u, & u &\in D_z := \phi_*(H_2(S)). \end{aligned}$$

($\phi_*(H_2(S))$ does not coincide with $H_2(S_0)$ in general.) Q_z and $\mathfrak{q}_z$ are related by the equation

$$(14.6.11) \quad \mathfrak{q}_z[u, v] = \ell_z[Q_z u, v], \quad u \in D_z, \; v \in H_1(S_0)$$

(cf. Proposition 14.2.5). Since ϕ_* is an isomorphism of Hilbert spaces, ϕ_* pulls back every closed operator on S to a closed operator on S_0 and every bounded operator on S to a bounded operator on S_0. In particular, Q_z is a closed operator (cf. Exercise 14.3.1). From the spectral theorem of the Laplacian (Theorem 7.2.6) we then have

14.6.12 Proposition. *Let $z \in \mathcal{Z}_g \subset \mathbf{R}^{6g-6}$, $S = S[z]$ and $Q_z \in \mathcal{C}(H_0(S_0))$ be as above. Then Q_z has a complete system of eigenfunctions $\varphi_0[z]$, $\varphi_1[z]$, ..., in $H_0(S_0)$ with corresponding eigenvalues $0 = \lambda_0(z) < \lambda_1(z) \leq \ldots$, where $\lambda_n(z) \to \infty$ as $n \to \infty$ and where each $\lambda_n(z)$ coincides with the n-th eigenvalue $\lambda_n(S)$ of the Laplacian of $S = S[z]$. If $\zeta \in \mathbf{C}$ is not one of these eigenvalues, then $(Q_z - \zeta)^{-1} \in \mathcal{B}(H_0(S_0))$.* ◇

Observe that the above sequence $\varphi_0[z]$, $\varphi_1[z]$, ... is orthonormal with respect to the inner product ℓ_z but not with respect to the inner product $(\,,\,)_0$ of $H_0(S_0)$, in general.

14.7 Reduction to Finite Dimension

We now prove the basic theorem of this chapter. It states that in Teichmüller space the eigenvalues of the Laplacian behave locally like the eigenvalues of an analytic family of matrices. Hence, the theorem gives access to the results from the analytic perturbation theory in finite dimensions (as, for instance, in Baumgärtel [1]). For better reference, the theorem is stated in a form which is independent of the description of Teichmüller space as given in the preceding section. The analytic structure in question is the standard real analytic structure of Teichmüller space as defined in Section 6.3. (By Theorem 6.8.13, the parameters for Teichmüller space commonly used in the literature are analytic with respect to this structure.)

We denote by $\mathbf{M}^{n+1}$ the set of all $((n+1) \times (n+1))$-matrices with coefficients in $\mathbf{C}$.

14.7.1 Theorem. *Let Ω be an open neighborhood in $\mathbf{R}^{6g-6}$ and let $z \mapsto S[z] \in \mathcal{T}_g$, $z \in \Omega$, be a real analytic diffeomorphism of Ω onto some open neighborhood in $\mathcal{T}_g$. We denote by $\lambda_k(z)$ the eigenvalues $\lambda_k(S[z])$ of the Laplacian of $S[z]$, $k = 0, \ldots$. Now consider $z_0 \in \Omega$ and let $n \in \mathbf{N}$ and $b \in \mathbf{R}$ be constants such that*

$$\lambda_n(z_0) < b < \lambda_{n+1}(z_0).$$

Then there exists an open neighborhood U of z_0 contained in Ω, and an open neighborhood V in $\mathbf{C}^{6g-6}$ satisfying $V \cap \mathbf{R}^{6g-6} = U$, and a matrix valued function $z \mapsto A_z = (a_{ij}(z)) \in \mathbf{M}^{n+1}$, $z \in V$, with the following properties.

(1) *For all $z \in U$ we have*

$$\lambda_n(z) < b < \lambda_{n+1}(z).$$

(2) *For $z \in U$ the sequence $\lambda_0(z), \ldots, \lambda_n(z)$ is the sequence of all eigenvalues (listed with multiplicities) of A_z.*

(3) *The functions $z \mapsto a_{ij}(z)$, $z \in V$, are holomorphic.*

Proof. For the proof we use the model of the preceding section with base surface S_0 and let $z_0 \in \mathcal{T}_g$ be the point with $S[z_0] = S_0$. By virtue of Proposition 14.6.12, we prove the theorem in terms of the eigenvalues $\lambda_0(z)$, $\lambda_1(z), \ldots,$ of the operators Q_z.

We proceed in the following steps. First we extend q_z, ℓ_z and Q_z to complex z. Then we show that for ζ_0 not an eigenvalue of Q_{z_0}, the bounded operator-valued function $(z, \zeta) \mapsto R_z(\zeta) = (Q_z - \zeta)^{-1} \in \mathcal{B}(H_0(S_0))$ is well defined and holomorphic in a neighborhood of (z_0, ζ_0) in $\mathbf{C}^{6g-6} \times \mathbf{C}$. Finally, we use projectors to interpret $\lambda_0(z), \ldots, \lambda_n(z)$ as the eigenvalues of a matrix A_z, where the matrix-valued function $z \mapsto A_z$ is holomorphic in a neighborhood of z_0.

We will from now on use H_ν to denote $H_\nu(S_0)$, $\nu = 0, 1, 2$.

We begin with the extension to complex z. Recall that $S_0 = S[z_0]$ is obtained by pasting together the sides of a polygon $\mathcal{P}$ which is tessellated by quadrilaterals $\mathcal{Q}_1, \ldots, \mathcal{Q}_{2g-1}$. The interior of each $\mathcal{Q}_\kappa$ is isometrically embedded in S_0, and for each κ we have the diffeomorphism

$$x_\kappa : \mathcal{Q}_\kappa \to \mathcal{R} = \{ (s, t) \in \mathbf{R}^2 \mid 0 \le s, t \le 1 \}.$$

For $F, G \in C^\infty(S_0)$ we denote by F_κ and G_κ the coordinate representations with respect to x_κ:

$$F_\kappa = F \circ (x_\kappa)^{-1}, \qquad G_\kappa = G \circ (x_\kappa)^{-1}.$$

The forms ℓ_z and q_z of (14.6.10) are given as follows

$$\ell_z[F, G] = \sum_{\kappa=1}^{2g-1} \int_{\mathcal{R}} F_\kappa(s, t) \overline{G}_\kappa(s, t) \theta_\kappa(s, t; z) \, ds \, dt,$$

(14.7.2)
$$q_z[F, G] = \sum_{\kappa=1}^{2g-1} \sum_{j,k=1}^{2} \int_{\mathcal{R}} \partial_j F_\kappa(s, t) \, \partial_k \overline{G}_\kappa(s, t) \, g_\kappa^{jk}(s, t; z) \theta_\kappa(s, t; z) \, ds \, dt,$$

where $\partial_1 = \partial/\partial s$, $\partial_2 = \partial/\partial t$ and

$$\theta_\kappa(s, t; z) = \sqrt{\det(g_{\kappa, ij}(s, t; z))}.$$

To simplify the notation we abbreviate

$$\ell = \ell_{z_0}, \qquad q = q_{z_0}, \qquad Q = Q_{z_0}.$$

Observe that ℓ is the inner product $(\,,\,)_0$ of H_0, q is the quadratic form as in (14.1.1) for $M = S_0$ and Q is the Friedrichs extension of the Laplacian on S_0.

By Proposition 14.6.8 there exists $\varepsilon > 0$ and an open neighborhood U_1 of z_0 in $\mathbf{R}^{6g-6}$ such that the functions $g_{\kappa, ij}(s, t; z)$ and $g_\kappa^{jk}(s, t; z)$ are real analytic on $\mathcal{R}_\varepsilon \times U_1$. Note that in particular $\det(g_{\kappa, ij}(s, t; z)) > 0$ on $\mathcal{R}_\varepsilon \times U_1$. There is therefore an open neighborhood V_1 of z_0 in $\mathbf{C}^{6g-6}$ such that the functions $g_{\kappa, ij}(s, t; z)$, $g_\kappa^{jk}(s, t; z)$ and $\theta_\kappa(s, t; z)$ can be extended to smooth complex valued functions on $\mathcal{R}_\varepsilon \times V_1$ which are holomorphic in z for each fixed $(s, t) \in \mathcal{R}_\varepsilon$.

We now let $g_{\kappa, ij}$, g_κ^{jk} and θ_κ denote these extensions.

It then follows from Weierstrass' theorem (Narasimhan [1], p. 7) that for given $F, G \in C^\infty(S_0)$ the functions $z \mapsto \ell_z[F, G]$ and $z \mapsto q_z[F, G]$ are holomorphic on V_1. We shall extend this to elements of H_0 and H_1.

For the next lemma we fix a small open neighborhood V of z_0 in $\mathbf{C}^{6g-6}$ whose compact closure is contained in V_1. We again use the short notation $b[u, u] = b[u]$ for any sesquilinear form b. Recall that $\ell[u] = \|u\|_0^2$ for $u \in H_0$.

14.7.3 Lemma. *Let $z' \in V$. For any $\delta > 0$ there exists an open neighborhood $V(z', \delta)$ of z' in V such that*

$$|\ell_z[F, G] - \ell_{z'}[F, G]| \leq \delta \|F\|_0 \|G\|_0$$

and

$$|q_z[F, G] - q_{z'}[F, G]| \leq \delta q^{1/2}[F] q^{1/2}[G]$$

for all $F, G \in C^\infty(S_0)$ and for all $z \in V(z', \delta)$.

Ch.14, §7]   Reduction fo Finite Dimension   391

Proof. We first consider the case $F = G$.

Fix $\delta > 0$. By the uniform continuity of θ_κ on $\mathcal{R} \times V$ there exists an open neighborhood W of z' in V such that $|\theta_\kappa(s, t; z) - \theta_\kappa(s, t; z')| \leq \frac{1}{2}\delta\theta_\kappa(s, t; z_0)$ for all $(s, t, z) \in \mathcal{R} \times W$, $\kappa = 1, \ldots, 2g - 1$. For $z \in W$

$$|\ell_z[F] - \ell_{z'}[F]| \leq \sum_{\kappa=1}^{2g-1} \int_\mathcal{R} |\theta_\kappa(s, t; z) - \theta_\kappa(s, t; z')| |F_\kappa(s, t)|^2 \, ds \, dt$$

$$\leq \tfrac{1}{2}\delta \|F\|_0^2.$$

By continuity, and since the matrix $(g_\kappa^{jk}(s, t; z_0))$ is symmetric and positive definite for all $(s, t) \in \mathcal{R}$ and all κ, there exists a positive constant μ such that

$$\sum_{j,k=1}^{2} g_\kappa^{jk}(s, t; z_0) a_j \bar{a}_k \geq \mu(|a_1|^2 + |a_2|^2)$$

for all $(a_1, a_2) \in \mathbf{C}^2$ and all $(s, t) \in \mathcal{R}$. We let W' be an open neighborhood of z' in V such that

$$|g_\kappa^{jk}(s, t; z)\theta_\kappa(s, t; z) - g_\kappa^{jk}(s, t; z')\theta_\kappa(s, t; z')| \leq \tfrac{1}{4}\delta\mu \, \theta_\kappa(s, t; z_0)$$

for all $(s, t, z) \in \mathcal{R} \times W'$, $\kappa = 1, \ldots, 2g - 2$ and $1 \leq j, k \leq 2$. Then

$$|\mathfrak{q}_z[F] - \mathfrak{q}_{z'}[F]| \leq \tfrac{1}{4}\delta\mu \sum_{\kappa=1}^{2g-1} \sum_{j,k=1}^{2} \int_\mathcal{R} |\partial_j F_\kappa(s, t)| |\partial_k F_\kappa(s, t)| \theta_\kappa(s, t; z_0) \, ds \, dt$$

$$\leq \tfrac{1}{2}\delta\mu \sum_{\kappa=1}^{2g-1} \int_\mathcal{R} (|\partial_1 F_\kappa(s, t)|^2 + |\partial_2 F_\kappa(s, t)|^2)\theta_\kappa(s, t; z_0) \, ds \, dt$$

$$\leq \tfrac{1}{2}\delta \sum_{\kappa=1}^{2g-1} \sum_{j,k=1}^{2} \int_\mathcal{R} \partial_j F_\kappa(s, t) \, \partial_k \bar{F}_\kappa(s, t) g_\kappa^{jk}(s, t; z_0)\theta_\kappa(s, t; z_0) \, ds \, dt$$

$$= \tfrac{1}{2}\delta\mathfrak{q}[F].$$

We take $V(z', \delta) = W \cap W'$.

In the next step we "sesquilinearize" the inequalities $|\ell_z[F] - \ell_{z'}[F]| \leq \tfrac{1}{2}\delta\|F\|_0^2$ and $|\mathfrak{q}_z[F] - \mathfrak{q}_{z'}[F]| \leq \tfrac{1}{2}\delta\mathfrak{q}[F]$ using the following identity, in which β may be any sesquilinear form in a complex vector space.

(14.7.4) $\beta[x, y] = \tfrac{1}{4}(\beta[x + y] - \beta[x - y] + i\beta[x + iy] - i\beta[x - iy]).$

We only consider $\mathfrak{q}_z$; the procedure for ℓ_z is the same. Let s and t be positive real parameters. Applying (14.7.4) to $\beta = \mathfrak{q}_z - \mathfrak{q}_{z'}$ and using the identity $\mathfrak{q}[x + y] + \mathfrak{q}[x - y] = 2\mathfrak{q}[x] + 2\mathfrak{q}[y]$ (recall that $\mathfrak{q} = \mathfrak{q}_{z_0}$ is positive semi-definite), we see that

$$|q_z[F, G] - q_{z'}[F, G]| = \frac{1}{st}|q_z[sF, tG] - q_{z'}[sF, tG]|$$
$$\leq \frac{\delta}{2st}(q[sF] + q[tG])$$
$$= \tfrac{1}{2}\delta(\tfrac{s}{t}q[F] + \tfrac{t}{s}q[G]).$$

If $q[F] = 0$ or $q[G] = 0$, then the inequality of the lemma follows with $t/s \to 0$, or with $s/t \to 0$, respectively. If $q[F]$ and $q[G]$ are different from 0, then the inequality follows with $s/t = (q[G]/q[F])^{1/2}$. ◊

By restricting V to a smaller neighborhood of z_0 if necessary, we may assume, in the particular case $z' = z_0$ and $\delta = 1$, that

$$V(z_0, 1) = V.$$

Lemma 14.7.3 together with Schwarz's inequality applied to the positive semi-definite form q then implies

(14.7.5) $|\ell_z[F, G]| \leq 2\|F\|_0 \|G\|_0$, $|q_z[F, G]| \leq 2q^{1/2}[F]q^{1/2}[G]$

for all $z \in V$ and all $F, G \in C^\infty(S_0)$. We may therefore extend the forms ℓ_z and q_z from $C^\infty(S_0)$ to H_0 and H_1:

(14.7.6) $\ell_z[u, v] = \lim_{k \to \infty} \ell_z[F_k, G_k]$, $q_z[u', v'] = \lim_{k \to \infty} q_z[F_k', G_k']$,

where $u, v \in H_0$ are the 0-equivalence classes of 0-Cauchy sequences $\{F_k\}$ and $\{G_k\}$ in $C^\infty(S_0)$, and $u', v' \in H_1$ are the 1-equivalence classes of 1-Cauchy sequences $\{F_k'\}$ and $\{G_k'\}$ in $C^\infty(S_0)$.

In the next lemma the neighborhoods $V(z', \delta)$ are as in Lemma 14.7.3.

14.7.7 Lemma. (i) *The function $z \mapsto \ell_z[u, v]$ is holomorphic on V for all $u, v \in H_0$.*
(ii) *Suppose $z' \in V$ and $\delta > 0$. Then*

$$|\ell_z[u, v] - \ell_{z'}[u, v]| \leq \delta \|u\|_0 \|v\|_0$$

for all $z \in V(z', \delta)$ and all $u, v \in H_0$.
(iii) *The function $z \mapsto q_z[u, v]$ is holomorphic on V for all $u, v \in H_1$.*
(iv) *Suppose $z' \in V$ and $\delta > 0$. Then*

$$|q_z[u, v] - q_{z'}[u, v]| \leq \delta q^{1/2}[u] q^{1/2}[v]$$

for all $z \in V(z', \delta)$ and all $u, v \in H_1$.

Proof. If $u = [F_k]$ and $v = [G_k]$, where $\{F_k\}$ and $\{G_k\}$ are 0-Cauchy sequences in $C^\infty(S_0)$, then the functions $z \mapsto \ell_z[F_k, G_k]$, $k = 1, 2, \ldots$, are holomorphic as mentioned a few lines before Lemma 14.7.3. By (14.7.5), $\ell_z[F_k, G_k] \to \ell_z[u, v]$ *uniformly* on V. By Weierstrass' theorem, the function $z \mapsto \ell_z[u, v]$ is holomorphic. The same proof applies to q_z. The inequalities follow immediately from Lemma 14.7.3. ◇

14.7.8 Proposition. *There exists an open neighborhood $V(z_0)$ in $\mathbf{C}^{6g-6}$ and a family of operators $z \mapsto \theta_z \in \mathscr{C}(H_0)$, $z \in V(z_0)$, with the following properties.*

(i) *The domain $D(\theta_z)$ is 1-dense in H_1.*
(ii) $q_z[u, v] = \ell_z[\theta_z u, v]$ *for all $u \in D(\theta_z)$ and $v \in H_1$.*
(iii) $(\theta_z + 1)^{-1} \in \mathscr{B}(H_0)$ *for all $z \in V(z_0)$.*
(iv) *If $z \in V(z_0) \cap \mathbf{R}^{6g-6}$, then $D(\theta_z) = D_z$ and $\theta_z = Q_z$, where D_z and Q_z are as in (14.6.10).*

If ζ_0 is a complex number different from the eigenvalues of $Q = Q_{z_0}$, then there exists an open neighborhood $V[\zeta_0] \subset V(z_0)$ of z_0 and an open neighborhood $U(\zeta_0)$ of ζ_0 in $\mathbf{C}$ such that the following additional properties hold.

(v) $(\theta_z - \zeta)^{-1} \in \mathscr{B}(H_0)$ *for all $z \in V[\zeta_0]$ and all $\zeta \in U(\zeta_0)$.*
(vi) *The bounded operator-valued function $(z, \zeta) \mapsto (\theta_z - \zeta)^{-1}$ is holomorphic on $V[\zeta_0] \times U(\zeta_0)$.*

Proof. We let $V'(z_0)$ be the neighborhood $V(z_0, \frac{1}{2})$ as in Lemma 14.7.3 and Lemma 14.7.7. By Lemma 14.7.7(iv),

$$|q_z[u, v] - q[u, v]| \le \tfrac{1}{2} q^{1/2}[u] q^{1/2}[v]$$

for $u, v \in H_1$ and any $z \in V'(z_0)$. By Theorem 14.3.5, there exists a closed operator $T_z \in \mathscr{C}(H_0)$ with properties (1) - (4) below.

(1) *The domain $D(T_z)$ is 1-dense in H_1.*
(2) $q_z[u, v] = (T_z u, v)_0$ *for all $u \in D(T_z)$ and all $v \in H_1$.*
(3) $(T_z + 1)^{-1} = K(B_z + K^2)^{-1} K \in \mathscr{B}(H_0)$, *where $B_z \in \mathscr{B}(H_0)$ is a bounded operator satisfying*

$$(B_z u, v)_0 = q_z[Ku, Kv] \text{ for all } u, v \in H_0.$$

To state the next property of the operators T_z, we let λ_k be the eigenvalues of Q, $\lambda_k = \lambda_k(z_0)$, $k = 0, 1, \ldots$, and ζ_0 a complex number different from these. We define

$$\rho = \tfrac{1}{2} \min_k |\lambda_k - \zeta_0|, \quad U'(\zeta_0) = \{\zeta \in \mathbf{C} \mid |\zeta - \zeta_0| < \rho\}.$$

There is then a constant $c(\zeta_0) \in {]0, 1[}$ such that

$$c(\zeta_0) < \inf_k \frac{|\lambda_k - \zeta|}{\lambda_k + 1} \text{ for all } \zeta \in U'(\zeta_0).$$

By Lemma 14.7.7 there exists an open neighborhood $V'[\zeta_0]$ of z_0 in $V'(z_0)$ such that for $z \in V'[\zeta_0]$ and for $u, v \in H_1$

$$|q_z[u, v] - q[u, v]| \le c(\zeta_0) q^{1/2}[u] q^{1/2}[v].$$

By points (v) and (vi) of Theorem 14.3.5

(4) $\qquad (T_z - \zeta)^{-1} = K(B_z - \zeta K^2)^{-1} K \in \mathcal{B}(H_0)$

for any $z \in V'[\zeta_0]$ and any $\zeta \in U'(\zeta_0)$.

We first prove that the family $z \mapsto B_z \in \mathcal{B}(H_0)$, $z \in V'(z_0)$, is continuous. For this we fix $z' \in V'(z_0)$, let δ be a positive number and consider any $z \in V(z', \delta)$, where $V(z', \delta)$ is as in Lemma 14.7.7. We take δ so small that $V(z', \delta) \subset V'(z_0)$. By Lemma 14.7.7 and by (14.2.10), for any $u, v \in H_0$ with $\|u\|_0 = \|v\|_0 = 1$,

$$|((B_z - B_{z'})u, v)_0| = |(q_z - q_{z'})[Ku, Kv]|$$
$$\le \delta q^{1/2}[Ku] q^{1/2}[Kv] \le \delta.$$

Hence, $\|B_z - B_{z'}\| \le \delta$. This proves the continuity of the family $z \mapsto B_z$.

By Lemma 14.7.7 and by Theorem 14.5.4, the family $z \mapsto B_z$ is holomorphic on $V'(z_0)$. By Theorem 14.5.6 and by Exercise 14.5.2, we have this.

(5) *The family* $(z, \zeta) \mapsto (T_z - \zeta)^{-1}$ *is holomorphic on* $V'[\zeta_0] \times U'(\zeta_0)$.

We now introduce θ_z which represents q_z with respect to ℓ_z rather than to ℓ. (Recall that $\ell = \ell_{z_0} = (\,,\,)_0$.) By Lemma 14.7.7 applied to $z' = z_0$ and to $\delta = \frac{1}{2}$, ℓ_z is a bounded sesquilinear form on H_0 for any $z \in V'(z_0)$. By the Riesz representation theorem there exists a bounded linear operator $L_z \in \mathcal{B}(H_0)$ satisfying

$$\ell_z[u, v] = (L_z u, v)_0, \text{ for } u, v \in \mathcal{B}(H_0).$$

By Lemma 14.7.7 and by Theorem 14.5.4, the family $z \mapsto L_z$ is holomorphic on $V'(z_0)$, the proof of the continuity of the family L_z being the same as for B_z. By Lemma 14.7.7,

$$\|L_z - 1\| = \|L_z - L_{z_0}\| = \sup_1 (\ell_z - \ell_{z_0})[u, v] \le \tfrac{1}{2}.$$

(The supremum is over all $u, v \in H_0$ with $\|u\|_0 = \|v\|_0 = 1$.) This shows that $L_z^{-1} \in \mathcal{B}(H_0)$. By Theorem 14.5.6, the family $z \mapsto L_z^{-1}$ is holomorphic on $V'(z_0)$. We define

$$D(\theta_z) = D(T_z) \qquad \theta_z = L_z^{-1} T_z$$

for any $z \in V'(z_0)$. By Exercise 14.3.1(iii), θ_z is a closed operator. Together with (1) and (2) we obtain (i) and (ii) of Proposition 14.7.8.

We let $V(z_0) \subset V'(z_0)$ be an open neighborhood of z_0 such that

(6) $$\|(T_z + 1)^{-1}\| \|L_z - 1\| < 1$$

for $z \in V(z_0)$. For ζ_0 as above we let $V[\zeta_0] \subset V(z_0) \cap V'[\zeta_0]$ be an open neighborhood of z_0 and let $U(\zeta_0) \subset U'(\zeta_0)$ be an open neighborhood of ζ_0 such that for any $z \in V[\zeta_0]$ and any $\zeta \in U(\zeta_0)$

(7) $$|\zeta| \|(T_z - \zeta)^{-1}\| \|L_z - 1\| < 1.$$

Now consider the pair (z, ζ) where either $z \in V(z_0)$ and $\zeta = -1$, or else $z \in V[\zeta_0]$ and $\zeta \in U(\zeta_0)$. Lemma 14.3.2 implies that the operator $(1 - \zeta(T_z - \zeta)^{-1}(L_z - 1))$ has a bounded inverse, and we check that this operator restricted to $D(T_z)$ is a one-to-one mapping of $D(T_z)$ onto itself. It follows that

$$\theta_z - \zeta = L_z^{-1}(T_z - \zeta)(1 - \zeta(T_z - \zeta)^{-1}(L_z - 1)),$$

and that $\theta_z - \zeta$ has the bounded inverse

$$(\theta_z - \zeta)^{-1} = (1 - \zeta(T_z - \zeta)^{-1}(L_z - 1))^{-1}(T_z - \zeta)^{-1}L_z.$$

When $(z, \zeta) \in V(z_0) \times \{-1\}$ this is property (iii) by virtue of (6); when $(z, \zeta) \in V[\zeta_0] \times U(\zeta_0)$ this is (v) by virtue of (7). By Theorem 14.5.6, by (5) and by Exercise 14.5.2, the family $(z, \zeta) \mapsto (\theta_z - \zeta)^{-1}$ is holomorphic on $V[\zeta_0] \times U(\zeta_0)$. This is (vi).

To prove (iv) we let $z \in V(z_0) \cup \mathbf{R}^{6g-6}$. Then (H_0, ℓ_z) is a Hilbert space which is isometric to $H_0(S[z])$ under the isomorphism $\phi_* : H_0(S[z]) \to H_0(S_0) = H_0$ as in Proposition 14.6.9. Using the same proposition note that ϕ_* is also a one-to-one mapping from $H_1(S[z])$ onto $H_0(S_0)$. Together with (14.6.10) we translate point (iii) of Proposition 14.2.5 as follows.

$$D_z = D(Q_z) = \{u \in H_1 \mid \text{there exists } u' \in H_0 \text{ satisfying}$$
$$\ell_z[u', v] = q_z[u, v] \text{ for all } v \in H_1\}.$$

By point (ii) of Proposition 14.7.8, $D(\theta_z) \subset D(Q_z)$. Moreover, this with (14.6.11) shows that $\ell_z[\theta_z u, v] = \ell_z[Q_z u, v]$ for all $v \in H_1$. Since H_1 is dense in the Hilbert space (H_0, ℓ_z) (cf. Proposition 14.6.9 and (14.6.10)), it follows that $\theta_z u = Q_z u$ for all $u \in D(\theta_z)$. Now Q_z is the pull-back on S_0 of the Friedrichs extension of the Laplacian of $S[z]$. Therefore, -1 belongs to the resolvent set of Q_z. In particular, $Q_z + 1$ is one-to-one. Since $\theta_z + 1$ has range H_0, we conclude that $D(\theta_z) = D(Q_z)$. Proposition 14.7.8 is now proved. ◇

We turn to the proof of Theorem 14.7.1. Let Γ be an admissible Jordan curve in the resolvent set $\rho(Q) = \mathbf{C} - \{\lambda_0(z_0), \lambda_1(z_0), \dots\}$ such that $\lambda_\nu(z_0) \in \text{int } \Gamma$ for $\nu = 0, \dots, n$, and $\lambda_\nu(z_0) \in \text{ext } \Gamma$ for $\nu \geq n + 1$.

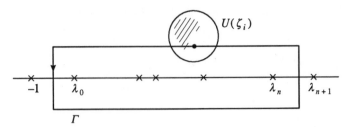

Figure 14.7.1

We first work with complex z. Consider the family of operators θ_z, $z \in V(z_0)$ as in Proposition 14.7.8. For every ζ_0 on Γ we have the neighborhoods $V[\zeta_0] \subset V(z_0) \subset \mathbf{C}^{6g-6}$ of z_0 and $U(\zeta_0) \subset \mathbf{C}$ of ζ_0 as in Proposition 14.7.8. By the compactness of Γ, there exist finitely many such points, say $\zeta_1, \ldots, \zeta_N$ such that the corresponding neighborhoods $U(\zeta_1), \ldots, U(\zeta_N)$ cover Γ. We define

$$V = V[\zeta_1] \cap \ldots \cap V[\zeta_N], \quad U_\Gamma = U(\zeta_1) \cup \ldots \cup U(\zeta_N).$$

By Proposition 14.7.8,

$$U_\Gamma \text{ is contained in } \rho(\theta_z)$$

for any θ_z with $z \in V$, and the family $(z, \zeta) \mapsto R_z(\zeta) := (\theta_z - \zeta)^{-1} \in \mathscr{B}(H_0)$ is holomorphic on $V \times U_\Gamma$. Letting

$$P_z = -\frac{1}{2\pi i} \int_\Gamma R_z(\zeta) \, d\zeta$$

as in Section 14.4, we obtain a holomorphic family $z \mapsto P_z \in \mathscr{B}(H_0)$, $z \in V$. Note that the integral is holomorphic by virtue of Theorem 14.5.5. By Theorem 14.4.8, P_z is a projector with range $P_z(H_0) \subset D(\theta_z)$ and with the property

$$\theta_z P_z u = P_z \theta_z P_z u = -\frac{1}{2\pi i} \int_\Gamma \zeta R_z(\zeta) u \, d\zeta$$

for all $u \in H_0$. By restricting V to a smaller neighborhood of z_0, if necessary, we obtain

$$\|P_z - P_{z_0}\| \le \tfrac{1}{2} \text{ for all } z \in V.$$

By Lemma 14.4.6, $\dim P_z(H_0) = \dim P_{z_0}(H_0)$ for any $z \in V$. By Theorem 14.4.12, $P_{z_0}(H_0) = E(\lambda_0, Q) + \ldots + E(\lambda_n, Q)$ (recall that $Q = Q_{z_0} = \theta_{z_0}$). In particular, if $\{\varphi_0, \varphi_1, \ldots\}$ is the complete sequence of eigenfunctions of Q with respect to $\lambda_0(z_0), \lambda_1(z_0), \ldots$, then

$\varphi_0, \ldots, \varphi_n$ *is a vector space basis of* $P_{z_0}(H_0)$,

and by Lemma 14.4.6,

$P_z\varphi_0, \ldots, P_z\varphi_n$ *is a vector space basis of* $P_z(H_0)$,

for any $z \in V$. By Exercise 14.5.2, the products $(P_z\varphi_i, P_z\varphi_k)_0$ are holomorphic functions of z, for $i, k = 0, \ldots, n$. By Theorem 14.4.8(v), the family $z \mapsto P_z\theta_z P_z$ is holomorphic. We obtain therefore by Cramer's rule that the coefficients $a_{ij}(z)$ in the linear combination

$$P_z\theta_z P_z(P_z\varphi_i) = \sum_{j=0}^{n} a_{ij}(z) P_z\varphi_j$$

are holomorphic for $z \in V$. By Theorem 14.4.8, θ_z restricted to $P_z(H_0)$ coincides with $P_z\theta_z P_z$ and is given by the matrix $A_z = (a_{ij}(z))$. Now let

$$U = V \cap \mathbf{R}^{6g-6}$$

and let $z \in U$. Then, by Proposition 14.7.8(iv), $\theta_z = Q_z$, and by Proposition 14.6.12, Q_z has a complete sequence of eigenfunctions. Only finitely many eigenvalues of Q_z are contained in int Γ, and since Γ is contained in $\rho(\theta_z)$, all the remaining eigenvalues of Q_z are contained in ext Γ. By Theorem 14.4.12, $P_z(H_0)$ is spanned by the eigenspaces corresponding to the eigenvalues of $\theta_z = Q_z$ which are contained in int Γ. Since dim $P_z(H_0) = $ dim $P_{z_0}(H_0)$ and since the eigenvalues are listed in increasing order, these eigenvalues are $\lambda_0(z), \ldots, \lambda_n(z)$. Hence, for any $z \in U$, the inequality $\lambda_n(z) < b < \lambda_{n+1}(z)$ holds, and $\lambda_0(z), \ldots, \lambda_n(z)$ are the eigenvalues of A_z. Theorem 14.7.1 is now proved. ◊

14.8 Holomorphic Families of Laplacians

In this section (it will not be needed afterwards) we shall interpret θ_z occurring in Proposition 14.7.8 as a holomorphic family of Laplacians in the sense of Kato [1], p. 366.

Let $z \mapsto T_z$ with $T_z : \mathsf{X} \to \mathsf{X}$ be a family of operators in a Banach space X, where $z \in W$ and W is an open subset of $\mathbf{C}^m$. The family T_z is called *holomorphic* if there exist bounded holomorphic families $A_z \in \mathcal{B}(\mathsf{X})$ and $B_z \in \mathcal{B}(\mathsf{X})$, $z \in W$, such that for all $z \in W$, $\mathsf{D}(T_z) = \mathsf{R}(A_z)$ and $T_z A_z = B_z$. The domains $\mathsf{D}(T_z)$ depend on z in general.

Consider now the family θ_z as in Proposition 14.7.8. For $z \in V$ as in the proposition, $(\theta_z + 1)^{-1}$ is bounded holomorphic, and

$$\theta_z(\theta_z + 1)^{-1} = (\theta_z + 1)(\theta_z + 1)^{-1} - (\theta_z + 1)^{-1} = 1 - (\theta_z + 1)^{-1}$$

is bounded holomorphic. Hence, θ_z is holomorphic in the more general sense.

We can restate the preceding remark together with Proposition 14.7.8 in a global form. For this we pull back the structures from $S[z]$ to a fixed base surface - for instance the surface S^* of Section 14.6 - for all $z \in \mathcal{L}_g$. We use the homeomorphism $\psi[z] : S^* \to S[z]$ as in (14.6.6). Note from the definition of the homeomorphisms $\phi = \phi[z] : S_0 = S[z_0] \to S[z]$ that

(14.8.1) $$\phi[z] \circ \psi[z_0] = \psi[z].$$

For any $z \in \mathcal{L}_g$ we now let $Q_z \in \mathcal{C}(H_0(S^*))$ denote the pull-back on S^* of the Friedrichs extension of the Laplacian of $S[z]$. By abuse of notation we may take the family Q_z, $z \in \mathcal{L}_g$, as the family of the Laplacians of the compact Riemann surfaces in the Teichmüller space $\mathcal{T}_g$.

14.8.2 Theorem. *There exists an open neighborhood $\tilde{\mathcal{L}}_g$ of $\mathcal{L}_g$ in $\mathbf{C}^{6g-6}$ and a holomorphic family of closed operators $\theta_z \in \mathcal{C}(H_0(S^*))$, $z \in \tilde{\mathcal{L}}_g$, with the following properties.*

(i) *If $z \in \mathcal{L}_g$ then $\theta_z = Q_z$.*

(ii) *$(\theta_z + 1)^{-1} \in \mathcal{B}(H_0(S^*))$ for all $z \in \tilde{\mathcal{L}}_g$ and the family $(\theta_z + 1)^{-1}$ is bounded holomorphic on $\tilde{\mathcal{L}}_g$.*

(iii) *Let $z' \in \tilde{\mathcal{L}}_g$ and let W be a connected open neighborhood of z' in $\tilde{\mathcal{L}}_g$. If Σ_z, $z \in W$, is a family of closed operators such that $\Sigma_z = Q_z$ for all $z \in W \cap \mathcal{L}_g$ and such that $(\Sigma_z + 1)^{-1} \in \mathcal{B}(H_0(S^*))$ is bounded holomorphic for $z \in W$, then $\Sigma_z = \theta_z$ for all $z \in W$.*

(iv) *Let $z_0 \in \tilde{\mathcal{L}}_g$ and let $\zeta_0 \in \rho(Q_{z_0})$. Then there exists an open neighborhood V of z_0 in $\tilde{\mathcal{L}}_g$ and an open neighborhood U of ζ_0 in $\mathbf{C}$ such that $(\theta_z - \zeta)^{-1} \in \mathcal{B}(H_0(S^*))$ for all $(z, \zeta) \in V \times U$ and such that the family $(\theta_z - \zeta)^{-1}$ is bounded holomorphic on $V \times U$.*

Proof. Let $z_0 \in \mathcal{L}_g$ and let $V(z_0) \subset \mathbf{C}^{6g-6}$ be the neighborhood of z_0 as in Proposition 14.7.8. By restricting $V(z_0)$ to a smaller neighborhood of z_0, if necessary, we may assume that $V(z_0)$ is connected and that for any $z = (z_1, \ldots, z_{6g-6}) \in V(z_0)$ the real part $\mathcal{R}e\, z := (\mathcal{R}e\, z_1, \ldots, \mathcal{R}e\, z_{6g-6})$ belongs to $V(z_0)$.

Let θ_z with $z \in V(z_0)$ be the family of operators on $S_0 = S[z_0]$ as in Proposition 14.7.8. If $z \in V(z_0) \cap \mathcal{L}_g$, then (14.8.1) implies that the pull-back of θ_z on S^* via $\psi[z_0]$ coincides with Q_z. We may therefore remain notationally consistent and denote the pull-backs of all θ_z again by θ_z. Since $\psi[z_0]$ induces isomorphisms between the Hilbert spaces $H_\nu(S_0)$ and $H_\nu(S^*)$

for $\nu = 0, 1$ (Proposition 14.6.9), the statements of Proposition 14.7.8 remain valid on S^*.

We let $\widetilde{\mathcal{L}}_g$ be the union set of all $V(z_0)$ for $z_0 \in \mathcal{L}_g$. For the proof of the theorem it remains to show that for overlapping neighborhoods $V(z_0)$ and $V(z_0')$ the pull-backs via $\psi[z_0]$ and $\psi[z_0']$ coincide, and that we have (iii). Both statements follow from the following uniqueness argument.

Let $z' \in V(z_0)$ and let W be a connected open neighborhood of z' in $\mathbf{C}^{6g-6}$ together with a family of operators Σ_z in $H_0(S^*)$ such that $(\Sigma_z + 1)^{-1}$ is bounded holomorphic and such that $\Sigma_z = \theta_z$ for $z \in V(z_0) \cap W \cap \mathcal{L}_g$. Assume also that $\mathcal{R}e\, z \in W$ for all $z \in W$. Then for $u, v \in H_0(S^*)$, the function $z \mapsto (((\theta_z + 1)^{-1} - (\Sigma_z + 1)^{-1})u, v)_0$ is holomorphic and vanishes on $V(z_0) \cap W \cap \mathbf{R}^{6g-6}$. It vanishes therefore on $V(z_0) \cap W$, meaning that $(\theta_z + 1)^{-1} = (\Sigma_z + 1)^{-1}$ and thus $\Sigma_z = \theta_z$ for all $z \in V(z_0) \cap W$. ◊

In view of the preceding theorem we may take the operators θ_z, $z \in \widetilde{\mathcal{L}}_g$ as a holomorphic family of Laplacians.

14.9 Analytic Properties of the Eigenvalues

In this section we draw conclusions from Theorem 14.7.1 concerning the eigenvalues. (Section 14.8 will not be needed.) We show that they are real analytic functions (of several real variables) in neighborhoods where they are simple. Then we prove that along real analytic curves in Teichmüller space the eigenvalues are analytic functions (now of one variable) even at those points where they have multiplicities.

To simplify the notation in the first theorem, we let $\lambda_{-1}(\ ) = -\infty$. The theorem is stated with respect to the standard real analytic structure of Teichmüller space as defined in Section 6.3.

14.9.1 Theorem. *Let $S_0 \in \mathcal{T}_g$, let a and b be real numbers, $a < b$, and let m and n be integers, $m < n$, such that*

$$\lambda_{m-1}(S_0) < a < \lambda_m(S_0) \leq \ldots \leq \lambda_n(S_0) < b < \lambda_{n+1}(S_0).$$

Then there exists an open neighborhood U of S_0 in $\mathcal{T}_g$ such that

$$\lambda_{m-1}(S) < a < \lambda_m(S) \leq \ldots \leq \lambda_n(S) < b < \lambda_{n+1}(S)$$

for all $S \in U$ and such that for any symmetric function σ of $(n + 1 - m)$ variables the function $S \mapsto \sigma(\lambda_m(S), \ldots, \lambda_n(S))$ is real analytic on U. In particular, $\lambda_n(S)$ is an analytic function in a neighborhood of S_0 if $\lambda_n(S_0)$ is a simple eigenvalue.

Proof. It suffices to consider the case $m = 0$. We let the neighborhoods $U \subset \mathbf{R}^{6g-6}$, $V \subset \mathbf{C}^{6g-6}$ and the matrices $A_z \in \mathbf{M}^{n+1}$, $z \in V$, be as in Theorem 14.7.1. The eigenvalues are now the eigenvalues of A_z, and we get for any integer k,

$$\lambda_0^k(z) + \ldots + \lambda_n^k(z) = \text{trace } A_z^k.$$

This shows that for any k the function $z \mapsto \lambda_0^k(z) + \ldots + \lambda_n^k(z)$, $z \in U$, is the restriction to U of a holomorphic function on V. Since any symmetric function is a polynomial expression of power sums, this proves the theorem. ◇

As a supplement to Theorem 14.9.1 we show that the eigenvalues $\lambda_n(S)$ vary with uniformly controlled speed as S varies through $\mathcal{T}_g$. The statement is in terms of the Teichmüller distance δ as defined in Section 6.4. We recall that this distance function is compatible with the standard topology of $\mathcal{T}_g$.

14.9.2 Theorem. *Let $M, S \in \mathcal{T}_g$ at distance $\delta = \delta(M, S)$. Then for any $n \in \mathbf{N}$,*

$$e^{-4\delta}\lambda_n(M) \leq \lambda_n(S) \leq e^{4\delta}\lambda_n(M).$$

Proof. Let $\phi : S \to M$ be a piecewise smooth q-quasi isometry. By the definition of the distance function δ, it suffices to show that

$$\lambda_n(S) \leq q^4 \lambda_n(M).$$

We use that ϕ induces a mapping $f \mapsto f \circ \phi$ from $C^L(M)$ to $C^L(S)$ (cf. Definition 14.1.5 and Lemma 14.1.6). We let $\varphi_0, \varphi_1, \ldots$ and $\psi_0, \psi_1, \ldots$ be complete orthonormal sequences of eigenfunctions of the Laplacian on M and S respectively. We denote by R the Rayleigh quotient

$$R(f, M) = \frac{\int_M \|\text{grad} f\|^2 \, dM}{\int_M f^2 \, dM},$$

where $f \in C^L(M)$ and $\| \ \|$ and grad are taken with respect to the given metric. Since ϕ is a q-quasi isometry, we have $\|\text{grad}(f \circ \phi)\| \leq q \|\text{grad} f\|$. An elementary argument from linear algebra shows that we have the stronger inequality $\|\text{grad}(f \circ \phi)\|^2 dS \leq q^2 \|\text{grad} f\|^2 dM$. This yields

$$R(f \circ \phi, S) \leq q^4 R(f, M).$$

Now let f be any linear combination $f = c_0\varphi_0 + \ldots + c_n\varphi_n$. Then, by Green's formula, $R(f, M) = (\lambda_1(M)c_1^2 + \ldots + \lambda_n(M)c_n^2)/(c_1^2 + \ldots + c_n^2) \leq \lambda_n$. There exists a choice of coefficients $c_0, \ldots, c_n$ such that $f \circ \phi$ is a non-zero function on S which is orthogonal to $\psi_0, \ldots, \psi_{n-1}$. For this choice we have, again by Green's formula, $R(f \circ \phi, S) \geq \lambda_n(S)$. This gives us the string of inequalities $\lambda_n(S) \leq R(f \circ \phi, S) \leq q^4 R(f, M) \leq q^4 \lambda_n(M)$. ◇

We like to mention that Theorems 14.9.1 and 14.9.2 do not allow us to prove the analyticity of the *individual* eigenvalues. Although we have not constructed explicit counterexamples in Teichmüller space for this, the following example of an analytic family of positive definite symmetric matrices shows that the analyticity of the individual eigenvalues is likely to fail at points where they occur with multiplicities. The example is due to Rellich [1], the variables are $(x_1, x_2) \in \mathbf{R}^2$.

$$A_x = \begin{pmatrix} 1 + 2x_1 & x_1 + x_2 \\ x_1 + x_2 & 1 + 2x_2 \end{pmatrix}.$$

The eigenvalues of these matrices are given by the two-valued function

$$\lambda_{1,2}(x) = 1 + x_1 + x_2 \pm \sqrt{2(x_1^2 + x_2^2)}.$$

The graph of this function is a cone, and thus there exists no pair of analytic functions which would represent it in a neighborhood of $x = 0$.

We observe, however, that if we restrict this function to analytic curves through the origin, for example the line $t \mapsto (\alpha t, \beta t) \in \mathbf{R}^2$, $t \in \mathbf{R}$, then the restriction of $\lambda_{1,2}$ to this line is indeed represented by two analytic functions; in our example these functions are

$$\ell_{1,2}(t) = 1 + t(\alpha + \beta \pm \sqrt{2(\alpha^2 + \beta^2)}).$$

Hence, the eigenvalues may become more regular if we restrict ourselves to one variable. This has been shown by Rellich [1] for operators in finite dimensional spaces and later in more generality. In conjunction with Theorem 14.7.1 we translate this to analytic curves in Teichmüller space to obtain the following improvement of Theorem 14.9.1 in the case of one variable.

14.9.3 Theorem. *Let $I \subset \mathbf{R}$ be an interval and let $t \mapsto S(t) \in \mathcal{T}_g$, $t \in I$, be a real analytic curve in Teichmüller space. Then there exist real analytic functions $f_k : I \to \mathbf{R}$, $k = 0, 1, \ldots$ such that for any $t \in I$ the sequence $f_0(t)$, $f_1(t), \ldots$ is the sequence of all eigenvalues of the Laplacian of $S(t)$ (listed with multiplicities, though not in increasing order).*

Proof. In view of Theorem 14.9.2, it suffices to prove that if $t_0 \in I$, $n \in \mathbf{N}$ and $b > 0$ are such that $\lambda_n(S(t)) < b < \lambda_{n+1}(S(t_0))$, then there exists an open interval I_0 containing t_0 and real analytic functions $g_0, \ldots, g_n : I_0 \to \mathbf{R}$ such that for any $t \in I_0 \cap I$ the inequality $\lambda_n(S(T)) < b < \lambda_{n+1}(S(t))$ holds and such that, up to a permutation depending on t, the sequence $g_0(t), \ldots, g_n(t)$ coincides with the sequence $\lambda_0(S(t)), \ldots, \lambda_n(S(t))$. This statement, in turn, is a consequence of Theorem 14.7.1 and the next theorem. ◇

The theorem in question is Rellich's theorem [1], which we state in a slightly restricted form; for the general form we refer to the original paper, or to Baumgärtel [1], section 3.5, or to Kato [1], chapter II, §6.

We let G be an open subset of the complex plane $\mathbf{C}$. A matrix valued function $s \mapsto A_s = (a_{ij}(s)) \in \mathbf{M}^N$, $s \in G$, is called a *holomorphic family* if the coefficients a_{ij} are holomorphic functions on G.

14.9.4 Theorem. *Let $s \mapsto A_s \in \mathbf{M}^N$, $s \in G$, be a holomorphic family of matrices. Let $s_0 \in G$ and assume that for any $s = s_0 + t \in G$ with $t \in \mathbf{R}$ the eigenvalues of A_s are real. Then there exists an open neighborhood G' of s_0 in G and holomorphic functions $h_1, \ldots, h_N : G' \to \mathbf{C}$ such that for $s \in G'$ the eigenvalues of A_s (listed with multiplicities) are $h_1(s), \ldots, h_N(s)$.*

Proof. We denote by $H(G)$ the ring of all holomorphic functions and by $M(G)$ the field of all meromorphic functions on G. The eigenvalues of A_s are the roots of the characteristic polynomial

$$\chi(s; \lambda) = \lambda^N + \alpha_{N-1}(s)\lambda^{N-1} + \ldots + \alpha_0(s),$$

where $\alpha_k \in H(G)$, $k = 0, \ldots, N-1$. We shall prove Theorem 14.9.4 solely by looking at the roots of χ as functions of s. For this we first prove some general facts about polynomials.

We use the following conventions and notations. When speaking of roots of a polynomial or of eigenvalues, then we shall always take them to be listed according to their multiplicities. For $r > 0$ and for $w \in \mathbf{C}$ we denote by $B_r(w)$ the open disk $B_r(w) = \{ s \in \mathbf{C} \mid |s - w| < r \}$ and by $C_r(w)$ its boundary, $C_r(w) = \{ s \in \mathbf{C} \mid |s - w| = r \}$.

Now let

$$p = p(s; \lambda) = \lambda^m + a_{m-1}(s)\lambda^{m-1} + \ldots + a_0(s)$$

be an arbitrary polynomial of degree $\deg(p) = m$ with coefficients in $H(G)$ and leading coefficient 1. If $p(s', \lambda') = 0$, we shall say that λ' is a *root of p for s'*. If, for fixed s' the polynomial $\lambda \mapsto p(s', \lambda)$ has a k-fold root at λ', then λ' will be called a *k-fold root of p for s'*. In (1) - (5) we prove a number of general facts about p. We begin with the continuity of the roots.

(1) *Let $s' \in G$, $\lambda' \in \mathbf{C}$ and assume that $p(s'; \lambda') = 0$. If λ' is a k-fold root of p for s' and if $\varepsilon > 0$, then there exists $\delta > 0$ such that for any $s \in B_\delta(s')$, p has k roots for s contained in $B_\varepsilon(\lambda')$.*

Proof. We write p in the following form

$$p(s; \lambda) = (\lambda - \lambda')^m + b_{m-1}(s)(\lambda - \lambda')^{m-1} + \ldots + b_0(s).$$

By assumption, $b_0(s') = \ldots = b_{k-1}(s') = 0$, $b_k(s') \neq 0$. For s in a sufficiently small open neighborhood of s' we may thus write

$$p(s; \lambda) = b_k(s)(\lambda - \lambda')^k(1 + u + v),$$

where

$$u = u(s; \lambda) = \frac{b_{k+1}(s)}{b_k(s)}(\lambda - \lambda') + \ldots + \frac{b_m(s)}{b_k(s)}(\lambda - \lambda')^{m-k},$$

$$v = v(s; \lambda) = \frac{b_0(s)}{b_k(s)}(\lambda - \lambda')^{-k} + \ldots + \frac{b_{k-1}(s)}{b_k(s)}(\lambda - \lambda')^{-1}.$$

There exists $\rho > 0$ and $\sigma > 0$ such that for any $s \in B_\sigma(s')$ and any $\lambda \in B_\rho(\lambda')$ the inequality $|u(s; \lambda)| < \frac{1}{2}$ holds. Now let $\varepsilon \in \,]0, \rho[$. Then there exists $\delta > 0$ such that for any $s \in B_\delta(s')$ and any $\lambda \in C_\varepsilon(\lambda')$ the inequality $|v(s; \lambda)| < \frac{1}{2}$ holds. For any fixed $s \in B_\delta(s')$, Rouché's theorem applied to the holomorphic function $\lambda \mapsto p(s; \lambda)$, $\lambda \in B_\varepsilon(\lambda') \cup C_\varepsilon(\lambda')$ tells us that this function has k roots in $B_\varepsilon(\lambda')$. ◇

(2) *Let $s' \in G$, $\lambda' \in \mathbf{C}$ and assume that $p(s'; \lambda') = 0$. If λ' is a simple root of p for s', then there exist open neighborhoods $B_\delta(s')$, $B_\varepsilon(\lambda')$ and a holomorphic function $\ell : B_\delta(s') \to B_\varepsilon(\lambda')$ such that for any $s \in B_\delta(s')$ and for any $\lambda \in B_\varepsilon(\lambda')$ the equation $p(s; \lambda) = 0$ holds if and only if $\lambda = \ell(s)$.*

Proof. By (1) there exist $\varepsilon > 0$, $\delta > 0$ and a *continuous* function $\ell : B_\delta(s') \to B_\varepsilon(\lambda')$ with these properties, and it remains to show that ℓ is holomorphic. Denote by p_1 and p_2 the partial derivatives of p with respect to s and λ respectively. Fix $s \in B_\delta(s')$ and let $\lambda = \ell(s)$. For ζ and ω small in absolute value and satisfying $\ell(s + \zeta) = \lambda + \omega$ we then have

$$0 = p(s + \zeta; \lambda + \omega) - p(s; \lambda) = \zeta p_1(s; \lambda) + \omega p_2(s; \lambda) + r,$$

where $|r| \leq c(|\zeta|^2 + |\omega|^2 + |\zeta\omega|)$ for some constant $c > 0$. By the known continuity of ℓ, such ω exists, and we have $\omega \to 0$ as $\zeta \to 0$. By the simplicity of the root λ for s (recall the properties of ℓ which are already established), we have $p_2(s; \lambda) \neq 0$. Therefore, the limit of ω/ζ for $\zeta \to 0$ exists, that is, ℓ is complex differentiable at s. ◇

(3) *Assume that $p = p'p''$, where p' and p'' are polynomials with coefficients in* M(G) *and leading coefficients* 1. *Then the coefficients are in* H(G).

Proof. Let $s^* \in G$. We show that the coefficients of p' and p'' have no poles in s^*.

There exists a neighborhood W of s^* whose closure is compact and contained in G such that the coefficients are holomorphic in $W - \{s^*\}$. By (1), there exists a constant $K > 0$ such that whenever $s \in W$ and $\lambda \in \mathbf{C}$ with $p(s; \lambda) = 0$, then $|\lambda| \leq K$.

Now fix $s \in W - \{s^*\}$ temporarily. Then p, p' and p'' become polynomials with coefficients in $\mathbf{C}$ and decay into linear factors. Any such factor $(\lambda - c)$ (where now $c \in \mathbf{C}$) of p' or p'' is also a factor of p and has therefore the upper bound $|c| \leq K$. This shows that, now again for variable s, the coefficients of p' and p'' are bounded functions on $W - \{s^*\}$, that is, they have no poles in s^*. ◇

According to (3), p (with leading coefficient 1) is irreducible in the ring of polynomials with coefficients in H(G) if and only if it is irreducible in the ring of polynomials with coefficients in M(G). We shall thus simply say that p is *irreducible* if either is the case.

(4) *Let q be another polynomial with coefficients in* H(G) *and leading coefficient* 1. *Assume that* $1 \leq \deg(q) \leq \deg(p) - 1$. *If there exists $s' \in G$ and a sequence* $\{s_n\}_{n=1}^{\infty}$ *in* $G - \{s'\}$ *with $s_n \to s'$ such that for each n there exists $\lambda(n) \in \mathbf{C}$ satisfying $p(s_n; \lambda(n)) = q(s_n; \lambda(n)) = 0$, then p and q have a nontrivial common divisor with coefficients in* H(G). *In particular, p is not irreducible.*

Proof. By virtue of (3) it suffices to prove this statement with H(G) replaced by M(G). We can then dispense with the condition about the leading coefficients and, since M(G) is a field, apply the Euclidean algorithm: write $p = hq + r$, where h and q are polynomials with coefficients in M(G) and $\deg(r) \leq \deg(q) - 1$. If $\deg(r) = 0$, that is, if $r \in$ M(G), then the hypothesis implies that $r(s_n) = 0$ for all n, and consequently $r = 0$. In this case q is a common divisor. If $\deg(r) \geq 1$, then q and r satisfy the same hypothesis as p and q, and we can repeat the argument. We thus obtain the common divisor in a finite number of steps. ◇

(5) *If p is irreducible and if $s' \in G$, then there exists an open neighborhood $B_\delta(s')$ such that for any $s \in B_\delta(s') - \{s'\}$ the polynomial $\lambda \mapsto p(s; \lambda)$ has only simple roots.*

Proof. This follows from (4) by taking $q = \frac{1}{m}\partial p/\partial \lambda$ ◇

We return to the characteristic polynomial $\chi = \chi(s; \lambda)$ of A_s. In the ring of polynomials with coefficients in M(G), χ may be decomposed into irreducible factors, and by (3) we can write the decomposition such that each factor P is a polynomial with coefficients in H(G) and leading coefficient 1. It suffices to show that the roots of any such P are holomorphic functions in a neighborhood of s_0. (This is no longer a general property of polynomials, and we shall use the assumptions about A_s in the proof.)

Let λ_0 be a k-fold root of P for s_0. We must find neighborhoods $B_\delta(s_0)$ and $B_\varepsilon(\lambda_0)$ and k holomorphic functions $h_1, \ldots, h_k : B_\delta(s_0) \to B_\varepsilon(\lambda_0)$ such that for any $s \in B_\delta(s_0)$ and any $\lambda \in B_\varepsilon(\lambda_0)$ the equation $P(s; \lambda) = 0$ holds if and only if $\lambda = h_\kappa(s)$ for some $\kappa \in \{1, \ldots, k\}$.

For $k = 1$ this was shown in (2) and we may thus assume that $k \geq 2$. We let $\varepsilon > 0$ be so small that all roots of P for s_0 different from λ_0 are further apart from λ_0 than 2ε. By (1), there exists $\delta > 0$ such that for any $s \in B_{2\delta}(s_0)$, P has exactly k roots for s contained in $B_\varepsilon(\lambda_0)$. By (5) we may take this δ such that, in addition, all roots of P for $s \in B_{2\delta}(s_0) - \{s_0\}$ are simple. We fix $s' = s_0 + \delta$. The polynomial P has k pairwise different roots $\lambda_1(s'), \ldots, \lambda_k(s')$ in $B_\varepsilon(\lambda_0)$ for s'. Let us first consider $\lambda_1(s')$.

By (2), there exists a holomorphic function ℓ_1 defined in a neighborhood of s' satisfying $\ell_1(s') = \lambda_1(s')$ and $P(s; \ell_1(s)) = 0$ for all s in this neighborhood. Now we go once around $C_\delta(s_0)$, starting at s', and construct an analytic continuation of ℓ_1 around $C_\delta(s_0)$. This is possible because the roots are continuous functions of s, for $s \in B_{2\delta}(s_0)$. As we come around, back to s', we obtain a holomorphic function ℓ_2 defined in a neighborhood of s' satisfying $P(s; \ell_2(s)) = 0$ for s in this neighborhood. If ℓ_2 is different from ℓ_1, then, as P has only simple roots for s', we also have $\ell_2(s') \neq \ell_1(s')$. If this is the case, we go around $C_\delta(s_0)$ a second time to obtain ℓ_3, and so on. Eventually we find an integer n such that for the first time $\ell_{n+1} = \ell_1$. For the functions $\ell_1, \ldots, \ell_n$ thus obtained, analytic continuation along $C_\delta(s_0)$ induces the cyclic permutation $\ell_1 \mapsto \ell_2, \ell_2 \mapsto \ell_3, \ldots, \ell_n \mapsto \ell_1$. We uniformize these functions using the mapping $\zeta \mapsto s_0 + \delta\zeta^n$ which maps a tubular neighborhood of the unit circle $C_1(0)$ onto a tubular neighborhood of $C_\delta(s_0)$. Thus we get a holomorphic function L defined in a tubular neighborhood of $C_1(0)$ with the property that for $j = 1, \ldots, n$ and for ζ in a small neighborhood of $\zeta_j = e^{2\pi i(j-1)/n}$ we have

(6) $$L(\zeta) = \ell_j(s_0 + \delta\zeta^n).$$

Using analytic continuation we see that L has a holomorphic extension onto the punctured disk $B_1(0) - \{0\}$ satisfying

(7) $$P(s_0 + \delta\zeta^n; L(\zeta)) = 0.$$

By (1), L has a continuous extension into $\zeta = 0$, and thus L has a holomorphic extension onto the full disk $B_1(0)$. We now show that $n = 1$. This will imply that the function ℓ_1 has a holomorphic extension h_1 on the disk $B_\delta(s_0)$ satisfying $h_1(s') = \lambda_1(s')$ and $P(s; h_1(s)) = 0$ for all $s \in B_\delta(s_0)$. Since the same arguments may also be applied to $\lambda_2(s'), \ldots, \lambda_k(s')$, this will conclude the proof of Theorem 14.9.4.

To prove that $n = 1$, we use that, by the hypothesis of the theorem, all roots of P for $s_0 + t \in G$ are real if t is real.

In view of (7), this hypothesis implies that $L(\tau e^{2\pi i(j-1)/n}) \in \mathbf{R}$ for $0 < \tau < 1$ and for $j = 1, \ldots, n$. For the coefficients $c_0, c_1, \ldots$ in the power series expansion

$$L(\zeta) = c_0 + c_1\zeta + c_2\zeta^2 + \ldots$$

this implies that $c_j \in \mathbf{R}$ if j is an integer multiple of n and that $c_j = 0$ if j is not an integer multiple of n. In particular,

$$L(\zeta) = c_0 + c_n\zeta^n + c_{2n}\zeta^{2n} + \ldots .$$

Together with (6), applied to the case $\zeta = \zeta_2 = e^{2\pi i/n}$, we obtain the sequence of equations $\ell_2(s') = \ell_2(s_0 + \delta\zeta_2^n) = L(\zeta_2) = L(1) = \ell_1(s_0 + \delta) = \ell_1(s')$, that is, $n = 1$. Theorem 14.9.4 is now proved. ◊

14.10 Finite Parts of the Spectrum

We denote by $\mathcal{R}_g = \mathcal{T}_g/\mathcal{M}_g$ the space of all compact Riemann surfaces of genus g. For any $\varepsilon > 0$ we denote by $\mathcal{R}_g(\varepsilon)$ the subset of all $S \in \mathcal{R}_g$ with injectivity radius $\geq \varepsilon$. As an application of the analyticity of the eigenvalues given in the preceding section, we prove the analog of Theorem 10.1.4 for the eigenvalues (Buser-Courtois [1]). As in the proof of Theorem 10.1.4, we use that analytic varieties are Noetherian. A difficulty arises from the fact that, in contrast to the geodesic length functions, the eigenvalues are not globally defined analytic functions on Teichmüller space. The difficulty is bypassed by using a theorem of Bruhat-Cartan [1], which states that, under suitable conditions, points on a real analytic variety may be joined by analytic curves on the variety. For the full statement of the theorem and its proof we refer the reader to Bruhat-Cartan [1].

14.10.1 Theorem. *For any $\varepsilon > 0$ there exists an integer $m(g, \varepsilon)$ with the following property. If $S, F \in \mathcal{R}_g(\varepsilon)$ and if $\lambda_n(S) = \lambda_n(F)$ for $n = 0, \ldots, m(g, \varepsilon)$, then S and F are isospectral.*

Proof. By Theorem 6.6.5, there exists a compact subset $Q(\varepsilon) \subset \mathcal{T}_g$ such that any $S \in \mathcal{R}_g(\varepsilon)$ is represented by an element in $Q(\varepsilon)$. It suffices therefore to fix any pair $(S_0, F_0) \in \mathcal{T}_g \times \mathcal{T}_g$, and to find an open neighborhood U_0 of (S_0, F_0) in $\mathcal{T}_g \times \mathcal{T}_g$ together with a positive integer N such that the following holds: if $(S, F) \in U_0$ and if $\lambda_i(S) = \lambda_i(F)$, for $i = 0, \ldots, N$, then S and F are isospectral.

If S_0 and F_0 have different spectra, then there exists N such that $\lambda_N(S_0) \neq \lambda_N(F_0)$. By Theorem 14.9.2, there exists an open neighborhood U_0 of (S_0, F_0) in $\mathcal{T}_g \times \mathcal{T}_g$ such that $\lambda_N(S) \neq \lambda_N(F)$ for any $(S, F) \in U_0$. Hence, nothing has to be proven in this case, and we may from now on assume that S_0 and F_0 have the same spectrum.

Let $L_1, L_2, \ldots$ with $L_1 < L_2 < \ldots$ be an infinite sequence of positive real numbers such that no L_k is an eigenvalue of S_0 and such that $L_k \to \infty$ as $k \to \infty$. By Theorem 14.9.2, there exists, for each L_k, a neighborhood U_k of (S_0, F_0) in $\mathcal{T}_g \times \mathcal{T}_g$ such that if $(S, F) \in U_k$, then the spectra of S and F do not contain L_k. For simplicity we take these neighborhoods in such a way that

$$U_1 \supset U_2 \supset \ldots,$$

and such that the boundary of each U_k is a smooth connected real analytic subvariety of $\mathcal{T}_g \times \mathcal{T}_g$. For each k there exists an integer k^* such that

(1) $\qquad \lambda_{k^*}(S) < L_k < \lambda_{k^*+1}(S), \quad \lambda_{k^*}(F) < L_k < \lambda_{k^*+1}(F),$

for any pair $(S, F) \in U_k$. We define

(2) $\qquad A_k = \{ (S, F) \in U_k \mid \lambda_i(S) = \lambda_i(F), i = 0, \ldots, k^* \}.$

By Theorem 14.9.1, A_k is a real analytic subvariety of U_k. Since the germs of analytic functions at a given point form a Noetherian ring (Grauert-Remmert [1]), and since for $j > k$ we always have $A_j \subset A_k \cap U_j$, this implies that there exists an integer K satisfying

(3) $\qquad A_j = A_K \cap U_j \text{ for all } j \geq K.$

By the theorem of Bruhat-Cartan [1], there exists an open neighborhood U_0 of (S_0, F_0) in U_K such that for any $(S, F) \in A_K \cap U_0$ an analytic curve can be found which is contained in $A_K \cap U_0$ joining (S_0, F_0) to (S, F).

Now let $(S, F) \in U_0$ and assume that $\lambda_i(S) = \lambda_i(F)$ for $i = 0, \ldots, N$, where $N = K^*$ (cf. (1)). We want to prove that S and F are isospectral. For this we recall that $(S, F) \in A_K$ and let

$$t \mapsto C(t) = (S(t), F(t)) \in A_K \cap U_0, \quad t \in [0, 1],$$

be a real analytic curve, with $C(0) = (S_0, F_0)$ and $C(1) = (S, F)$. Then we let b be any (large) positive real number. By Theorem 14.9.3, there exists an integer M and two sequences of real analytic functions $f_m : [0, 1] \to \mathbf{R}$, $g_m : [0, 1] \to \mathbf{R}$, $m = 0, \ldots, M$, such that for any $t \in [0, 1]$ the $f_m(t)$ and $g_m(t)$ are eigenvalues of $S(t)$ and $F(t)$, respectively, and such that the subsequence of all $f_m(t) \leq b$ (respectively, all $g_m(t) \leq b$) coincides, up to a permutation, with the subsequence of all eigenvalues $\leq b$ of $S(t)$ (respectively, $F(t)$).

In view of (2) and (3) we can renumber the functions g_m such that for some $\delta > 0$ we get

$$f_m(t) = g_m(t), \quad m = 0, \ldots, M,$$

for all $t \in [0, \delta]$. By the real analyticity of the functions, these equations then hold for all $t \in [0, 1]$. Hence, the eigenvalues $\leq b$ of S coincide with the eigenvalues $\leq b$ of F. As b was chosen arbitrarily, this proves Theorem 14.10.1. ◇

We remark that the property of having a small injectivity radius can be seen in a finite part of the spectrum: if ρ is the injectivity radius and A the number of eigenvalues in $[\frac{1}{4}, \frac{1}{2}]$ of a compact Riemann surface S of genus g, then

$$c_1 \log \frac{1}{\delta} \leq A \leq c_2 \log \frac{1}{\delta},$$

where c_1 and c_2 are constants depending only on g (Buser [5]).

As in Buser-Courtois [1], we conjecture that Theorem 14.10.1 holds with a constant which depends only on g.

Appendix

Curves and Isotopies

Teichmüller theory hinges strongly on the fact that homotopies of simple curves can be carried out by surface homeomorphisms. The necessary theorems have been studied by Baer [1], Epstein [1], Zieschang [1] and others. We reprove these theorems for completeness. The proofs are similar to those in the literature but are not based on piecewise linear mappings. At certain points we use hyperbolic geometry. In the final part we give an application to the mapping class group of the 3-holed sphere and to length decreasing homotopies.

The Theorems of Baer-Epstein-Zieschang

Throughout the appendix we make the following assumption.

A.0 Assumption. S is a smooth compact connected orientable two dimensional manifold with possibly empty piecewise smooth boundary. At any boundary point p of S where the boundary is not smooth, we assume that the adjacent arcs do not have parallel tangent vectors.

Since all compact surfaces of the same signature are homeomorphic, this is no restriction of generality. Moreover, all compact hyperbolic surfaces satisfy this assumption.

If two surfaces S and S' satisfy Assumption A.0, and if both have the same signature, then they are piecewise diffeomorphic (see for instance Massey [1] or Hirsch [1]). We may therefore assume, when this is convenient, that S carries a metric of constant curvature in which the boundary is piecewise geodesic with all interior angles $\leq \pi$. Thus, in most cases S is a surface with a hyperbolic structure and we shall use results of Chapter 1.

A.1 Definition. Let S and S' be given as above. Two homeomorphisms $\varphi_0, \varphi_1 : S \to S'$ are called *isotopic* if there exists a continuous mapping $J : [0, 1] \times S \to S'$ with the following properties.

(i) $J(0, \ ) = \varphi_0$, $J(1, \ ) = \varphi_1$.
(ii) $J(\rho, \ ) : S \to S'$ is a homeomorphism for any $\rho \in [0, 1]$.

Now assume that $S = S'$. If $A \subset S$ is a subset, then the homeomorphisms $\varphi_0, \varphi_1 : S \to S$ are called *isotopic with A pointwise fixed* if $\varphi_0(x) = x$ and $\varphi_1(x) = x$ for any $x \in A$, and if there exists a mapping J with the above properties such that in addition

(iii) $J(\rho, x) = x$ for any $x \in A$ and any $\rho \in [0, 1]$.

A homeomorphism $h : S \to S$ is called a *1-homeomorphism* if it is isotopic to the identity with ∂S pointwise fixed. h is an *A-1-homeomorphism* if it is isotopic to the identity with $A \cup \partial S$ pointwise fixed.

The mapping J in Definition A.1 is sometimes called an isotopy or a level-preserving isotopy.

A.2 Theorem. (Alexander-Tietze). *Let $D \subset S$ be a domain homeomorphic to the compact unit disk in $\mathbf{R}^2$ and let $A = S - \mathrm{int}\, D$. If $\varphi : S \to S$ is a homeomorphism which is the identity on A, then φ is an A-1-homeomorphism.*

Proof. It suffices to prove the theorem for the case $\varphi : E \to E$, where $E = \{x \in \mathbf{R}^2 \mid |x| \leq 1\}$, and where $\varphi(x) = x$ for all $x \in \partial E$. We define, for $x \in E$ and $\rho \in [0, 1]$,

$$J(\rho, x) = \begin{cases} x & \text{if } |x| \geq \rho \\ \rho\varphi(x/\rho) & \text{if } x < \rho. \end{cases}$$

Then $J : E \times [0, 1] \to E$ is a continuous mapping and satisfies the conditions (i)-(iii) of Definition A.1 with $\varphi_0 = \mathrm{id}$ and $\varphi_1 = \varphi$. ◇

We now state the theorems of Baer, Epstein and Zieschang in a form which is adapted to the canonical curve systems as in Chapter 6. Afterwards we shall look at some examples to explain the lengthy hypotheses of the theorems. The proofs are be based on the Jordan Schoenflies theorem and will begin after Corollary A.7.

We introduce the following notation. Curves α and β are said to be *non ± homotopic* if α is neither homotopic to β nor to β^{-1}, where β^{-1} is the curve parametrized in the opposite sense. (If non-oriented curves are considered then "*non ± homotopic*" may be replaced by "non-homotopic".)

In the first theorem, homotopy means *free homotopy*.

A.3 Theorem. (Baer-Zieschang). *Let $c_1, \ldots, c_m$ be pairwise disjoint homotopically non-trivial simple closed curves in the interior of S which are pairwise non ± homotopic. If $\gamma_1, \ldots, \gamma_m$ in the interior of S is another set of curves with these properties, and if c_μ is freely homotopic to γ_μ for $\mu = 1, \ldots, m$, then there exists a 1-homeomorphism $\phi : S \to S$ such that $\phi \circ c_\mu = \gamma_\mu, \mu = 1, \ldots, m$.*

The next theorem concerns simple *loops* that is, simple closed curves with a given base point. In this theorem, "homotopy" means *homotopy with fixed base point*.

A.4 Theorem. (Epstein-Zieschang). *Let $p \in \text{int } S$ and let $c_1, \ldots, c_m$ be homotopically non-trivial simple loops in int S with base point p which pairwise intersect each other only at p and which are pairwise non ± homotopic. If $\gamma_1, \ldots, \gamma_m$ in int S with base point p is another set of loops with these properties, and if c_μ is homotopic to γ_μ (with p fixed) for $\mu = 1, \ldots, m$, then there exists a p-1-homeomorphism $\phi : S \to S$ such that $\phi \circ c_\mu = \gamma_\mu, \mu = 1, \ldots, m$.*

The third theorem will be needed in the induction proof of Theorem A.4. In itself, it is interesting mainly in the case $m = 1$ and for arcs c_1, γ_1 with two different endpoints. In this theorem, "homotopy" means *homotopy with fixed endpoints*.

A.5 Theorem. (Epstein-Zieschang). *Let $p_1, \ldots, p_m, q_1, \ldots, q_m \in \partial S$ and let for $\mu = 1, \ldots, m$, c_μ be an arc from p_μ to q_μ which, except for the endpoints, lies in the interior of S. Assume the following hypotheses.*

(i) *If $p_\mu \neq q_\mu$, then c_μ is a simple arc oriented from p_μ to q_μ; if $p_\mu = q_\mu$, then c_μ is a simple homotopically non-trivial loop with base point p_μ; $\mu = 1, \ldots, m$.*

(ii) *For $\mu \neq \nu$ the curves c_μ and c_ν are non ± homotopic (in the homotopy class with p_μ and q_μ fixed).*

(iii) *The curves $c_1, \ldots, c_m$ pairwise intersect each other at most at their endpoints.*

If $\gamma_1, \ldots, \gamma_m$ is another set of curves with these properties, where γ_μ is oriented from p_μ to q_μ, $\mu = 1, \ldots, m$, and if each c_μ is homotopic (with fixed endpoints) to γ_μ, then there exists a 1-homeomorphism $\phi : S \to S$ such that $\phi \circ c_\mu = \gamma_\mu$, $\mu = 1, \ldots, m$.

We first consider a few examples.

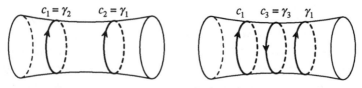

Figure A.1

Fig. A.1 shows that Theorem A.3 fails if homotopic curves are admitted: the closed curves c_1 and c_2 shown on the cylinder cannot be interchanged by a homeomorphism of the cylinder which keeps the boundary fixed.

The second example in Fig. A.1 shows that the theorem also fails if c_1 is homotopic to some c_ν^{-1}. Similar counterexamples can be given in the case of Theorems A.4 and A.5.

The next example is more difficult and we give two figures, one showing the surface S (an annulus) and one showing its universal covering.

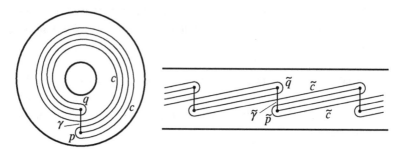

Figure A.2

The example shows that Theorem A.5 cannot be extended to arcs with different endpoints in the interior of S. One checks in the universal covering that c and γ are homotopic with fixed endpoints. However, any homotopy between c and γ moves c over one of the endpoints. Therefore this homotopy cannot be carried out by a $\{p, q\}$-1-homeomorphism.

The final example is due to Epstein-Zieschang [1] and shows that Theorem A.4 needs certain modifications if S is a non-orientable surface.

The example is on the Möbius band. In the first figure, the Möbius band is shown cut open into a rectangle. The second figure shows again the universal covering. Note that the group of covering transformations is generated by a

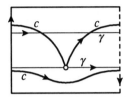
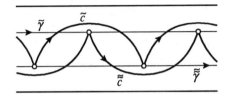

Figure A.3

glide reflection which maps the covering curve $\tilde{\gamma}$ of γ onto $\tilde{\tilde{\gamma}}$ and $\tilde{c}$ onto $\tilde{\tilde{c}}$. In the proof of Theorem A.4 the reader will indeed find an argument which uses the fact that in the orientable case no covering transformation is a glide reflection.

For the proofs of the theorems we need the Jordan-Schoenflies theorem and corollaries which we state in the following condensed form. For the proof we refer to Dostal-Tindell [1], Stillwell [1] and Moise [1].

A.6 Theorem. (Jordan-Schoenflies). (i) *Let c and γ be simple closed curves in the interior of the closed unit disk E. Then there exists a 1-homeomorphism $\phi : E \to E$ such that $\phi \circ c = \gamma$ or $\phi \circ c = \gamma^{-1}$.*

(ii) *Let $p, q \in E$ (possibly with $p = q$) and let c and γ be simple arcs connecting p to q which, except possibly for the endpoints, are contained in the interior of E. Then there exists a $\{p, q\}$-1-homeomorphism $\phi : E \to E$ such that $\phi \circ c = \gamma$ or $\phi \circ c = \gamma^{-1}$.*

We note the following corollary.

A.7 Corollary. *Let c be a homotopically trivial simple closed curve in S. Then c bounds a disk.*

Proof. We assume that S has signature (g, n) with $g + n \geq 2$ and leave the remaining cases to the reader. We may assume that S carries a hyperbolic structure and let $\pi : \tilde{S} \to S$ be a universal covering with $\tilde{S} \subset \mathbf{D}$, where $\mathbf{D}$ is the unit disk model of the hyperbolic plane.

Since c is simple and homotopically trivial, c has a simple closed lift $\tilde{c}$ in $\tilde{S} \subset \mathbf{D}$ which bounds a disk $D \subset \mathbf{D}$. No boundary point of $\tilde{S}$ is in the interior of D for otherwise $\partial \tilde{S}$ would intersect $\tilde{c}$ and then ∂S would have to intersect c. Hence, $D \subset \tilde{S}$. If $T \neq id$ is a covering transformation with respect to π, then neither $T(D) \subset D$ nor $D \subset T(D)$. Since both components of $\mathbf{D} - \tilde{c}$ and both components of $\mathbf{D} - T(\tilde{c})$ are connected, we conclude that if $D \cap T(D) \neq \emptyset$, then $\tilde{c}$ intersects $T(\tilde{c})$; but then c would have a self-intersection. Hence,

$D \cap T(D) = \emptyset$ for any covering transformation $T \neq id$. It follows that π restricted to D is one-to-one, and c bounds the disk $\pi(D)$. ◇

We now prove Theorems A.3 - A.5. In the first step we show that we may restrict ourselves to "nice" curves.

A.8 Lemma. *Let $c_1, \ldots, c_m$ be as in Theorem A.3, or A.4 or A.5. Then there exists a 1-homeomorphism $\varphi : S \to S$ such that $\varphi \circ c_\mu$ is a piecewise geodesic curve for $\mu = 1, \ldots, m$. In the case of Theorem A.4, φ may be chosen to be a $\{p\}$-1-homeomorphism.*

Proof. We let $c_1, \ldots, c_m$ be as in Theorem A.4, the remaining cases have similar proofs.

Let $p \in \text{int } S$ be the base point of the loops $c_1, \ldots, c_m$ and let ε be a small positive number such that the set $D = \{ x \in S \mid \text{dist}(p, x) \leq \varepsilon \}$ is a disk and such that each c_μ contains points outside D. Each c_μ is a product of three arcs: $c_\mu = a_\mu v_\mu b_\mu$, where a_μ goes from p along c_μ to the first intersection point of c_μ with ∂D, v_μ is the middle section, and b_μ goes from the last intersection point of c_μ with ∂D back to p.

By Theorem A.6, there exists a $\{p\}$-1-homeomorphism $D \to D$ mapping a_1 onto a geodesic arc with initial point p. Now cut D open along a_1 and apply Theorem A.6 to modify this homeomorphism such that also the image of b_1 is a geodesic arc (with endpoint p). We may now assume that a_1 and b_1 themselves are these arcs. The arcs a_1 and b_1 separate D into two topological disks. Applying Theorem A.6 to one of these disks we find another $\{p\}$-1-homeomorphism fixing a_1 and b_1 and mapping a_2 onto a geodesic arc; and so on.

We may from now on assume that $a_1, \ldots, a_m$ and $b_1, \ldots, b_m$ are geodesic arcs.

Since $c_1, \ldots, c_m$ intersect each other only at p, there exist disjoint open neighborhoods $U_1, \ldots, U_m$ in S such that the arc v_μ is contained in U_m for $\mu = 1, \ldots, m$. It remains, for each $c = c_\mu$, to find a 1-homeomorphism of $U = U_\mu$ onto itself which maps the middle section $v = v_\mu$ of $c = avb$ onto a piecewise geodesic curve such that the geodesic arcs a and b are kept fixed.

For this we consider the universal covering $\pi : \tilde{S} \to S$, where $\tilde{S} \subset \mathbf{D}$. We shall use the fact that $\tilde{S}$ is a subset of $\mathbf{R}^2$ with the induced topology. Let $\tilde{c} = \tilde{a}\tilde{v}\tilde{b}$ be a lift of the loop $c = avb$. Then $\tilde{c}$ is a simple arc in $\mathbf{R}^2$ contained in the interior of some large rectangle $\Delta \subset \mathbf{R}^2$ (Δ need not be contained in $\tilde{S}$). By Theorem A.6, there exists a homeomorphism $\psi : \Delta \to \Delta$ which maps $\tilde{c}$ onto a straight line segment $c' = a'v'b'$ (where a' is the image of $\tilde{a}$, etc.).

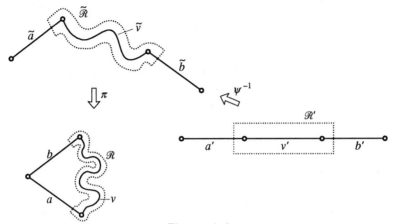

Figure A.4

Let $\mathcal{R}' \subset \mathbf{R}^2$ be a rectangle as in Fig. A.4 parallel to c' containing v' in the interior. We take $\mathcal{R}'$ so small that $\tilde{\mathcal{R}} := \psi^{-1}(\mathcal{R}')$ is contained in $\tilde{S}$ and such that π restricted to $\tilde{\mathcal{R}}$ is one-to-one with image $\mathcal{R} := \pi(\tilde{\mathcal{R}}) \subset U$. (Recall that π restricted to $\tilde{v}$ is one-to-one.) There exists a piecewise geodesic curve $\hat{v}$ in the interior of $\mathcal{R}$ which connects the endpoints of v.

The existence proof for $\hat{v}$ need not yield a simple $\hat{v}$, but if $\hat{v}$ is not simple, then it forms finitely many loops such as loop λ in Fig. A.5. Removing them successively we end up with a simple $\hat{v}$. In a similar way we achieve that $\hat{v}$ intersects a and b only at the endpoints. In view of Theorem A.6 there exists a 1-homeomorphism of $\mathcal{R}$ mapping v onto $\hat{v}$. This proves the lemma. ◇

Before proceeding with the proofs of Theorems A3 - A5, we draw a conclusion of the preceding proof. We use the following terminology. Let $D \subset S$ be a topological disk and let v be a simple arc in D with two different endpoints on ∂D and all remaining points in the interior of D. Then v will be called a *topological diameter of D*.

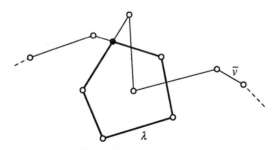

Figure A.5

A.9 Lemma. *Let v be a simple arc in int S and let U be an open neighborhood in S containing v. Then there exists a simple closed curve $\rho \subset U$ which bounds a disk $\mathfrak{R} \subset U$ such that v is a topological diameter of $\mathfrak{R}$.* ◊

In the proof of Theorem A.3 - A.5 we shall successively map c_μ to γ_μ for $\mu = 1, \ldots, m$. To distinguish the various cases, we introduce the following classes of pairs α, β of curves in S.

Types of Curves.

(1) α and β are simple closed homotopically non-trivial freely homotopic curves in the interior of S.

(2) α and β are homotopically non-trivial simple loops in the interior of S with the same base point and homotopic to each other with a homotopy which keeps the base point fixed.

(3) α and β are homotopic arcs with endpoints on ∂S and all other points in the interior of S. If the two endpoints are different, then α and β are simple arcs. If the two endpoints coincide, then α and β are simple homotopically non-trivial loops.

Two subarcs a on α and b on β are said to form an *overlap* if $a \cup b$ is a simple closed curve and if there exists a closed topological disk $D \subset S$ with $\partial D = a \cup b$. The overlap D is *proper* if there exists a topological disk D' with $D \subset D'$ such that $D' \cap \alpha = a$, $D' \cap \beta = b$, and such that a and b are topological diameters of D'.

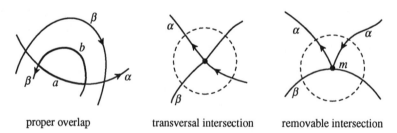

proper overlap   transversal intersection   removable intersection

Figure A.6

Let $D \subset S$ be a topological disk such that $D \cap \alpha$ and $D \cap \beta$ are topological diameters of D which intersect each other in exactly one point $m \in \text{int } D$. If α and β are of type (2) or (3) we also assume that m is not the common base point of α and β. We say that m is a *transversal intersection* if $D \cap \alpha$ has points on either side of $D \cap \beta$; otherwise m is called a *removable intersection*. By Theorem A.6, there exists a 1-homeomorphism $\varphi : D \to D$ such that $\varphi(D \cap \alpha) \cap \beta = \emptyset$ if m is a removable intersection.

A.10 Lemma. *Let α, β be a pair of curves of type (1), (2) or (3). Assume that α intersects β only in finitely many points and that each intersection point different from the endpoints (in case of type (2) or (3)) is a transversal intersection. If α and β have at least one transversal intersection, then there exist proper subarcs a of α and b of β which form a proper overlap.*

Proof. We restrict ourselves again to the case that S has signature (g, n) with $g + n \geq 2$ and leave the remaining cases to the reader. We may assume that S is a surface with a hyperbolic structure and work with a universal covering $\pi : \tilde{S} \to S$, where $\tilde{S} \subset \mathbf{D}$. Every covering transformation $T \neq id$ is the restriction to $\tilde{S}$ of a hyperbolic element of $\mathrm{Is}^+(\mathbf{D})$ and has a unique invariant geodesic $a_T \subset \tilde{S}$ (the axis of T in $\mathbf{D}$).

Every homotopically non-trivial closed curve c on S is freely homotopic to a unique closed geodesic γ. If $\tilde{c}$ is a lift of c in $\tilde{S}$, then the homotopy between c and γ lifts to a homotopy in $\tilde{S}$ between $\tilde{c}$ and a covering geodesic $\tilde{\gamma}$ of γ. The curves $\tilde{c}$ and $\tilde{\gamma}$ have the same endpoints at infinity. A covering transformation T leaves $\tilde{c}$ invariant if and only if it leaves $\tilde{\gamma}$ invariant. By Theorem A.6, $\tilde{c}$ separates $\mathbf{D} - \tilde{c}$ into two connected components, and since T preserves orientation, these components are T-invariant if $\tilde{c}$ is T-invariant

Let now α, β be a pair of type (1) with at least one transversal intersection point m. Let $\tilde{\alpha}$ and $\tilde{\beta}$ be (not necessarily homotopic) lifts in $\tilde{S}$ of α and β respectively which intersect each other in a lift $\tilde{m}$ of m. We first claim that $\tilde{\alpha}$ and $\tilde{\beta}$ have an overlap in $\tilde{S}$.

Let a and b denote the closed geodesics freely homotopic to α and β respectively, and let $\tilde{a}$ and $\tilde{b}$ be lifts of a and b homotopic to $\tilde{\alpha}$ and $\tilde{\beta}$ respectively. Then $\tilde{\alpha}$ and $\tilde{a}$ have the same endpoints at infinity and the same holds for $\tilde{\beta}$ and $\tilde{b}$.

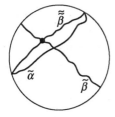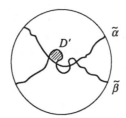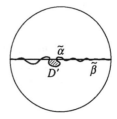

Figure A.7

If $\tilde{a}$ intersects $\tilde{b}$ transversally as suggested in the first example of Fig. A.7, then the endpoints at infinity of $\tilde{\alpha}$ lie on different sides of $\tilde{\beta}$ (here we are using Theorem A.6). There exists a lift $\tilde{\tilde{\beta}}$ of β which has the same endpoints at infinity as $\tilde{\alpha}$. It follows that $\tilde{\tilde{\beta}}$ intersects $\tilde{\beta}$, and this implies that β has a self-intersection, a contradiction. Hence, $\tilde{a}$ does not intersect $\tilde{b}$ transversally.

If $\tilde{a} \cap \tilde{b} = \emptyset$, as in the second example of Fig. A.7, then $\tilde{\beta}$ has both endpoints at infinity on the same side of $\tilde{\alpha}$, and so $\tilde{\beta}$ and $\tilde{\alpha}$ have at least two intersection points. Since all intersection points are transversal and since there are only finitely many intersection points, there exist subarcs α' of $\tilde{\alpha}$ and β' of $\tilde{\beta}$ which form an overlap in **D**. Since every connected component of $\partial \tilde{S}$ has two endpoints at infinity, no boundary point of $\tilde{S}$ is in the interior of the overlap. Therefore the overlap is contained in $\tilde{S}$. If $\tilde{a} = \tilde{b}$ as in the third example of Fig. A.7, then we also find such an overlap in $\tilde{S}$ because then $\tilde{\alpha}$ and $\tilde{\beta}$ are T-invariant under some covering transformation $T \neq id$, and so $\tilde{\beta}$ crosses $\tilde{\alpha}$ periodically.

Now let $D' \subset \tilde{S}$ be the disk bounded by α' and β'. If D' contains a point x in its interior lying on a lift $\bar{\alpha}$ of α (not necessarily different from $\tilde{\alpha}$), then there exists an arc α'' in D' with endpoints on β'. By Theorem A.6, there exists a subarc β'' of β' such that α'' and β'' form an overlap $D'' \subset D'$. Repeating this argument finitely many times we end up with an innermost overlap D^*, and we claim that its image $\pi(D^*)$ in S is a proper overlap. For this we let α^* and β^* denote the arcs which form the boundary of D^*, where α^* lies on some lift of α and β^* lies on some lift of β. Then we let m^* and n^* be the common endpoints of α^* and β^*. We claim that for any covering transformation $T \neq id$, $T(D^*) \cap D^* = \emptyset$.

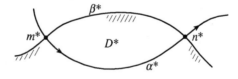

Figure A.8

For this we first observe that neither $T(D^*) \subset D^*$ nor $D^* \subset T(D^*)$. Since D^* and $T(D^*)$ contain no points in the interior which lie on a lift of α and β, and since all intersection points of α and β are transversal, it follows that the only possible intersection point of D^* and $T(D^*)$ is one of the points m^* and n^*. For the same reason as before, our intersection point must also be one of the points $T(m^*)$ and $T(n^*)$. Since T has no fixed points, it follows that either $T(m^*) = n^*$ or $T(n^*) = m^*$. In both cases, T leaves the lift $\bar{\beta}$ of β invariant which carries the arc β^*. Since T preserves orientation it follows that D^* and $T(D^*)$ lie on the same side of $\bar{\beta}$. Since the lift $\bar{\alpha}$ of α which carries α^*

changes sides at m^* and n^*, this is impossible. We have thus proved that $T(D^*) \cap D^* = \emptyset$ for any covering transformation $T \neq id$.

We remark here that for non-orientable S, the above covering transformation T may be a glide reflection and then it is indeed possible that $T(D^*) \cap D^* = \{n^*\}$.

Since $T(D^*) \cap D^* = \emptyset$ for all $T \neq id$, it follows that π restricted to D^* is one-to-one, and since π is a covering map, there exists an open neighborhood containing D^* on which π is one-to-one. Hence, $\pi(D^*)$ is a proper overlap in S. This proves the lemma for pairs of type (1).

Next let α, β be a pair of type (2) and let p be the common base point of α and β. As before, we find an overlap D^* in $\tilde{S}$ as in Fig. A.8 bounded by arcs α^* and β^* which lie on lifts of α and β respectively, and which intersect each other at points m^* and n^*. Moreover, we find this D^* such that it contains no further points in the interior which lie on lifts of α and β. Again, if T is a covering transformation, $T \neq id$, then $T(D^*)$ intersects D^* at most in m^* or n^*, and this only by sending either n^* to m^* or m^* to n^*.

If m^* and n^* are transversal intersections, then the earlier argument (for pairs of type (1)) is still valid and implies that $T(D^*) \cap D^* = \emptyset$.

If m^* (or n^*) is not transversal, then it is a lift of p, and since $n^* = T(m^*)$, n^* is also a lift of p. It follows that $\pi(\alpha^*) = \alpha$. Since α intersects β transversally at some point different from p, α^* intersects a lift of β at some point different from m^* and n^*, a contradiction. Hence, this last case cannot occur and the lemma is proved for pairs of type (2).

Finally, let α, β be a pair of type (3). If the endpoints p, q of α and β coincide, the proof for type (2) works without modification. Let us therefore assume that $p \neq q$.

Denote by u the boundary component of S containing p and by v the boundary component containing q (we may, of course, have $u = v$). Consider homotopic lifts $\tilde{\alpha}$ and $\tilde{\beta}$ in $\tilde{S}$ with initial point $\tilde{p}$ on lift $\tilde{u}$ of u and endpoint $\tilde{q}$ on lift $\tilde{v}$ of v as shown in Fig. A.9.

We let α' be the arc on $\tilde{\alpha}$ from $\tilde{p}$ to the first intersection point r (after $\tilde{p}$) of $\tilde{\alpha}$ with $\tilde{\beta}$, and we let β' be the arc on $\tilde{\beta}$ from $\tilde{p}$ to r. Then α' and β' form an

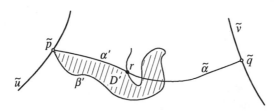

Figure A.9

overlap D'. We now claim the following (this is the argument which fails in the example of Fig. A.2).

D' contains no lift of p and q in its interior.

In fact, if p^* is a lift of p in int D', then p^* lies on some lift u^* of u. Now u^* has two endpoints at infinity, and therefore u^* intersects the boundary of D' at some point different from $\tilde{p}$ and $\tilde{q}$. This is impossible because α and β intersect ∂S only at p and q. This proves the claim.

From the claim follows that if $x \in$ int D' lies on a lift of α or β, then x lies on an arc on this lift which has *both* endpoints on $\partial D'$ and all other points in int D'. We find therefore a smaller overlap and eventually an innermost overlap D^*.

The boundary arcs of D^* and their intersection points are again denoted by α^*, β^*, m^* and n^*. If T is a covering transformation, $T \neq id$, then, as before, the intersection $T(D^*) \cap D^*$ is non-empty only if either $T(m^*) = n^*$ or $T(n^*) = m^*$. Neither is possible because in the present case, π is injective on any lift of α. Lemma A.10 is now proved. ◇

In the final argument of the preceding proof we did not use the fact that α and β have a transversal intersection. Hence, α and β form a simple overlap if $\alpha \cap \beta = \{p, q\}$.

The counterpart of this for curves of type (1) is the following.

A.11 Proposition. *Let α and β be homotopically non-trivial simple closed curves on S. If α and β are freely homotopic and if $\alpha \cap \beta = \emptyset$, then α and β bound an embedded annulus in S.*

Proof. By attaching annuli along the boundary of S if necessary, we may assume that α and β lie in the interior of S so that they form a pair of type (1). We assume again that S has a hyperbolic structure and leave the remaining case of the torus to the reader.

Let $\tilde{\alpha}$ and $\tilde{\beta}$ be homotopic lifts of α and β in $\tilde{S}$. They have the same endpoints at infinity and by Theorem A.6, they bound a strip Σ in $\mathbf{D}$ that is, a domain homeomorphic to $[0, 1] \times \mathbf{R}^1$. If T is a covering transformation, $T \neq id$, then either $T(\tilde{\alpha}) = \tilde{\alpha}$ and $T(\tilde{\beta}) = \tilde{\beta}$, or else $T(\tilde{\alpha}) \cap \tilde{\alpha} = \emptyset$ and $T(\tilde{\beta}) \cap \tilde{\beta} = \emptyset$. It follows that either $T(\Sigma) = \Sigma$ or $T(\Sigma) \cap \Sigma = \emptyset$. If there were a boundary component δ of $\partial \tilde{S}$ which intersects the interior of Σ, then δ would be contained in Σ and have the same endpoints at infinity as $\tilde{\alpha}$ and $\tilde{\beta}$. But then the lift of the homotopy between α and β (which is a homotopy in $\tilde{S}$ between $\tilde{\alpha}$ and $\tilde{\beta}$) would take place in the closure of one of the two half-planes of

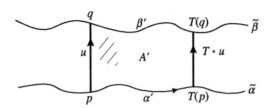

Figure A.10

$D - \delta$, a contradiction. Hence, $\Sigma \subset \tilde{S}$.

Let T be a generator of the cyclic subgroup of all covering transformations which map Σ onto itself.

Let u be a shortest geodesic arc in Σ from $\tilde{\alpha}$ to $\tilde{\beta}$ and let $p \in \tilde{\alpha}$ and $q \in \tilde{\beta}$ be its endpoints. We denote by α' and β' the arcs along $\tilde{\alpha}$ and $\tilde{\beta}$ from p to $T(p)$ and from q to $T(q)$. By the minimizing properties of u, $\alpha'(T \circ u)(\beta')^{-1}u^{-1}$ is a simple closed curve. By Theorem A.6, this curve encloses a closed disk A'. Now $T(\beta') \cap \beta' = T(q)$ and $T(\alpha') \cap \alpha' = T(p)$ so that $T(A') \cap A' = T(u)$. Similarly $T^{-1}(A') \cap A' = u$. For any other covering transformation $\tau \neq id$ we have $\tau(A') \cap A' = \emptyset$. It follows that π restricted to $A' - T(u)$ is one-to-one, and $\pi(A')$ is an embedded annulus. ◊

We turn to the proof of Theorems A.3 - A.5. The structure of the proof is as follows. By Lemma A.8, there exist 1-homeomorphims $\varphi, \psi: S \to S$ such that $\varphi \circ c_1, \ldots, \varphi \circ c_m$ and $\psi \circ \gamma_1, \ldots, \psi \circ \gamma_m$ are piecewise geodesic curves. It is easy to see that φ can be modified such that in addition all intersection points of $\varphi \circ c_\nu$ and $\psi \circ \gamma_\mu$, $\nu, \mu = 1, \ldots, m$, are transversal except for the possible endpoints. We thus assume that $c_1, \ldots, c_m$ and $\gamma_1, \ldots, \gamma_m$ themselves have these properties and then only admit homeomorphisms which preserve these properties. We do not change the names of the curves when they are deformed.

In the first step, c_1 will be mapped to γ_1. Then S will be cut open along $c_1 = \gamma_1$ and we shall have to prove that on the new surface the curve systems $c_2, \ldots, c_m$ and $\gamma_2, \ldots, \gamma_m$ satisfy again the hypotheses of one of the Theorems A.3 - A.5. We then have a 1-homeomorphism of the new surface mapping c_2 onto γ_2. Since this homeomorphism keeps the boundary pointwise fixed, it can be interpreted as a 1-homeomorphism of S itself which keeps $c_1 = \gamma_1$ pointwise fixed and mapping c_2 onto γ_2. After that we cut along $c_2 = \gamma_2$ and so on.

Let us start with Theorem A.5 and assume that c_1 and γ_1 (which form a pair of type (3)) have two different endpoints p and q on ∂S. If c_1 and γ_1 have

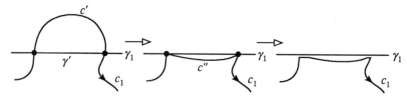

Figure A.11

a transversal intersection, then there exist subarcs c' of c_1 and γ' of γ_1 forming a proper overlap.

By definition, there exists a disk $D \subset S$ such that we get $c' = D \cap c_1$ and $\gamma' = D \cap \gamma_1$, and such that c' and γ' are topological diameters of D. By Theorem A.6, there exists a 1-homeomorphism $\varphi : D \to D$ which maps c' to a piecewise geodesic arc c'' with c' and c'' on different sides of γ'. This removes the overlap and creates two removable intersection points (or one, if one of the endpoints of c' coincides with p or q). After removing the removable intersection points we have at least one intersection point less. After finitely many such steps, c_1 and γ_1 intersect each other only in p and q. By the remark preceding Proposition A.11, c_1 and γ_1 form a proper overlap, and we find a final 1-homeomorphism ϕ which maps c_1 onto γ_1.

Remark. Under the homeomorphism ϕ described in this way, $c_2, \ldots, c_m$ may have lost the property of being piecewise geodesics. We may fix this by applying Lemma A.7 again. In view of other approaches to surface topology, for instance based on piecewise smooth homeomorphisms, we do not want to apply Lemma A.7 but rather improve the construction of ϕ as follows.

Let again D' be the above proper overlap formed by the arcs c' on c_1 and γ' on γ_1, and let x be a point in the interior of D' lying on one of the curves c_ν, $\nu = 2, \ldots, m$. Then x belongs to a subarc c'' of c_ν contained in D' which has both endpoints on γ'. If there exist more than one such arcs, then we find an innermost one which together with a subarc of γ' forms a proper overlap contained in D'. With the above procedure we then first remove this innermost overlap *without affecting the remaining curves*. Then we continue in this way until all overlaps in D' are removed. After this has been done we shall say that D' is *free* (with respect to $c_2, \ldots, c_m$) and then finally remove D'. The homeomorphism ϕ modified in this form now maps $c_2, \ldots, c_m$ again onto piecewise geodesic curves.

Next, let c_1 and γ_1 be either of type (3) with the same endpoint p on ∂S, or of type (2) with base point $p \in \text{int } S$. With the above procedure we remove the transversal intersection points of c_1 and γ_1 also in this case.

We now have to deform c_1 into γ_1. For this we use the universal covering $\pi : \tilde{S} \to S$. Let c' and γ' be homotopic lifts of c_1 and γ_1 in $\tilde{S}$. Then c' and γ' connect lifts p_1 and p_2 of p and π restricted to $c' - \{p_2\}$ and π restricted to $\gamma' - \{p_2\}$ is one-to-one onto c_1 and γ_1 respectively. By hypothesis, c_1 is homotopically non-trivial and therefore $p_1 \neq p_2$.

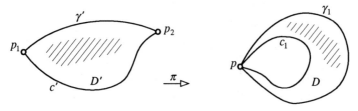

Figure A.12

In $\tilde{S}$, c' and γ' form a proper overlap D' and, as in the earlier cases, any covering transformation $T \neq id$ satisfying $T(D') \cap D' \neq \emptyset$ maps p_1 to p_2 or inversely and satisfies $T(D') \cap D' = \{p_1\}$ or $T(D') \cap D' = \{p_2\}$. Hence, π restricted to $D' - \{p_2\}$ is one-to-one onto a domain $D \subset S$ which is a limiting case of an overlap. If D is not yet free, that is, if there are points in int D lying on one of the curves $c_2, \ldots, c_m$, then there exist overlaps which we remove as in the earlier cases until D is free. Now there exists a disk $\tilde{D}$ in $\tilde{S}$ containing D' with the following properties.

π restricted to $\tilde{D} - \{p_2\}$ is one-to-one.

c' and γ' are topological diameters of $\tilde{D}$.

$\pi(\tilde{D})$ intersects $c_2, \ldots, c_m$ at most at p.

By Theorem A.6, there exists a $\{p_2\}$-1-homeomorphism $\tilde{D} \to \tilde{D}$ which maps c' to γ'. Since π restricted to $\tilde{D}$ is one-to-one, this induces a 1-homeomorphism $\pi(\tilde{D}) \to \pi(\tilde{D})$ which maps c_1 to γ_1 without affecting $c_2, \ldots, c_m$.

Finally, we consider the case where c_1 and γ_1 are of type (1). Here we remove all overlaps so that by Lemma A.10, c_1 and γ_1 are disjoint. By Proposition A.11, c_1 and γ_1 bound an embedded annulus A. As before, we remove all intersections of A with $c_2, \ldots, c_m$ (the hypothesis of Theorem A.3 is such that no c_ν is completely contained in A, $\nu = 2, \ldots, m$). Now there exists a larger annulus A' containing A in its interior, and together with Theorem A.6 we find a 1-homeomorphism of A' which deforms c_1 into a piecewise geodesic which intersects γ_1 in exactly one point p. We may do this in such a way that $c_1(0) = \gamma_1(0) = p$. The pair c_1, γ_1 is now of type (2) and we continue as in the preceding case.

We have achieved the first step of the proof of Theorems A1 - A3.

Now that $c_1 = \gamma_1$, we proceed by induction. We begin with Theorem A.3. The idea is to cut S open along c_1 into a new surface S' and then to apply the induction hypothesis to the remaining $m - 1$ pairs of curves on S'. Since the 1-homeomorphism on S' which will map c_ν to γ_ν, $\nu = 2, \ldots, m$, fixes $\partial S'$ pointwise and since $c_1 = \gamma_1 \subset \partial S'$, this homeomorphism can be interpreted as a 1-homeomorphism of S which fixes c_1 pointwise, and we are done. The problem, however, is to show that on S' the curves still satisfy the hypothesis of Theorem A.3 that is, we have to show that on S', c_ν remains freely homotopic to γ_ν, $\nu = 2, \ldots, m$.

For this we use hyperbolic geometry. By Theorem 1.6.6, c_1 is freely homotopic to a simple closed geodesic, and by what we have proved so far, this homotopy can be carried out by a 1-homeomorphism of S. We may therefore assume that c_1 is a closed geodesic so that S' is again a surface with a hyperbolic structure. (S' may consist of two connected components.) We let c_1' and c_1'' be the two boundary geodesics of S' which on S together form c_1.

Fix $\nu \in \{2, \ldots, m\}$. By Theorem 1.6.6, c_ν and γ_ν are freely homotopic in S' to closed geodesics $\bar{c}_\nu$ and $\bar{\gamma}_\nu$. The corresponding homotopies can also be carried out in S. By the uniqueness of the closed geodesic in a free homotopy class we have $\bar{c}_\nu = \bar{\gamma}_\nu$ on S. By the hypothesis of Theorem A.3, c_ν is not $\pm$homotopic to c_1 (cf. the definition preceding Theorem A.3), and therefore $\bar{c}_\nu$ and $\bar{\gamma}_\nu$ are different (as point sets) from c_1' and c_1''. Hence, we also have $\bar{c}_\nu = \bar{\gamma}_\nu$ on S' that is, c_ν is homotopic to γ_ν on S, $\nu = 2, \ldots, m$, as claimed. Theorem A.3 is now proved.

The same argument goes through in the case of Theorem A.5. For Theorem A.4 we may again assume that c_1 is a smooth closed geodesic. On S' the curves $c_2, \ldots, c_m$ and $\gamma_2, \ldots, \gamma_m$ no longer satisfy the hypotheses of Theorem A.4, but they satisfy the hypothesis of Theorem A.5 and we are done. Theorems A.3 - A.5 are now proved. ◊

An Application to the 3-Holed Sphere

We give an application to the mapping class groups of the annulus and the 3-holed sphere. All homeomorphisms will keep the boundary pointwise fixed. The following terminology is used.

We let $\mathcal{A}$ be the *standard annulus*

$$\mathcal{A} = \{(\rho, \sigma) \mid 0 \leq \rho \leq 1, \sigma \in \mathcal{S}^1\},$$

where $\mathcal{S}^1 = \mathbf{R}^1/[s \mapsto s + 1]$. On $\mathcal{A}$ we have the *standard twists* τ^n, $n \in \mathbf{Z}$, defined by

$$\tau^n(\rho, \sigma) = (\rho, \sigma + n\rho).$$

For the next definition we recall point (iii) of Definition A.1.

A.12 Definition. Let M be a surface with boundary $\partial M \neq \emptyset$. Two homeomorphisms $\varphi_0, \varphi_1 : M \to M$ which coincide on ∂M are called *∂-isotopic* if they are isotopic with an isotopy which fixes ∂M pointwise.

A.13 Proposition. *Let $\phi : \mathcal{A} \to \mathcal{A}$ be a homeomorphism which fixes $\partial \mathcal{A}$ pointwise. Then there exists a unique $n \in \mathbf{Z}$ such that ϕ is ∂-isotopic to τ^n.*

Proof. We look how ϕ acts on the arc $\rho \mapsto a(\rho) := (\rho, 0) \in \mathcal{A}, \rho \in [0, 1]$.

Let $\tilde{\mathcal{A}} = \{(\rho, s) \mid 0 \le \rho \le 1, -\infty < s < +\infty\}$ be the universal covering surface with covering map $\pi(\rho, s) = (\rho, s \bmod(1))$.

The curve $\phi \circ a$ has a lift in $\tilde{\mathcal{A}}$ which connects point $(0, 0)$ with point $(1, n)$ for some $n \in \mathbf{Z}$. Hence, $\tau^{-n} \circ \phi \circ a$ has a lift in $\tilde{\mathcal{A}}$ which connects $(0, 0)$ with $(1, 0)$. By Theorem A.6, this lift is homotopic with fixed endpoints to the curve $\rho \mapsto \tilde{a}(\rho) = (\rho, 0) \in \tilde{\mathcal{A}}, \rho \in [0, 1]$. Hence, $\tau^{-n} \circ \phi \circ a$ is homotopic with fixed endpoints to a. By Theorem A.5, there exists a 1-homeomorphism $J : \mathcal{A} \to \mathcal{A}$ satisfying $J \circ \tau^{-n} \circ \phi \circ a = a$. By Theorem A.2 (applied to the rectangle obtained by cutting $\mathcal{A}$ open along a), $J \circ \tau^{-n} \circ \phi$ is ∂-isotopic to the identity. Hence, $\tau^{-n} \circ \phi$ is ∂-isotopic to the identity. The uniqueness of n is left as an exercise. ◇

Proposition A.13 allows us to define boundary twists. For this we consider again an arbitrary surface S (satisfying Assumption A.0) and let γ be a boundary component of S. We assume for convenience that γ is smooth.

A.14 Definition. A homeomorphism $\varphi : S \to S$ is called a *boundary twist along γ* if there exists an embedded annulus $A_\varphi \subset S$ satisfying $A_\varphi \cap \partial S = \gamma$, together with a homeomorphism $\varphi' : S \to S$ such that the following hold.
(i) φ' is the identity mapping on $S - A_\varphi$,
(ii) φ and φ' are ∂-isotopic.

We insert here the definition of Dehn twists.

A.15 Definition. Let c be a homotopically non-trivial simple closed curve in the interior of S which is not homotopic to a boundary component of S. A homeomorphism $\Delta : S \to S$ is called a *Dehn twist* along c if there exists an embedded annulus $A \subset S$ containing c together with a homeomorphism $\Delta' : S \to S$ such that the following hold.

(i) Δ' is the identity mapping on $S - A$,
(ii) Δ and Δ' are ∂-isotopic.

A.16 Proposition. *Let A be a fixed embedded annulus in S satisfying $A \cap \partial S = \gamma$, and let $h : \mathcal{A} \to A$ be a homeomorphism. Define*
$$\vartheta^n = \begin{cases} h \circ \tau^n \circ h^{-1} & \text{on } A \\ id & \text{on } S - A. \end{cases}$$
Then any boundary twist along γ is ∂-isotopic to a unique ϑ^n.

Proof. Let φ' and A_φ be as in Definition A.14 and let $j : \mathcal{A} \to A_\varphi$ be a homeomorphism. We may assume that $j(\{0\} \times \mathbf{S}^1) = h(\{0\} \times \mathbf{S}^1) = \gamma$. There exists a positive $\varepsilon < 1$ so small that $j([0, \varepsilon] \times \mathbf{S}^1) \subset A$. There exists a ∂-isotopy changing $\phi := j^{-1} \circ \varphi \circ j$ into a homeomorphism of $\mathcal{A}$ which is the identity mapping on $[\varepsilon, 1] \times \mathbf{S}^1$. This is obtained as follows. First we define
$$f_s(\rho, \sigma) = ((1 - s(1 - \varepsilon))\rho, \sigma), \quad 0 \leq s \leq 1.$$
The desired isotopy is then given by
$$\phi_s(\rho, \sigma) = \begin{cases} f_s \circ \phi \circ f_s^{-1}(\rho, \sigma) & \text{if } 0 \leq \rho \leq 1 - s(1 - \varepsilon) \\ (\rho, \sigma) & \text{if } 1 - s(1 - \varepsilon) \leq \rho \leq 1. \end{cases}$$
This shows that φ is ∂-isotopic to a homeomorphism which is the identity mapping outside $j([0, \varepsilon] \times \mathbf{S}^1)$ and in particular outside A. The proposition follows now from Proposition A.13. ◇

For our next application we let Y be a hyperbolic surface of signature $(0, 3)$ with smooth boundary geodesics γ_1, γ_2 and γ_3 that is, a pair of pants as in Chapter 3. For $k = 1, 2, 3$ we fix an annulus A_k together with a homeomorphism $j_k : \mathcal{A} \to A_k$ such that $A_k \cap A_\kappa = \emptyset$ for $k \neq \kappa$ and $A_k \cap Y = \gamma_k$, $k = 1, 2, 3$. As A_k we may, for instance, take the half-collars as in Proposition 3.1.8. For $k = 1, 2, 3$, we let ϑ_k denote the boundary twists
$$\vartheta_k = j_k \circ \tau \circ j_k^{-1}.$$

A.17 Proposition. *Any homeomorphism $\phi : Y \to Y$ fixing ∂Y pointwise is ∂-isotopic to a product $\vartheta_1^{n_1} \circ \vartheta_2^{n_2} \circ \vartheta_3^{n_3}$ with uniquely determined n_1, n_2, n_3.*

Proof. We let $\rho \mapsto a(\rho) \in Y$, $\rho \in [0, 1]$, with $a(0) \in \gamma_2$ and $a(1) \in \gamma_3$ be a simple arc, for instance the common simple perpendicular. By Proposition A.18 below, $\phi \circ a$ is homotopic to a with endpoints gliding on γ_2 and γ_3.

Hence, there are boundary twists $\vartheta_2^{-n_2}$ and $\vartheta_3^{-n_3}$ such that $\vartheta_2^{-n_2} \circ \vartheta_3^{-n_3} \circ \phi \circ a$ is homotopic to a with fixed endpoints. By Theorem A.5, there exists a 1-homeomorphism $J : Y \to Y$ such that $J \circ \vartheta_2^{-n_2} \circ \vartheta_3^{-n_3} \circ \phi \circ a = a$. Cutting Y open along a we get an annulus. By Proposition A.17, $J \circ \vartheta_2^{-n_2} \circ \vartheta_3^{-n_3} \circ \phi$ is ∂-isotopic to $\vartheta_1^{n_1}$ for some $n_1 \in \mathbf{Z}$. Hence, ϕ is ∂-isotopic to $\vartheta_2^{n_2} \circ \vartheta_3^{n_3} \circ \vartheta_1^{n_1} = \vartheta_1^{n_1} \circ \vartheta_2^{n_2} \circ \vartheta_3^{n_3}$. We leave the uniqueness as an exercise. ◇

A.18 Proposition. *Let $a, b : [0, 1] \to Y$ with $a(0), b(0) \in \gamma_2$ and $a(1), b(1) \in \gamma_3$ be simple arcs. Then a and b are homotopic with endpoints gliding on γ_2 and γ_3.*

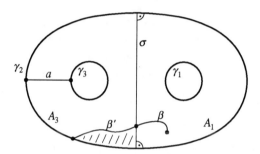

Figure A.13

Proof. We may assume that Y is a standard Y-piece as in Section 3.1 and that a is the the common perpendicular a_1 from γ_2 to γ_3. We let σ denote the perpendicular from γ_2 to γ_2 which separates γ_1 from γ_3 as in Fig. A.13. Let β be any simple arc in the homotopy class of b with endpoints gliding on γ_2 and γ_3. By Lemma A.8, there exists a 1-homeomorphism $Y \to Y$ which transforms β into a curve which intersects σ only finitely many times and only with transversal intersections. Let m be the minimal possible number of such intersection points in the homotopy class of b (with gliding endpoints). Then $m = 0$. In fact, take β with m intersection points and assume $m \geq 1$. Then there exists an arc β' from $\beta(0)$ to the first intersection point p with σ. This arc runs in one of the annuli A_1 and A_3 separated by σ and decomposes this annulus into an annulus and a disk D. If some other parts of β are contained in the interior of D, then there are overlaps with σ which we may remove as in the proof of Theorems A.3 - A.5. If no other parts of β are contained in the interior of D, then we can remove D itself. In both cases we end up with a curve β^* which is again in the homotopy class of b. Now β^* intersects σ in less than m points, a contradiction. Hence, $m = 0$.

We may now assume that b is contained in the annulus A_3 (Fig. A.13) and that b has the same endpoints as a. There exists a boundary twist $\vartheta_3^{n_3}$ such that $b' := \vartheta_3^{n_3} \circ b$ is homotopic to a with fixed endpoints. Finally, b' is homotopic to b with endpoint gliding on γ_3, and we are done. ◊

In Chapter 6 we used Proposition A.17 to show that if $\phi : S \to S$ is a homeomorphism of a compact surface S of genus $g \geq 2$ which fixes each γ_ν in a pants decomposition $\gamma_1, \ldots, \gamma_{3g-3}$ pointwise and without interchanging the two sides of γ_ν, then ϕ is isotopic to a product of Dehn twists along $\gamma_1, \ldots, \gamma_{3g-3}$. We mention here without proof that any orientation-preserving homeomorphism of S is isotopic to a product of Dehn twists. Hence, the Dehn twists are generators of the orientation-preserving mapping class group. This has first been shown by Dehn [1]. Later, Lickorish [1] has given a generating set of Dehn twists along only $3g - 1$ curves. A simplified proof may be found in Birman [1]. Humphries [1] has shown that the minimal possible number of Dehn twist generators is $2g + 1$ and that the Dehn twists about the following simple closed geodesics $a_1, \ldots, a_g, c_1, \ldots, c_{g-1}, m_1, m_2$ generate the orientation-preserving mapping class group.

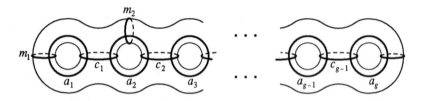

Figure A.14

Length-Decreasing Homotopies

To prove the existence of geodesics in a given homotopy class in Chapter 1, we used the Arzelà-Ascoli theorem. In view of Lemma A.8 we describe here a procedure proposed by K.-D. Semmler which is entirely based on geodesic polygons. An important property of the procedure is that if the original polygon is simple, then the polygon remains simple at all steps. In this section, *polygon* means geodesic polygon curve.

For the sake of completeness we first recall the theorem of Arzelà-Ascoli. We let X and Z be metric spaces with the following properties: X has a countable dense subset $D = \{d_1, d_2, \ldots\}$ and Z is compact (usually a compact subset of some larger space). In both spaces we let dist denote the distance function. A mapping $\gamma : X \to Z$ is *L-Lipschitzian* if

$$\text{dist}(\gamma(x), \gamma(y)) \leq L \, \text{dist}(x, y)$$

for any $x, y \in X$. The Arzelà-Ascoli theorem is this.

A.19 Theorem. *Let $\{\gamma_n\}_{n=1}^{\infty}$ be a sequence of L-Lipschitzian mappings $\gamma_n : X \to Z$. Then there exists a subsequence of $\{\gamma_n\}_{n=1}^{\infty}$ which converges uniformly on compact sets to an L-Lipschitzian mapping $\gamma : X \to Z$.*

Proof. We outline the well-known argument. Since Z is compact, there exists a subsequence $\{\gamma_{1n}\}_{n=1}^{\infty}$ of $\{\gamma_n\}_{n=1}^{\infty}$ such that $\{\gamma_{1n}(d_1)\}_{n=1}^{\infty}$ converges. Then there exists a subsequence $\{\gamma_{2n}\}_{n=1}^{\infty}$ of $\{\gamma_{1n}\}_{n=1}^{\infty}$ such that $\{\gamma_{2n}(d_2)\}_{n=1}^{\infty}$ converges, then a subsequence $\{\gamma_{3n}\}_{n=1}^{\infty}$ of $\{\gamma_{2n}\}_{n=1}^{\infty}$ such that $\{\gamma_{3n}(d_3)\}_{n=1}^{\infty}$ converges, and so on. The diagonal sequence $\{\gamma_{nn}\}_{n=1}^{\infty}$ has the property that $\{\gamma_{nn}(d_k)\}_{n=1}^{\infty}$ converges for any $d_k \in D$. Since D is dense in X and since all mappings are L-Lipschitzian, $\{\gamma_{nn}(x)\}_{n=1}^{\infty}$ converges for any $x \in X$. Here we use the triangle inequality. The triangle inequality also shows that the limiting mapping is L-Lipschitzian and that the convergence is uniform on any compact subset of X. ◊

If the mappings γ_n are smooth curves on a compact subset Z of a Riemannian manifold M and if all curves are defined on the same interval (or on S^1) and with uniformly bounded speed, then the limiting mapping γ is a rectifiable curve. If the curves γ_n belong to a given homotopy class H, and if the boundary of M is sufficiently regular (for instance the piecewise geodesic boundary of a surface), then γ belongs to H. If in addition the lengths $\ell(\gamma_n)$ converge to the infimum $\lambda = \inf\{\ell(c) \mid c \in H\}$ and if $\lambda > 0$, then any arc of γ in the interior of M is a geodesic arc.

We now restate this in terms of polygons. For simplicity we restrict ourselves to closed curves and assume that M is a compact Riemannian manifold without boundary.

There exists $\delta > 0$ such that for any $p \in M$, $U = \{x \in M \mid \text{dist}(x, p) < \delta\}$ is diffeomorphic to an open Euclidean ball and has the property that all x, $y \in U$ can be joined in U by a unique length minimizing geodesic arc (cf. e.g. Cheeger-Ebin[1]). Distance sets with these properties are called *normal neighborhoods*.

We consider the free homotopy class of a piecewise smooth curve $c : S^1 \to M$. The curve can be subdivided into small arcs such that each arc is contained in a normal neighborhood. Each small arc may be replaced by a minimizing geodesic arc in the normal neighborhood with the same endpoints. Hence, c is homotopic to a closed geodesic polygon. To simplify the notation, we denote this polygon again by c. The geodesic arcs of c are called the *sides* of c.

Let now c_1 be the geodesic polygon which goes from the midpoint of the first side of c to the midpoint of the second side, from there to the midpoint of the third side and so on until it returns to the midpoint of the first side. In this way c has been changed, within its homotopy class, into a shorter geodesic polygon c_1 with the same number of sides. One can prove that if this procedure is repeated indefinitely, then a sequence $c_1, c_2, \ldots$ is obtained which converges to a geodesic γ freely homotopic to c. This is the usual length-decreasing homotopy.

Let now M be two-dimensional. Then the difference between simple and non-simple curves is essential, and we want to improve the above procedure in such a way that simple curves remain simple. If c is piecewise smooth and simple, then the above subdivision of c can be made fine enough so that the approximating geodesic polygon is again simple. We may therefore assume that the original curve $c : S^1 \to M$ is a geodesic polygon.

In the following, $\delta > 0$ is a constant such that for all $p \in M$ the distance set $U = \{x \in M \mid \text{dist}(x, p) < 2\delta\}$ is a normal neighborhood. Semmler's algorithm is this.

A.20 Algorithm.

- Let $c^0 = c$ and choose $p^0 = q^0 \in c^0$.
- Assume that c^k and p^k, $q^k \in c^k$ have been defined, where the segment $p^k q^k =: \sigma^k$ is an oriented geodesic arc on c^k of length $\leq \delta$.

If c^k lies in a normal neighborhood of radius 2δ about p^k, then replace c^k by the point curve p^k and *stop*.

- Otherwise let p^{k+1} be the midpoint of σ^k, and define $q^{k+1} \in c^k$ by the condition that the oriented arc on c^k from p^{k+1} via q^k to q^{k+1} has length δ. Replace this arc by the minimizing geodesic arc σ^{k+1} from p^{k+1} to q^{k+1}. Then "clean up" as explained below, call the new curve c^{k+1} and continue.

The procedure is illustrated in Fig. A.15. The dashed line is the new geodesic arc $\sigma = \sigma^{k+1}$. The *"cleaning"* is optional: If the original polygon c is simple

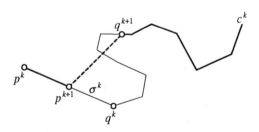

Figure A.15

and we want all c^k to be simple, we have to remove the intersections with σ which may occur. Since c^k is piecewise geodesic, there are at most finitely many intersection points, and if c^k is simple, all intersections are caused by arcs in a normal neighborhood U of σ and having both endpoints on σ, as for example the arcs ζ_1 and ζ_2 in Fig. A.16. This follows from the Jordan curve theorem and from the fact that c^k is a simple *closed* curve which has points outside U.

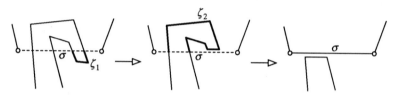

Figure A.16

The "cleaning" is illustrated in Fig. A.16: each arc ζ with the two endpoints on σ can be replaced by a shorter arc ζ' which lies *on* σ. After that, an infinitesimal homotopy removes all self-intersections. Clearly this can be carried out so that c^{k+1} is again piecewise geodesic and such that still $\ell(c^{k+1}) \leq \ell(c^k)$. The curves c^k and c^{k+1} are freely homotopic.

If the procedure stops, then c is homotopic to a point. Let us therefore assume that the procedure does not stop. Then all c^k have length bounded below by 4δ, and there exists $\ell \geq 4\delta$ such that $\ell(c^k) \to \ell$ as $k \to \infty$. From this follows that $\ell(\sigma^k) \to \delta$, and that the angle between σ^k and σ^{k+1} converges to π. If we now extract a subsequence $\{k_n\}_{n=1}^{\infty}$ such that $p^{n_k} \to p$ and $q^{n_k} \to q$ for suitable $p, q \in M$, then c^{n_k} converges to the geodesic γ through p and q, and it follows that γ is a closed geodesic which is freely homotopic to c. Moreover, if c is simple and the cleaning option is on, then γ is also simple. (Observe that no diagonal argument has been used.)

A.21 Remarks. (i) Unless M has negative curvature, the above algorithm does not necessarily produce the shortest closed geodesic in the given homotopy class.

(ii) It seems possible, although we do not know a proof, that the cleaning option does not come into action if the original polygon c has sufficiently small sides.

(iii) The example in Fig. 1.5.2 shows that the cleaning option cannot be defined for homotopy classes with fixed endpoints.

(iv) If M is a surface as above but with piecewise smooth boundary, and if $A, B \subset \partial M$ are closed and connected, then Algorithm A.20 can be modified (including the cleaning option) for the homotopy classes of curves with endpoints gliding freely on A and B. The modifications are straightforward and we omit the details. If c is a curve with endpoints on A and B, the algorithm produces a geodesic γ from A to B. If A is a smooth closed boundary geodesic or a smooth side which meets its adjacent sides under an angle less or equal $\pi/2$, then γ is orthogonal to A, and the same holds for B. If c is simple then γ is simple.

(v) The reader will convince himself that if all curves c^k lie in a given homotopy class, then the limit γ lies also in this class.

Bibliography

Abikoff, W.
[1] *The real analytic theory of Teichmüller space*, Springer Lecture Notes in Math. **820** (1980).

Accola, R. D. M.
[1] *On the number of automorphisms of a closed Riemann surface*, Trans. Amer. Math. Soc. **131** (1968), 398-408.

Anker, J.-P.
[1] *Le noyau de la chaleur sur les espaces symétriques $U(p, q)/U(p) \times U(q)$*, in: Harmonic analysis, Proc. Int. Symp. Luxembourg, Luxembourg 1987, Springer Lecture Notes in Math. **1359** (1988), 60-82.

Apostol, T. M.
[1] *Introduction to analytic number theory*, Springer-Verlag, Berlin–Heidelberg–New York (1976).

Baer, R.
[1] *Isotopie von Kurven auf orientierbaren, geschlossenen Flächen und ihr Zusammenhang mit der topologischen Deformation der Flächen*, J. Reine Angew. Math. **159** (1928), 101-111.

Ballmann, W.
[1] *Doppelpunktfreie geschlossene Geodätische auf kompakten Flächen*, Math. Z. **161** (1978), 41-46.

Ballmann, W., Gromov, M. and Schroeder, V.
[1] *Manifolds of nonpositive curvature*, Birkhäuser Verlag, Basel–Boston–Stuttgart (1985).

Basmajian, A.
[1] *The stable neighborhood theorem*, to appear in Proc. Amer. Math. Soc. (1992).

Baumgärtel, H.
[1] *Analytic perturbation theory for matrices and operators*, Birkhäuser Verlag, Basel–Boston–Stuttgart (1985).

Beardon, A. F.
[1] *The geometry of discrete groups*, Springer-Verlag, Berlin–Heidelberg–New York (1983).
[2] *A primer on Riemann surfaces*, Cambridge University Press, Cambridge–London–New York (1984).

Benedetti, R. and Dedò, M.
[1] *Una introduzione alla geometria e topologia delli varietà di dimensione tre*, Pitagora Editrice, Bologna (1984).

Bérard, P. and Berger, M.
[1] *Le spectre d'une variété riemannienne en 1982*, in: Spectra of riemannian manifolds, Proc. France-Japan Siminar, Kyoto (M. Berger et al. eds.), Kaigai Publications, Tokio (1983), 139-194, (reprinted in Bérard [1]).

Bérard, P. H.
[1] *Spectral geometry: direct and inverse problems*, Springer Lecture Notes in Math. **1207** (1986).
[2] *Variétés riemanniennes isospectrales non isométriques*, Sém. Bourbaki (1989), exposé n⁰ 705, Astérisque **177-178** (1989), 127-154.
[3] *Transplantation et isospectralité* I, Séminaire de théorie spectrale et géométrie n⁰ 7, Institut Fourier, Grenoble (1990-1991), 153-175; II, ibid. 177-188.

Bérard-Bergery, L.
[1] *Laplacien et géodésiques fermées sur les formes d'espace hyperbolique compactes*, in: Sém. Bourbaki (1971-1972), exposé n⁰ 406, Springer Lecture Notes in Math. **317** (1973).

Berger, M.
[1] *A l'ombre de Loewner*, Ann. Sci. Ecole Norm. Sup. (4) **5** (1972), 241-260.
[2] *Geometry of the spectrum* I, in: Differential geometry, Proc. Sympos. Pure Math. (Amer. Math. Soc.) **27** (1975), 129-152.

Berger, M., Gauduchon, P. and Mazet, E.
[1] *Le spectre d'une variété riemannienne*, Springer Lecture Notes in Math. **194** (1971).

Berry, J. P.
[1] *Tores isospectraux en dimension* 3, C. R. Acad. Sci. Paris Sér. I Math. **292** (1981), 163-166.

Bers, L.
[1] *Quasiconformal mappings and Teichmüller's theorem*, in: Analytic functions (R. Nevanlinna et al., eds.), Princeton Univ. Press, Princeton, NJ (1960), 89-119.
[2] *A remark on Mumford's compactness theorem*, Israel J. Math. **12** (1972), 400-407.
[3] *Spaces of degenerating Riemann surfaces*, in: Discontinuous groups and Riemann surfaces (L. Greenberg, ed.), Annals of Mathematics Studies, vol. **79**, Princeton Univ. Press, Princeton, NJ (1974), 43-55.
[4] *An inequality for Riemann surfaces*, in: Differential geometry and complex analysis (I. Chavel and H. M. Farkas, eds.), Springer-Verlag, Berlin–Heidelberg–New York (1985).

Besson, G.
[1] *Sur la multiplicité de la première valeur propre des variétés riemanniennes*, Ann. Inst. Fourier (Grenoble) **30** (1980), 109-128.

Birman, J.
[1] *The algebraic structure of surface mapping class groups*, in: Discrete groups and automorphic functions (W. J. Harvey, ed.), Academic Press, London–New York–San Francisco (1977), 163-198.

Blatter, Chr.
[1] *Über Extremallängen auf geschlossenen Flächen*, Comment. Math. Helv. **35** (1961), 153-168.

Bollobás, B.
[1] *Graph theory, an introductory course*, Springer-Verlag, Berlin–Heidelberg–New York (1979).
[2] *A probabilistic proof of an asymptotic formula for the number of labelled regular graphs*, Europ. J. Combinatorics **1** (1980), 311-316.

Brooks, R.
[1] *Constructing isospectral manifolds*, Amer. Math. Monthly **95** (1988), 823-839.

Brooks, R. and Matelski, J. P.
[1] *Collars in Kleinian groups*, Duke Math. J. **49** (1982), 163-182.

Brooks, R. and Tse, R.
[1] *Isospectral surfaces of small genus*, Nagoya Math. J. **107** (1987), 13-24.

Bruhat, F. and Cartan, H.
[1] *Sur la structure des sous-ensembles analytiques réels*, C. R. Acad. Sci. Paris Sér. I Math. **244** (1957), 988-996.

Burger, M.
[1] *Asymptotics of small eigenvalues of Riemann surfaces*, Bull. Amer Math. Soc. **18** (1988), 39-40.
[2] *Small eigenvalues of Riemann surfaces and graphs*, Math. Z. **205** (1990), 395-420.

Burger, M. and Schroeder, V.
[1] *Volume, diameter and the first eigenvalue of locally symmetric spaces of rank one*, J. Differential Geom. **26** (1987), 273-284.

Buser, P.
[1] *Riemannsche Flächen mit Eigenwerten in* $(0, \frac{1}{4})$, Comment. Math. Helv. **52** (1977), 25-34.
[2] *Cubic graphs and the first eigenvalue of a Riemann surface*, Math. Z. **162** (1978), 87-99.
[3] *Riemannsche Flächen mit grosser Kragenweite*, Comment. Math. Helv. **53** (1978), 395-407.
[4] *The collar theorem and examples*, Manuscripta Math. **25** (1978), 349-357.
[5] *Dichtepunkte im Spektrum Riemannscher Flächen*, Comment. Math. Helv. **54** (1979), 431-439.
[6] *On Cheeger's inequality* $\lambda_1 \geq h^2/4$, in: Proc. Sympos. Pure Math. (Amer. Math. Soc.) **36** (1980), 29-77.
[7] *Riemannsche Flächen und Längenspektrum vom trigonometrischen Standpunkt aus*, Habilitationsschrift, Bonn (1980).
[8] *Isospectral Riemann surfaces*, Ann. Inst. Fourier (Grenoble) **36** (1986), 167-192.
[9] *Cayley graphs and planar isospectral domains*, in: Geometry and analysis on manifolds (T. Sunada, ed.), Proc. Taniguchi Symp., Springer Lecture Notes in Math. **1339** (1988), 64-77.

Buser, P. and Courtois, G.
[1] *Finite parts of the spectrum of a Riemann surface*, Math. Ann. **287** (1990), 523-530.

Buser, P. and Sarnak, P.
[1] *On the period matrix of a Riemann surface of large genus*, preprint, Princeton (1991).

Buser, P. and Semmler, K.-D.
[1] *The geometry and spectrum of the one holed torus*, Comment. Math. Helv. **63** (1988), 259-274.

Buser, P. and Seppälä, M.
[1] *Symmetric pants decompositions of Riemann surfaces*, to appear in Duke Math. J. (1991).

Chavel, I.
[1] *Eigenvalues in riemannian geometry*, Academic Press, Orlando-San Diego-New York (1984).

Chavel, I. and Feldman, E. A.
[1] *Cylinders on surfaces*, Comment. Math. Helv. **53** (1978), 439-447.

Cheeger, J.
[1] *A lower bound for the smallest eigenvalue of the Laplacian*, in: Problems in analysis, a symposium in honor of S. Bochner, Princeton Univ. Press, Princeton, NJ (1970), 195-199.

Cheeger, J. and Ebin, D. G.
[1] *Comparison theorems in riemannian geometry*, North-Holland Publishing Comp., Amsterdam–Oxford (1975).

Colbois, B.
[1] *Petites valeurs propres du laplacien sur une surface de Riemann compacte et graphes*, C. R. Acad. Sci. Paris Sér. I Math. **301** (1985), 927-930.
[2] *Sur la multiplicité de la première valeur propre non nulle du Laplacien des surfaces à courbure -1*, doctoral thesis, Lausanne (1987).

Colbois, B. and Colin de Verdière, Y.
[1] *Sur la multiplicité de la première valeur propre d'une surface de Riemann à courbure constante*, Comment. Math. Helv. **63** (1988), 194-208.

Colbois, B. and Courtois, G.
[1] *Les valeurs propres inférieures à 1/4 des surfaces de Riemann de petit rayon d'injectivité*, Comment. Math. Helv. **64** (1989), 349-362.
[2] *Convergence de variétés et convergence du spectre du Laplacien*, Ann. Sci. Ecole Norm. Sup. (4) **24** (1991), 507-518.

Colin de Verdière, Y.
[1] *Spectre du Laplacien et longueurs des géodésiques périodiques* I, Compositio Math. **27** (1973), 83-106; II, Compositio Math. **27** (1973), 159-184; see also: Sém. Goulaouic-Schwartz (1973-1974), exposé n⁰ 14.
[2] *Pseudo-Laplaciens* I, Ann. Inst. Fourier (Grenoble) **32** (1982), 275-286; II, Ann. Inst. Fourier (Grenoble) **33** (1983), 87-113.

Conder, M.
[1] *Hurwitz groups: a brief survey*, Bull. Amer Math. Soc. (N.S.) **23** (1990), 359-370.

Conway, J. H., Curtis, R. T., Norton, S. P., Parker, R. A. and Wilson, R. A.
[1] *Atlas of finite groups*, Clarendon Press, Oxford (1985).

Coxeter, H. S. M.
[1] *Non-euclidean geometry*, University of Toronto Press (1968).

Davies, E. B.
[1] *Heat kernels and spectral theory*, Cambridge University Press, Cambridge–New York–New Rochelle (1989).

Davies, E. B. and Mandouvalos, N.
[1] *Heat kernel bounds on hyperbolic space and Kleinian groups*, Proc. London Math. Soc. (3) **57** (1988), 182-208.

De Rham, G.
[1] *Sur les polygones générateurs de groupes fuchsiens*, Enseignement Math. **17** (1971), 49-61.

De Turck, D. M. and Gordon, C. S.
[1] *Isospectral metrics and finite Riemannian coverings*, in: The legacy of Sonya Kovalevskaya, Proc. Symp., Radcliffe Coll. 1985, Contemp. Math. **64** (1987), 79-92.
[2] *Isospectral riemannian metrics and potentials*, Bull. Amer Math. Soc. (N.S.) **17** (1987), 137-139.

de la Vallée Poussin, Ch.
[1] *Recherches analytiques sur la théorie des nombres premier*, Ann. Soc. Sci. Bruxelles 20_2 (1896), 183-256 and 281-297.

Dehn, M.
[1] *Die Gruppe der Abbildungsklassen*, Acta Math. **69** (1938), 135-206.

Delsarte, J.
[1] *Sur le gitter fuchsien*, C. R. Acad. Sci. Paris Sér. I Math. **214** (1942), 147-149.
[2] *Le gitter fuchsien*, Oeuvres de Jean Delsarte, Vol. II, Editions du Centre National de la Recherche Scientifique, Paris (1971), 829-845.

Dodziuk, J.
[1] *Eigenvalues of the Laplacian and the heat equation*, Amer. Math. Monthly **88** (1981), 686-695.

Dodziuk, J. and Randol, B.
[1] *Lower bounds for λ_1 on a finite volume hyperbolic manifold*, J. Differential Geom. **24** (1986), 133-139.

Dodziuk, J., Pignataro, T., Randol, B. and Sullivan, D.
[1] *Estimating small eigenvalues of Riemann surfaces*, Contemp. Math. (Amer. Math. Soc.) **64** (1987), 93-121.

Dostal, M. and Tindell, R.
[1] *The Jordan curve theorem revisited*, Jahresber. Deutsch. Math.-Verein. **80** (1978), 111-128.

Duistermaat, J. J. and Guillemin, V. W.
[1] *The spectrum of positive elliptic operators and periodic bicharacteristics*, Invent. Math. **29** (1975), 39-79.
[2] *The spectrum of positive elliptic operators and periodic geodesics*, in: Differential geometry, Proc. Sympos. Pure Math. (Amer. Math. Soc.) **27**, part. II (1975), 205-209.

Earle, C. J. and Eells, J.
[1] *A fibre bundle description of Teichmüller theory*, J. Differential Geom. **3** (1969), 19-43.

Edwards, H. M.
[1] *Riemann's zeta function*, Academic Press New York–London (1974).

Ejiri, N.
[1] *A construction of non-flat, compact irreductible riemannian manifolds which are isospectral but not isometric*, Math. Z. **168** (1979), 207-212.

Elstrodt, J.
[1] *Die Resolvente zum Eigenwertproblem der automorphen Formen in der hyperbolischen Ebene*, I, Math. Ann. **203** (1973), 295-330; II, Math. Z. **132** (1973), 99-134; III, Math. Ann. **208** (1974), 99-132.
[2] *Die Selbergsche Spurformel für kompakte Riemannsche Flächen*, Jahresber. Deutsch. Math.-Verein. **83** (1981), 45-77.

Epstein, D. B. A.
[1] *Curves on 2-manifolds and isotopies*, Acta Math. **115** (1966), 83-107.

Epstein, D. B. A. and Zieschang, H.
[1] *Curves on 2-manifolds: a counter example*, Acta Math. **115** (1966), 109-110.

Euler, L.
[1] *Variae observationes circa series infinitas*, Commentarii Academiae Scientiarum Imperialis Petropolitanae **9** (1737), 160-188; reprinted in: Opera Omnia (1), **14**, 216-244.

Farkas, H. M. and Kra, I.
[1] *Riemann surfaces*, Springer-Verlag, Berlin–Heidelberg–New York (1980).

Fathi, A., Laudenbach, F. and Poénaru, V. (eds.).
[1] *Travaux de Thurston sur les surfaces*, Séminaire Orsay, Astérisque **66-67** (1979).

Fejes Tóth, L.
[1] *Kreisausfüllungen der hyperbolischen Ebene*, Acta Math. Acad. Sci. Hungar. **4** (1953), 103-110.
[2] *Lagerungen in der Ebene, auf der Kugel und im Raum*, Springer-Verlag, Berlin–

Heidelberg–New York (1972).

Fenchel, W.
[1] *Elementary geometry in hyperbolic space*, Walter de Gruyter, Berlin–New York (1989).

Fenchel, W. and Nielsen, J.
[1] *Discontinuous groups of non-euclidean motions*, unpublished manuscript (1948).
[2] *On discontinuous groups of isometric transformations of the non-euclidean plane*, Studies and Essays Presented to R. Courant, Interscience, New York (1948).

Fischer, A. E. and Tromba, A. J.
[1] *On a purely "riemannian" proof of the structure and dimension of the unramified moduli space of a compact Riemann surface*, Math. Ann. **267** (1984), 311-345.
[2] *A new proof that Teichmüller space is a cell*, Trans. Amer. Math. Soc. **303** (1987), 257-262.

Freedman, M., Hass, J. and Scott, P.
[1] *Closed geodesics on surfaces*, Bull. London Math. Soc. **14** (1982), 385-391.

Fricke, R. and Klein, F.
[1] *Vorlesungen über die Theorie der automorphen Funktionen*, Vol. I: die gruppentheoretischen Grundlagen, Vol. II: Die funktionentheoretischen Ausführungen und die Anwendungen, R. G. Teubner, Leipzig (1987/1926); Reprint: Johnson Reprint Corp. New York; B. G. Teubner Verlagsgesellschaft, Stuttgart(1965).

Fricker, F.
[1] *Ein Gitterpunktproblem im dreidimensionalen hyperbolischen Raum*, Comment. Math. Helv. **43** (1968), 402-416.
[2] *Eine Beziehung zwischen der hyperbolischen Geometrie und der Zahlentheorie*, Math. Ann. **191** (1971), 293-312.

Gangolli, R.
[1] *The length spectra of some compact manifolds of negative curvature*, J. Differential Geom. **12** (1977), 403-424.

Gardiner, F. P.
[1] *Teichmüller theory and quadratic differentials*, John Wiley & Sons, New York–Chichester–Brisbane (1987).

Gassmann, F.
[1] *Bemerkungen zur vorstehenden Arbeit von Hurwitz*, Math. Z. **25** (1926), 665-675.

Gel'fand, I. M.
[1] *Automorphic functions and the theory of representations*, in: Proc. Internat. Congr. Math. Stockholm, 1962, Almqvist & Wiksell, Uppsala (1963), 74-85.

Gerst, I.
[1] *On the theory of n-th power residues and a conjecture of Kronecker*, Acta Arith. **17** (1970), 121-139.

Gilman, J.
[1] *A geometric approach to Jørgensen's inequality*, Bull. Amer Math. Soc. **16** (1987), 91-92.

Gordon, C. S.
[1] *Riemannian manifolds isospectral on functions but not on 1-forms*, J. Differential Geom. **24** (1986), 79-96.
[2] *The Laplace spectrum versus the length spectra of riemannian manifolds*, in: Nonlinear problems in geometry (D. M. De Turck, ed.), Contemp. Math. (Amer. Math. Soc.) **51** (1986), 63-80.

Gordon, C. S. and Wilson, E. N.
[1] *Isospectral deformations of compact solvmanifolds*, J. Differential Geom. **19** (1984), 241-256.

Grauert, H. and Remmert, R.
[1] *Analytische Stellenalgebren*, Springer-Verlag, Berlin–Heidelberg–New York (1971).

Grauert, K. and Fritzsche, K.
[1] *Several complex variables*, Springer-Verlag, Berlin–Heidelberg–New York (1976).

Greenberg, L.
[1] *Maximal Fuchsian groups*, Bull. Amer Math. Soc. **69** (1963), 569-573.
[2] *Maximal groups and signatures*, in: Discontinuous groups and Riemann surfaces (L. Greenberg, ed.), Annals of Mathematics Studies, vol. **79**, Princeton Univ. Press, Princeton, NJ (1974), 207-226.
[3] *Finiteness theorems for Fuchsian and Kleinian groups*, in: Discrete groups and automorphic functions (W. J. Harvey, ed.), Academic Press, London–New York–San Francisco (1977), 199-257.

Gromov, M.
[1] *Filling riemannian manifolds*, J. Differential Geom. **18** (1983), 1-147.

Gromov, M., Lafontaine, J. and Pansu, P.
[1] *Structures métriques pour les variétés riemanniennes*, Editions CEDIC, Fernand Nathan, Paris (1981).

Guillemin, V. W. and Kazhdan, D.
[1] *Some inverse spectral results for negatively curved 2-manifolds*, Topology **19** (1980), 301-312.

Guillopé, L.
[1] *Sur la distribution des longueurs des géodésiques fermées d'une surface compacte à bord totalement géodésique*, Duke Math. J. **53** (1986), 827-848.

Günther, P.
[1] *Poisson formula and estimations for the length spectrum of compact hyperbolic space forms*, Studia Sci. Math. Hungar. **14** (1979), 105-123.
[2] *Problèmes de réseaux dans les espaces hyperboliques*, C. R. Acad. Sci. Paris Sér. I Math. **288** (1979), 49-52.
[3] *Gitterpunktprobleme in symmetrischen Riemannschen Räumen vom Rang 1*, Math. Nachr. **94** (1980), 5-27.

Guralnick, R. M.
[1] *Subgroups inducing the same permutation representation*, J. Algebra **81** (1983), 312-319.

Haas, A.
[1] *Length spectra as moduli for hyperbolic surfaces*, Duke Math. J. **5** (1985), 922-935.

Hadamard, J.
[1] *Sur la distribution des zéros de la fonction ζ(s) et ses conséquences arithmétiques*, Bull. Soc. Math. France **24** (1896), 199-220.

Halpern, N. A.
[1] *A proof of the collar lemma*, Bull. London Math. Soc. **13** (1981), 141-144.

Hebda, J. J.
[1] *Some lower bounds for the area of surfaces*, Invent. Math. **65** (1982), 485-490.

Hejhal, D. A.
[1] *The Selberg trace formula and the Riemann zeta function*, Duke Math. J. **43** (1976), 441-482.
[2] *The Selberg trace formula for* PSL(2, **R**), Vol. I, Springer Lecture Notes in Math. **548** (1976); Vol II, ibid. **1001** (1983).

Hempel, J.
[1] *Traces, lengths, and simplicity of loops on surfaces*, Topology and its Applications **18** (1984), 153-161.

Hersch, J.
  [1] *Contribution à la théorie des fonctions pseudo-analytiques*, Comment. Math. Helv. **30** (1955), 1-19.

Hirsch, M. W.
  [1] *Differential topology*, Springer-Verlag, Berlin–Heidelberg–New York (1976).

Horowitz, R. D.
  [1] *Characters of free groups represented in the two-dimensional special linear group*, Comm. Pure Appl. Math. **25** (1972), 635-649.

Huber, H.
  [1] *Über eine neue Klasse automorpher Funktionen und ein Gitterpunktproblem in der hyperbolischen Ebene*, Comment. Math. Helv. **30** (1955), 20-62.
  [2] *Zur analytischen Theorie hyperbolischer Raumformen und Bewegungsgruppen* I, Math. Ann. **138** (1959), 1-26; II, Math. Ann. **142** (1961), 385-398; Nachtrag zu II, Math. Ann. **143** (1961), 463-464.
  [3] *Über den ersten Eigenwert des Laplace-Operators auf kompakten Riemannschen Flächen*, Comment. Math. Helv. **49** (1974), 251-259.
  [4] *Über die Eigenwerte des Laplace-Operators auf kompakten Riemannschen Flächen*, Comment. Math. Helv. **51** (1976), 215-231.
  [5] *Über die Darstellungen der Automorphismengruppe einer Riemannschen Fläche in den Eigenräumen des Laplace-Operators*, Comment. Math. Helv. **52** (1977), 177-184.
  [6] *Über die Dimension der Eigenräume des Laplace-Operators auf Riemannschen Flächen*, Comment. Math. Helv. **55** (1980), 390-397.
  [7] *Über den ersten Eigenwert des Laplace-Operators auf Flächen vom Geschlecht zwei*, J. Reine Angew. Math. **408** (1990), 202-218.

Humphries, S. P.
  [1] *Generators for the mapping class group*, in: Topology of low dimensional manifolds (R. Fenn, ed.), Springer Lecture Notes in Math. **722** (1979), 44-47.

Hurwitz, A.
  [1] *Über algebraische Gebilde mit eindeutigen Transformationen in sich*, Math. Ann. **41** (1893), 403-442; see also: Hurwitz, A., Mathematische Werke, Vol I, Birkhäuser Verlag, Basel–Boston–Stuttgart (1932/1962), 391-430.
  [2] *Über Beziehungen zwischen den Primidealen eines algebraischen Körpers und den Substitutionen seiner Gruppe*, Math. Z. **25** (1926), 661-665; see also: Hurwitz, A., Mathematische Werke, Vol 2, Birkhäuser Verlag, Basel–Boston–Stuttgart (1962), 733-737.

Huxley, M. N.
  [1] *Cheeger's inequality with a boundary term*, Comment. Math. Helv. **58** (1983), 347-354.
  [2] *Exceptional eigenvalues and congruence subgroups*, in: The Selberg Trace Formula and Related Topics, Proc. AMS-IMS-SIAM, Contemp. Math. (Amer. Math. Soc.) **53** (1986), 341-349.

Ikeda, A.
  [1] *On lens spaces which are isospectral but not isometric*, Ann. Sci. Ecole Norm. Sup. (4) **13** (1981), 303-315.

Ikeda, A. and Yamamoto, Y.
  [1] *On the spectra of 3-dimensional lens spaces*, Osaka J. Math. **16** (1979), 447-469.

Jenni, F. W.
  [1] *Über das Spektrum des Laplace-Operators auf einer Schar kompakter Riemannscher Flächen*, doctoral thesis, Basel (1981).
  [2] *Über den ersten Eigenwert des Laplace-Operators auf ausgewählten Beispielen kompakter Riemannscher Flächen*, Comment. Math. Helv. **59** (1984), 193-203.

Jörgens, K.
  [1] *Lineare Integraloperatoren*, B. G. Teubner Verlagsgesellschaft, Stuttgart (1970).

Jørgensen, T.
[1]   *On discrete groups of Möbius transformations*, Amer. J. Math. **98** (1976), 739-749.

Jost, J.
[1]   *Two-dimensional geometric variational problems*, John Wiley & Sons, Chichester–New York–etc. (1991).

Kato, T.
[1]   *Perturbation theory for linear operators*, Springer-Verlag, Berlin–Heidelberg–New York (1976).

Keen, L.
[1]   *Canonical polygons for finitely generated Fuchsian groups*, Acta Math. **115** (1966), 1-16.
[2]   *Intrinsic moduli on Riemann surfaces*, Ann. of Math. (2)**84** (1966), 404-420.
[3]   *On Fricke moduli*, in: Advances in the theory of Riemann surfaces (L. V. Ahlfors et al., eds.), Ann. of Math. Stud. **66**, Princeton Univ. Press, Princeton, NJ (1971), 205-224; Correction: Proc. Amer. Math. Soc. **30** (1973), 618.
[4]   *Collars on Riemann surfaces*, in: Discontinuous groups and Riemann surfaces (L. Greenberg, ed.), Ann. of Math. Stud. **79**, Princeton Univ. Press, Princeton, NJ (1974), 263-268.
[5]   *On fundamental domains and the Teichmüller modular group*, in: Contributions to analysis, a collection of papers dedicated to Lipman Bers (L. V. Ahlfors et al., eds.), Academic Press, London–New York–San Francisco (1974), 185-194.
[6]   *The modular group and Riemann surfaces of genus two*, Math. Z. **142** (1975), 205-219.
[7]   *A rough fundamental domain for Teichmüller spaces*, Bull. Amer Math. Soc. **83** (1977), 1199-1226.

Klingenberg, W.
[1]   *Eine Vorlesung über Differentialgeometrie*, Springer-Verlag, Berlin–Heidelberg–New York (1973).
[2]   *A course in differential geometry* (transl. by D. Hoffman), Springer-Verlag, Berlin–Heidelberg–New York (1978).

Koebe, P.
[1]   *Riemannsche Mannigfaltigkeiten und nichteuklidische Raumformen*, dritte Mitteilung, Sitzungsber. Preuss. Akad. Wiss. phys. math. Kl. **XXIII** (1928).

Komatsu, K.
[1]   *On the adele ring of algebraic number fields*, Kodai math. Sem. Reports **28** (1976), 78-84.

Koornwinder, T. H.
[1]   *Jacobi functions and analysis on noncompact semisimple Lie groups*, in: Special functions: group theoretical aspects and applications (R. A. Askey, T. H. Koornwinder and W. Schempp, eds.), D. Reidel Publishing Company, Dordrecht–Boston–Lancaster (1984).

Kra, I.
[1]   *On lifting of Kleinian groups to* SL(2, C), in: Differential geometry and complex analysis (I. Chavel and H. M. Farkas, eds.), Springer-Verlag, Berlin–Heidelberg–New York (1985), 181-193.
[2]   *Non-variational global coordinates for Teichmüller spaces*, in: Holomorphic functions and moduli II, Proc. Workshop, Berkeley 1986 (D. Drasin et al., eds.), Publ. Math. Sci. Res Just. **11**, Springer-Verlag, Berlin–Heidelberg–New York (1988), 221-249.
[3]   *Horocyclic coordinates for Riemann surfaces and moduli spaces*, I: Teichmüller and Riemann spaces of Kleinian groups, preprint, Stony Brook (1989).

Kronecker, L.
[1]   *Über die Irreductibilität von Gleichungen*, Monatsber. d. Berl. Akad. d. Wiss. (1880), 155-163.

Kuwabara, R.
[1] *On isospectral deformations of riemannian metrics*, Compositio Math. **40** (1980), 319-324.
[2] *On the characterization of flat metrics by the spectrum*, Comment. Math. Helv. **55** (1980), 427-444, correction: ibid. **56** (1981) 196-197.

Lagrange, J. L.
[1] *Recherches sur la nature et la propagation du son*, Miscellanea Taurinensia t. **I** (1759); reprinted in: Oeuvres de Lagrange, Gauthier-Villars, Paris (1867) and: Georg Olms Verlag, Hildesheim–New York (1973).

Laplace, P. S.
[1] *Traité de mécanique céleste*, Livre III, Duprat Librairie, Paris (1807); reprinted in: Oeuvres complètes de Laplace, Gauthier-Villars, Paris (1878).

Lehto, O.
[1] *Univalent functions and Teichmüller spaces*, Springer-Verlag, Berlin–Heidelberg–New York (1987).

Lehto, O. and Virtanen, K. I.
[1] *Quasiconformal mappings in the plane*, Springer-Verlag, Berlin–Heidelberg–New York (1973).

Lickorish, W. B. R.
[1] *A finite set of generators for the homeotopy group of a surface*, Math. Proc. Cambridge Philos. Soc. **60** (1964), 769-778; Corrigendum: ibid. **62** (1966), 679-681.

Löbell, F.
[1] *Die überall regulären unbegrenzten Flächen fester Krümmung*, doctoral thesis, Tübingen (1927).

Maaß, H.
[1] *Über eine neue Art von nichtanalytischen automorphen Funktionen und die Bestimmung Dirichletscher Reihen durch Funktionalgleichungen*, Math. Ann. **121** (1949), 141-183.

Macbeath, A. M.
[1] *On a theorem of Hurwitz*, Proc. Glasgow Math. Assoc. **5** (1961), 90-96.
[2] Proceedings of the Summer School in Geometry and Topology, Queens College, Dundee (1961).
[3] *Topological background*, in: Discrete groups and automorphic functions (W. J. Harvey, ed.), Academic Press, London–New York–San Francisco (1977), 1-45.

Maclachlan, C.
[1] *A bound for the number of automorphisms of a compact Riemann surface*, J. London Math. Soc. **44** (1969), 265-272.

Magnus, W.
[1] *Noneuclidean tesselations and their groups*, Academic Press New York–London (1974).

Magnus, W., Oberhettinger, F. and Soni, R. P.
[1] *Formulas and theorems for the special functions of mathematical physics*, Springer-Verlag, Berlin–Heidelberg–New York (1966).

Maskit, B.
[1] *Moduli of marked Riemann surfaces*, Bull. Amer Math. Soc. **80** (1974), 773-777.
[2] *Kleinian groups*, Springer-Verlag, Berlin–Heidelberg–New York (1988).

Massey, W. S.
[1] *Algebraic topology: an introduction*, Harcourt, Brace & World, Inc., New York–Chicago–San Francisco (1967).

Matelski, J. P.
[1] *A compactness theorem for Fuchsian groups of the second kind*, Duke Math. J. **43**

(1976), 829-840.

McKean, H. P.
[1] *Selberg's trace formula as applied to a compact Riemann surface*, Comm. Pure Appl. Math. **25** (1972), 225-246, Correction: ibid. **27** (1974), 134.

Meschkowski, H.
[1] *Noneuclidean geometry*, Academic Press, New York (1964).

Milnor, J.
[1] *Eigenvalues of the Laplace operator on certain manifolds*, Proc. Nat. Acad. Sci. U.S.A. **51** (1964), 542.
[2] *Hyperbolic geometry: The first 150 years*, Bull. Amer Math. Soc. (N.S.) **6** (1982), 9-24.

Minakshisundaram, S. and Pleijel, Å.
[1] *Some properties of the eigenfunctions of the Laplace operator on riemannian manifolds*, Canad. J. Math. **1** (1949), 242-256.

Moise, E. E.
[1] *Geometric topology in dimensions 2 and 3*, Springer-Verlag, Berlin–Heidelberg–New York (1977).

Mumford, D.
[1] *A remark on Mahler's compactness theorem*, Proc. Amer. Math. Soc. **28** (1971), 289-294.

Nag, S.
[1] *The complex analytic theory of Teichmüller spaces*, John Wiley & Sons, New York–Chichester–Brisbane (1988).

Nakanishi, T.
[1] *The lengths of the closed geodesics on a Riemann surface with self-intersections*, Tôhoku Math. J. (2) **41** (1989), 527-541.

Narasimhan, R.
[1] *Several complex variables*, The University of Chicago Press, Chicago–London (1971).
[2] *Analysis on real and complex manifolds*, Masson & Cie, Editeur-Paris, North-Holland Publishing Comp., Amsterdam (1986).

Patterson, S. J.
[1] *A lattice-point problem in hyperbolic space*, Mathematika **22** (1975), 81-88; Correction: ibid. **23** (1976), 227.
[2] *Spectral theory and Fuchsian groups*, Math. Proc. Cambridge Philos. Soc. **81** (1977), 59-75.
[3] *An introduction to the theory of the Riemann zeta-function*, Cambridge Studies in Advanced Mathematics, **14** Cambridge University Press, Cambridge–New York–New Rochelle (1988).

Perlis, R.
[1] *On the equation $\zeta_K(s) = \zeta_{K'}(s)$*, J. Number Theory **9** (1977), 342-360.
[2] *On the class numbers of arithmetically equivalent fields*, J. Number Theory **10** (1978), 489-509.

Perron, O.
[1] *Nichteuklidische Elementargeometrie der Ebene*, B. G. Teubner Verlagsgesellschaft, Stuttgart (1962).

Pesce, H.
[1] *Borne explicite du nombre de tores plats isospectraux à un tore donné*, Prépublication de l'Institut Fourier, Grenoble (1990).

Randol, B.
[1] *Small eigenvalues of the Laplace operator on compact Riemann surfaces*, Bull.

Amer Math. Soc. **80** (1974), 996-1000.

[2] *On the asymptotic distribution of closed geodesics on compact Riemann surfaces*, Trans. Amer. Math. Soc. **233** (1977), 241-247.

[3] *The Riemann hypothesis for Selberg's zeta-function and the asymptotic behavior of eigenvalues of the Laplace operator*, Trans. Amer. Math. Soc. **236** (1978), 209-223.

[4] *Cylinders in Riemann surfaces*, Comment. Math. Helv. **54** (1979), 1-5.

[5] *The length spectrum of a Riemann surface is always of unbounded multiplicity*, Proc. Amer. Math. Soc. **78** (1980), 455-456.

[6] *A remark on the multiplicity of the discrete spectrum of congruence groups*, Proc. Amer. Math. Soc. **81** (1981), 339-340.

[7] *The Selberg trace formula*, chapter XI in: Eigenvalues in riemannian geometry (I. Chavel), Academic Press, Orlando–San Diego–New York (1984), 266-302.

[8] *A remark on λ_{2g-2}*, Proc. Amer. Math. Soc. (1991).

Rayleigh, J. W. S.

[1] *On the character of the complete radiation at a given temperature*, Phil. Mag. **27** (1889); see also: Scientific papers, Cambridge University Press, Cambridge 1902 and Dover 1964, Vol. 3 p. 273.

Read, R. C.

[1] *Some enumeration problems in graph theory*, doctoral thesis, University of London (1958).

Rees, E. G.

[1] *Notes on geometry*, Springer-Verlag, Berlin–Heidelberg–New York (1983).

Rellich, F.

[1] *Störungstheorie der Spektralzerlegung* I, Math. Ann. **113** (1937), 600-619.

Riesz, F. and Sz.-Nagy, B.

[1] *Functional analysis*, (transl. by L. F. Boron), Frederick Ungar Publ. Co., New York (1972).

Riggenbach, H.

[1] *Freie Homotopieklassen und das Eigenwertspektrum des Laplace-Operators bei hyperbolischen Raumformen*, doctoral thesis, Basel (1975).

Roelcke, W.

[1] *Über die Wellengleichung bei Grenzkreisgruppen erster Art*, Sitzungsber. Heidelb. Akad. Wiss. Math.-Natur. Kl. 1953/55, 4. Abh. (1956).

[2] *Das Eigenwertproblem der automorphen Formen in der hyperbolischen Ebene* I, Math. Ann. **167** (1966), 292-337; II, ibid. **168** (1967), 261-324.

Saccheri, G.

[1] *Euclides ab omni naevo vindicatus*, Paolo Antonio Mastano (1733); edited and translated by G. B. Halsted, Chelsea Publishing Company, New York (1986).

Sarnak, P.

[1] *Prime geodesic theorems*, doctoral thesis, Stanford (1980).

[2] *The arithmetic and geometry of some hyperbolic three-manifolds*, Acta Math. **151** (1984), 253-295.

Sarnak, P. and Xue, X.

[1] *Bounds for multiplicities of automorphic representations*, Duke Math. J. **64** (1991), 207-227.

Schmutz, P.

[1] *Zur Anzahl kleiner Eigenwerte auf Riemannschen Flächen*, doctoral thesis, University of Zürich (1989).

[2] *Small eigenvalues on Y-pieces and on Riemann surfaces*, Comment. Math. Helv. **65** (1990), 603-614.

[3] *Small eigenvalues on Riemann surfaces of genus 2*, Invent. Math. **106** (1991), 121-138.

Schoen, R., Wolpert, S. and Yau, S.-T.
[1] *Geometric bounds on the low eigenvalues of a compact surface*, in: Geometry of the Laplace operator, Proc. Sympos. Pure Math. (Amer. Math. Soc.) **36** (1980), 279-285.

Selberg, A.
[1] *Harmonic analysis*, part 2. Lecture note, library of the mathematics department, Univ. of Göttingen, Summer Semester (1954).
[2] *Harmonic analysis and discontinuous groups in weakly symmetric riemannian spaces with applications to Dirichlet series*, J. Indian Math. Soc. **20** (1956), 47-87.
[3] *Discontinuous groups and harmonic analysis*, in: Proc. Internat. Congr. Math. Stockholm, 1962, Almqvist & Wiksell, Uppsala (1963), 177-189.

Semmler, K.-D.
[1] *A fundamental domain for the Teichmüller space of compact Riemann surfaces of genus 2*, doctoral thesis, University of Lausanne (1988).

Seppälä, M.
[1] *Real algebraic curves in the moduli space of complex curves*, Compositio Math. **74** (1990), 259-283.
[2] *Moduli spaces of stable real algebraic curves*, Ann. Sci. Ecole Norm. Sup. (4) **24** (1991), 519-544.

Seppälä, M. and Sorvali, T.
[1] *Parametrization of Möbius groups acting in a disk*, Comment. Math. Helv. **61** (1986), 149-160.
[2] *Parametrization of Teichmüller spaces by geodesic length functions*, in: Holomorphic functions and moduli II, Proc. Workshop, Berkeley 1986 (D. Drasin et al., eds.), Publ. Math. Sci. Res Just. **11**, Springer-Verlag, Berlin–Heidelberg–New York (1988), 267-283.
[3] *Traces of commutators of Möbius Transformations*, Math. Scand. **68** (1991), 53-58.
[4] *Geometry of Riemann surfaces and Teichmüller spaces*, North-Holland Mathematical Studies **169**, North-Holland–Amsterdam–etc. (1992).

Siegel, C. L.
[1] *Über einige Ungleichungen bei Bewegungsgruppen in der nichteuklidischen Ebene*, Math. Ann. **133** (1957), 127-138.

Simon, U. and Wissner, H.
[1] *Geometrische Aspekte des Laplace-Operators*, Jahrbuch Überblicke Mathematik (1982), 73-92.

Singerman, D.
[1] *Finitely maximal Fuchsian groups*, J. London Math. Soc. (2) **6** (1972), 29-38.
[2] *Symmetries of Riemann surfaces with large automorphism group*, Math. Ann. **210** (1974), 17-32.

Stillwell, J.
[1] *Classical topology and combinatorial group theory*, Springer-Verlag, Berlin–Heidelberg–New York (1980).

Strebel, K.
[1] *Vorlesungen über Riemannsche Flächen*, studia Mathematica, Vandenhoek Ruprecht, Göttingen–Zürich (1980).

Sullivan, D.
[1] *Hyperbolic geometry and homeomorphisms*, in: Geometric topology (J. C. Cantrell, ed.), Academic Press, London–New York–San Francisco (1979), 543-555.

Sunada, T.
[1] *Spectrum of compact flat manifolds*, Comment. Math. Helv. **53** (1978), 613-621.
[2] *Trace formula and heat equation asymptotics for a non-positively curved manifold*, Amer. J. Math. **104** (1982), 795-812.

[3] *Riemannian coverings and isospectral manifolds*, Ann. of Math. **121** (1985), 169-186.
[4] *Gelfand's problem on unitary representations associated with discrete subgroups of PSL(2, R)*, Bull. Amer Math. Soc. (N.S.) **12** (1985), 237-238.

Tanaka, S.
[1] *Selberg's trace formula and spectrum*, Osaka J. Math. **3** (1966), 205-216.

Terras, A.
[1] *Harmonic analysis on symmetric spaces and applications* I, Springer-Verlag, Berlin–Heidelberg–New York (1985).

Thurston, W.
[1] *The Geometry and topology of 3-manifolds*, Lecture Notes, Dep't. of Math., Princeton Univ. Press, Princeton, NJ (1979).

Tse, R. M.
[1] *A lower bound for the number of isospectral surfaces of arbitrarily large genus*, doctoral thesis, University of Southern California (1988).
[2] *A lower bound for the number of isospectral surfaces*, in: Recent developments in geometry (S.-Y. Cheng et al., eds.), Contemp. Math. (Amer. Math. Soc.) **101** (1989), 161-164.

Urakawa, H.
[1] *Bounded domains which are isospectral but not isometric*, Ann. Sci. Ecole Norm. Sup. (4) **15** (1982), 441-456.

Venkov, A. B.
[1] *Spectral theory of automorphic functions, the Selberg zeta-function and some problems of analytic number theory and mathematical physics*, Russian Math. Surveys **34**, n⁰ 3 (1979), 79-153; transl. from Uspekhi Mat. Nauk **34**, n⁰ 3 (207) (1979), 69-135.

Vignéras, M.-F.
[1] *Exemples de sous-groupes discrets non conjugués de PSL(2, R) qui ont la même fonction zêta de Selberg*, C. R. Acad. Sci. Paris Sér. I Math. **287** (1978), 47-49.
[2] *Variétés riemanniennes isospectrales et non isométriques*, Ann. of Math. **112** (1980), 21-32.
[3] *Arithmétique des algèbres de quaternions*, Springer Lecture Notes in Math. **800** (1980).
[4] *Quelques remarques sur la conjecture $\lambda_1 \geq 1/4$*, in: Sém. Théorie des nombres, Paris 1981-1982, (M.-J. Bertin, ed.) Birkhäuser Verlag, Basel–Boston–Stuttgart (1983), 321-343.

Weinstein, M.
[1] *Examples of groups*, Polygonal Publishing House, Passaic (NJ) (1977).

Weyl, H.
[1] *Über die asymptotische Verteilung der Eigenwerte*, Nachr. d. Königl. Ges. d. Wiss. zu Göttingen (1911), 110-117; see also: Weyl, H., Gesammelte Abhandlungen, Vol. I, Springer-Verlag, Berlin–Heidelberg–New York (1968), 368-375.
[2] *Das asymptotische Verteilungsgesetz der Eigenwerte linearer partieller Differentialgleichungen (mit einer Anwendung auf die Theorie der Hohlraumstrahlung*, Math. Ann. **71** (1912), 441-479; see also: Weyl, H., Gesammelte Abhandlungen, Vol. I, Springer-Verlag, Berlin–Heidelberg–New York (1968), 393-430.

Whittaker, E. T. and Watson, G. N.
[1] *A course of modern analysis*, fourth edition reprinted, Cambridge University Press, Cambridge (1969).

Widder, D. V.
[1] *The Laplace transform*, Princeton Univ. Press, Princeton, NJ (1946).

Wiener, N.

[1] *The Fourier integral and certain of its applications*, Cambridge University Press, Cambridge (1933).

[2] *Tauberian theorems*, Ann. of Math. **33** (1932), 1-100; see also: Collected Works, Vol II, The MIT Press, Cambridge Massachussets and London, England (1979), 519-618.

Wolfe, W.

[1] *The asymptotic distribution of lattice points in hyperbolic space*, J. Funct. Anal. **31** (1979), 333-340.

Wolpert, S. A.

[1] *The eigenvalue spectrum as moduli for compact Riemann surfaces*, Bull. Amer Math. Soc. **83** (1977), 1306-1308.

[2] *The eigenvalue spectrum as moduli for flat tori*, Trans. Amer. Math. Soc. **244** (1978), 313-321.

[3] *The length spectra as moduli for compact Riemann surfaces*, Ann. of Math. **109** (1979), 323-351.

[4] *Riemann surfaces, moduli and hyperbolic geometry* (Trieste Lectures), mimeographed lecture notes, Trieste (1979).

[5] *Thurston's riemannian metric for Teichmüller space*, J. Differential Geom. **23** (1986), 143-174.

[6] *Geodesic length functions and the Nielsen problem*, J. Differential Geom. **25** (1987), 275-296.

Wormwald, N.

[1] *Enumeration of labelled graphs II: cubic graphs with a given connectivity*, J. London Math. Soc. (2) **20** (1979), 1-7.

[2] *Enumeration of cyclically 4-connected cubic graphs*, J. Graph Theory **9** (1985), 563-573.

Yamada, A.

[1] *On Marden's universal constant of Fuchsian groups, II*, Journal d'Analyse Mathémathique **41** (1982), 234-248.

Yau, S.-T.

[1] *Isoperimetric constants and the first eigenvalue of a compact Riemannian manifold*, Ann. Sci. Ecole Norm. Sup. (4) **8** (1975), 487-507.

Zieschang, H.

[1] *Finite groups of mapping classes of surfaces*, Springer Lecture Notes in Math. **875** (1981).

Zieschang, H., Vogt, E. and Coldewey, H.-D.

[1] *Flächen und ebene diskontinuierliche Gruppen*, Springer Lecture Notes in Math. **122** (1970).

[2] *Surfaces and planar discontinuous groups*, Springer Lecture Notes in Math. **835** (1980).

Zimmer, R. J.

[1] *Ergodic theory and semisimple groups*, Birkhäuser Verlag, Basel–Boston–Stuttgart (1984).

Zograf, P. G.

[1] *Fuchsian groups and small eigenvalues of the Laplace operator*, Russian studies in topology, IV, Zap Naučn. Sem. Leningrad, Otdel, Mat. Inst. Steklov (LOMI) **122** (1982), 24-29.

[2] *Fuchsian groups and small eigenvalues of the Laplace operator*, J. Soviet Math. **26** (1984), 1618-1621.

[3] *Small eigenvalues of automorphic Laplacians in spaces of parabolic forms*, J. Soviet Math. **36** (1987), 106-114.

Index

1-homeomorphism 83, 134, 410
1-product 363

A

Abel transform 194, 196, 200
admissible
  Jordan curve 376
  list for graph 79
  sequence for isospectrality 335
  set for homotopy 19
  transform pair 252
Alexander-Tietze theorem 410
almost conjugate 285
angle 34
  of cone-like singularity 313
  subsequent 34
angle type 35
Arzelà-Ascoli theorem 429
asymptotic equality 235
atlas
  conformal 29
  hyperbolic 6
axis 171, 227

B

$\mathcal{B}_g$ 172
Baer-Zieschang theorem 411
base surface for markings 71, 138, 140, 336, 384
Bers coordinates 172, 174
Bers' constant 123
Bers' theorem 123, 124, 125, 130

boundary orientation 11
boundary point 4
  at infinity 4, 22, 50
  ordinary 6
boundary twist 425
building block 30, 63, 302, 305

C

$\mathcal{C}(M)$ 229
canonical
  curve system 141
  dissection 166
  generators of Fuchsian group 171
  polygon 168
  polygon, normal 383
Cayley graph 297
cell structure
  of Teichmüller space 144, 334
Chebyshev functions 243, 245, 246
Cheeger's constant 215
circle sector 6
cleaning option 430
closed geodesic 22
  iterate 22
  non-oriented 269, 291
  norm 245
  one-sided 26
  oriented 22, 227, 229
  prime 22, 228, 245
  primitive 22, 228, 291
co-area formula 217
collar 70, 94, 343

width of 70
collar theorems 94, 97, 106, 112
color 298
commensurable 336
complete sequence 379
completely multiplicative 242
complex differentiable 380
cone-like singularity 159, 313
conformal structure 29, 139, 156
conjugation in quaternion algebra 53
const 204
convolution 259
coordinate maps
  for Teichmüller space 147
coordinates
  Bers 172, 174
  Fenchel-Nielsen 29, 82, 147
  Fermi 4
  Fricke 179
  horocyclic 5, 198
  polar 3
  quadrilateral 385
  Z-V-C 169
correlated 292
counterclockwise 58
curve
  piecewise regular 375
curve system
  canonical 141
cusp 25, 111
cut open 15
cycle 313
defining relation 164, 171, 179

D

Dehn twist 72, 425
Dehn-Nielsen theorem 171
direct 156
Dirichlet
  domain 221
  fundamental domain 205
Dirichlet series 225, 247
displacement length 171, 175, 227, 292
dissection
  canonical 166
  normal canonical 166
distance
  hyperbolic 2, 38, 59, 198
  in graph 221
  in Teichmüller space 152, 153

domain
  of operator 370
  of sesquilinear form 370

E

edge
  of graph 78, 298
eigenfunction 187
  radial 198, 236
eigenspace 191, 295, 296
eigenvalue 187
  isolated 378
  small 210
endpoint
  at infinity 22
  of geodesic 109
Epstein-Zieschang theorem 411
equivalence
  Gassmann 285
  of canonical dissections 166
  of closed geodesics 229
  of homeomorphisms 154
  of marked Fuchsian groups 171, 334
  of marked Riemann surfaces 139, 145, 147
  of marked Y-pieces 67
  of normal canonical polygons 168
  of parametrized geodesics 21, 22
  of vectors 49
  of vertices 313
  of words 92
  relation for pasting 8, 9
  weak 269, 291
Euler product 242, 243
exponent 287

F

$\mathcal{F}_g$ 171
Fenchel-Nielsen parameters 29, 82, 147
Fermi coordinates 4, 70
  distance formula 38
  on cylinder 10
  sign conventions 4
figure eight geodesic 86, 100, 159, 272
fixed point
  attractive 172
  repulsive 172
Fourier transform 252, 258, 259
Fricke-coordinates 179
Friedrichs extension 368

Fuchsian group 170
  marked 170, 171
fundamental domain 230
  Dirichlet 205
  in Teichmüller space 67
  rough 161

G

Gassmann equivalent 285
Gel'fand conjecture 183, 269
general sine and cosine law 61
generalized triangle 36
generating function 205
  decay condition 205
  induced kernel 206
geodesic
  figure eight 272
  in the quaternion model 49
  loop 26
  polygon (curve) 34
  polygon (domain) 3
  prime 228, 245, 291
geodesic-like 49
geometric spectrum 235
gradient 185, 363
graph
  3-regular 79
  connected 79
  cubic 78
  marked 79
  quotient 298
Green's formula 185, 363

H

hairy torus 131
half-disk 6
half-edge 79
heat equation 186, 188
  fundamental solution 186
  solution 188
heat kernel 185, 186
Heisenberg group 287
hexagon 40
  self-intersecting 42
Hilbert-Schmidt theorem 187
holomorphic 380
holomorphic family
  of bounded operators 381, 393
  of closed operators 397
  of Laplacians 399

  of matrices 402
homotope (as verb) 18, 73
homotopically trivial 21
homotopy
  free 21
  length-decreasing 430
  with gliding endpoints 18
horocycle 5, 25
Huber's theorem 229
Hurwitz' theorem 159
hyperbolic
  atlas 6
  cylinder 9, 10
  distance 2, 38, 59, 198
  structure 7
  surface 7
hyperbolic plane
  disk model 2
  Poincaré model 2
  quaternion model 52
hyperboloid 32, 49
hypergeometric function 263, 264

I

incidence relations
  in the quaternion model 50
induced kernel 206
infinity-like 49
injectivity radius 96
integral logarithm 257
interior 6
intersection
  removable 417
  transversal 416
irreducible 404
isomorphism
  of graphs 80, 298
  strong 298
isoperimetric
  constant 215
  quotient 212
isospectral 183
isotopy 83, 410, 425
  level-preserving 410
  with pointwise fixed set 410
iterate
  of closed geodesic 22, 228

J

Jordan-Schoenflies theorem 413

L

Lsp 269
Laplace operator 184, 366
  holomorphic family 399
lattice point problem 225
length
  of cycle 314
  of prime number 284
length distortion 68, 152
length spectrum 183, 226, 229, 269
  primitive 229
length trace formula 230
let go 18
lift
  of prime geodesic 291
Lipschitzian 428
local
  analytic variety 270
loop 20
  geodesic 26
  simple 20

M

$\mathfrak{M}_g$ 154
$\mathcal{M}_g$ 154
M-O-S 248
Mahler compactness theorem 163, 273
mapping class 154
mapping class group 154
marking
  of Fuchsian group 170, 171
  of Riemann surface 139
  of X-piece 70
  of Y-piece 67
marking equivalence
  of Fuchsian groups 171, 334
  of Riemann surfaces 139, 145, 147
  of Y-pieces 67
marking homeomorphism 139, 143
Mercer's theorem 187
metric tensor 184, 185
minimax principles 213
moduli problem 67, 160
moduli space 161, 211

N

ν-Cauchy sequence 363
ν-convergence 363
ν-equivalent 363

Neumann problem 215
non-solitary 284
non±homotopic 411
norm
  in quaternion algebra 53
  of closed geodesic 245, 253
  of operator 370
normal
  neighborhood 429
normal canonical 337
  dissection 166
  generators of Fuchsian group 172
  polygon 168, 383
normal form
  presentation of surface 164

O

one-sided 26
operator
  bounded 370
  closed 370
  domain of 370
  linear 370
  range of 370
  sum 370
  bounded inverse 370
order
  of cone-like singularity 313
  of cycle 314
orientation
  of edge in graph 299
  of graph 298
  of surface 7
overlap 416
  free 422
  proper 416

P

$\mathcal{P}(M)$ 229
$\mathcal{P}(N)$ 291
pair of pants 63
  degenerate 109
  in standard form 64
parallel curve 135
parameter geodesic 82, 341
parity of configuration 59
partition 123
pasting 8
  of closed geodesics 10
pasting condition 8, 9, 64

pasting scheme 9, 82, 143, 164
pentagon 39
piecewise circular 135
point
  in the quaternion model 49
  singular 313
point at infinity 4, 22
  in the hyperboloid model 50
point-like 49
polygon
  canonical 168
  normal canonical 168, 383
  oriented 34
pre-trace formula 238
prime geodesic 22, 228, 245, 291
prime lift 291
prime number theorem
  for compact Riemann surfaces 246
  in number theory 241
  with error terms 257
primitive
  closed curve 23
  closed geodesic 228, 291
  transformation 228
projector 375
puncture 112

Q

quasi isometry 68
quasiconformal mapping 153
quaternion
  algebra 52
  model of hyperbolic plane 52
quotient graph 298

R

$\mathcal{R}_g$ 211
$\mathcal{R}^{6g-6}$ 142, 147
range
  of operator 370
Rayleigh quotient 400
resolvent 373
resolvent equation 374
resolvent set 373
Riemann space 156, 161
  of 3-holed sphere 67
Riemann surface
  compact 30
  marked 139
  of signature (g, n) 30

  with cone-like singularities 159
Riemann zeta function 243
Roelcke problem 223
root 402

S

S^1 21
Schwartz space 195, 196
  modified 195
Selberg's zeta function 246
semi group property 185, 190
sesquilinear form 370
side 6
  of polygon 34, 429
  of trigon 116
  subsequent 34
sign conventions in figures 4
signature 30, 111, 333
  of Fuchsian group 333
simple
  curve 19
  loop 20
simplicity
  of the length spectrum 85
small eigenvalues 210
smoothing operator 365
spectrum
  of closed operator 373
standard annulus 424
Stieltjes integral 244
stretch 68, 143
strong isomorphism 298
structure
  conformal 29
  hyperbolic 7
subsequent (sides and angles) 34
Sunada's theorem 291, 295
support 213
surface 5
  hyperbolic 7

T

$\mathcal{T}_g$ 139
$\mathcal{T}_G$ 29, 144, 147
Tauberian theorem 243
Teichmüller mapping 154
Teichmüller modular group 154
  of 3-holed sphere 67
Teichmüller space 139, 383, 384
  arbitrary signature 334

Index   453

cell structure   144, 334
  of 3-holed sphere   67
  real analytic structure   148
tessellation   119
thick and thin decomposition   96, 108, 109
thick part   117
topological diameter   415
trace
  in quaternion algebra   53
  of kernel operator   227
transform pair   252
transplantation
  of closed geodesics   308
  of functions   305
triangulation   116
trigon   116
triple trace theorem   56, 57
trirectangle   37
twist   143
twist homeomorphism   72, 77, 143
twist parameter   11, 29, 69, 77
  of X-piece   72
type
  of edge   312
  of edge in graph   298, 299
  of geodesic   305

U

unique lifting property   17

V

vector product   50
vertex
  of boundary of surface   6
  of graph   78
vertex cycle   12
volume element   363
von Mangoldt function   243
von Neumann series   374

W

Ω_G 141
weakly equivalent   269, 291
Weyl's asymptotic law   182, 235
width
  of collar   70, 95, 117, 343
  of half-collar   66
Wiener-Ikehara theorem   243

Wolpert's theorem   270
word   178
  cyclic   92
  in generators of a group   90
  reduced   92

X

X-piece   69
  immersion of   140
  marked   70

Y

Y-piece   63
  degenerate   109
  marked   67
  standard form   64

Z

$\mathcal{T}_g$ 169
Z-V-C   166
Z-V-C parameters   169
zeta function   183
  Riemann   243
  Selberg   246

Formula Glossary

2.2.2 Right-angled triangles

(i) $\cosh c = \cosh a \cosh b$,
(ii) $\cosh c = \cot\alpha \cot\beta$,
(iii) $\sinh a = \sin\alpha \sinh c$,
(iv) $\sinh a = \cot\beta \tanh b$,
(v) $\cos\alpha = \cosh a \sin\beta$,
(vi) $\cos\alpha = \tanh b \coth c$.

2.3.1 Trirectangles

(i) $\cos\varphi = \sinh a \sinh b$,
(ii) $\cos\varphi = \tanh\alpha \tanh\beta$,
(iii) $\cosh a = \cosh\alpha \sin\varphi$,
(iv) $\cosh a = \tanh\beta \coth b$,
(v) $\sinh\alpha = \sinh a \cosh\beta$,
(vi) $\sinh\alpha = \coth b \cot\varphi$.

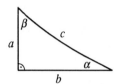

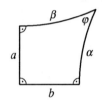

2.3.4 Right-angled pentagons

(i) $\cosh c = \sinh a \sinh b$,
(ii) $\cosh c = \coth\alpha \coth\beta$.

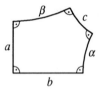

2.4.1 Right-angled hexagons

(i) $\cosh c = \sinh a \sinh b \cosh\gamma - \cosh a \cosh b$,
(ii) $\sinh a : \sinh\alpha = \sinh b : \sinh\beta = \sinh c : \sinh\gamma$,
(iii) $\coth\alpha \sinh\gamma = \cosh\gamma \cosh b - \coth a \sinh b$.

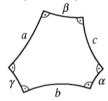

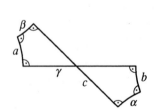

2.4.4 Crossed right-angled hexagons

$$\cosh c = \sinh a \sinh b \cosh\gamma + \cosh a \cosh b.$$

2.2.1 Triangles

(i) $\cosh c = -\sinh a \sinh b \cos\gamma + \cosh a \cosh b$,
(ii) $\cos\gamma = \sin\alpha \sin\beta \cosh c - \cos\alpha \cos\beta$,
(iii) $\sinh a : \sin\alpha = \sinh b : \sin\beta = \sinh c : \sin\gamma$.

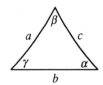

Edited by:

J. Oesterlé
Département de Mathématiques
Université de Paris VI
4, Place Jussieu
75230 Paris Cedex 05, France

A. Weinstein
Department of Mathematics
University of California
Berkeley, CA 94720
U.S.A.

Progress in Mathematics is a series of books intended for professional mathematicians and scientists, encompassing all areas of pure mathematics. This distinguished series, which began in 1979, includes authored monographs and edited collections of papers on important research developments as well as expositions of particular subject areas.

We encourage preparation of manuscripts in some form of TeX for delivery in camera-ready copy which leads to rapid publication, or in electronic form for interfacing with laser printers or typesetters.

Proposals should be sent directly to the editors or to: Birkhäuser Boston, 675 Massachusetts Avenue, Cambridge, MA 02139, U. S. A.

A complete list of titles in this series is available from the publisher.

53 LAURENT. Théorie de la Deuxième Microlocalisation dans le Domaine Complexe
54 VERDIER/LE POTIER. Module des Fibres Stables sur les Courbes Algébriques: Notes de l'Ecole Normale Supérieure, Printemps, 1983
55 EICHLER/ZAGIER. The Theory of Jacobi Forms
56 SHIFFMAN /SOMMESE. Vanishing Theorems on Complex Manifolds
57 RIESEL. Prime Numbers and Computer Methods for Factorization
58 HELFFER/NOURRIGAT. Hypoellipticité Maximale pour des Opérateurs Polynomes de Champs de Vecteurs
59 GOLDSTEIN. Séminaire de Théorie des Nombres, Paris 1983–84
60 PROCESI. Geometry Today: Giornate Di Geometria, Roma. 1984
61 BALLMANN/GROMOV/SCHROEDER. Manifolds of Nonpositive Curvature
62 GUILLOU/MARIN. A la Recherche de la Topologie Perdue

63 GOLDSTEIN. Séminaire de Théorie des Nombres, Paris 1984–85
64 MYUNG. Malcev-Admissible Algebras
65 GRUBB. Functional Calculus of Pseudo-Differential Boundary Problems
66 CASSOU-NOGUES/TAYLOR. Elliptic Functions and Rings and Integers
67 HOWE. Discrete Groups in Geometry and Analysis: Papers in Honor of G.D. Mostow on His Sixtieth Birthday
68 ROBERT. Autour de L'Approximation Semi-Classique
69 FARAUT/HARZALLAH. Deux Cours d'Analyse Harmonique
70 ADOLPHSON/CONREY/GHOSH/YAGER. Analytic Number Theory and Diophantine Problems: Proceedings of a Conference at Oklahoma State University
71 GOLDSTEIN. Séminaire de Théorie des Nombres, Paris 1985–86

72 VAISMAN. Symplectic Geometry and Secondary Characteristic Classes
73 MOLINO. Riemannian Foliations
74 HENKIN/LEITERER. Andreotti-Grauert Theory by Integral Formulas
75 GOLDSTEIN. Séminaire de Théorie des Nombres, Paris 1986–87
76 COSSEC/DOLGACHEV. Enriques Surfaces I
77 REYSSAT. Quelques Aspects des Surfaces de Riemann
78 BORHO /BRYLINSKI/MACPHERSON. Nilpotent Orbits, Primitive Ideals, and Characteristic Classes
79 MCKENZIE/VALERIOTE. The Structure of Decidable Locally Finite Varieties
80 KRAFT/PETRIE/SCHWARZ. Topological Methods in Algebraic Transformation Groups
81 GOLDSTEIN. Séminaire de Théorie des Nombres, Paris 1987–88
82 DUFLO/PEDERSEN/VERGNE. The Orbit Method in Representation Theory: Proceedings of a Conference held in Copenhagen, August to September 1988
83 GHYS/DE LA HARPE. Sur les Groupes Hyperboliques d'après Mikhael Gromov
84 ARAKI/KADISON. Mappings of Operator Algebras: Proceedings of the Japan-U.S. Joint Seminar, University of Pennsylvania, Philadelphia, Pennsylvania, 1988
85 BERNDT/DIAMOND/HALBERSTAM/ HILDEBRAND. Analytic Number Theory: Proceedings of a Conference in Honor of Paul T. Bateman
86 CARTIER/ILLUSIE/KATZ/LAUMON/ MANIN/RIBET. The Grothendieck Festschrift: A Collection of Articles Written in Honor of the 60th Birthday of Alexander Grothendieck. Vol. I
87 CARTIER/ILLUSIE/KATZ/LAUMON/ MANIN/RIBET. The Grothendieck Festschrift: A Collection of Articles Written in Honor of the 60th Birthday of Alexander Grothendieck. Volume II
88 CARTIER/ILLUSIE/KATZ/LAUMON/ MANIN/RIBET. The Grothendieck Festschrift: A Collection of Articles Written in Honor of the 60th Birthday of Alexander Grothendieck. Volume III
89 VAN DER GEER/OORT / STEENBRINK. Arithmetic Algebraic Geometry
90 SRINIVAS. Algebraic K-Theory
91 GOLDSTEIN. Séminaire de Théorie des Nombres, Paris 1988–89
92 CONNES/DUFLO/JOSEPH/RENTSCHLER. Operator Algebras, Unitary Representations, Enveloping Algebras, and Invariant Theory. A Collection of Articles in Honor of the 65th Birthday of Jacques Dixmier
93 AUDIN. The Topology of Torus Actions on Symplectic Manifolds
94 MORA/TRAVERSO (eds.) Effective Methods in Algebraic Geometry
95 MICHLER/RINGEL (eds.) Representation Theory of Finite Groups and Finite Dimensional Algebras
96 MALGRANGE. Equations Différentielles à Coefficients Polynomiaux
97 MUMFORD/NORI/NORMAN. Tata Lectures on Theta III
98 GODBILLON. Feuilletages, Etudes géométriques
99 DONATO /DUVAL/ELHADAD/TUYNMAN. Symplectic Geometry and Mathematical Physics. A Collection of Articles in Honor of J.-M. Souriau
100 TAYLOR. Pseudodifferential Operators and Nonlinear PDE
101 BARKER/SALLY. Harmonic Analysis on Reductive Groups
102 DAVID. Séminaire de Théorie des Nombres, Paris 1989-90
103 ANGER /PORTENIER. Radon Integrals
104 ADAMS /BARBASCH /VOGAN. The Langlands Classification and Irreducible Characters for Real Reductive Groups
105 TIRAO/WALLACH. New Developments in Lie Theory and Their Applications
106 BUSER. Geometry and Spectra of Compact Riemann Surfaces